TERRESTRIAL ECOSYSTEMS

D. Muench/H. Armstrong Roberts

H. Armstrong Roberts

Donald Paterson/Photo Researchers

TERRESTRIAL ECOSYSTEMS

John D. Aber
University of New Hampshire

Jerry M. Melillo
Marine Biological Laboratory
Woods Hole, Massachusetts

Saunders College Publishing

Philadelphia Fort Worth Chicago
San Francisco Montreal Toronto
London Sydney Tokyo

THIS BOOK IS PRINTED ON **ACID-FREE, RECYCLED** PAPER 

Text Typeface: Palatino
Compositor: Digitype
Acquisitions Editor: Julie Alexander
Developmental Editor: Gabrielle Goodman
Managing Editor: Carol Field
Project Editor: Mary Patton
Copy Editor: Becca Gruliow
Manager of Art and Design: Carol Bleistine
Art Director: Doris Bruey
Text Designer: Doris Bruey
Cover Designer: Doris Bruey
Text Artwork: Larry Ward
Director of EDP: Tim Frelick
Production Manager: Bob Butler
Marketing Manager: Marjorie Waldron

Cover: Burning pasture for renewal of vegetation near Rio Claro, São Paulo, Brazil. © Georg Gerster/Comstock.

Printed in the United States of America

TERRESTRIAL ECOSYSTEMS

ISBN 0-03-047443-4

Library of Congress Catalog Card Number: 90-053331

7 8 9 0 1 2 016 9 8 7 6 5 4

PREFACE

From the clearing of tropical forests to the deposition of air pollutants, humans are altering the way the Earth looks and functions. The cumulative effects of our actions are clearly visible in the changing chemistry of the global atmosphere and alterations in forest health and agricultural production.

Managing this increasing human presence on the Earth requires an understanding of how terrestrial ecosystems, the basic units of the landscape, function. It is from these systems that we obtain food and fiber, clean air and water, and the esthetic and social benefits of contact with the natural world. Climate change, acid rain, increasing concentrations of carbon dioxide in the atmosphere, all of these affect, and are affected by, the metabolism of the biotic communities that cover the land surface.

Our purpose in writing this book has been to integrate information from several different traditional disciplines in order to present a holistic view of ecosystem function. Topics as diverse as photosynthesis, soil chemistry, and population dynamics of herbivores are discussed as they relate to the movement of energy, water, and elements through units of the landscape. We believe that this interdisciplinary approach is required for the successful application of ecosystem concepts to the pressing problems resulting from human use of the biosphere.

The book has been written for advanced undergraduates and beginning graduate students in a wide range of disciplines. To achieve this, we assume only an introductory-level knowledge of the workings of soils, plants, and animal populations and have made every effort to minimize the use of specialized terms (scientific "jargon"). Concepts and processes are presented in ways that should be accessible to a wide audience. Figures, photographs, and diagrams are used liberally to illustrate the major points in each chapter, and references are provided for those who wish to pursue particular topics in greater detail.

The structure of the book reflects our goal of providing an integrative, interdisciplinary discussion of the function of terrestrial ecosystems. Part I introduces basic concepts of ecosystem studies, including a focus on the movement of energy and matter through the landscape, and some first steps in measuring function at the ecosystem level. A brief introduction to systems analysis reemphasizes the system-level viewpoint of the text.

Part II deals with the key processes that control ecosystem function. The transfers of carbon and nutrients are used to structure the presentation of factors controlling net photosynthesis and water use at the ecosystem level, nutrient uptake and release through decomposition, the role of soil chemistry in controlling nutrient availability, and the importance of herbivory and fire as alternate "sinks" for fixed carbon and other elements.

Part III returns to the ecosystem level, presenting an integrated view of how the processes described in Part II interact to produce the patterns of change and distribution visible in three very different landscapes. The role of human use of each of these three systems, and the interaction of management with natural processes, is an important part of this discussion.

In Part IV, the growing dominance of human activity in controlling ecosystem distribution and function is dealt with in detail. Forest management, agricultural practices, and the indirect effects of pollution loading, which represent three of the spatially dominant uses of the landscape, are presented. The final chapter deals with terrestrial ecosystems as components of the global system and describes methods for measuring and predicting the net effect of terrestrial ecosystem function on the regional, continental, and global environment.

For Lynn, Lalise, Ted, Patrick, Colleen, and Caitlin

ACKNOWLEDGMENTS

Many people have contributed generously of their time and enthusiasm to help bring this book to completion. Foremost of these is Dan Binkley, who has read through the entire text twice, and who provided insightful, objective, and knowledgeable criticism at every step. Dick Waring and Keith Van Cleve provided both partial reviews and much encouragement to finish the task. Additional substantial reviews were provided by: Herb Bormann, Wally Covington, Lowell Getz, Henry Gholz, Jim Gosz, Calvin Meier, John Pastor, Francis Putz, Peter Reich, Steve Running, Dan Schiller, Dave Schimel, Nellie Stark, James Vilkitis, Peter Vitousek, Kristiina Vogt, Neil West, and several additional reviewers who chose to remain anonymous. We also owe thanks to many students who worked conscientiously through early versions of the text and provided valuable commentary. These include: Carol Wessmann, Steve McNulty, Mary Martin, Alice Cialella, and many others enrolled in the Terrestrial Ecosystems course at the University of Wisconsin between 1982 and 1987. Critical figures and references were provided by Wally Covington, Marylin Jordan, Steve McNulty, John Pastor, Barry Rock, Dave Skole, Lloyd Swift, Jim Vogelmann, Diane Wickland, and Trish and Peter Wolter. Finally, we would like to thank Julie Alexander at Saunders for her encouragement to complete the project, and Mary Patton, Gabrielle Goodman, and their co-workers for orchestrating the production of the text.

John Aber and Jerry Melillo
January, 1991

CONTENTS OVERVIEW

CONTENTS

Part I

INTRODUCTION The Nature of Ecosystem Studies

H. Armstrong Roberts

The study of ecosystems has developed rapidly over the last three decades, yet the field is still a relatively new one. The ecosystem approach has been used with increasing frequency in the analysis of large-scale environmental problems because it deals explicitly with large units of the landscape, the size of units dealt with by forest and range managers, and the agencies concerned with environmental protection. Increasingly, the ecosystem is the basic landscape unit of study of regional and global environmental problems.

For all of this, it is often difficult to visualize the scope and purpose of ecosystem studies. Just what are ecosystems? What are their important characteristics? What distinguishes one system from another? Perhaps one of the most troubling questions is: How do you measure the function of such large landscape units?

The purpose of this first section is to present an introduction to the study of ecosystems. The five chapters present a very brief historical view of the development of the ecosystem concept and some basic definitions (Chapter 1), a discussion of how ecosystems vary in structure and function under different climates (Chapter 2), an introduction to the measurement of carbon, water, and nutrient balances over terrestrial ecosystems (Chapters 3 and 4), and a brief introduction to concepts and definitions borrowed from the field of systems analysis that are frequently applied to ecosystems (Chapter 5).

Chapter 1 Development of Concepts in Ecosystem Studies

P. Kreson/H. Armstrong Roberts

WHY STUDY ECOSYSTEMS?

The term **ecosystem** is often encountered in media presentations of current environmental problems. Air pollution is often said to be damaging forest ecosystems. Water pollution may be threatening lake ecosystems. New approaches may be discussed for managing wetland ecosystems. These presentations may convey a sense that ecosystems are something more than the plants and animals residing in an area. They may also suggest that something will go wrong with the system, that some important function will be damaged or altered. Both of these ideas, indeed, are related to the definition of ecosystem studies. An ecosystem is the sum of all of the biological and nonbiological parts of an area that interact to cause plants to grow and decay, soils or sediments to form, and the chemistry of water to change. The study of terrestrial ecosystems can be thought of as the study of the metabolism of units of the landscape—their energy, chemical, water, and mass balance. Ecosystem studies can be defined, then, as the study of the movement of energy and materials, including water, chemicals, nutrients, and pollutants, into, out of, and within ecosystems.

This sounds pretty dry. Yet the study of ecosystems has grown rapidly over the last 30 years as we have become aware of the extent to which human society depends on the continuing health of ecosystems. We rely on ecosystem function to clean polluted water and air, as well as to provide food and fiber. The approaches and concepts of ecosystem studies discussed in this book apply just as well to intensively managed systems, such as cornfields and forest plantations, as they do to wild, unmanaged systems. Scholarly studies have even been done on suburban lawns as ecosystems! In fact, one of the values of ecosystem studies is in determining just how much human use of the landscape changes the energy, water, and chemical balances of ecosystems. In this way, we monitor the "health" of managed systems or their effect on adjacent urban or wild ecosystems. Studies of individual ecosystems are also increasingly combined into larger, regional units as part of the examination of the entire Earth. Many environmental problems are not local or even regional in scope, but rather global. The link between the metabolism of individual ecosystems and the metabolism of the Earth will be made at several points in this book.

From a purely intellectual point of view, the field of ecosystem studies has offered a new and unique challenge to natural scientists. It is most fundamentally an interdisciplinary or holistic science, involving experts from several traditional fields. For example, an atom of nitrogen may, in a single growing season, move from an organic compound (such as protein) in the soil to an inorganic one (such as ammonium), into the soil solution, and then into the plant root, stem, and leaf. From the leaf it may be washed by rainfall back into the soil, down past the roots, and into a stream. So far we have mentioned processes that traditionally lie in the provinces of soil science, microbiology, plant physiology, water chemistry, micrometeorology, and hydrology. Others could easily be included. Integration of these fields has been a continuing challenge in ecosystem studies.

Another and more recent characteristic of ecosystem studies is an attempt to determine where, in the vast array of biotic (living) and abiotic (nonliving) components that make up an ecosystem, the most significant interactions lie. This is a point that will be central throughout this book. Everything that happens in nature happens within ecosystems. This does not mean that all processes are equally important in terms of controlling ecosystem function. For example, plants take both nitrogen and sodium out of soils and incorporate them into plant tissues. However, nitrogen is a required element while sodium generally is not. Low availability of nitrogen frequently limits overall growth of plants in terrestrial ecosystems by limiting the total amount of leaf biomass or chlorophyll. Thus, the movement of nitrogen will receive more attention than the movement of sodium. In this book, we will emphasize the most important interactions—those that control the overall rates of energy, water, and nutrient flux.

DEVELOPMENT OF ECOSYSTEM CONCEPTS

The concept of an ecosystem as a natural entity involving the interrelationships of biotic and abiotic factors in nature is an ancient one. However, the analysis of entire natural systems as a rigorous science in terms of factors controlling structure and function is a recent phenomenon derived from several disciplines. Two important characteristics of ecosystem studies are the use of large areas as the unit of study (such as whole lakes or large blocks of forest or prairie) and an attempt to study each of the system's components (soils, sediments, plants, animals, microbes) with an intensity reflecting their overall importance in the system's energy and material movement.

Tansley, a plant ecologist, was the first to coin the term "ecosystem" in 1935. He recognized, and attempted to minimize, the tendency in his time to emphasize the study of organisms (such as algae) rather than the processes they perform (such as photosynthesis). He tried to encourage the study of interactions between biotic and abiotic components.

> *The more fundamental conception is . . . the whole system (in the sense of physics), including not only the organism-complex, but also the whole complex of physical factors forming what we call the environment . . . the habitat factors in the widest sense. . . . Our natural human prejudices force us to consider the organisms . . . as the most important parts of these systems, but certainly the inorganic "factors" are also parts, . . . and there is constant interchange of the most various kinds within each system, not only between the organisms but between the organic and inorganic. These ecosystems, as we may call them, are of the most various kinds and sizes. (Tansley 1935)*

While several other similar terms conveying similar meanings have surfaced at different times throughout the world, "ecosystem" has become most widely accepted. As discussed later, this may be at least partly due to the rapid growth of "systems analysis" as a method for expressing interactions between components of any system, whether biological, electrical, or mechanical. This method has been applied frequently to ecological data (see Chapters 5 and 21).

Acceptance of the ecosystem as a new unit of study creates some serious problems. If all parts of the system are to be treated in a similar way, what is the common denominator that expresses their interdependence? A classic paper by Lindeman, a young aquatic ecologist, or limnologist, suggested one factor that is common to all life processes: energy. He used the storage and movement of energy to express interdependence between organisms within a system, in this case a lake, and also to compare rates of important processes between very different types of ecosystems.

> *A lake, for example, might be considered by a botanist as containing several distinct plant aggregations. . . . The associated animals would be "biotic factors" of the plant environment, tending to limit or modify the development of the aquatic plant communities. To a strict zoologist, on the other hand, a lake would seem to contain animal communities, although the*

> *"associated vegetation" would be considered merely as a part of the environment of the animal community. . . . The trophic dynamic viewpoint, as adopted in this paper, emphasizes the relationship of trophic or "energy-availing" relationships within the community unit. From this viewpoint . . . a lake is considered as a primary ecological unit in its own right, since all the lesser "communities" . . . are dependent upon other components of the lacustrine food cycle for their very existence. (Lindeman 1942)*

Lindeman emphasized that such a complex set of interactions between living and dead organic material existed in any ecosystem, so that separating spatially distinct "communities" such as the "plankton community" was meaningless. The key to the function of the system was the transfer of energy from green plants (producers) to animals (consumers) and microbes (decomposers). This occurred throughout the lake ecosystem. Thus the energy transfers within the lake ecosystem became the focus of his study, and different spatial or organismal components were treated as important in their contribution to this total energy flow. Lindeman also produced one of the first energy flow diagrams, again for a lake ecosystem (Figure 1.1). Such diagrams have become the most common method for expressing interrelationships between components of ecosystems. Many will be presented in this text. Energy has thus become a "universal currency" of ecosystems and ecosystem studies.

Another point mentioned by Lindeman but not fully developed in his paper is the inseparable relationship between the flow of energy and the flow of nutrient elements (refer again to the quotation above). Plants require several chemical elements in order to grow, photosynthesize, and reproduce. Pioneer work and early synthesis in this area was carried out by J. D. Ovington, an English forester and silviculturist. In another classic paper he set forth the basic outline of terrestrial ecosystem studies that remains largely unaltered to this day.

As a forester, Ovington was concerned with the quantity of water and nutrients required to grow a given amount of tree biomass. He pulled together all existing data on forest production, water use, and nutrient content and tried to express nutrient and energy movements as yearly rates. Transfers within the system were examined, such as those detailed by Lindeman for lakes. However, Ovington realized that in an era of increasing demand for wood products, the continued productivity of forests might depend on their input–output balances. Removing wood from the forest also removes the nutrients contained in that wood. These must eventually be balanced by inputs in rainfall or by other processes. Ovington thus began to consider all the possible inputs and outputs of nutrients in forests. His concept of the movement of energy, water, and nutrients within and through ecosystems was therefore more complete (Figure 1.2). His statement of the need for more basic understanding of ecosystem function was one of the first to envision the importance of ecosystem analysis in the wise management of natural resources.

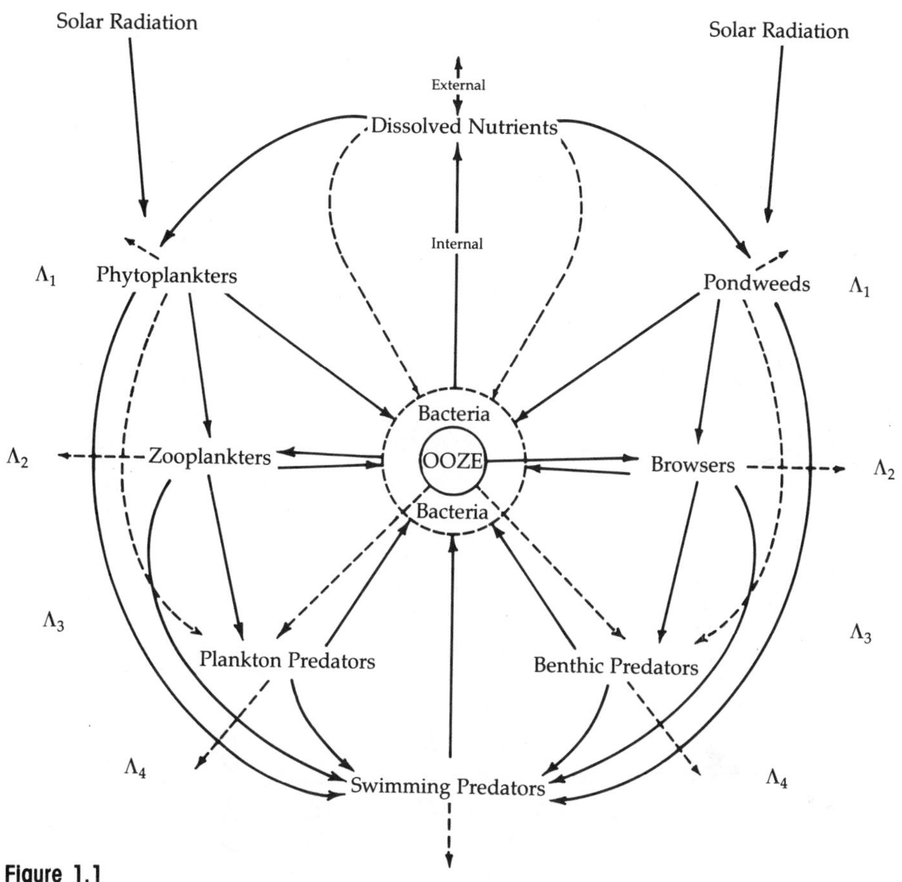

Figure 1.1
Early diagram of energy exchanges among components of a lake ecosystem. (Lindeman 1942)

> *Of all terrestrial communities, woodlands are probably the most complex and massive, and not unexpectedly woodland ecologists have attempted to restrict their research to whatever features they felt were of greatest importance or could be recorded most readily. . . . The resultant lack of integration has caused an oversimplified approach to woodland ecology which has further hindered the effective co-ordination of results from different disciplines and areas. Multiple and more intensive use of forest land is inevitable as the world population multiplies, and the more thorough and comprehensive our knowledge of woodland ecology, the better will be the prospect for wise use and long-term conservation of the woodland resource. Some unifying concept embracing all aspects of woodland ecology is needed to bring forth a deeper understanding of woodland systems and to serve as the basis for their more rational utilization. It seems that the concept of ecosystem may provide the universal backcloth against which to show woodlands in all their patterned complexity. (Ovington 1962)*

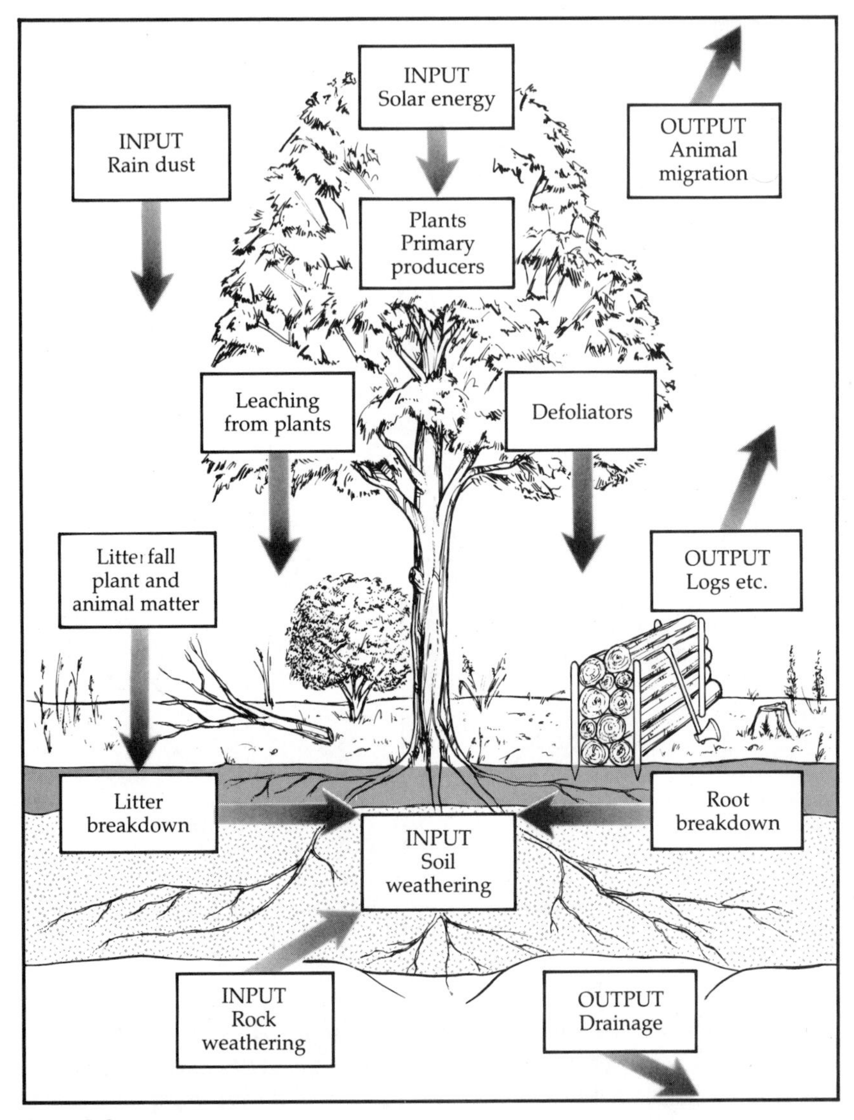

Figure 1.2
An early diagram of the movement of energy, water, and nutrients through forest ecosystems. (Ovington 1962)

Thus Ovington largely defined the focus of terrestrial ecosystem studies as the movement of energy, water, and nutrients within and through ecosystems, with particular emphasis on factors controlling those rates of flow. He admitted that the necessary data were not available at the time of his paper to answer the crucial questions on controlling factors or to predict effects of management on ecosystems.

In the 30 years since Ovington's paper, there has been an explosion in the amount of information available on the structure and function of terrestrial ecosystems. However, we still have not realized the predictive capability Ovington sought. We shall see that it is far easier to conceptualize the components of an ecosystem than it is to take all the measurements necessary to accurately describe its function. Still, definite progress has been made. That progress is the subject of this volume.

DELIMITING THE ECOSYSTEM

Thus far, ecosystems have been discussed in the abstract as combinations of biotic and abiotic components, using examples such as lakes or forests. In order to develop a rigorous set of methods for measuring processes, particularly inputs and outputs of energy, water, and nutrients, a definite boundary must be placed around the system. How are the boundaries of an ecosystem determined?

To begin with a general and not very satisfying answer, the boundaries of the system are determined by the purposes of a study or the questions posed. All of the ecosystems of the world, however defined, are linked by their inputs and outputs to all other ecosystems, so the largest definition is the Earth itself. This has been shown most dramatically by studies on the global distribution of pollutants such as dichlorodiphenyltrichloroethane (DDT) or radioisotopes from nuclear bomb tests.

For example, large-scale application of the chlorinated hydrocarbon DDT began in the 1940s. DDT was one of several potent synthetic pesticides that were to rid the world of insect-borne diseases and revolutionize agriculture. However, it was quickly learned that serious side effects resulted from its use (for a popular account of this experience, see Rachel Carson's *Silent Spring*). Among these was the disruption of reproduction among predatory birds, including some of the rarest and most valued, in whose tissues the chemical became concentrated. Surveys of animals worldwide revealed measurable concentrations of DDT even in Antarctic penguins who were far removed from areas of direct application. Thus, the Earth's intertwined food chains had carried this new chemical to the farthest corners, indicating that all of the distinct ecosystems were indeed linked.

At the other extreme, researchers have found that the ecosystem concept is applicable to areas as small as a single leaf. Enclosing these in small chambers and measuring gas balances (oxygen, carbon dioxide) allows calculations of photosynthesis and respiration within the leaf (see Chapter 3).

In terrestrial ecosystem studies, two definitions are most common: the watershed and the stand. A **watershed** is a topographically defined area such that all the precipitation falling into the area leaves in a single stream.

Figure 1.3
A watershed ecosystem at the Coweeta Hydrologic Laboratory in North Carolina. Note the use of the complete hydrologic basin (watershed) for an experiment on the effects of disturbing the plant community. (Courtesy of the Coweeta Hydrologic Laboratory, U.S. Forest Service)

Figure 1.3 clearly shows the outline of a small watershed used for hydrologic research in a forested region. Watersheds can range in size from a few to hundreds of thousands of hectares. The watershed of the Mississippi River, for example, includes roughly one-third of the 48 contiguous United States.

The watershed concept has been used extensively because of the importance of water balances in the study of ecosystems. The water balance itself is of interest because of the role of water availability in limiting plant growth and because of the importance of water yield from ecosystems for urban and agricultural uses. Precipitation inputs and stream water outputs are also important in nutrient balances over ecosystems. Watersheds can allow accurate measurements, particularly of nutrient losses, as the stream acts as a sampling device, giving an average concentration of nutrients in water leaving the watershed ecosystem. Most watersheds used in ecosystem studies are relatively small (10–50 hectares) and occur in areas of extreme topography (for example, see Figure 1.3; further discussion of the use of watersheds is presented in Chapter 4).

The second important spatial definition of an ecosystem is the stand, a term borrowed from forestry or agriculture but applicable to any form of terrestrial vegetation. A **stand** can be defined as an area of sufficient homogeneity with regard to vegetation, soils, topography, microclimate, and past disturbance history to be treated as a single unit. This is clearly a less precise definition than that for watershed, as variation over small distances is the rule in natural systems, and a strict definition of "sufficient homogeneity" is lacking. However, stands can be much smaller than watersheds (usually an area less than 5 hectares is actually sampled), which reduces the effort required to measure processes within the system. In areas of little or no topographic variation it is the only practical definition. In arid areas such as deserts or short-grass prairies, where soil water percolation and streamflow are minimal, the watershed loses its value in sampling exports from the system. Thus, while being less satisfying than the watershed, the stand definition is the only practical approach in many situations.

COMPONENTS OF TERRESTRIAL ECOSYSTEMS

What do grasslands, deserts, forests, shrublands, tundra, and all other terrestrial ecosystems have in common? If we want to devise a general structure for terrestrial ecosystems, what components do all these systems share? They are very different in terms of plant species, soil structure, and other visible characteristics (see Chapter 2). Building on the early efforts of Lindeman and Ovington, can we discern a common set of functional components?

For the flow of energy, associated in terrestrial systems with carbon-based compounds (mainly sugars, proteins, cellulose, lignin), components can be separated according to their source of energy (Figure 1.4*a*). Green plants are called autotrophs (*auto* = self, *troph* = energy), or primary producers. Animals that graze on plants, ingesting living tissues, are called heterotrophs (*hetero* = other), secondary producers or consumers. Consumption of live plants is generally less important quantitatively than the pathway through saprotrophs (*sapro* = dead), which feed on and decompose dead organic matter shed by plants, animals, and other decomposers. More simply stated, more plant production dies and is shed as litter (such as with the falling of leaves in the autumn) than is consumed live. Notable exceptions to this generalization are discussed in Chapters 15 and 16.

The flow of energy and carbon compounds through plants, consumers, and decomposers is complicated by the wide array of complex biochemical compounds formed by plants from the products of photosynthesis (Chapter 12). This is further complicated by an even more complex set of organic compounds that are produced by decomposer organisms and which accumulate in soils in terrestrial ecosystems. Separating relatively fresh litter (such as this year's leaves or roots) from that which has been decomposed

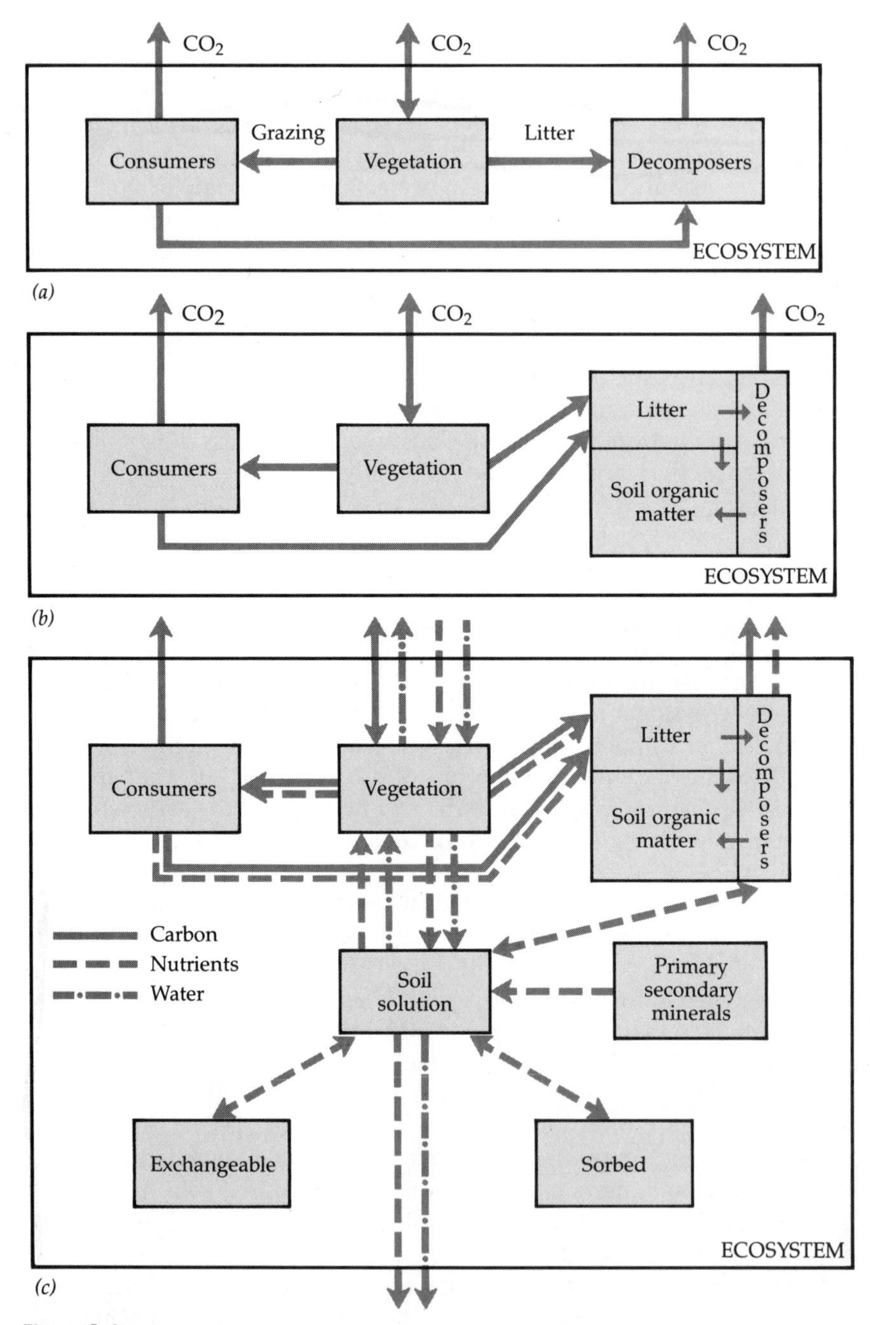

Figure 1.4
Components common to all terrestrial ecosystems. *(a, b)* Components important in the carbon cycle. *(c)* Enlarged model containing components that are important in nutrient and water dynamics.

to some extent results in the situation illustrated in Figure 1.4*b*, in which energy and carbon in the ecosystem reside either in plants, consumers, decomposers, or in fresh or partially decomposed organic matter. The last three are spatially combined within the soil subsystem.

Each of these categories could be divided further. Vegetation consists of different species, perhaps of different growth forms (trees, shrubs, grasses), and of different tissue types (such as leaves, stems, and roots). Soil organic matter is really a composite of the simple and complex carbon substances produced by plants and decomposers. The decomposer component can be a vast and diverse assemblage of microorganisms and soil animals operating as an integrated community or can be dominated by a few major organisms. The number of compounds recognized within an ecosystem is much like the boundaries of the system in that they can be selected to reflect the goals of a study. At different points in the following chapters, it will be necessary to dissect the basic components presented here to achieve greater understanding and accuracy. For now, we will use this simpler model.

Production of organic matter cannot occur without nutrients essential to plant function. Thus the flow of energy and carbon through an ecosystem depends on a parallel flow of essential nutrients. Unlike carbon (and the hydrogen and oxygen associated with it), which is exchanged with the atmosphere or derived from precipitation, mineral nutrients, such as nitrogen, phosphorus, and calcium, reside largely within the system and are recycled. Inputs and outputs are relatively small compared to the total amount cycled annually.

In addition, most nutrients can occur in numerous forms within the system. They can be part of the vegetation, litter, soil organic matter, animals, and microbes—components already described in Figure 1.4.*b*. However, plants take up nutrients mainly in simple inorganic forms, not directly from organic matter. Microbes have to "digest" organic material and release the inorganic forms. Thus nutrients can be present in the soil water solution in these simple inorganic forms (see Figure 1.4*c*).

Mineral nutrients in solution can have several fates. They can be taken up and used either by plants or by microbes. Also, depending on the element, they can participate in several strictly chemical interactions. They can be held on the surfaces of soil particle surfaces either through simple electrostatic attraction (exchangeable) or by stronger surface complexation reactions (sorbed). Both of these transfers are reversible, though exchangeable elements are more readily available to plants. Elements previously held in completely unavailable forms in rocks can be released to the soil solution as simple inorganic forms through weathering. Finally, rainwater washing down through the soil can leach nutrient below the rooting zone, out of the biologically active area, and intro streams or groundwater. Under certain conditions, selected elements can be lost from the system as gases. Thus our list of components and transfers is much larger than for carbon alone (Figure 1.4*c*).

The movement of water is less complex conceptually (Figure 1.4*c*), though no easier to measure. Precipitation can first contact plant surfaces such as leaves or stems or fall directly on the soil. In the former case, it can either evaporate or fall to the soil surface. Soil water becomes part of the soil solution and will either evaporate, be taken up by plants and evaporated from leaf surfaces, or percolate below the rooting zone. Water carries nutrients throughout these pathways and exchanges them with plant and soil surfaces contacted.

What emerges in Figure 1.4*c* is a fairly complete but not very detailed diagram or model of the pathways by which energy, water, and nutrient elements are transported into, out of, and within terrestrial ecosystems. We will return frequently to this diagram in later chapters.

Contained in this diagram is a conception of the function of ecological systems that was unknown 30 years ago. As we will see, rates of important ecosystem processes, such as plant production and decomposition, are largely determined by the relative availabilities of energy, water, and nutrients and the efficiency with which they are used. Thus plants and decomposers have evolved to optimize rates of function under the constraints of the availability of these resources. As environments vary widely over the face of the globe, organisms and communities have developed that, although they carry out the same functions, are vastly different in structure and appearance. These differences, which produce the great diversity of terrestrial communities on Earth, are the subject of Chapter 2.

REFERENCES CITED

Lindeman, R. L. 1942. The trophic–dynamic aspects of ecology. *Ecology* 23:399–418.

Ovington, J. D. 1962. Quantitative ecology and the woodland ecosystem concept. *Advances in Ecological Research* 1:103–192.

Tansley, A. G. 1935. The use and abuse of vegetational concepts and terms. *Ecology* 16:284–307.

ADDITIONAL REFERENCES

Bormann, F. H., and G. E. Likens. 1967. Nutrient Cycling. *Science* 155:424–429.

Carson, R. 1962. *Silent Spring*. Houghton Mifflin, Boston.

Kormondy, E. J. 1969. The nature of ecosystems. In *Concepts of Ecology*. Prentice-Hall, Englewood Cliffs, New Jersey.

Odum, E. P. Principles and Concepts Pertaining to the Ecosystem. Chapter 2 in *Fundamentals of Ecology*. W. B. Saunders, Philadelphia.

Odum, E. P. 1969. The strategy of ecosystem development. *Science* 164:262–270.

Swift, M. J., O. W. Heal, and M. Anderson. 1979. Decomposition in processes in terrestrial ecosystems. In *Decomposition in Terrestrial Ecosystems.* University of California Press, Berkeley.

Woodwell, G. M. 1967. Toxic substances and ecological cycles. *Scientific American* 216(3):24–31.

Chapter 2 Structure of Terrestrial Ecosystems

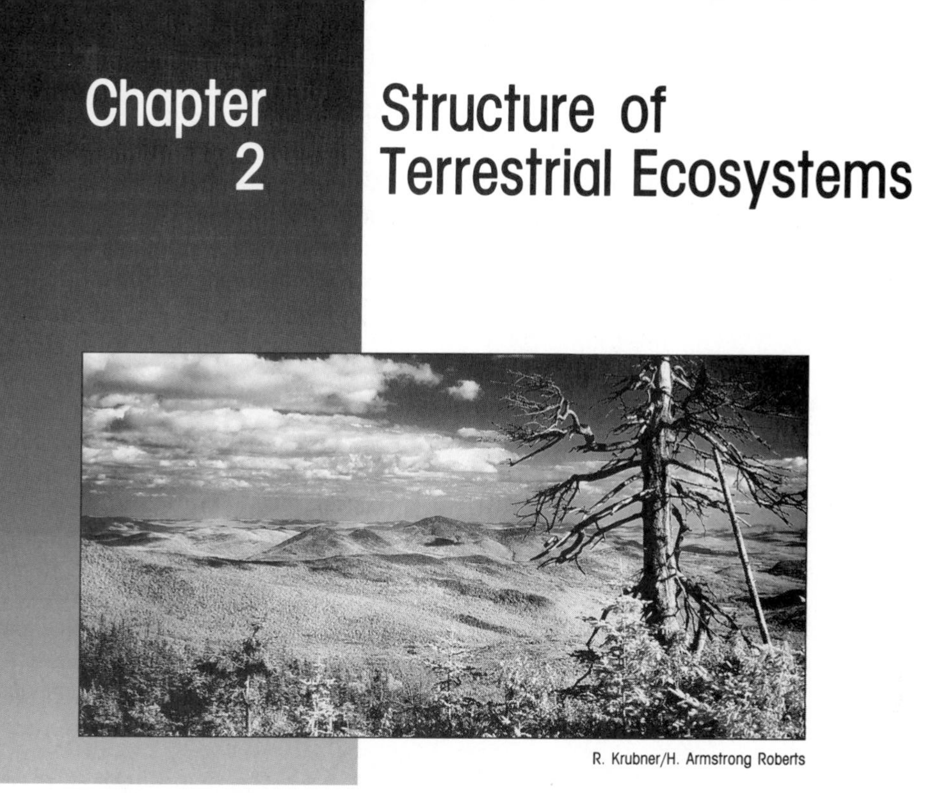

R. Krubner/H. Armstrong Roberts

INTRODUCTION

In Chapter 1, a single conceptual model was developed that emphasized major processes common to all terrestrial ecosystems. While thinking about different systems in a uniform way is useful in drawing comparisons between them, this model masks the tremendous variation that exists in the physical structure of ecosystems. The purpose of this chapter is to provide a brief, descriptive tour of some major terrestrial ecosystems, presenting differences in the structure of both vegetation and soil, along with a brief description of the factors accounting for those differences. Differences in key processes such as plant growth (primary production), water use (transpiration), and soil formation are also described.

On a global scale, climate plays the largest role in determining the structure of both vegetation and soils in natural ecosystems. Temperature and the balance between precipitation and evaporation (including evaporation through plant surfaces, known as transpiration) are particularly important, since these largely determine the rate at which biological and chemical reactions occur. Both the production of new organic matter by plants and decomposition of dead organic matter by microbes are temperature- and moisture-dependent processes. The amount of water available for use by plants, relative to the potential for that water to evaporate, determines the amount of water stress encountered by plants and often limits the length of the growing season. The amount of water percolating down through the soil profile and the chemistry of that water play a large role in determining soil structure and nutrient content.

The precise mechanisms controlling vegetation structure and soil development are numerous and complex. The simplified discussion presented here is intended only to describe trends in vegetation and soil resulting from the large changes in climate occurring over the Earth. Greater detail will be provided in later chapters.

IMPORTANT CHARACTERISTICS OF TERRESTRIAL ECOSYSTEMS

Figure 2.1*a* describes the occurrence of major terrestrial ecosystem types, or biomes, identified by the growth form of the dominant vegetation, in relation to climate. The shaded area represents the range of mean precipitation and mean temperature values on the surface of the Earth. Figure 2.1*b* maps the geographic distribution of these biomes.

Figure 2.2 summarizes differences in structure and function between the major ecosystem types identified in Figure 2.1. Three characteristics

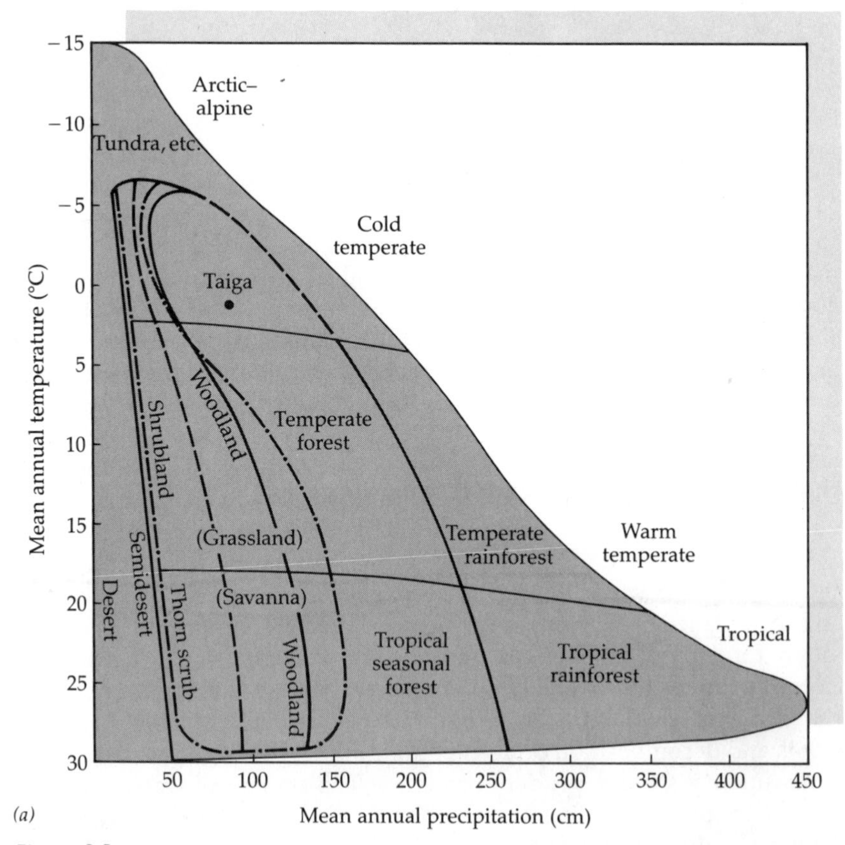

Figure 2.1

(a) Distribution of major vegetation types, or biomes, in relation to mean annual precipitation and temperature (Whittaker 1975). *(b, next page)* Geographic distribution of major vegetation types of the world (adapted from Villee et al., *Biology*, 2nd ed., Saunders College Publishing, Philadelphia, 1989).

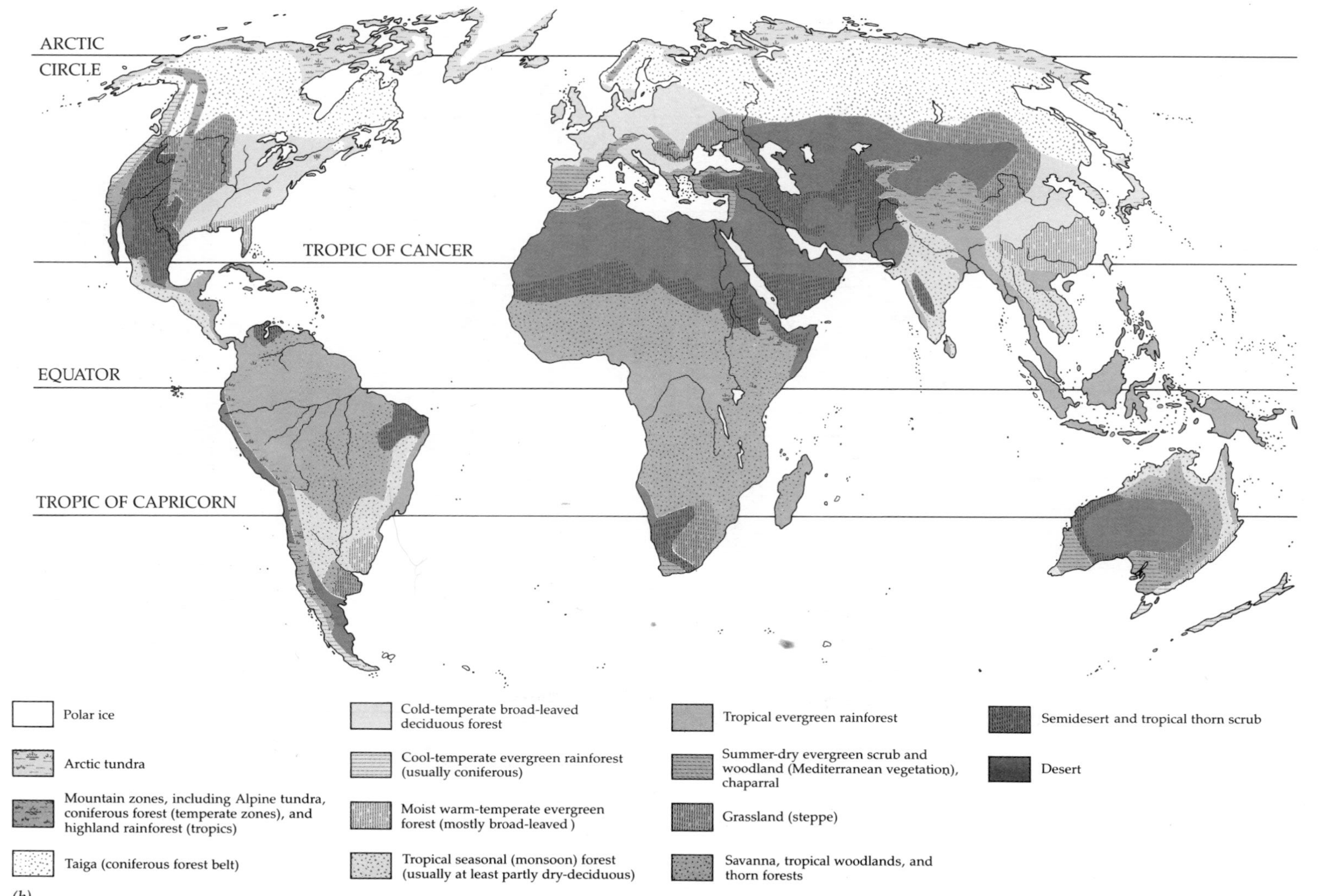
ARCTIC CIRCLE
TROPIC OF CANCER
EQUATOR
TROPIC OF CAPRICORN
Polar ice
Arctic tundra
Mountain zones, including Alpine tundra, coniferous forest (temperate zones), and highland rainforest (tropics)
Taiga (coniferous forest belt)
Cold-temperate broad-leaved deciduous forest
Cool-temperate evergreen rainforest (usually coniferous)
Moist warm-temperate evergreen forest (mostly broad-leaved)
Tropical seasonal (monsoon) forest (usually at least partly dry-deciduous)
Tropical evergreen rainforest
Summer-dry evergreen scrub and woodland (Mediterranean vegetation), chaparral
Grassland (steppe)
Savanna, tropical woodlands, and thorn forests
Semidesert and tropical thorn scrub
Desert

(b)

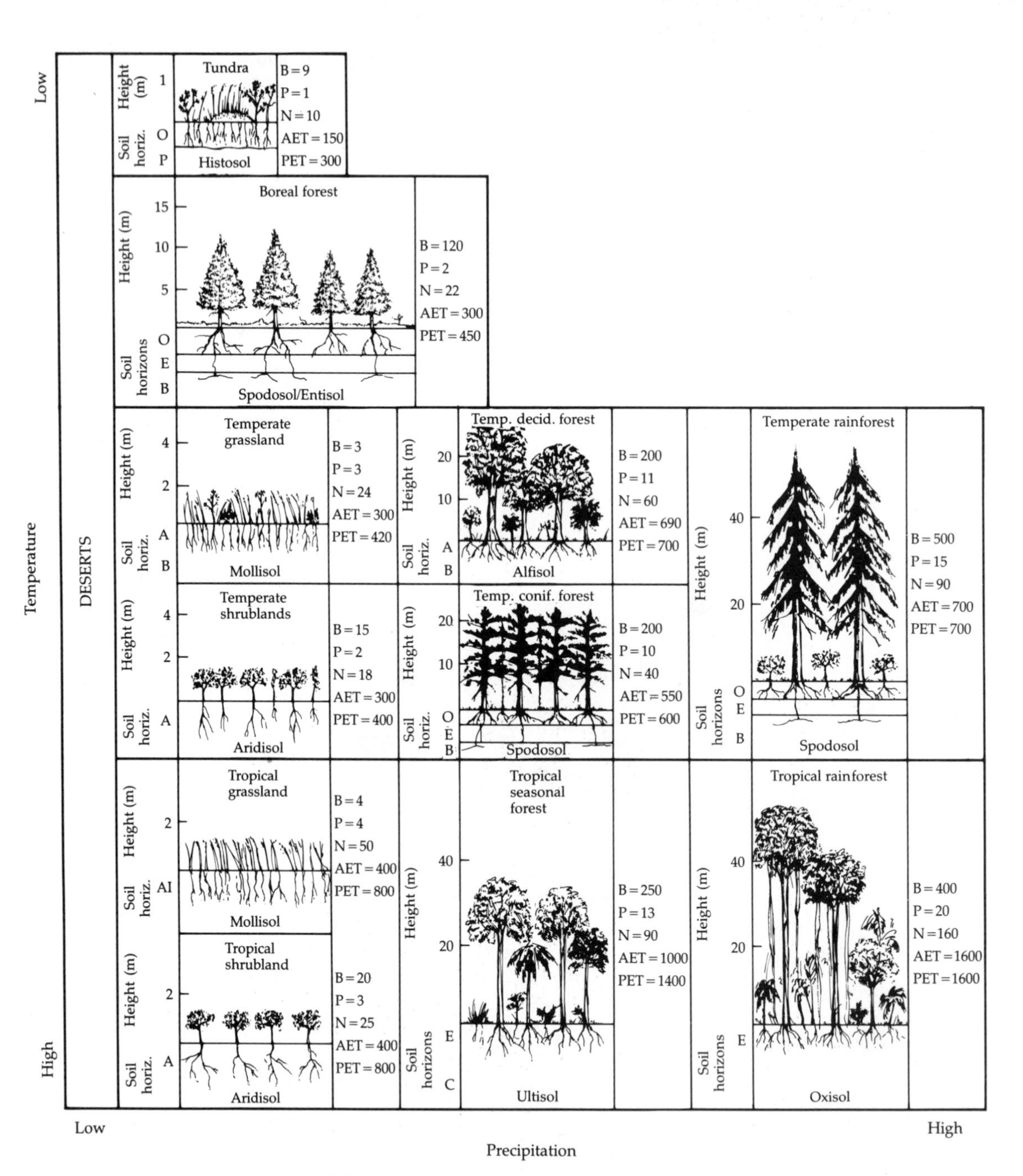

Figure 2.2
Modified diagram of the distribution of major ecosystem types in relation to precipitation and temperature. Values for elements of structure and function that are used to characterize ecosystems are given for each type. B = total plant biomass in Mg/ha, P = total plant production above ground in Mg/ha · yr, N = nitrogen uptake by plants in kg/ha · yr, AET = actual evapotranspiration in mm of water per year, PET = potential evapotranspiration in mm of water per year. (Bliss et al. 1973, Bokhari and Singh 1975, Gray and Schlesinger 1981, Murphy and Lugo 1986, Nadelhoffer et al. 1985, Pastor et al. 1984, Sims et al. 1978, Sinclair and Norton-Griffiths 1979, Van Cleve et al. 1983, Vorosmarty et al. 1989, Vitousek and Sanford 1986)

are used to describe the structure of the vegetation: (1) total height, (2) the distribution of the fine roots, which are most active in nutrient and water uptake, and (3) total above-ground biomass (B), or the dry weight of all live plants.

Three important processes by which ecosystems are compared are also included. The first is annual net primary production above ground (P). This is the summation of the dry weight of all the new plant growth except roots and other below-ground parts. Second is annual nitrogen uptake or use by above-ground vegetation (N). This is determined by measuring the concentration of nitrogen in the different kinds of tissues produced (leaves, wood, fruits) and multiplying by the weight of those tissues. The third process is actual evapotranspiration (AET). This is the sum of water lost to the atmosphere from the soil surface (evaporation) and through plant tissues (transpiration) over the course of a year. It is often compared with potential evapotranspiration (PET), or the amount of evapotranspiration that would occur with unlimited supplies of water. In hot, dry climates, AET is limited by the amount of water available to be evaporated and may be much lower than PET. Regular droughts occur in such areas, and little water moves down through the soil. In cool, rainy climates, less solar energy and more humid air reduce evaporative demand and PET is lower. AET may then be equivalent to PET, and drought and water stress on plants is lower. When AET is limited by evaporative demand (PET), more water is available for movement down through the soil.

Soils are described according to the depths of the different soil horizons or layers. There are five major soil horizons (Figure 2.3).

Horizon	Characteristics
O	Surface accumulation of partially decayed organic matter
A	Mineral soil in which mixing of surface organic matter and root growth masks effects of chemical leaching
E	Mineral soil in which weathering and leaching dominate horizon characteristics
B	Mineral soil zone affected by chemical deposition of material leached from A horizon
C	Parent material or mineral horizon largely unaffected by soil development

Figure 2.3
Major soil horizons and their characteristics.

The O or organic horizon is a surface accumulation of partially decomposed litter and organic matter that has not been mixed into the mineral soil. The A horizon is the upper layer of mineral soil altered mainly by the mixing of organic matter from the O horizon and the deposition of organic matter by root production and death. Chemical weathering also occurs in this horizon, but the results of weathering are masked by organic matter content. The E horizon is defined as that mineral soil horizon the chemistry of which has been altered mainly by the weathering and leaching of minerals. Many soils will show *either* an A or an E horizon, depending on the relative importance of chemical weathering versus biological incorporation of organic matter in determining the characteristics of the top layer of mineral soil.

The B horizon is the part of the soil profile that has been altered by the chemical deposition or precipitation of material leached from the overlying horizons. The C horizon is mineral material that has not been affected by soil development. In some cases it represents the original material from which the A, E, and B horizons have developed.

Which substances are leached from the E horizon is determined largely by climate but also partly by changes in water chemistry caused by the presence of different types of vegetation and their decay products. Four basic processes can be defined.

1 **Podzolization** is the leaching of iron, aluminum, and organic matter along with very little formation of clays, producing an E horizon rich in silicon (sand) and of very low fertility. Podzolization is most effective under cool, moist, and acidic conditions. A sizable O horizon is commonly found over podzol soils, and organic acids leaching from the O horizon can increase the rate of podzolization. The B horizon can be enriched in iron and aluminum oxides leached from the E horizon.

2 **Laterization** is the converse of podzolization. Under hot, moist conditions, silicon is weathered and leached more readily than iron and aluminum oxides or clays, and the E horizon becomes enriched in the latter. The leached silicon is not redeposited lower in the soil profile, and there may be no visible B horizon.

3 **Melanization/lessivage** occurs in temperate areas with fine-textured soils and less acidic conditions. The leaching of clays (lessivage) is important in determining the structure of the B horizon, while the A horizon tends to be well mixed by the activity of soil animals and is enriched and darkened by the addition of organic matter (melanization). The boundary between A and B horizons can be very hard to distinguish, and there may be no visible E horizon.

4 **Calcification/salinization** occurs in arid regions where percolation of water through the soil profile is minimal. Even very soluble com-

pounds such as calcium carbonate (lime) and sodium chloride and other salts remain within the soil column, although they may be moved down from the surface to an extent dependent on precipitation.

In many soils, some combination of these processes is evident. In all but the driest conditions, the E horizon increases in depth with time. Its depth at any one time reflects the intensity of soil-forming processes and the length of time they have been at work.

MAJOR VEGETATION AND SOIL TYPES OF THE EARTH

Tropical rainforests (Figure 2.2) occur where temperatures are warm and constant and where rainfall is high and occurs throughout the year. The growing season lasts all year and plants grow quickly and continuously, leading to high levels of production and biomass accumulation within a very tall and vertically stratified canopy. A tremendous diversity of growth forms is present, including vines and epiphytes, as well as the greatest diversity of species. High production levels require the uptake and cycling of large amounts of nitrogen. Although evapotranspiration is high, rainfall still exceeds evaporative demand and leaching of water down through the soil column is intense. Many soils under this vegetation type are also very old, so the E horizon can be exceptionally deep. Laterization is the dominant soil-forming process, and the E horizon is high in iron and aluminum content and clays. Conditions are also optimal for decomposition, so little or no surface litter accumulates, yet there is an abundance of organic matter in the mineral soil. The occurrence of large stores of soil organic matter and nitrogen in almost all terrestrial ecosystems is a fundamental and perplexing property, which will receive more attention in later chapters.

Such fully weathered soils with large accumulations of clays and iron and aluminum oxides are called **Oxisols**. They are very nutrient-poor soils. A paradox of the tropical rainforest is that it supports the richest and most luxuriant vegetation on the poorest soils. Most of the nutrient capital is present in the vegetation at any one time and is quickly recycled to the vegetation after litterfall occurs. This is reflected in the concentration of fine roots at the soil surface, which take up nutrients as they are made available by decomposers from litter, and little nutrient matter is absorbed from lower soil horizons.

Tropical areas in which regular dry seasons occur support tropical seasonal forests. Periods of drought reduce plant growth, so total plant production and associated litterfall and nitrogen cycling over a full year are lower than in the rainforest, as is evapotranspiration. Soils are less severely leached but otherwise are similar to the Oxisols and are called

Ultisols. The major soil-forming process is still laterization, which results in a red clay E horizon rich in iron and aluminum. This soil type may also be found in subtropical forests such as the pine and hardwood stands in the southeastern United States.

In a still drier tropical area, forest gives way to grassland/savanna or shrub "thorn" forest. Both are shown in the same climatic square since both can occur under similar mean annual climatic conditions. As discussed in more detail for temperate grasslands and shrublands (below), disturbance, particularly fire and grazing, might play a role in determining which growth form predominates. The severely limiting moisture conditions lead to much lower rates of evapotranspiration, although evaporative demand (PET) in this hot, dry climate is very high. Total production and nitrogen cycling are also reduced. Growth form reflects these limitations; there is a shift from trees to shorter vegetation and lower total biomass. Roots, especially under shrub vegetation, may penetrate very deeply into the soil to reach groundwater stored well below the soil surface.

The potential for evaporation in these hot and relatively dry areas reduces leaching; thus soil-forming factors are less active. Under shrub vegetation on moderate to steep slopes, occasional heavy rainstorms may cause extensive erosion, which removes material from the top, most developed horizon. Root masses are lower than for grasslands, and so soils are relatively low in organic matter. In grasslands, fine root masses are larger, creating a deep A horizon. This increases water percolation into the soil, decreases runoff, and reduces erosion. The profile shown for the shrubland in Figure 2.2 is called an **Aridisol**, while the grassland soil is called a **Mollisol.** Calcification and salinization are dominant in the aridisol, while melanization tends to mask these processes in the Mollisol.

A similar sequence of vegetation types may be described for temperate regions. Temperate rainforests occupy a much smaller area than their tropical counterparts, usually occurring on the windward side of coastal mountains or in other coastal areas. The vegetation is totally different from the tropical systems in species composition, but structure and function are generally quite similar (Figure 2.2). In North America, the temperate rainforests are dominated by evergreen conifers rather than broadleaved trees, but vines and epiphytes and a deep, stratified canopy are all similar to tropical conditions. Production is somewhat lower than in the tropics as temperatures are less favorable, but AET is close to PET due to abundant rainfall. Litter production is somewhat lower than in the tropical forests, while nitrogen cycling is considerably lower because the dominant conifers have lower concentrations of nitrogen in the biomass produced, including leaf litter. This low nitrogen content, along with lower temperatures, reduces decomposition and soil animal activity, creating large accumulations of surface organic matter (O horizon) and a reduced A horizon. With the cooler climate and more acidic litter, podzolization is the predominant soil-forming process. There is often a distinct E horizon.

Accumulations of iron, aluminum, and to some extent organic matter occur in the B horizon. The soil type is a **Spodosol.** Root distribution is similar to that of organic matter. Roots are concentrated in the O horizon, very low in the E, and somewhat higher in the organically enriched top of the B horizon.

Temperate seasonal forests may be either deciduous or evergreen. The two types occupy climatic regions that may appear very similar in terms of annual mean temperature and precipitation (Figure 2.2). However, the timing of precipitation and annual temperature extremes are important in determining the occurrence of one type over the other. Deciduous forests occupy areas where summer precipitation is high and winter temperatures are low. Evergreen conifers are dominant in areas where winters are warm and summers are dry, as the evergreen habit allows significant photosynthesis during periods when deciduous trees would be leafless. Conifers also occur on poor sites within the areas normally dominated by hardwoods, largely because of lower demands for nutrients and water for production.

Conifers and hardwoods can have large differences in the chemical constituents from which they are made. When living tissues are shed as litter (for example, by the loss of leaves and needles in autumn), these chemical differences can affect soil formation and fertility. In general, deciduous litter decomposes more quickly than coniferous and tends to support higher levels of soil animal activity, leading to the creation of an A horizon, with moderately dense rooting throughout. Leaching is primarily of clays, so the B horizon is enriched in these and may contain some fine roots. The soil is called an **Alfisol.** Production is lower than in other forests described previously but higher than in tropical grasslands and shrublands, as is biomass. AET is close to PET, and drought stress is relatively slight. Nitrogen cycling can be higher than in temperate rainforest because of higher nitrogen concentrations in the biomass produced by deciduous trees.

Productivity in temperate coniferous forests may be similar to that in deciduous forests, but less of it may take the form of leaf litter, since needles are replaced only every two to five years. Longer needle retention may result in higher total foliar biomass in evergreen stands, even when foliage production is lower than in deciduous stands. Nitrogen cycling is significantly less in most cases, again because less production goes into relatively nitrogen-rich needles and more into nitrogen-poor wood. Needle litter also has less nitrogen than broadleaf litter, which reduces decomposition and mixing. A substantial O horizon can develop. The acidifying effect of conifer litter fosters podzolization, resulting in the formation of an E horizon. Rooting is concentrated in the O and upper B horizons.

In drier temperate regions, grasslands and shrublands occur that are somewhat similar to their tropical counterparts. Again, the reason for the predominance of one growth form over the other is not always clear, but the timing of precipitation and the occurrence of disturbance seem important. For example, the shrub-dominated chaparral vegetation type occupies coastal areas of the southwestern United States, the Mediterranean region, and western South America, where winters are warm and wet. In central North America, where winters are cold and most of the precipitation occurs in the summer, grasslands were predominant in presettlement times. However, fire exclusion and excessive grazing have encouraged a transition to forest at the humid end of the prairie region and to semiarid shrubland at the dry end. In either case, AET is far below PET and the length and timing of the growing season are determined mainly by the timing and quantity of precipitation.

Differences in growth form produce large differences in soil properties under shrubs and grasses. Although production may be similar for the two, grasses form much more extensive fine root systems, which incorporate much more organic matter into lower soil levels. This creates a very deep A horizon with no clear line between the A and B. Soil animal activity is high, surface organic accumulations are minimal, due perhaps as much to fire as to rapid decomposition, and the soil type is called a **Mollisol**.

Under shrub communities, as discussed for tropical shrublands, the soil-forming factors of calcification and salinization are more active, causing the accumulation of carbonates and salts.

To the north of the temperate forests and grasslands in the northern hemisphere is a circumpolar region of coniferous forest known as the boreal forest, or taiga. Even though precipitation can be much lower than in the tropical or temperate forest zones, cold temperatures reduce evaporation and transpiration so that excess moisture is nearly always present. Production is lower than for other forest types because of low temperatures, and nitrogen cycling is further reduced by the coniferous vegetation's low demand for this nutrient. The latter in turn creates very poor litter quality, which, along with cold temperatures, reduces decomposition rates. Soil animal activity is minimal and podzolization is the dominant process. However, soil development has generally not proceeded very far due to the cold temperatures and because the soils have been in place for only a few thousand years since the last glaciation. Such soils have large O horizons but less pronounced E horizons and are called **Entisols or Inceptisols** (actually, these two classes of soils can be found in any region where the soil-forming processes have been interrupted by such factors as glaciation, flooding, volcanic flow, erosion, etc.). Mature or old-growth forests, or those over shallow water tables, are susceptible to invasion by the ground moss *Sphagnum*, which can lead to creation of histosols (discussed below). Forests similar to the taiga in structure and growth form also occur in the temperate zone at high elevation. These are called subalpine forests.

The lowland tundra is a mixture of grass-like sedges, low shrubs, lichens, and mosses growing at the northern limit of vegetation in the northern hemisphere or above timberline in mountains elsewhere (alpine tundra). Production, decomposition, and evaporation are severely limited by low temperatures. Soils can be water-logged during the short growing season, further reducing decomposition and plant root activity. Large amounts of organic matter accumulate over a layer of rock or frozen water ("permafrost"). This soil type is called a **Histosol** (again, this soil class can be found in any region in wetlands where decomposition is restricted and large amounts of organic matter accumulate). Rooting occurs throughout the thawed organic horizon but is concentrated toward the top.

Under the driest conditions at all mean temperatures occur the sparsely vegetated "deserts." These can range from the moving sand dunes of the Sahara region to the uniquely vegetated Sonoran region in North America, to the moss and lichen systems of the high arctic and the Antarctic. Deserts mingle with dry grasslands and shrublands under slightly wetter conditions, and the boundary between them is difficult to define. It is at this boundary, however, that human impact on ecosystems can have its most profound effects. The term **desertification** is applied to the creation of deserts out of dry grasslands and shrublands through human disturbance. The climatic extremes in deserts, particularly water stress, allow plants only a tenuous existence. Ecosystems develop very slowly and are easily disrupted. Once disturbed, recovery may be slow or may not occur at all. Overgrazing and fuelwood harvesting are two of the main causes of desertification. Relatively few ecosystem-type studies have been done in deserts, where physical rather than biological processes are the dominant shapers of the landscape.

CORRELATIONS BETWEEN CLIMATE AND ECOSYSTEM STRUCTURE AND FUNCTION

Some fairly clear and not-too-surprising trends emerge from this brief survey of terrestrial vegetation and soils. The first is that total plant growth, expressed as primary production, is strongly related to temperature and moisture. These two variables are combined in actual evapotranspiration (AET). Figure 2.4 shows a relationship between AET and total above-ground production for several sites around the world. Total nitrogen cycling shows a similar pattern but with significant differences between coniferous and deciduous forest types, particularly in the temperate zone, due to differences in nitrogen use per unit production.

Total above-ground biomass in terrestrial ecosystems is also closely related to both AET and production. Figure 2.5 shows this relationship for mature systems only. Any disturbed or young systems will have less biomass than their mature counterparts.

Soil profiles show the effects of the relative intensity of the four major soil-forming processes. Laterization becomes less important with cooler temperatures. Podzolization becomes less important either with warmer

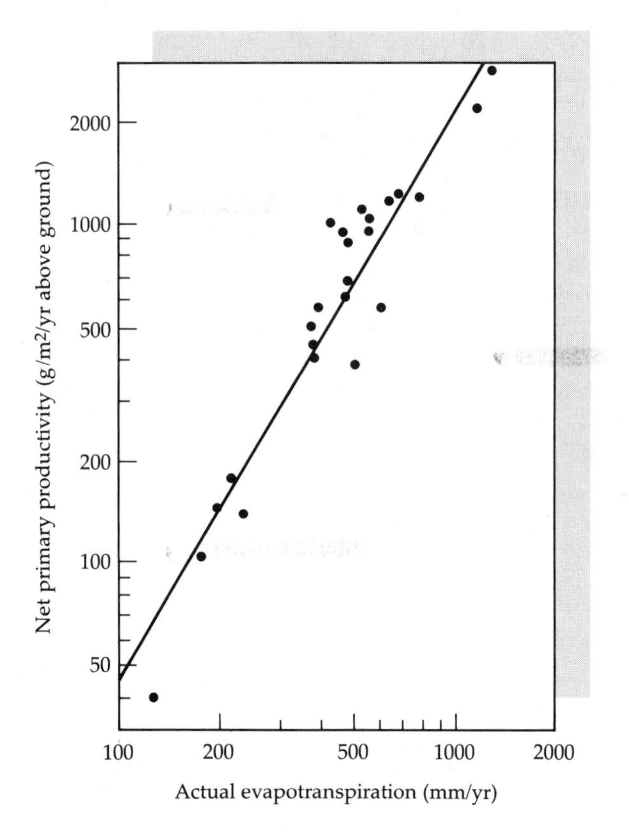

Figure 2.4
Net primary productivity of terrestrial ecosystems in relation to actual evapotranspiration. (Rosenzweig 1968, cited in Whittaker 1975)

temperatures or with decreasing precipitation. Lessivage/melanization increases in importance under near-neutral soil pH and where soil and vegetation allow large and active populations of soil-mixing animals. Calcification/salinization is important where the potential for evaporation exceeds precipitation and soil-water leaching is minimal, leading to the accumulation of calcium-rich minerals and salts. Surface organic accumulations tend to increase with decreasing temperature, being a minor component of total soil organic matter in the tropical rainforest and representing much of the active soil profile in the tundra Histosol.

VARIATION WITHIN LARGE CLIMATIC REGIONS

As we stated at the beginning of this chapter, this is a very oversimplified view of the interactions between climate, plants, and soils. Exceptions to every generalization presented here can be found. Spodosols can be found

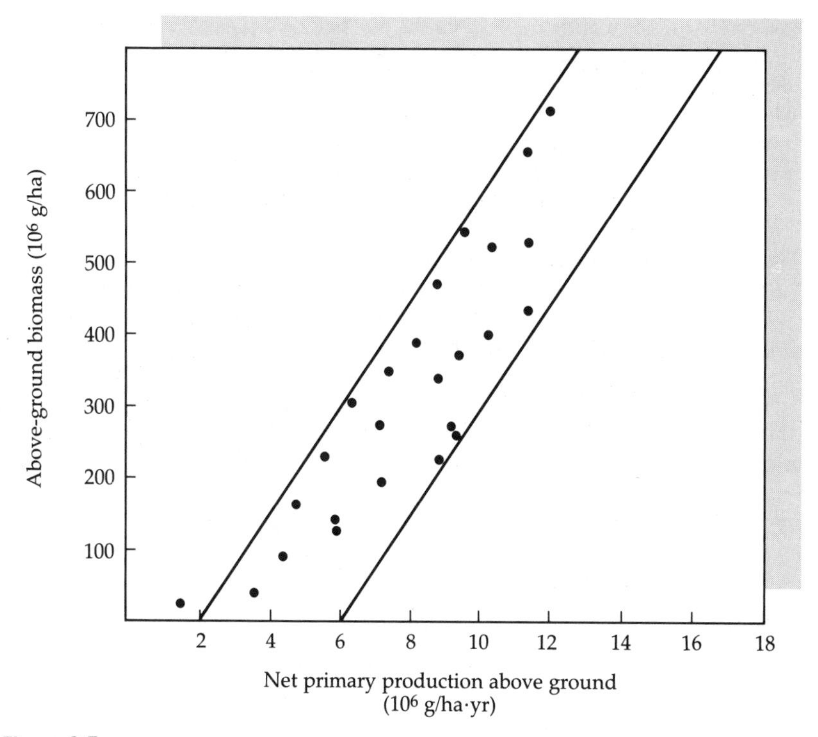

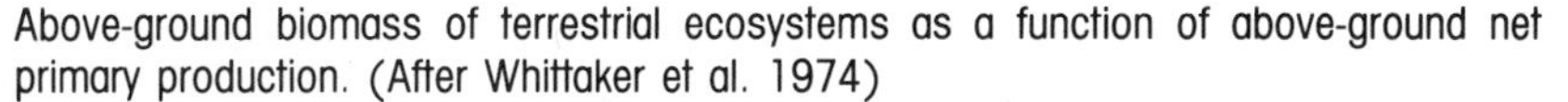

Figure 2.5
Above-ground biomass of terrestrial ecosystems as a function of above-ground net primary production. (After Whittaker et al. 1974)

in the tropics. Shrublands can be found in the heart of the deciduous forest. Sites with very low productivity occur in tropical rainforest regions. Without these exceptions, there would be little reason to study ecosystems in detail. If rates of important processes such as production, decomposition, and nutrient cycling could be predicted simply on the basis of AET, the subject would be dull indeed!

A closer examination of Figures 2.2 and 2.4 shows significant variation in both soil structure and production for a given climatic condition. For example, production ranges from 300 to 1000 $g/m^2 \cdot yr$ at 500 mm AET in Figure 2.4. Temperate forest soils show very different structures, ranging from spodosols to Alfisols, depending on soil texture, bedrock geology, species present, and other factors. In Parts II and III we will discuss the extent to which the interaction of soil, plants, consumers, decomposers, and disturbances can alter ecosystem structure and function within the limits established by climate.

REFERENCES CITED

Bliss, L. C., G. M. Courtin, D. L. Pattie, R. R. Riewe, D. W. A. Whitfield, and P. Widden. 1973. Arctic tundra ecosystems. *Annual Review of Ecology and Systematics* 4:359–399.

Bokhari, U. G., and J. S. Singh. 1975. Standing state and cycling of nitrogen in soil-vegetation components of prairie ecosystems. *Annals of Botany* 39:27–285.

Gray, J. T., and W. H. Schlesinger. 1981. Biomass, production and litterfall in the coastal sage scrub of Southern California. *American Journal of Botany* 68:24–33.

Murphy, P. G., and A. E. Lugo. 1986. Ecology of dry tropical forests. *Annual Review of Ecology and Systematics* 17:67–88.

Nadelhoffer, K. J., J. D. Aber, and J. M. Melillo. 1985. Fine root production in relation to total net primary production along a nitrogen availability gradient in temperate forests: A new hypothesis. *Ecology* 66:1377–1390.

Pastor, J., J. D. Aber, C. A. McClaugherty, and J. M. Melillo. 1984. Above-ground production and N and P cycling along a nitrogen mineralization gradient on Blackhawk Island, Wisconsin. *Ecology* 65:256–268.

Rosenzweig, M. L. 1968. Net primary production of terrestrial communities: Prediction from climatological data. *American Naturalist* 102:67–74.

Sims, P. L., J. S. Singh, and W. K. Lauenroth. 1978. The structure and function of ten western North American grasslands. I. Abiotic and vegetational characteristics. *Journal of Ecology* 66:251–285.

Sinclair, A. R. E., and M. Norton-Griffiths (eds.). 1979. *Serengeti: Dynamics of an Ecosystem*. The University of Chicago Press, Chicago and London.

Van Cleve, K., L. Oliver, and R. Schlentner. 1983. Productivity and nutrient cycling in taiga forest ecosystems. *Canadian Journal of Forest Research* 13: 747–766.

Vorosmarty, C. J., B. Moore III, A. L. Grace, M. P. Gildea, J. M. Melillo, B. J. Peterson, E. B. Rastetter, and P. A. Steudler. 1989. Continental scale models of water balance and fluvial transport: An application to South America. *Global Biogeochemical Cycles* 3:241–265.

Vitousek, P. M., and R. L. Sanford, Jr. 1986. Nutrient cycling in moist tropical forests. *Annual Review of Ecology and Systematics* 17:137–168.

Whittaker, R. H. 1975. Production. Chapter 5 in *Communities and Ecosystems*. Macmillan, New York.

Whittaker, R. H., F. H. Bormann, G. E. Likens, and T. G. Siccama. 1974. The Hubbard Brook Ecosystem Study: Forest biomass and production. *Ecological Monographs* 44:233–254.

ADDITIONAL REFERENCES

Buol, S. W., F. D. Hole, and R. J. McCracken. 1980. *Soil Genesis and Classification*. Iowa State University Press, Ames.

Emanuel, W. R., H. H. Shugart, and M. P. Stevenson. 1985. Climatic change and the broad-scale distribution of terrestrial ecosystem complexes. *Climate Change* 7:29–43.

Jenny, H. 1980. *The Soil Resource*. Springer-Verlag, New York.

Lieth, H., and R. H. Whittaker. 1975. *Primary Production of the Biosphere*. Springer-Verlag, New York.

Chapter 3

Measurement of Ecosystem Function I: Carbon Balances

C. Bauer/H. Armstrong Roberts

INTRODUCTION

We have now described ecosystems in terms of the conceptual parts that they all have in common (Chapter 1) and the tremendous diversity of structure and function they can exhibit (Chapter 2). It may be difficult to imagine just how the kind of information presented in Figure 2.2 is obtained. Just how do you go about measuring biomass or production or nitrogen cycling in something as large as an ecosystem? How might inputs and outputs of elements or water be measured in something as difficult to enclose or define as a forest stand? For the ecosystem concept to be useful, methods must be developed for measuring the structure and function of these systems. In this chapter and the next, we will present several classic studies that demonstrate different approaches to the measurement of whole ecosystem function. These two chapters should help convey something about the nature of ecosystem studies and also help visualize ecosystems the way a research scientist might see them.

In this chapter, several methods are introduced for measuring the overall carbon balance between terrestrial ecosystems and the atmosphere. Each method works from a slightly different conceptual model of the system. Both model and method will be presented simultaneously.

THE CARBON BALANCE OF TERRESTRIAL ECOSYSTEMS

The exchange of carbon as gaseous CO_2 between ecosystems and the atmosphere reflects the balance between the complementary processes of photosynthesis by plants and respiration by plants, animals, and microbes. This balance is receiving considerable attention now due to the increase in CO_2 concentrations in the earth's atmosphere resulting from the combustion of fossil fuels and the clearing of tropical forests. Several large-scale computer models have been devised to predict the effects of human activities on future atmospheric CO_2 levels and hence on plant growth and climate (see Chapter 24). An important part of such models is the role of terrestrial ecosystems in storing or giving off CO_2.

Basic questions about the rate of carbon fixation are also being asked to determine how efficient photosynthesis is as an energy-capturing process. Just what percentage of solar energy striking an ecosystem is converted to chemical energy in the form of carbohydrates? This question has been raised in discussions of the relative value of plants versus physical systems such as photovoltaic cells for converting sunlight to usable forms of energy.

Addressing the CO_2 balance first provides a good example of how the question determines the components of the ecosystem studied. The question in this case is: What is the total carbon balance for a terrestrial ecosystem of a certain type? We do not need to know the fate of carbon within the system. Using Figure 1.4*b*, reproduced here as Figure 3.1*a*, we need only know the net flux of CO_2 across the boundary between the ecosystem and the atmosphere. We need not even know the total flow in

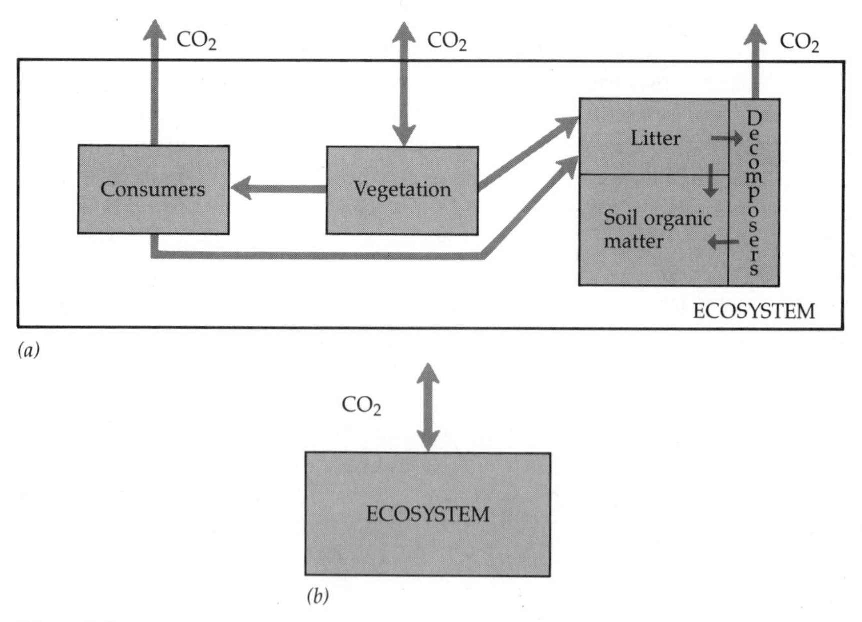

Figure 3.1
Conceptual models for the study of carbon balances in terrestrial ecosystems: *(a)* as developed in Chapter 1, and *(b)* as required for the measurement of total carbon balances only.

either direction, only the balance between them. Thus the conceptual model of system function is reduced to Figure 3.1*b*.

The method employed must reflect this boundary separating the terrestrial ecosystem from the atmosphere. Ideally, the method would allow accurate measurements while minimizing the effect of the method on the system. In other words, the method should not alter the function of the system significantly.

Three methods of measuring the carbon flux between ecosystems and the atmosphere have been developed. All depend on measuring changes in concentrations of CO_2 in air as it passes over parts or all of the system.

Method I. Whole System Enclosures

The method that produces boundaries for the system that are most similar to those in the conceptual diagram (Figure 3.1*b*) is simply to enclose the system in question, pump air through it, and measure the changes in CO_2 concentration between input and output air. The most spectacular example of this was carried out on a tropical rainforest in the El Verde forest of Puerto Rico (Figure 3.2). An entire stand (213 m^2) was enclosed in a giant cylinder of plastic over 20 meters tall. A fan was installed at the base of the

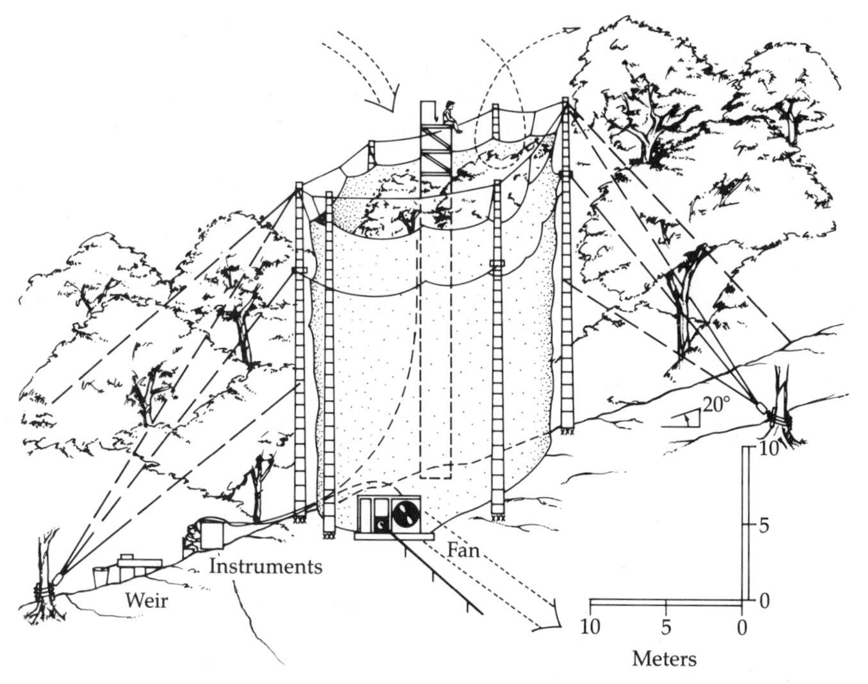

Figure 3.2

Diagram of the "giant cylinder" experiment with air exchange and measurement system for total carbon balance. (Odum and Jordan 1972)

cylinder to draw air down through it. A tall (21.5-m) instrument tower was constructed in the center to carry devices for measuring CO_2, moisture, and other parameters of incoming air. An identical set of measurements were made in the air leaving the fan, and differences were ascribed to photosynthesis, respiration, and transpiration by the enclosed vegetation and soil. Although the cylinder did not extend down into the soil, CO_2 generated within the soil would diffuse into the atmosphere within the cylinder before it could be lost. This would be measured at the fan.

Typical results of CO_2 balance for a single day are shown in Figure 3.3. Note that a net carbon gain by the system within the cylinder is recorded as a negative difference between the intake air and the air at the fan. When the air passing through the cylinder loses CO_2, the ecosystem has gained carbon. The mean change in concentration between intake and outlet air for this day is +3.33 parts per million CO_2. Concentration values can be multiplied by the rate of air movement through the fan to calculate the total CO_2 exchange for the day. Divided by the land area enclosed (213 m^2), this gives the rate of CO_2 taken up or given off per square meter of ground area per day ($g/m^2/day$). For the day recorded in Figure 3.3, there was a net carbon loss from the system of 10.2 g CO_2/m^2. To obtain a

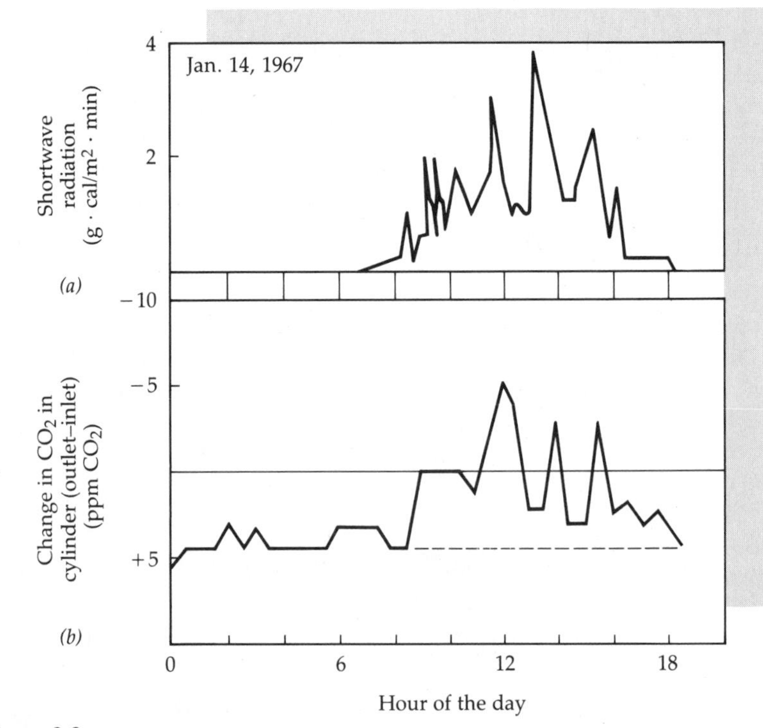

Figure 3.3
Data for a single day from the giant cylinder study: *(a)* net radiation, *(b)* carbon balance. (Odum and Jordan 1972)

balance for a full year, measurements can be taken daily for the full period, or a relationship can be obtained between the CO_2 balance and such factors as temperature, soil moisture, and time of year and a mathematical model developed to predict rates of CO_2 exchange for times when measurements are not made.

Two characteristics of ecosystem studies are apparent here. The first is the conversion of data from whatever original form (in this case CO_2 concentration and airflow rates) to rates expressed in units of mass per unit land area per unit time. The second is that a tremendous amount of work, time, and effort can be required to obtain a single number, in this case the CO_2 balance of a tropical rainforest.

The sample data set can also be used to calculate the efficiency of energy fixation, or the percentage of solar energy converted to chemical energy by photosynthesis. However, this now involves a different question and requires a different conceptual model of the system and a different set of calculations.

In this tropical rainforest ecosystem, only green plants can carry out photosynthesis and only during daylight hours. However, plants also respire, giving off CO_2, during metabolism in both light and dark. In addition, decomposers in both litter and soil respire CO_2 as well. Thus, we can redraw the conceptual model as in Figure 3.4. Note that conceptually and by measurement, respiration by plants and by microorganisms and animals are indistinguishable. A crucial assumption for the calculations is that total respiration of all parts of the ecosystem is the same in light and dark conditions. In Figure 3.3, nighttime respiration in the cylinder increases CO_2 concentration in the air passing through the cylinder by 5 parts per million. Assuming this rate also holds during the day, any decrease below this elevated CO_2 concentration during daylight hours (Figure 3.3) represents CO_2 fixation by photosynthesis (in Figure 3.3, the dashed line represents assumed rate of respiration during daylight hours).

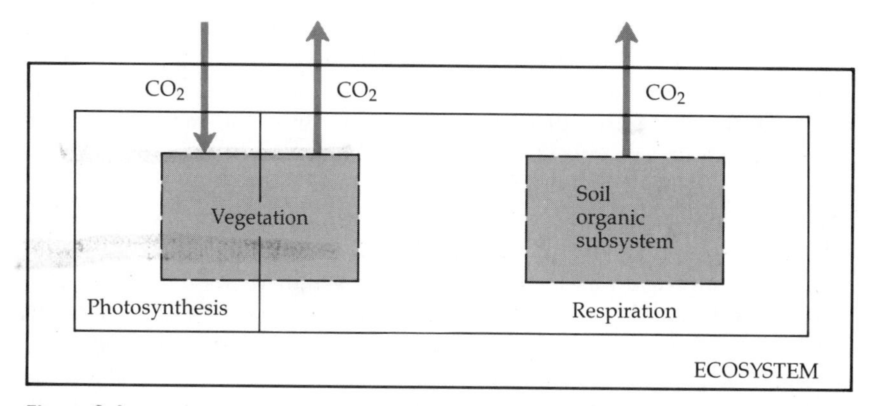

Figure 3.4
Conceptual model for the calculation of gross photosynthesis by plants and total respiration by plants, animals, and microbes, using data from the giant cylinder.

The calculations for the rate of CO_2 fixation during photosynthesis are similar to those for total CO_2 balance except that the concentration of CO_2 used is −1.67 rather than +3.33 parts per million. To derive an efficiency calculation, the quantity of CO_2 fixed must be converted to the amount of chemical energy in the bonds of the carbohydrate produced. This comes to 600 kcal/m^2 for this day. Dividing this value by the energy reaching this system in sunlight yields an efficiency of conversion of 2.3%. This value seems low, but it is not at all unusual. Very little of the energy in sunlight is actually converted to chemical energy in the bonds of carbohydrates by terrestrial ecosystems.

Results such as those provided by the giant cylinder lead directly to additional questions about the carbon dynamics within the system. What percentage of dark respiration is contributed by plants and what percentage by decomposers? How much plant respiration is associated with leaf function? Restating the second question, how much energy is left for export to the rest of the plant (stems, roots, flowers, and fruits) after leaf requirements are satisfied? Before considering these questions, we should introduce the terminology that describes the movement of carbon through ecosystems.

Production Terminology

A simple set of steps describes the fate of carbon fixed in photosynthesis (see Table 3.1). Gross photosynthesis is the total amount of carbon fixed by leaves, that total one-way transfer of carbon into foliage. Estimates of this value were obtained for the giant cylinder. Leaf respiration is the amount of carbon given off as a result of leaf metabolism. The difference between these two is the net carbon gain over the leaf, or net photosynthesis, the amount of carbon available to the rest of the plant. This value could not be obtained from the cylinder because it was not possible to separate leaf respiration from other sources of CO_2 in the system.

Part of the carbon exported by leaves to other parts of the plant is needed for respiration to keep cells in the stem and roots functioning. Subtracting this amount from net photosynthesis gives the total amount available for plant growth or the amount of structural material (leaves, wood, roots, etc.) produced. This is called **net primary production.**

Tissues produced can either be consumed by animals or retained until they fall to the ground as dead material (litterfall). Net primary production minus consumption and litterfall is termed **live biomass accumulation** and is the increase (or decrease!) in total weight of living plants. Live biomass accumulation can be negative, as for example, in a forest where a large blowdown of trees has occurred and a large quantity of living biomass suddenly becomes litter.

A similar set of steps can be described for carbon consumed by animals. Not all material eaten is digested. Subtracting that which is excreted leaves the amount assimilated. Of this carbon, much is respired; the remainder

Table 3.1 Terminology used to describe steps in the movement of fixed carbon and energy through terrestrial ecosystems

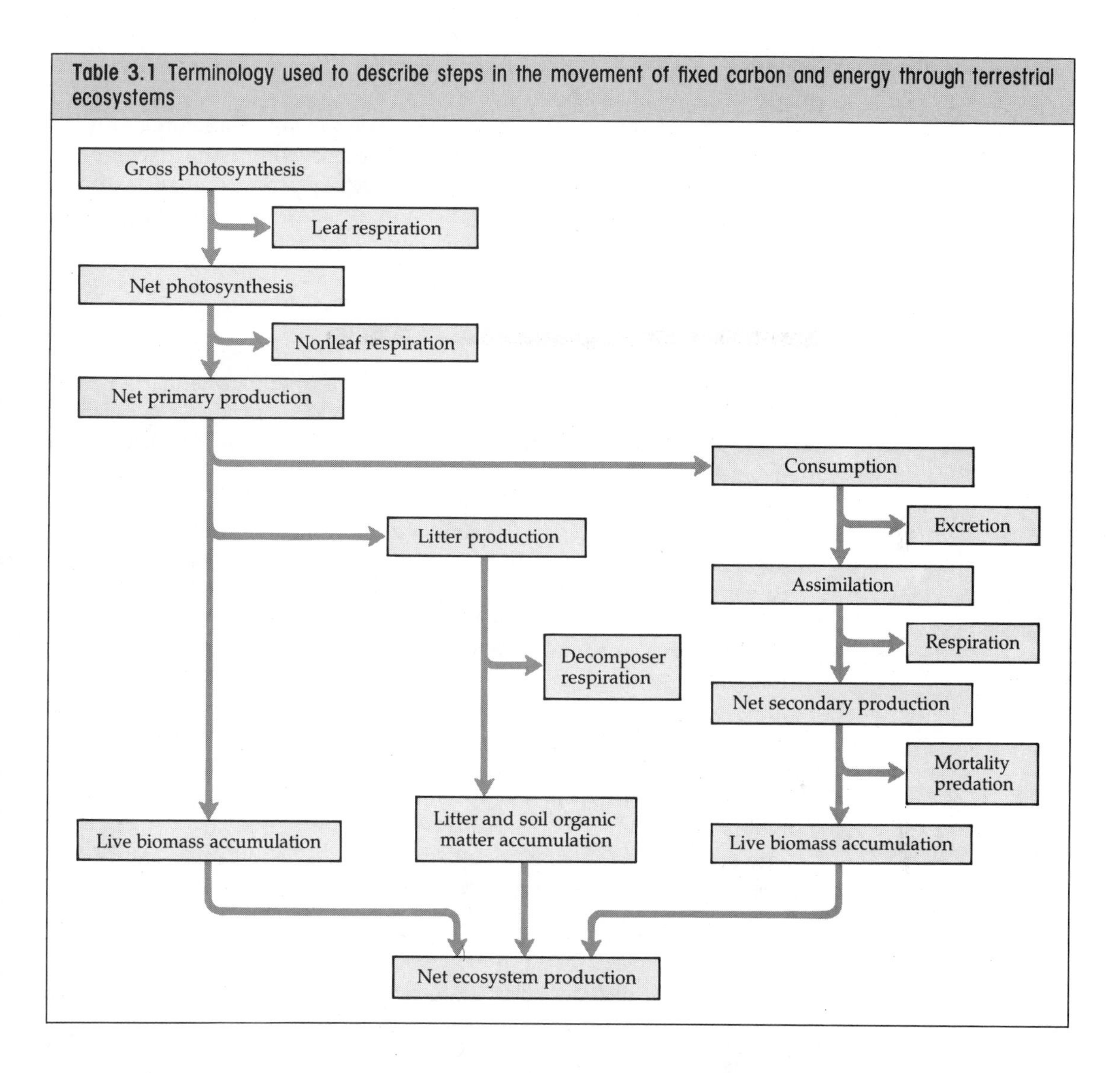

goes to growth and reproduction and is called **secondary production**. Subtracting mortality and consumption by other animals (predation) gives the change in standing crop, or the biomass accumulation (positive or negative). The portion consumed by predators goes through another, similar cycle.

Both fresh litter and accumulated soil organic matter are decomposed by soil microbes and animals with the generation of CO_2. Subtracting this from litter production gives the net accumulation (positive or negative) of total litter plus soil organic matter.

Net ecosystem production (NEP) is the sum of the changes in organic matter stored in live plant biomass, live animal biomass, and the soil organic matter pool. This is the total carbon balance of the ecosystem. It can be positive or negative.

Two more arrows could be drawn to create a complete picture of carbon flows. Consumption and predation both occur within the decomposer community, creating a second cycle of secondary production and biomass accumulation. In addition, animal mortality contributes to the pool of decomposable organic matter (litter) within the system.

Method II. Small-Chamber Enclosures

To begin to separate the overall carbon flux into components, the same approach of enclosing parts of the system and measuring CO_2 balances can be used, but the conceptual model and the methods of measurement used must be more detailed.

First, we need to separate leaf respiration from the rest of respiration. Next, it would be interesting to separate the rest of plant respiration and isolate plant function from that of decomposers, as in Figure 3.5.

To separate functions in this way, the cylinders must become much smaller and many more are required. While we can draw a box around photosynthesis conceptually, making a cylinder that would contain a large number of leaves without stems and branches is nearly impossible, except perhaps in the case of grasslands, where almost all above-ground biomass is photosynthetic. Where the small-cylinder technique has been applied in forests, the stand was separated into leaves, branches, stems, and soil and a subsample of each was enclosed in separate cylinders and monitored individually.

One example of this approach to measuring CO_2 balances and photosynthetic efficiency has been carried out in an oak–pine forest on sandy soils at the Brookhaven National Laboratory on Long Island, New York

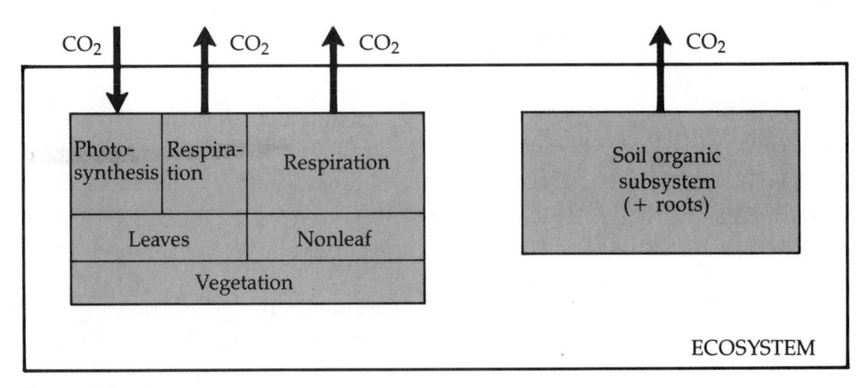

Figure 3.5
Conceptual model for measuring ecosystem carbon balances using the small chamber enclosure technique.

(Figure 3.6). Sampling problems arise here that were not important in the stand-size cylinder. If only a small portion of all the leaves in the stand are to be enclosed, care must be taken to sample the proper proportion of leaves of each species and also the right proportion in full sun, partial shade, and deep shade. Similar problems occur for enclosing stems. In soils, an additional problem is that plant roots and microbes cannot be physically separated without severely disturbing the soil system. Chambers placed over the soil to measure efflux of CO_2 actually include decomposer respiration plus plant root respiration. Thus, roots and the soil organic subsystem are measured together (see Figure 3.5). The difficulty of separating the functions of plants and microbes within soils is a recurring problem in ecosystem studies.

The Brookhaven carbon balance study involved a total of 75 cylinders each connected by a maze of plastic tubing to a central, computer-controlled sample lab, which obtained CO_2 exchange data from each cylinder once every 25 minutes. Computations were identical to those for the large cylinder but were carried out separately for each chamber and expressed per gram leaf weight or stem or soil surface area. Mean values for each cylinder type (for example, oak leaves or pine stem) were then obtained and multiplied by the total weight of leaves or surface area of stem in the stand (see the example of photosynthesis calculations in Table 3.2).

Total gross photosynthesis in this forest, assuming daytime respiration rates equal to dark respiration, was 5230 g/m^2/yr. This is equivalent to 13,546 kcal/m^2/yr of chemical energy produced as simple sugars. This is

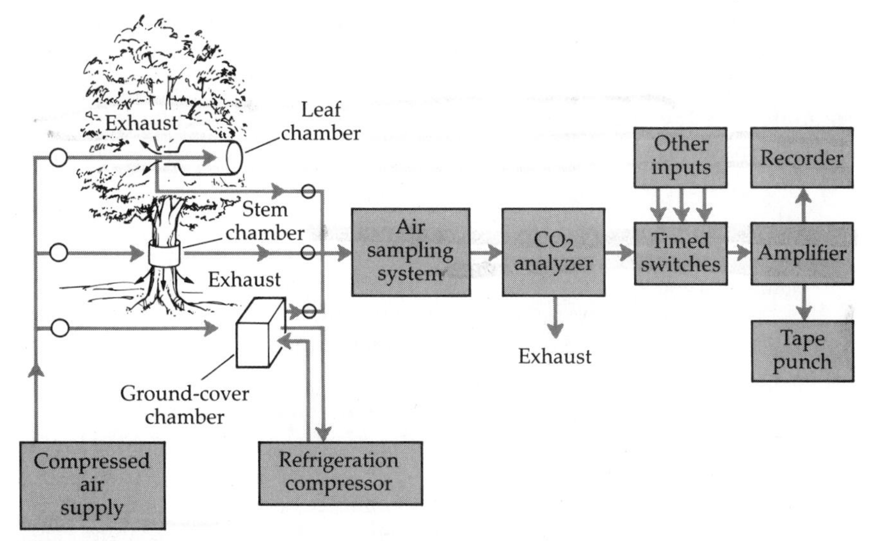

Figure 3.6
The Brookhaven small chamber system for monitoring CO_2 exchange rates of a forest. (Woodwell and Botkin 1970)

Table 3.2 Calculation of CO_2 balance in daylight and dark by leaves at the Brookhaven forest*

SPECIES	LEAF MASS (per land area, g/m²)	CO_2 UPTAKE		DARK RESPIRATION	
		(per leaf, g/g)	(per land area, g/m²)	(per leaf, g/g)	(per land area, g/m²)
White oak	110	16.8	1848	2.4	264
Scarlet oak	164	9.9	1624	1.0	164
Pitch pine	116	7.5	870	1.8	209
Other	—	—	228	—	23
Totals			4570		660

*From Botkin et al. 1970.

0.7% of total sunlight energy for the calendar year or 1.2% of total solar energy available during the growing season for this area.

For a total carbon balance, respiration from soil and from tree stems must be included. Chambers placed over both of these components always show increases in CO_2 concentration as respiration greatly predominates over photosynthesis. The annual pattern of respiration per square meter of bark surface largely reflects changes in temperature (Figure 3.7). Soil respiration shows similar trends.

Table 3.3 shows gross photosynthesis and the fate of the fixed carbon for the year studied at the Brookhaven forest. These results are more detailed than those obtained from the large cylinder in the tropical rainforest but also required more instrumentation and time. Despite the greater effort, the small-chamber method still cannot give an accurate estimate of either net primary production or biomass accumulation because respiration by roots and soil organisms cannot be separated.

Method III. Aerodynamic Analysis of Boundary Layer Atmosphere

It has been argued that enclosing all or part of the ecosystem in plastic containers alters the system enough to cause changes in function. Changes in temperature, wind speed, humidity, and CO_2 or O_2 concentrations could be particularly important. While researchers have put considerable effort into minimizing these differences, some certainly exist. Their importance is still subject to debate.

A third method for measuring carbon balances is to monitor changes in the concentration of CO_2 in the atmosphere above the ecosystem. This avoids the problem of enclosing the ecosystem but adds complexity, as the rate and direction of air movement over the study site must be known. Study of movements of air over ecosystems is a common practice in

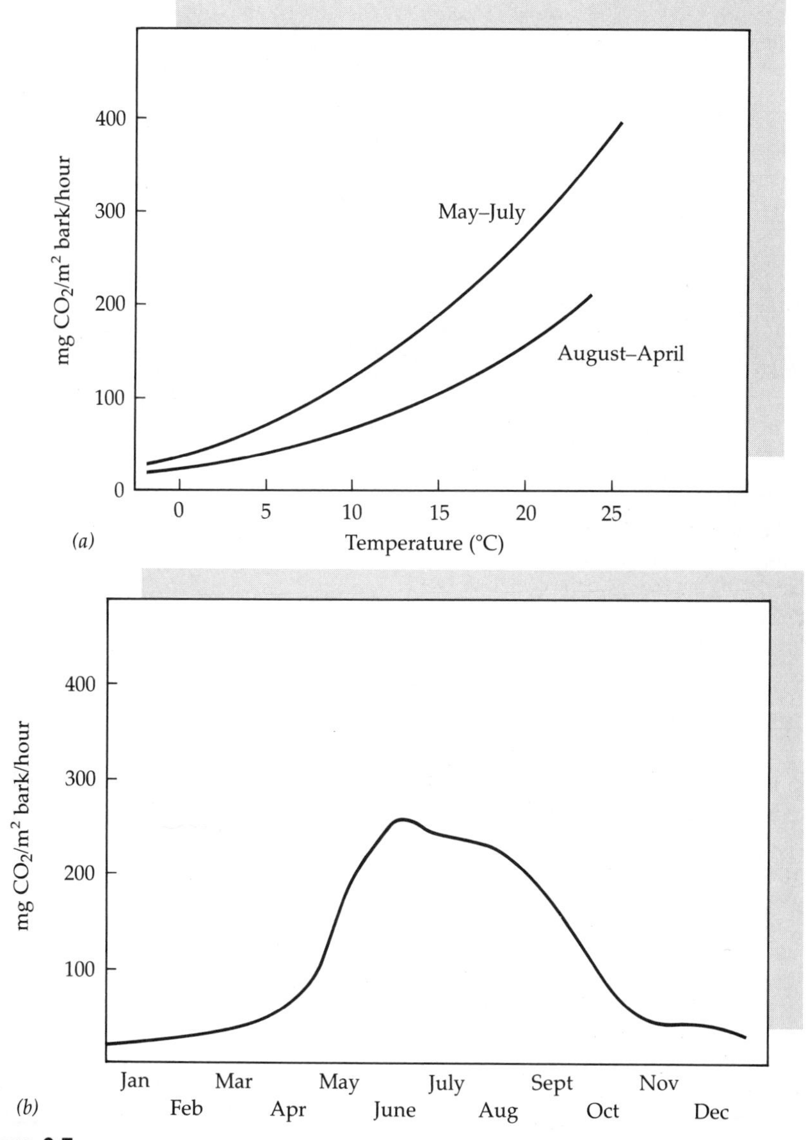

Figure 3.7

Respiration rates by tree stems in the Brookhaven study. *(a)* Rates as a function of air temperature. *(b)* Seasonal changes. (Woodwell and Botkin 1970)

meteorology, and the methods used are borrowed from that discipline. Because no measurements are made inside the ecosystems, only net CO_2 flux over the entire system can be measured, returning us to the conceptual model in Figure 3.1*b*.

Table 3.3 Gross photosynthesis and respiration by components of the Brookhaven forest*

		$g\ CO_2/m^2$
Gross photosynthesis (growing season)		5199
Leaf respiration	−	1524
Net photosynthesis	=	3675
Stem respiration	−	2049
Root and soil respiration	−	1724
Net ecosystem production	=	−98

*From Botkin et al. 1970, Woodwell and Botkin 1970.

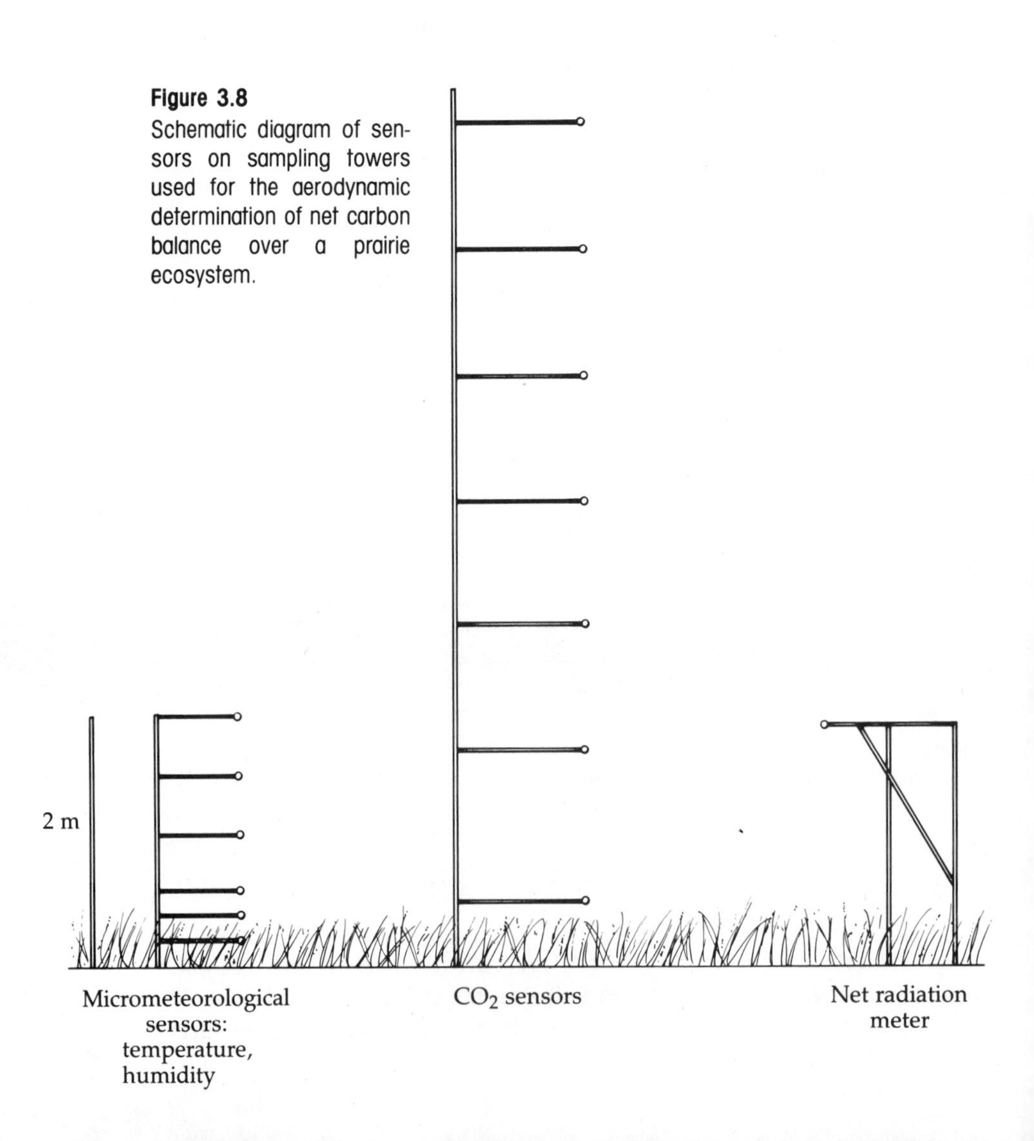

Figure 3.8
Schematic diagram of sensors on sampling towers used for the aerodynamic determination of net carbon balance over a prairie ecosystem.

A study of this type has been carried out over a short-grass prairie area in Saskatchewan (Figure 3.8). The instrumentation for the carbon balance measurement is a sampling mast 6.4 meters tall with several CO_2 analyzers at various heights (Figure 3.8*a*), combined with a second tower for measuring wind speeds and temperature (Figure 3.8*b*). From these data, the rate of turbulent airflow up from the canopy and away into the atmosphere can be calculated. Assigning CO_2 concentrations to the moving air masses allows calculation of total CO_2 flux. The CO_2 balance obtained by this meteorologic approach can again be separated into daylight and nighttime rates (Figure 3.9) to estimate gross photosynthesis as well as the total carbon balance.

A simplification of this system was used to measure nighttime respiration over the entire Brookhaven forest ecosystem. The same type of in-

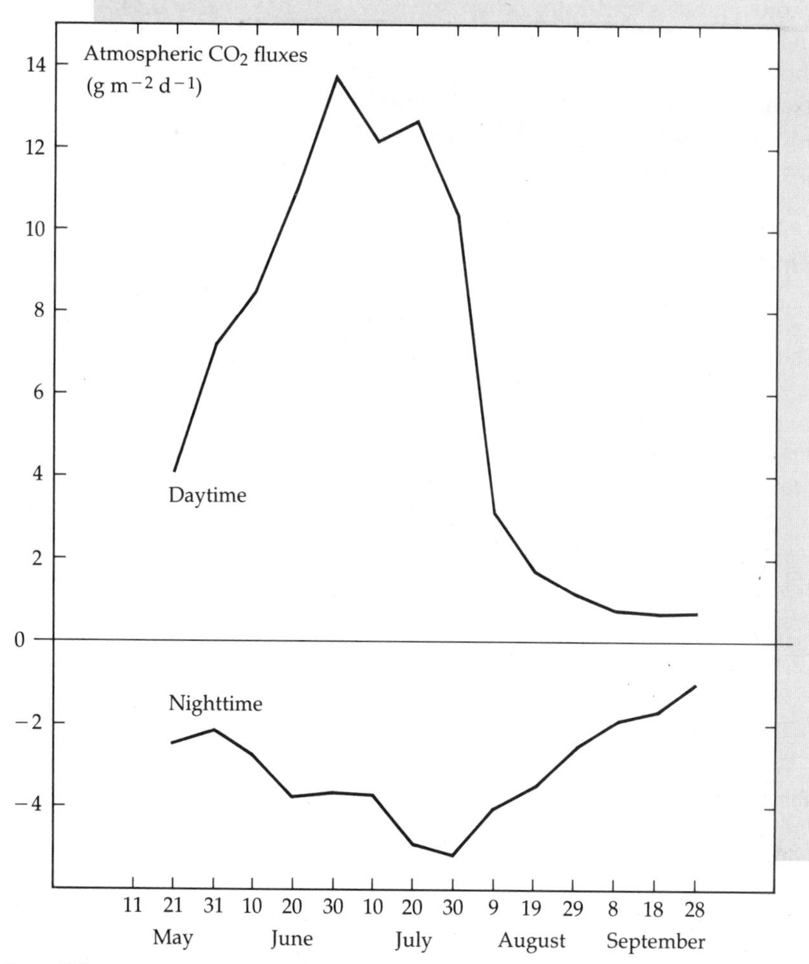

Figure 3.9
Seasonal variation in daytime and nighttime CO_2 balances over a prairie ecosystem. (Saugier and Ripley 1974)

strumentation was used, only on taller towers (21 meters) because of the taller vegetation. Concentrations of CO_2 were measured during the occurrence of nighttime temperature inversions, when warmer air overlies colder air and very little turbulent upward air movement occurs. Assuming stability of the air mass, the rate of total system respiration can be calculated as the change in total CO_2 content of the air mass. Unfortunately, this method is of limited value because temperature inversions on this scale are relatively rare.

COMPARISONS BETWEEN ECOSYSTEMS

Although different methods were used in each of the three systems described in this chapter, comparable data were obtained for both gross photosynthesis and total ecosystem CO_2 balance (Table 3.4). In accordance with our idealized discussion in Chapter 2, the tropical rainforest shows the largest gross CO_2 fixation, as a result of both faster rates of fixation and a longer growing season. The temperate forest at Brookhaven is intermediate in both daily and total gross photosynthesis, and the grassland is lowest. (Values in Table 3.4 and Figure 2.2 are not directly comparable because Table 3.4 lists gross photosynthesis as CO_2-fixed, while Figure 2.2 contains data on dry weight accumulation per year, which is gross photosynthesis minus plant respiration.)

Total carbon balances are quite different from gross photosynthesis. Respiration rates are also higher at higher temperatures and can balance high rates of CO_2 fixation. Both the Brookhaven and El Verde sites show near-zero net accumulation of carbon (actually a zero accumulation was assumed at El Verde in calculating gross photosynthesis). The grassland site shows a positive carbon balance, but data were collected only for the growing season (May–September). Respiration during the remainder of the year may have brought this balance much closer to zero.

Table 3.4 Gross photosynthesis and total carbon balance for three terrestrial ecosystems

SYSTEM	GROSS PHOTOSYNTHESIS			ANNUAL CO_2 BALANCE ($g/m^2/yr$)
	Mean Daily Rate for Growing Season ($g/m^2/day$)	Length of Season (days/yr)	Total ($g/m^2/yr$)	
Tropical rainforest	53	365	19,345	0
Temperate forest	29	180	5,200	197
Grassland	14	120	1,680	720*

*8 months only.

It may at first appear contradictory that ecosystems with such different rates of CO_2 fixation should all show similar and near-zero total CO_2 balances. However, this is a basic characteristic of mature (not recently disturbed) ecosystems. A constant, positive carbon balance would mean a continuous increase in total carbon within the system, approaching an infinite amount stored. This is impossible. Terrestrial ecosystems all tend toward a balance between CO_2 uptake and CO_2 evolution (this need not be the case for aquatic or wetland systems, where sediments or peat can accumulate for very long periods of time). The balance can be altered greatly by disturbances such as fire, harvesting, grazing, or plowing. Returning to the role of terrestrial ecosystems in the global atmospheric carbon balance, all mature and undisturbed ecosystems should have little net effect on atmospheric CO_2 content over long periods of time. However, disturbances, such as fire and clearing for agriculture, can cause releases of large amounts of carbon previously stored in plant biomass and soil organic matter. Currently there is much interest in the role of tropical forests on atmospheric CO_2. These forests have large amounts of stored carbon (see Figure 2.2) and are being harvested or converted to other uses at a rapid rate (see Chapters 22 and 24).

REFERENCES CITED

Botkin, D. B., G. M. Woodwell, and N. Tempel. 1970. Forest productivity estimated from carbon dioxide uptake. *Ecology* 51:1057–1060.

Odum, H. T., and C. Jordan. 1972. Carbon balance of the large cylinder. In Odum, H. T., and R. F. Pigeon (eds.), *A Tropical Rainforest*. Atomic Energy Commission, Washington, D.C.

Saugier, E. A., and B. Ripley. 1974. Microclimate and production of a native grassland. A micrometeorological study. *Oecologia Plantarum* 9:333–363.

Woodwell, G. M., and D. B. Botkin. 1970. Metabolism of terrestrial ecosystems by gas exchange techniques: The Brookhaven approach. In Reichle, D. (ed.), *Studies in Ecology*. Springer-Verlag, New York.

ADDITIONAL REFERENCES

More recent examples of CO_2 balances over enclosed systems can be found in:

Curtis, P. S., B. G. Drake, and D. F. Whigham. 1989. Nitrogen and carbon dynamics in C3 and C4 estuarine marsh plants grown under elevated CO_2 in situ. *Oecologia* 78:297–301.

Hillbert, D. W., and W. C. Oechel. 1987. Response of tussock tundra to elevated carbon dioxide regimes: Analysis of ecosystem CO_2 flux through non-linear modeling. *Oecologia* 72:466–472.

Tissue, D. T., and W. C. Oechel. 1987. Response of *Eriophorum vaginatum* to elevated CO_2 and temperature in the Alaskan tussock tundra. *Ecology* 68:401–410.

A more recent approach to measuring element balances using micrometeorological techniques is described in:

Hicks, B. B., D. R. Matt, and R. T. McMillen. 1989. A micrometeorological investigation of surface exchanges of O_3, SO_2 and NO_2: A case study. *Boundary-Layer Meteorology* 47:321–336.

Chapter 4

Measurement of Ecosystem Function II: Nutrient and Water Balances

R. Krubner/H. Armstrong Roberts

INTRODUCTION

The carbon balance discussed in Chapter 3 is only one part of the "metabolism" of ecosystems. Plants, animals, and microbes also require nutrients and water in order to function. The input and output balances for water and nutrients are linked to crucial global cycles that affect climate and environmental quality and are equally as important as the carbon cycle. Unlike carbon, inputs and outputs of mineral nutrients and other chemicals do not occur only in gaseous form. The movement of water through ecosystems provides a second important pathway, as both precipitation and streamflow can carry compounds either crucial or detrimental to life. Measuring the flow of water through ecosystems, and its nutrient or chemical content, has become a major activity in ecosystem studies.

The water balance of a terrestrial ecosystem is important in its own right. The amount of water available for transpiration plays a large role in determining the productivity of the plants in the system. The water that is not used by plants and finds its way to groundwater, or to a stream, becomes the source for human use. Some of the earliest ecosystem-level studies and experiments were carried out by hydrologists attempting to divert water from transpiration to streamflow. These experiments were conducted on whole watersheds so that the streams draining these watersheds could be used to measure total liquid water output.

The purpose of this chapter is to present an introduction to the watershed-ecosystem concept, to discuss its advantages and limitations, and to present important findings obtained by this and subsequent nonwatershed nutrient balance methods. The correlation between conceptual model and methods will again be presented, as in Chapter 3. Differences between different ecosystems will also be presented, although for watershed studies the spectrum is limited to areas with appreciable streamflow (mainly forested regions). One of the most powerful applications of the watershed technique is assessing the integrated response of all of the components of the ecosystem to human use, such as forest harvesting. This response can be "read out" as the change in the quantity and chemical quality of the water draining from the system. Results of this kind of research are directly applicable to questions of both forest productivity and water quality management.

NUTRIENT AND WATER BALANCES

An ecosystem's CO_2 balance reflects the difference between fixation through photosynthesis and respiration by plants, animals, and decomposers. The only major storage for carbon in the ecosystem is in some form of organic matter (actually a great deal of carbon can be found in limestone-type rocks within a system, but this carbon is part of a very different cycle; see Chapter 9). In contrast, mineral nutrients can exist in an array of inorganic forms with varying degrees of availability to plants. The increased complexity of transformations within the system is reflected in the differences between Figures 1.4*b* and 1.4*c* (the latter repeated here as Figure 4.1). The concentration of different nutrients in organic matter also

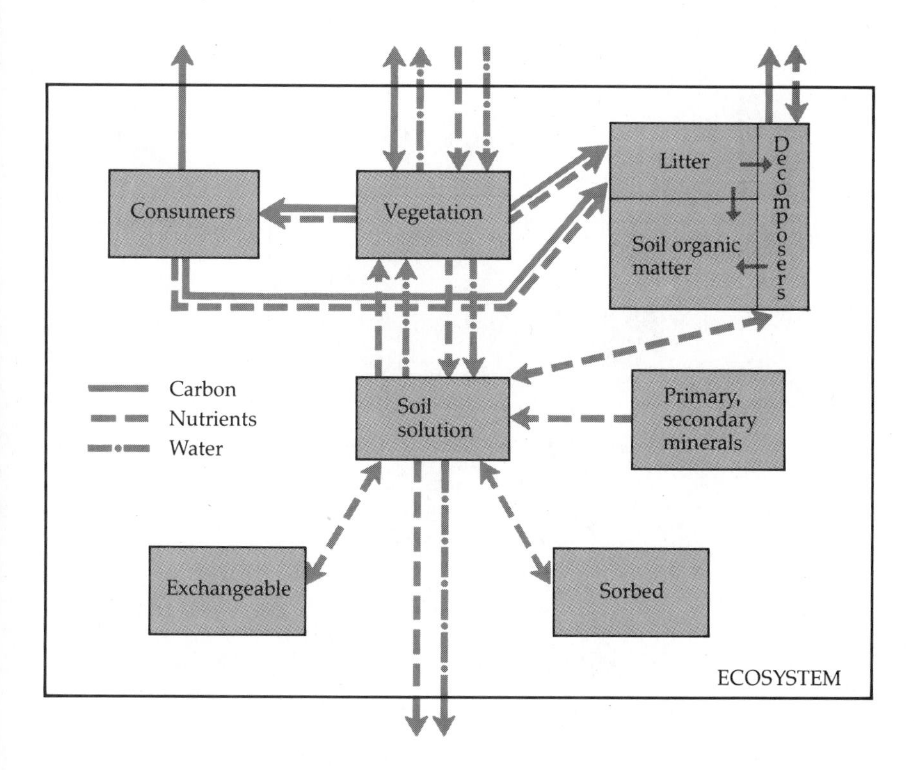

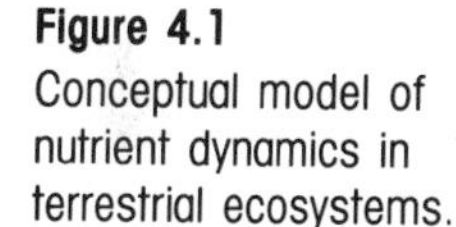
Figure 4.1
Conceptual model of nutrient dynamics in terrestrial ecosystems.

varies widely with the type of organic matter. For example, nitrogen concentration is very low in wood (about 0.1% of dry weight) and much higher in soil organic matter (2 to 4%). Inputs and outputs of nutrients need not parallel those for carbon.

Water balances can also be very different for different ecosystems. The importance of precipitation and evaporative demand (or potential evapotranspiration, PET) in determining the amount of actual evapotranspiration (AET), was mentioned in Chapter 2. Of particular importance for watershed-ecosystem studies is the division of the remaining water between surface flow (streams) and deep seepage (to groundwater). Storage within the system can also occur at different depths and changes with time.

These complexities will be treated at length in Part II of this book. They are mentioned here to introduce the idea that different ecosystems may show very different input–output balances and responses to disturbance even if total production or carbon balances are similar. At the watershed-ecosystem level, this internal complexity is initially ignored, and the conceptual model is greatly reduced to that in Figure 4.2.

METHODS IN WATERSHED-ECOSYSTEM STUDIES

There are two important criteria for a watershed study site: (1) that it be of reasonably small size so that conditions of vegetation, soils, geology, and

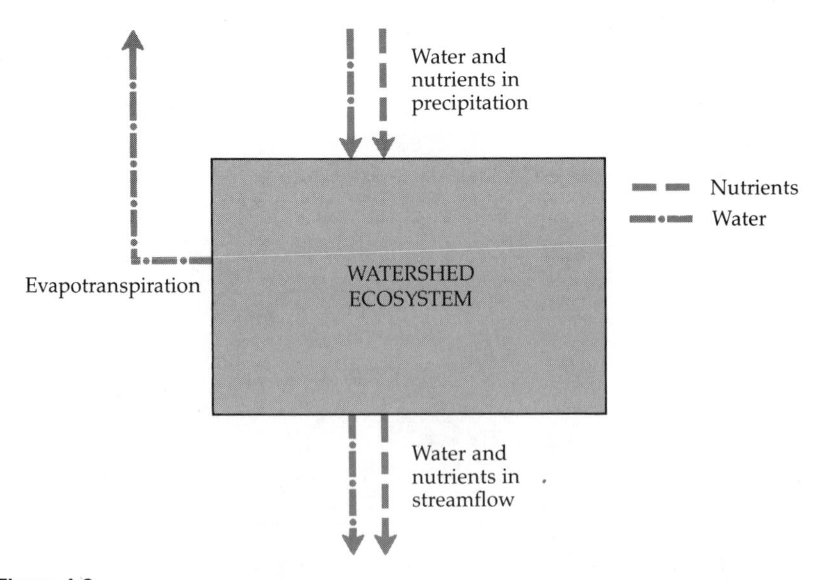

Figure 4.2
Simplified conceptual model for measuring water and nutrient balances over watershed ecosystems.

microclimate are sufficiently uniform and (2) that all liquid water leaving the system be measurable as streamflow. These two criteria together generally limit watershed-ecosystem studies to regions of extreme topography and relatively young, unweathered soils over hard rock formations such as granites. Extreme topography, as in mountainous areas, creates many small, distinct watersheds in a limited area. Shallow soils over hard bedrock keep water that leaches below the rooting zone near the surface, so it will appear in surface streamflow rather than be lost through deep seepage to groundwater.

Water and nutrient inputs to a watershed are measured with precipitation gauges for both rain and snow. From each collection, a subsample is taken to the laboratory and analyzed for nutrient concentration. Water quantity and concentration are multiplied together to give total nutrient inputs, usually expressed in grams per square meter per year ($g/m^2/yr$). Collectors are placed in small clearings at several locations within each watershed to account for variation in precipitation amounts and chemistry within the ecosystem, particularly with elevation.

Gauging water and nutrient losses from the system requires a weir of some type. The most common is the V-notch weir embedded in a concrete slab, which is set directly upon the bedrock at the base of the watershed. This forces all stream and subsurface water to flow up and over the weir (Figure 4.3). The height of the water column flowing through the weir is directly related to rate of streamflow in volume per unit time. The small structure next to the weir (Figure 4.3a) houses a recorder, which continuously monitors water height. A weighted mean flow rate for a month's time can be converted to rates of water loss in the same units as precipitation (depth/unit time).

Nutrient concentration in streamflow is measured by collecting a small amount of the water as it spills over the weir. However, this concentration will vary with the rate of flow and the time of year, so concentrations are measured under a wide range of flow rates throughout the year. Concentrations are again matched with the amount of water lost at different flow rates to yield a weighted mean concentration in stream water. Multiplying this times total streamflow gives total nutrient outputs. Again, the tremendous number of carefully collected and analyzed samples required to calculate this single number is apparent.

SOME RESULTS FROM WATERSHED-ECOSYSTEM STUDIES

One of the first watershed-ecosystem studies in the United States was begun at Coweeta in the 1950s. The work was not conceived as an ecosystem study, as that term was not widely known or used. Rather, it was one of several very applied studies in forest hydrology.

The Coweeta Experimental Forest (Figure 1.3) had been established in the Great Smoky Mountains of North Carolina as a site for hydrologic

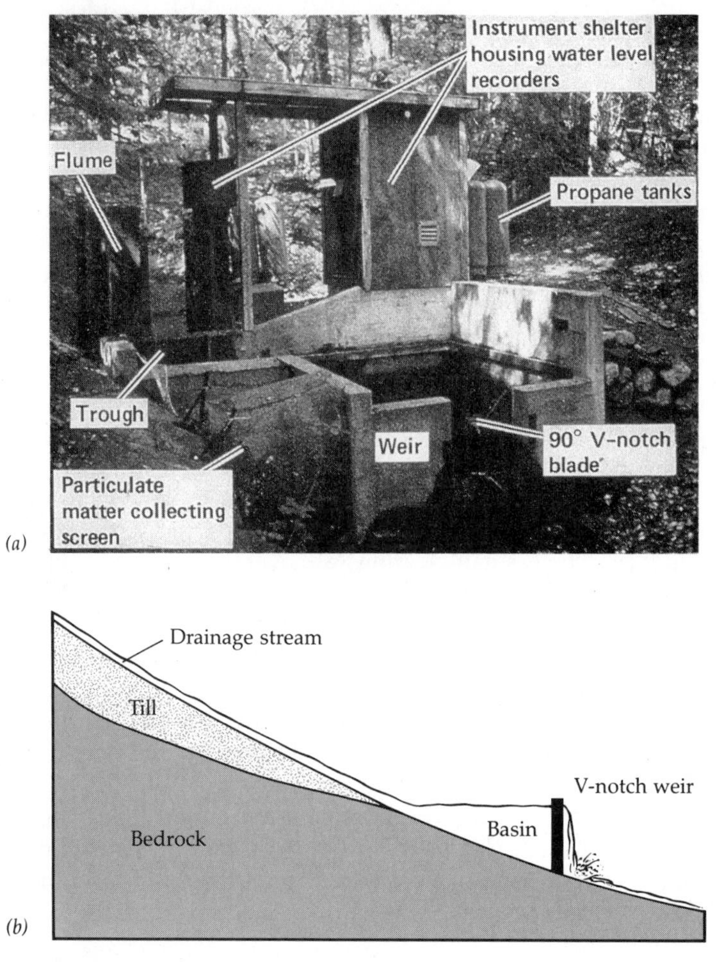

Figure 4.3
A stream gauging station for measuring stream-water losses of water and nutrients from a forested ecosystem. *(a)* The V-notch weir placed at the base of the watershed. *(b)* Schematic view of how the weir, when placed on watertight bedrock, forces all liquid water losses from the ecosystem to pass through the notch. (Likens et al. 1977)

research. It was known that forests use a considerable amount of water through transpiration and that increases in streamflow could be realized by cutting trees and removing the forest canopy. Unfortunately, at least for water yield, cutover forests regrow quickly in humid climates, and increases in water yield were short-lived. One idea for providing permanently higher yields was to convert the forest to grass, which might not transpire as much. A watershed-level experiment was established where the existing forest was cut and removed, and the area was replanted to grass. With the addition of a fertilizer treatment, this experiment actually became one of the first, at this scale, on the interactions between primary productivity and water and nutrient use in ecosystems.

Figure 4.4 shows the results of this experiment in terms of differences between expected and measured streamflow. The bars above 0.0 indicate

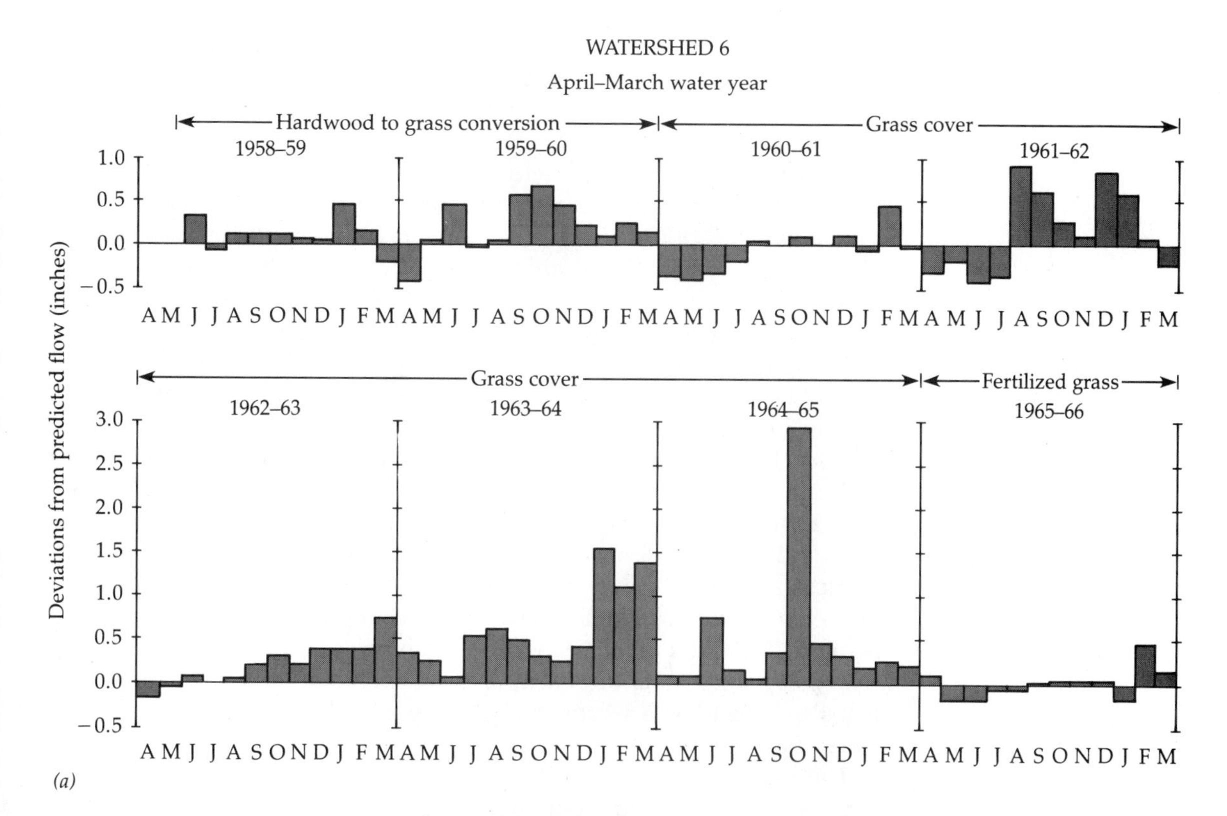

Oven-dry weight of grass
Difference in water yield
Difference in water yield (inches)
Dry weight of grass (tons/acre)
1960–61 61–62 62–63 63–64 64–65 65–66
Water year (April–March)
(b)

Figure 4.4

Results of forest-to-grass conversion at the Coweeta Hydrologic Laboratory. *(a)* Changes in seasonal streamflow due to conversion and the addition of fertilizer. *(b)* Relationship between increases in streamflow and the net primary production of grass. (Hibbert 1969)

increases in water yield due to the conversion to grass. In the first year (1960–1961), total water yield was roughly the same under grass as under forest, but more of it occurred in the fall and winter, with less in the spring (Figure 4.4*a*). In the four succeeding years, water yield rose. This coincided with a decline in the productivity of the grass cover (Figure 4.4*b*). To test the idea that this reduction in grass growth was due to reduced nutrient availability, a fertilizer treatment was carried out in 1965. Grass production returned to near first-year levels following the fertilization, and water yield declined markedly. These results showed that water use by grass could be just as high as by an intact forest when nutrient availability was sufficient to meet the plants' growth requirements. As nutrient availability declined, so did plant growth. This resulted in a decrease in transpiration and an increase in streamflow.

What happened on that converted watershed to cause nutrient losses and reduced grass growth? Increased water yield occurs because of decreased transpiration, which in turn results from reduced plant production. Is this reduced production somehow linked to the development of nutrient shortages? What role does productivity play in the retention of nutrients in an ecosystem? These kinds of questions on the interactions between important processes such as productivity, water use, and nutrient retention have been examined in detail at several other watershed-ecosystem sites, including a second research site initially established by the U.S. Forest Service.

The Hubbard Brook Experimental Forest in New Hampshire was also established as an outdoor hydrologic laboratory and used initially for forest hydrology research. However, since the early 1960s Hubbard Brook has been the site of one of the longest-running studies on the water and nutrient dynamics of forest ecosystems. Initially, nutrient concentration measurements were added to the regular monitoring of precipitation inputs and stream-water outputs. This allowed calculations of total nutrient balances over the watershed ecosystems much like the carbon balance calculations discussed in Chapter 3.

One of the major findings of the Hubbard Brook study is that undisturbed forests exhibit regularity and predictability in their input–output balances. This is particularly true for water. Figure 4.5 shows stream-water output at Hubbard Brook as a function of total precipitation. The difference between these two is evapotranspiration, which is nearly constant over this wide range of precipitation values. Most of this is transpiration by trees, indicating that water use by vegetation varies little from year to year in this forest. Water that is not used by plants appears in the stream. Thus an initial result of this study was an equation that predicts annual water yield from forests of this type as a function of precipitation. Such predictability can be important in the management of surface water resources.

The predictability at Hubbard Brook extends to the losses of certain chemical elements as well. Figure 4.6*a* shows losses for four major cations

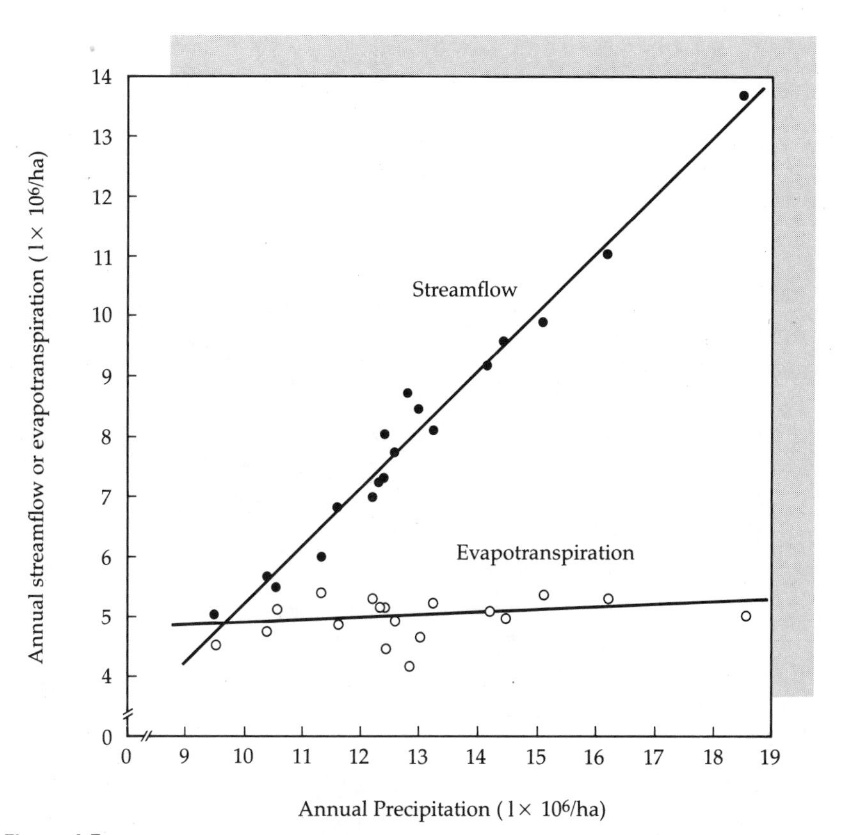

Figure 4.5
Relationship among precipitation, streamflow, and evapotranspiration for the control watershed at Hubbard Brook. (Likens et al. 1977)

as a function of annual streamflow. Nitrogen shows a more complex but still regular pattern of stream concentrations (Figure 4.6*b*). The seasonal changes in nitrogen losses have been linked to biological demand for nitrogen by plants and microbes, which keeps losses near zero during the growing season. Losses are higher in the dormant season, when biological activity is greatly reduced.

An early assumption of watershed studies, derived mainly from geologic methods, was that the storage of nutrients within the watershed ecosystem was not changing. This is called the **steady-state assumption**. If this were strictly true and precipitation and streamflow were the only significant inputs and outputs, then, according to the conceptual model in Figure 4.2, these inputs and outputs should be equal. They are not for most systems (Table 4.1).

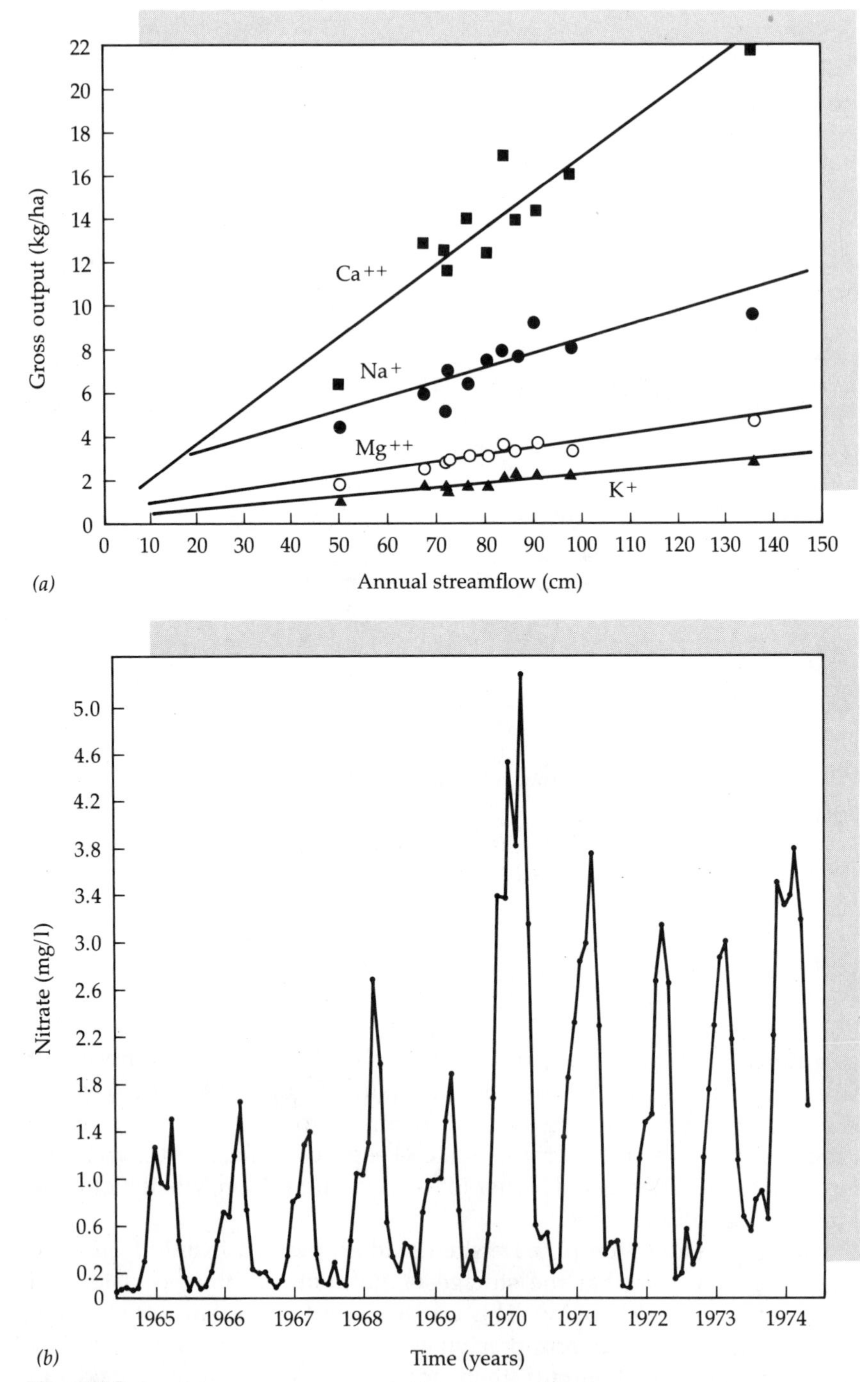

Figure 4.6

Nutrient outputs from the control watershed at Hubbard Brook. *(a)* Losses of calcium, sodium, magnesium, and potassium in relation to total annual streamflow. *(b)* Seasonal pattern of nitrate losses. (Likens et al. 1977)

Table 4.1 Element input–output balances for five forest ecosystems*

WATERSHED ECOSYSTEM	CALCIUM		MAGNESIUM		POTASSIUM		SODIUM		NITROGEN	
	In	Out	In	Out	In	Out	In	Out	In	Out
Hubbard Brook	2.2	13.7	0.6	3.1	0.9	1.9	1.6	7.2	6.5	3.9
Coweeta	4.5	7.5	1.0	3.8	2.5	5.5	4.5	11	4.9	0.2
Walker Branch	12	148	2.0	77	3.0	7.0	4.0	4.5	5.8	0.4
Andrews	4.5	55	.5	12	0.3	2.0	2.0	29	0.7	0.2
Tropical rainforest	28	3.9	3.0	0.7	24	4.6				

*All data given in kg/hectare-year.
From Likens et al. 1977, Henderson et al. 1978, Vitousek and Sanford 1986.

For several important ions (calcium, magnesium, sodium, and potassium), outputs are almost always larger than inputs. Geologists have long used this information with a slightly different conceptual model of the system to estimate rates of rock weathering by the difference between outputs and inputs of the elements. Large amounts of these and other ions are present in most systems as minor components of primary minerals or rocks. In limestone formations, calcium and occasionally magnesium are major constituents. Several reactions release these chemicals in simple ionic forms at a slow but continuous rate (see Chapter 9). By removing primary minerals from the ecosystem box as in Figure 4.7, measuring

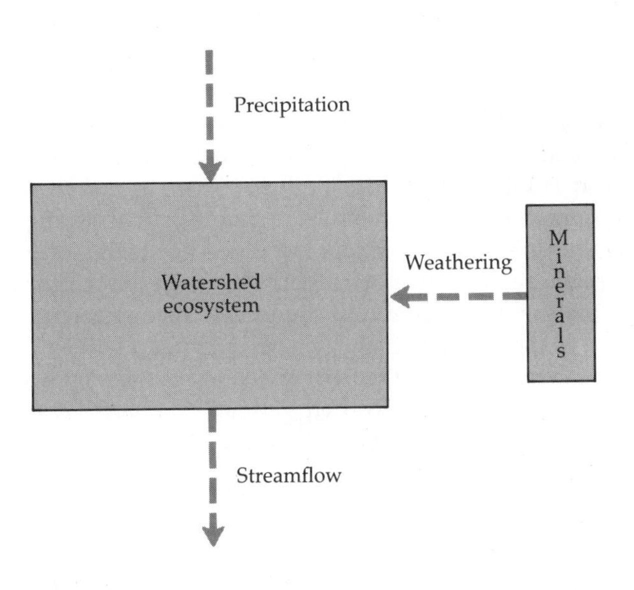

Figure 4.7
Modified conceptual model of nutrient balances over ecosystems, with primary minerals and weathering rates considered external to the ecosystem.

precipitation inputs and stream-water outputs, and assuming steady state for the remaining parts of the system within the box (soils and biological components), the weathering rate can be calculated as outputs minus inputs.

In general, results from this method coincide with expectations based on bedrock geology (Table 4.1). Streams draining the Hubbard Brook and Coweeta sites, both having granitic bedrock, which is low in calcium and magnesium and slow to weather, have low concentrations of these elements. In contrast, the Walker Branch watershed in Tennessee is on dolomitic limestone, which weathers more rapidly and is rich in both calcium and magnesium. The andesite substrate under the H. J. Andrews Forest in Oregon is intermediate in both content of calcium and magnesium and in rate of weathering.

These results can be compared with nutrient balances estimated by nonwatershed techniques (see below) for a tropical rainforest on a highly weathered soil (Oxisol, Table 4.1). Outputs below the rooting zone are actually less than inputs. The vegetation is taking up and storing nutrients and, in this case, the old, fully weathered soil does not provide significant new cations as weathering products.

Nitrogen balances in the temperate forest ecosystems are quite different from balances of other ions. All four ecosystems in Table 4.1 show net decreases in this nutrient between precipitation and streamflow, similar to those for other ions in the tropical system. The reasons are the same. All four are increasing in total nitrogen content, and weathering rates for nitrogen are negligible (there are no important rock types in which nitrogen is a significant component). From this limited analysis, which does not include gaseous exchanges, all five ecosystems are accumulating nitrogen.

STUDIES ON RESPONSES TO DISTURBANCE

One of the major advances made possible by the watershed-ecosystem approach is the experimental manipulation of entire landscape units and the measurement of the integrated response of all components of the system as judged by water and nutrient balances. This approach was first undertaken at Hubbard Brook in the mid-1960s, at a time when much controversy surrounded the use of clearcutting (the harvesting of all trees on a site) in the United States national forests. The results were relevant to that controversy because of the similarity between the size of the watersheds used for the experiment and the size of management units in the national forests and because the watershed method allowed the straightforward measurement of the effects of different practices on water quality.

However, the experiment carried out at Hubbard Brook was not designed to directly test the effects of clearcutting but rather to assess the effect of higher plant processes on ecosystem function. In particular, the question was asked: To what extent were the input–output balances in the undisturbed watershed (Table 4.1) and their predictability (Figures 4.5 and 4.6) controlled by plant function? Thus the experiment did not involve commercial clearcutting but rather a complete devegetation. One of the watersheds (watershed 2, or W2) that was determined to be nearly identical to a control, undisturbed watershed (W6) was selected for the devegetation. All trees on the watershed were cut, and the area was sprayed with herbicide to kill all plant regrowth. No logging roads were constructed into the watershed and all downed trees were left on-site.

Figure 4.8 shows that removal of vegetation had a marked effect on water and nutrient balances. Summer streamflow during the devegetation experiment, which lasted three years, was nearly four times higher than in the control watershed because evapotranspiration was now limited only to evaporation from the soil surface. (Note that all results in Figure 4.8 compare W2 with W6, which was left undisturbed. By this comparison, differences in hydrology due to annual differences in weather patterns are accounted for. The differences between the two lines in each panel express the effect of the experiment, not year-to-year variability.) While this increase in streamflow would, by itself, be expected to result in increases in total nutrient losses (see Figure 4.6*a*), increases in the concentration of nutrients within the stream also occurred, especially for nitrate (NO_3^-). These combined to yield increases in loss rates in g/m^2/yr of from 2.5 to 44 times that for undisturbed conditions. Nutrients such as nitrogen and potassium, which are used in large quantities by plants, showed the greatest increases in loss rates.

Nitrate concentrations in the stream water from the devegetated watershed exceeded U.S. Public Health Service standards for drinking water. This created concern over the effects of commercial cuttings as well. However, since this experiment was not an actual clearcut, and especially since all vegetation was killed for three years, it was argued that commercial cuttings would have a lesser effect. On the other hand, some argued that the removal of wood and the construction of logging roads into a real clearcut would make conditions worse.

To answer these arguments, concentrations of nitrate were measured in stream water draining several commercially clearcut watersheds. Results showed that concentrations were indeed lower in these streams than on watershed 2, most likely due to nitrogen uptake by the regrowing vegetation. Still, nitrate concentrations in some cases were above Health Service standards.

A question of both basic and applied interest immediately arose. Were high nitrate losses following cutting a general response of forest ecosystems? Would other systems show the same high concentrations of nitrate? Clearcutting experiments in other regions and forest types were soon conducted, with varied results. A Douglas fir forest in the Pacific North-

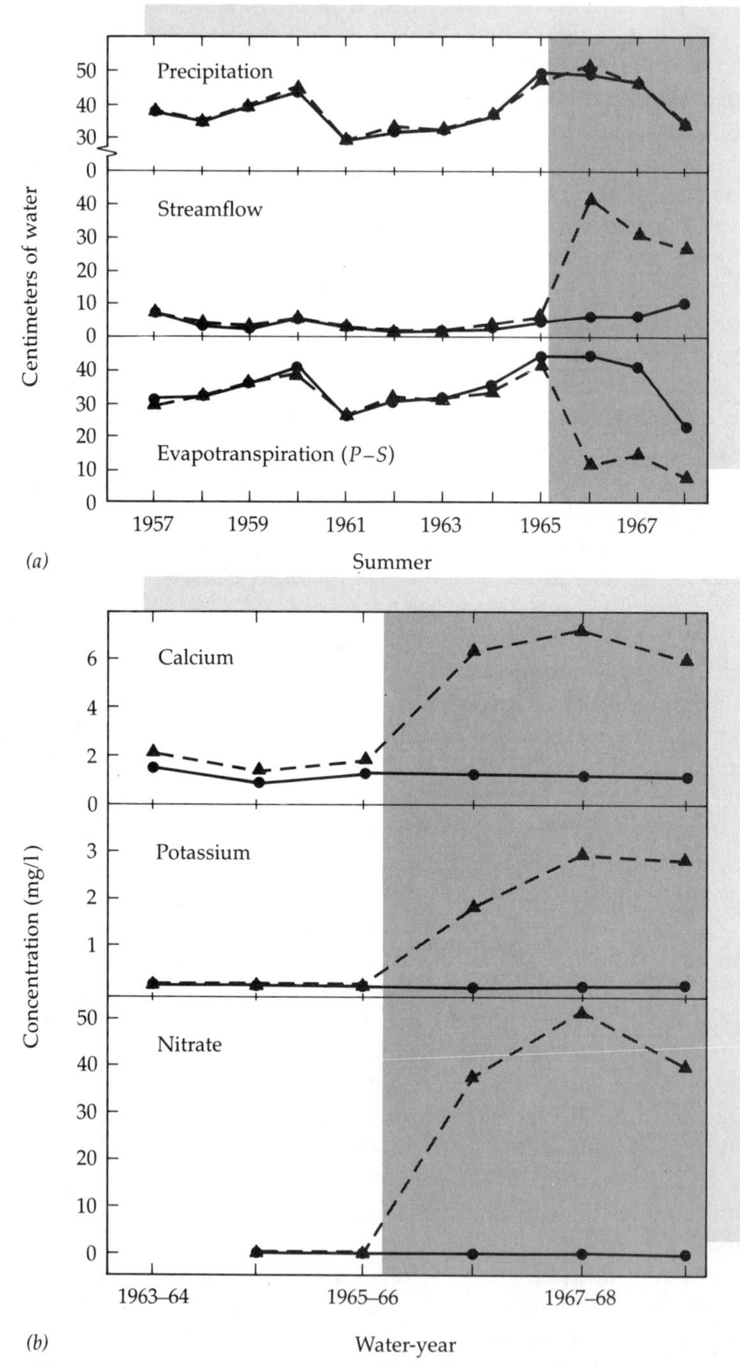

Figure 4.8

Effects of devegetation of watershed 2 at Hubbard Brook. *(a)* Changes in summer water balance. *(b)* Changes in stream-water chemistry. (Bormann and Likens 1979)

west showed negligible increases in nitrate loss. A hardwood forest in West Virginia showed slight but measurable increases. An alder forest in Alaska showed very high loss rates.

The low nitrogen loss response is the most difficult to explain. Looking again at Figure 4.1, nitrogen is cycling within the forest from litter to available condition due to decomposition and is then taken up by the vegetation. With uptake cut off, one of the other pathways within the system would have to become more important. The leaching of nitrogen to streams is the most obvious, and yet in many cases this did not happen. (Some possible mechanisms for this response will be discussed in Parts II and III.)

This fundamental difference in the response of forest ecosystems to similar types of disturbances led to a systematic, comparative study of a wide range of forest types. The goal of this study was to measure nitrate-leaching loss rates from as wide a range of forest and soil conditions as could be found in the United States, using identical methods of disturbance and measurement in each stand. However, it was not possible to establish watershed study areas for all of these sites because of the requirements of topography, size, and bedrock conditions listed above, as well as the cost. Instead, ten small study plots were established in each stand by digging a trench around small (1 – 6 m^2) blocks of soil to a depth of 1 meter, or below the rooting zone. Thus a good-sized block of soil was severed from the direct influence of living plants.

Without the benefit of whole watersheds with streams to be used for sampling, some other method for collecting water was required. In this case, a tube lysimeter, which is a small tube of plastic with a porous ceramic cup at the bottom and a rubber stopper on top, was installed in each soil block. The tube reaches below the rooting zone, and suction is created in the tube to draw water in through the porous cup. While measurements of water volume lost cannot be made by this method, relative concentrations of nutrients in this water can be obtained. The differences in methods of disturbance and measurement between the watershed approach and the trenched plot lysimeter approach indicate that methods of very different scale and requiring very different amounts of effort can be used with the same conceptual model.

Results from this study confirmed a wide range of nitrogen loss rates, even for the same forest type within the same region (see Table 4.2, especially the two Douglas fir sites in Washington). The general conclusion drawn from this study was that stands on richer or better sites, those in which nitrogen was presumably cycling at a faster rate, lost the most nitrogen following disturbance. However, it became clear from this type of work that more detailed information about the internal mechanisms of nitrogen cycling are required in order to understand, and possibly predict, the nitrogen output response for any particular forest ecosystem. We will return to the topic of nutrient losses following disturbance in Chapters 17 and 20.

Table 4.2 Concentrations of nitrate in water collected below the rooting zone in control and trenched soils from several forest ecosystems*

SITE	LYSIMETER NITRATE CONCENTRATIONS (μeq/liter)	
	Control	Trenched
Indiana		
Maple, beech	15	2150
Oak, hickory	12	1510
Shortleaf pine	20	175
Massachusetts		
Oak, pine	0	932
Red pine	0	263
Oak, red maple	1	140
New Hampshire		
Maple, beech	105	1055
Balsam fir	45	570
New Mexico		
Ponderosa pine	1	60
Mixed conifer	0	784
Aspen	0	645
Spruce, subalpine fir	1	24
North Carolina		
Mixed oak	0.5	434
White pine	1.9	610
Oregon		
Western hemlock	25	730
Washington		
Alder	371	1571
Douglas fir (low site quality)	1.4	114
Douglas fir (high site quality)	6.1	779
Pacific silver fir	6.2	5.6

*From Vitousek et al. 1979.

REFERENCES CITED

Bormann, F. H., and G. E. Likens. 1979. *Pattern and Process in a Forested Ecosystem.* Springer-Verlag, New York.

Henderson, G. S., W. T. Swank, J. B. Waide, and C. C. Grier. 1978. Nutrient budgets of Appalachian and Cascade region watersheds: A comparison. *Forest Science* 24:385–397.

Hibbert, A. R. 1969. Water yield changes after converting a forested catchment to grass. *Water Resources Research* 5:634–640.

Likens, G. E., F. H. Bormann, R. S. Pierce, J. S. Eaton, and N. M. Johnson. 1977. *Biogeochemistry of a Forested Ecosystem.* Springer-Verlag, New York.

Vitousek, P. M., and R. L. Sanford, Jr. 1986. Nutrient cycling in moist tropical forests. *Annual Reviews of Ecology and Systematics* 17:137–167.

Vitousek, P. M., J. R. Gosz, C. C. Grier, J. M. Melillo, W. A. Reiners, and R. L. Todd. 1979. Nitrate losses from disturbed ecosystems. *Science* 204:469–474.

ADDITIONAL REFERENCES

Binkley, D. 1986. Harvesting, site preparation and regeneration. In *Forest Nutrition Management.* Wiley Interscience, New York.

Vitousek, P. M., and W. A. Reiners. 1975. Ecosystem succession and nutrient retention: A hypothesis. *BioScience* 25:376–381.

Chapter 5 Concepts and Terminology in Ecosystem Studies

D. Muench/H. Armstrong Roberts

INTRODUCTION

Implicit in most of the following chapters is a "systems analysis" approach to the understanding of ecosystem function. Systems analysis is a method of describing and understanding complex interactions between large numbers of processes or components in a generalized way. It is concerned with the process of identifying the fundamental units of a system, defining how they interact and how they function in response to changing conditions.

As an example, consider the process of photosynthesis. Textbooks in botany and plant physiology deal with photosynthesis as a biochemical process, describing the organelles involved, the metabolic pathways, the intermediate products, enzyme systems, and so forth. Relatively less attention is paid to the rate of photosynthesis, how that rate is controlled by leaf condition, or how photosynthesis interacts with transpiration, leaf energy balance, light absorption by whole canopies, and other factors external to the leaf.

In contrast, a systems approach stresses factors controlling rates of photosynthesis, including the effects of water stress, nutrient availability, shading within the canopy, climate, and the genetic potential of the species. In addition, the further effects of photosynthesis on the movements of carbon, energy, water, and nutrients are of primary importance. The same emphasis and focus holds for other processes as well. For example, ecosystem analysis is more concerned with rates of organic matter decomposition and how this process interacts with plant uptake and nutrient cycling than with how it is achieved biochemically.

The purpose of this chapter is to introduce some of the concepts and jargon associated with systems analysis that have become part of the normal terminology of ecosystem analysis. The presentation will necessarily have a theoretical cast to it, as the terms are supposed to be useful in a most general way. Still, concrete examples are presented of the uses of each term in ecosystem studies. It is not necessary to feel completely comfortable with all of these concepts before proceeding to the rest of the book, but some familiarity with them will be helpful. This chapter can also be used as a reference section for terms used in later chapters.

SYSTEMS ANALYSIS AND ECOSYSTEMS

Systems analysis can be applied to any set of separate, definable components that interact. It was developed first, and has been used most extensively, in electrical engineering, particularly in designing computers, telephone systems, and other devices of similar complexity. However, our summary view of ecosystem function presented in Figure 1.4 also fits these criteria. Several separate components or parts of the system are defined, each enclosed in a separate box. Those components that interact with each other are connected by arrows showing possible directions of movement. For the application of systems analysis, the names on the box and arrows are not crucial. They could define traffic patterns in a city, currency flow through an economic system, or materials flow through a factory. What is of concern is how the outputs from any one component vary as a function of inputs and how all of the flows in the system affect overall system performance.

This does not mean that a meaningful analysis of a system can be carried out without a thorough knowledge of how the components in the system respond. Imagine trying to build a computer without knowing how the individual circuits or circuit boards would respond to input signals. The same can be said for ecosystems. An accurate analysis of any ecosystem must be built on a substantial database of solid information.

The concepts described here can be separated into three levels: (1) those that describe the current state of a component and its response to inputs; (2) those that describe interactions between components; and (3) those that describe the integrated response of the whole system to an external signal, perturbation, or disturbance.

Concepts Pertaining to Single Components

State The state of a component is a description of its current condition. The simplest mechanical example is a switch, which is either on or off. The on or off description describes the state of the switch. A car's gasoline tank can be described by the amount of gasoline it currently holds. The gasoline gauge in a car describes the state of the tank, in this case as a continuous variable, rather than a two-state on or off description.

Since ecosystem studies deals with the movement of energy and matter through ecosystems, the state of a given component is often described by the quantity it contains. For example, the state of the vegetation component in Figure 1.4 can be described by the amount of organic matter, nitrogen, or water it contains. The state of a component may also be described as its capacity to carry out a function. Again using the vegetation component, it can be described as actively photosynthesizing and growing, or after the end of the growing season it may be senescent or dormant. The ability of the vegetation to respond to a given input, say fertilization, may depend on whether it is in the active or dormant state.

The term "state" is also often applied at higher levels of organization. Whole ecosystems are at times described as being in "steady state," which means showing no net increase or decrease in total storage of nutrients, organic matter, and so forth over all of the components. Or, ecosystems may be described as "aggrading" (increasing in content) or "degrading" (decreasing in content). Another common description of state applied to whole ecosystems is "undisturbed" versus "disturbed." The changes in input/output balances at Hubbard Brook (Chapter 4) following disturbance emphasize the potential importance of this distinction.

Turnover Rate, Residence Time If the quantity of material in a component is a common descriptor of its state, then the relationships between inputs to, outputs from, and storage within components are key descriptors of the role that component plays in the system. The terms "turnover rate" and "residence time" describe these relationships. Turnover rate is the fraction of material in a component that enters or leaves in a specified time interval. For example, a reservoir that holds 10 million gallons of water and from which 1 million gallons are pumped out every month has a turnover rate of 1/10 or 0.1 (10%) per month. Similarly, if there are 200 tons of organic matter per hectare in a prairie soil, and plant death and senescence contribute 4 tons of new organic matter each year, the turnover rate for all soil organic matter is 4/200 or 0.02 (2%) per year. Residence time is just the inverse of turnover rate. The residence time for soil organic matter in the prairie example is 1.0/0.02, or 50 years.

Buffering (Damping) Versus Amplification If a component is in steady state (not changing in total content), then inputs must equal outputs. One application of this concept to ecosystems in order to estimate rates of rock weathering is described in Chapter 4. However, the steady state is rarely achieved over either individual components or whole systems. Much of the dynamic nature of ecosystems and their response to disturbance or management deals with how quickly and how strongly they respond to changing inputs, or changes in state. An example of changes in inputs would be responses to "acid rain" (Chapter 23), which has caused large changes in the chemistry of precipitation and inputs of several important elements to ecosystems. An example of change in state would again be the Hubbard Brook devegetation or a commercial harvest.

The internal mechanisms working within a compartment may "buffer" or "damp" outputs relative to inputs. Using the prairie soil example cited above, if litter inputs changed from 4 to 8 tons per hectare per year, but turnover rate remained at 2% per year, it would be quite a long time before decomposition outputs rose to be equal with inputs. At this point, total soil storage of organic matter would be double the initial value of 200 tons per hectare (Figure 5.1). In general, compartments with large storage and slow turnover rates tend to buffer a system against sudden or large changes in inputs. Indeed, soil organic matter is one of the major stabilizing forces in most terrestrial ecosystems.

The opposite of buffering is "amplification." This means that changes in inputs result in even larger changes in outputs. Examples of this are harder to come by in ecosystems. A simple mechanical example would be using a little bit of water to wet the valves in a hand pump, thus "priming

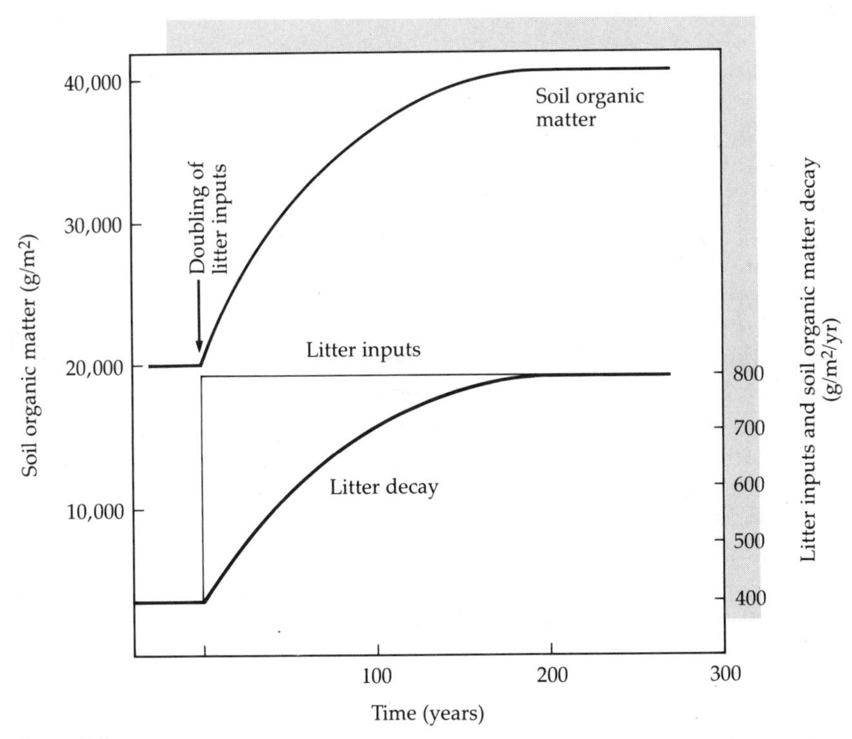

Figure 5.1

Effect of change in litter inputs on soil organic matter content. With litter production of 400 g/m^2/yr and turnover rate of 2%/yr, initial equilibrium soil organic matter content is 20,000 g/m^2. Changing litter inputs to 800 g/m^2/yr and keeping turnover rate at 2%/yr causes total soil organic matter to slowly rise to 40,000 g/m^2.

the pump" so that lots of water can be gotten out. One potential example in nature is the "nitrogen-priming" effect on soils. It has been noted that small applications of mineral nitrogen to soils can occasionally cause a large increase in the apparent decomposition rate and the release of mineral nitrogen from soil organic matter.

Concepts Pertaining to Interactions Among Components

Interactions among components within ecosystems include interactions governing the movement of mass or energy between compartments, interactions dealing with the ratio of two or more combinations of mass or energy through compartments, and control or feedback relationships.

Donor/Receptor Control The transfer of mass or energy between two compartments can be described as controlled either by the compartment from which (donor control), or the compartment to which (receptor control), the transfer occurs. A hydrologic system in which the amount of water entering a downslope portion of a stream is simply the amount flowing out of the next upslope portion is an example of a donor-controlled system. Active plant uptake from an existing soil nutrient pool, if determined by the energy status and nutrient requirements of the plant, represents a receptor-controlled system.

Resource Use Efficiency The concept of resource use efficiency is emerging as pivotal in understanding the function of ecosystems. It describes the quantity of water, energy, or nutrients required for an ecosystem process. It is most frequently used to describe photosynthesis or net primary production. For example, the temperate deciduous forest described in Figure 2.2 takes up, cycles, and uses more nitrogen per unit of aboveground productivity than does the temperate evergreen forest and so has a lower nitrogen use efficiency. In Chapter 6, several physiologic adaptations are described that reduce the amount of water required to fix a unit of carbon through photosynthesis or to increase water use efficiency.

Similar calculations of use efficiency can be made for light (Chapter 3) or any other resource required for growth. The relative abilities of different species or growth forms to use resources efficiently help to determine the types of environments in which they are the most effective competitors.

Optimization The concept of optimization is very closely allied to that of resource use efficiency. Matching resource use efficiencies with the relative availability of different resources in the environment "optimizes" total resource use and maximizes production. For example, abandoned agricultural land with fine-textured soils may have high water availability but low nutrient availability. Species that use nutrients efficiently but require lots of water might be expected to do well, and therefore dominate, in such an environment. Species with high nutrient demands would be less competitive. We will return to this subject again in Chapter 11. The idea that natural selection tends to "optimize" the physiology of different species relative to the full set of resources in the environment required for growth and reproduction is becoming an important cornerstone of ecosystem analysis.

Feedback "Feedback" is a term used to describe a control interaction, rather than a transfer of mass or energy. The classic example of feedback is an office heating/cooling system with a thermostat, an air conditioner, and a furnace. As the temperature falls in the office, a switch in the thermostat causes the furnace to turn on and release heat into the office. As the temperature rises past a set level, the thermostat switches the furnace off. If the temperature gets too high, then the air conditioner will come on. This is a negative feedback (Figure 5.2*a*). The cooling of the office causes the furnace to heat the office again, keeping it at a nearly constant temperature. Such responses are often called "homeostatic" mechanisms, meaning that they tend to keep the system in the same state. There are many examples of homeostatic mechanisms in nature. Humans and other mammals are "homeotherms" because several systems in the body work together to maintain a relatively constant temperature. Examples of homeostatic mechanisms at the ecosystem level are presented in detail for two ecosystem types in Part III.

The opposite of negative feedback is positive feedback. Imagine that the switches on the thermostat are reversed so that the furnace turns on when the office is too warm, and the air conditioner turns on when it is too cold. If the office is too warm, the furnace comes on and it gets even warmer. This is a weak positive feedback. If the furnace burns hotter the farther the office temperature gets from the temperature setting on the thermostat, then you have a strong positive feedback (Figure 5.2*b*). Positive feedbacks are very destabilizing and can have destructive effects. There are several examples of positive feedbacks in ecosystems, particularly in the dynamics of animal populations and the effects of site quality on net primary production and future site quality.

Concepts Pertaining to Whole Systems

A major goal of ecosystem analysis is the prediction of system responses to management, pollution, and other forms of disturbance, based on the interactions of the component parts. Many of the concepts discussed above, such as state, turnover rate, and resource use efficiency, can be applied to whole ecosystems as well. Turnover rates for whole systems are also often used to describe the system as "open" or "closed." Most terrestrial ecosystems in humid regions have a high turnover rate for water, losing more to streamflow and transpiration during the course of a year

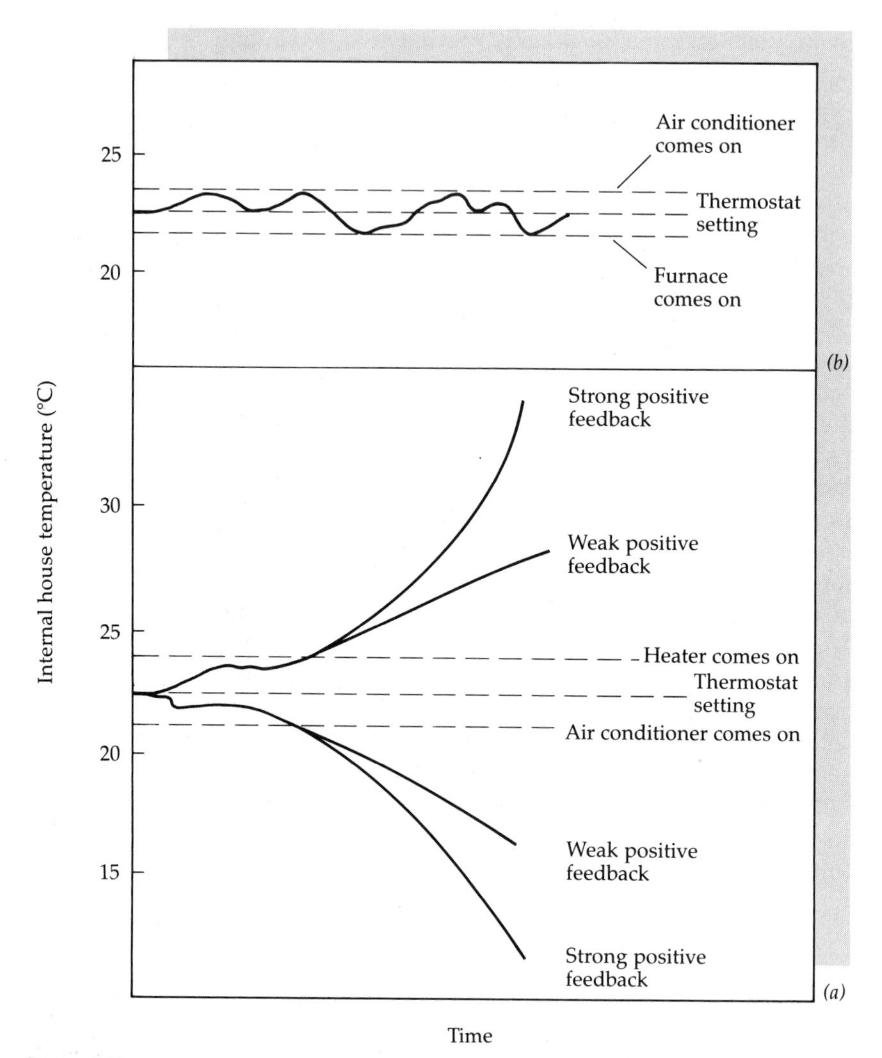

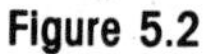

Figure 5.2
Effect of positive and negative feedbacks on system performance using the thermostat example. *(a)* Negative feedback. *(b)* Positive feedback.

than is stored in the system at any one time. In comparison, most terrestrial systems lose only a very small fraction of the total nutrient content of the system in any one year. Such systems can be described as "open" relative to water retention but "closed" relative to nutrient retention. The devegetated ecosystem at Hubbard Brook (Chapter 4) went from a relatively closed system of nitrogen recycling to a relatively open one of nitrogen loss.

There are, in addition, three frequently used terms that apply only at the whole-system level. All have to do with the response of the system to an external disturbance.

Resistance Systems that show relatively little response to disturbance are said to be "resistant." A severe disturbance is required to change the state of the system (Figure 5.3). Using the example of nitrate losses following clearcutting or trenching presented in Chapter 4, those systems that did not show elevated nitrate losses would be termed resistant. It is often the case that resistant systems will take a relatively long time to return to their initial state after a disturbance that is strong enough to alter that state.

Resilience Resilience is the opposite of resistance (Figure 5.3). Resilient systems can be altered relatively easily but will return to the initial state more rapidly. Again, using the nitrate loss example, the Hubbard Brook ecosystem and the others that showed high rates of nitrate loss after disturbance would not be called resistant. Resilience also requires a rapid return to the initial, low nitrate loss condition. At Hubbard Brook, the high nutrient availabilities evident in high stream-water concentrations cause

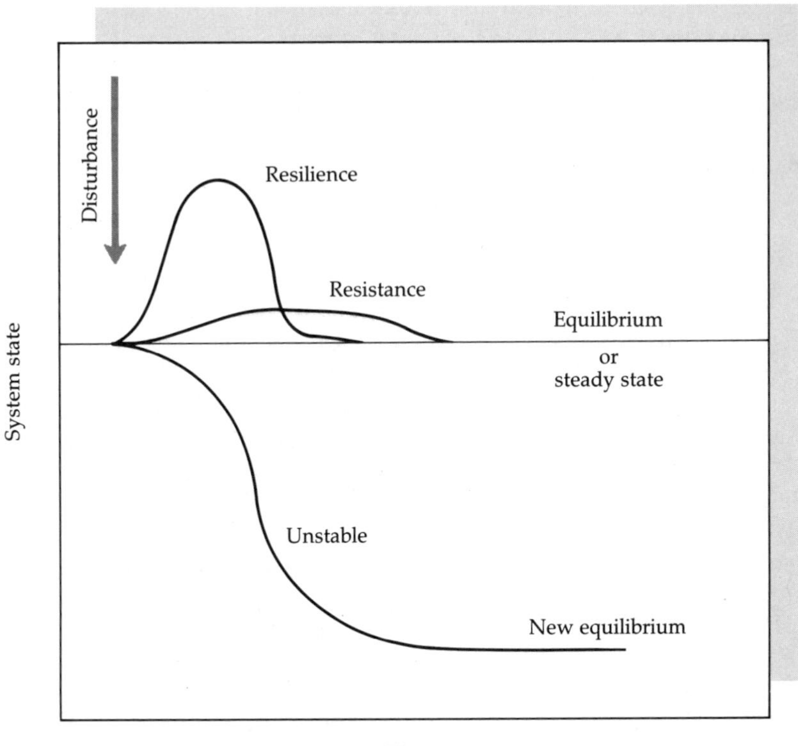

Figure 5.3
Ecosystem responses to disturbance: resistance, resilience, and unstable.

very rapid reestablishment of vegetation. High rates of nutrient uptake by this vegetation then reduces nitrate losses (an example of negative feedback). The elevated nitrate concentrations seen immediately after disturbance last for only three to four years. Unlike many of the previous concepts in this chapter, which can be quantified, resilience and resistance are strictly relative terms used generally in comparisons between systems.

Stability Suppose the Hubbard Brook system had not been either resistant or resilient, but had shown continuously high nitrate losses following disturbance, with no regrowth of vegetation and eventual deterioration to some severely degraded state. It would then be described as "unstable" with regard to devegetation. The initial state has been lost, and the system does not have the capacity to return to it (Figure 5.3). Some new state is reached, and the system continues in that state. Serious examples of this occur in "desertification," the conversion of semi-arid grasslands or shrublands to deserts by overgrazing or by agriculture, which disrupts the thin layer of vegetation that holds the soil in place.

Stability is a very difficult concept to pin down. While the desertification examples are clearly unstable, what of the resistant and resilient systems? Are they "stable" because they eventually return to the initial state? Is the resilient system more stable because it returns faster, or is the resistant system more stable because it is more difficult to dislodge from the initial state? Volumes have been written on the subject of ecosystem stability, how to define it, and what makes systems stable or unstable, without any real consensus being reached. We will use the term only in a relative and intuitive sense.

Additional Concepts

Gradients The concept of gradients does not fit neatly into a systems analysis framework, but it is central to many ideas in ecosystem studies. A gradient is any continuous change in an environmental parameter in time or space. There is a gradient of light availability from the top to the bottom of a forest canopy, caused by the interception of light by leaves overhead. There is a gradient of temperature from the outside to the inside of the wall of a heated house on a cold day. There is a gradient of carbon dioxide concentration from the outside air to that inside a leaf caused by the removal of CO_2 from the internal air space by photosynthesis.

The gradient concept is important because most changes in ecosystems are continuous and occur along gradients. Gradients in temperature cause heat to move through objects. The greater the difference in temperature, the faster the movement of heat. Water movement in plants is in response to gradients in water potential and again is faster as the gradients get steeper (Chapter 8). While there are some important "switches" in ecosystems that cause rapid changes in function over relatively small changes in environmental conditions, most changes are continuous and in response to gradual changes in conditions.

The Niche The concept of the niche is at the heart of much ecological research. It describes the types of environments for which species are best adapted. Figure 5.4 shows a two-dimensional field of light and nitrogen availability. Different species might be best adapted to different combinations of availability of these two resources due to specific morphologic or physiologic characteristics. For example, species A grows best under high-light and high-nitrogen conditions, while species B performs best with high light and low nitrogen. This definition involves both the concepts of gradients of resource availability and of resource use efficiency. A species' particular ratio of resource use efficiencies will determine where it will perform best over gradients of resource availability.

Niches are described in two ways. The fundamental niche is that part of the resource availability field in which the species can survive in the

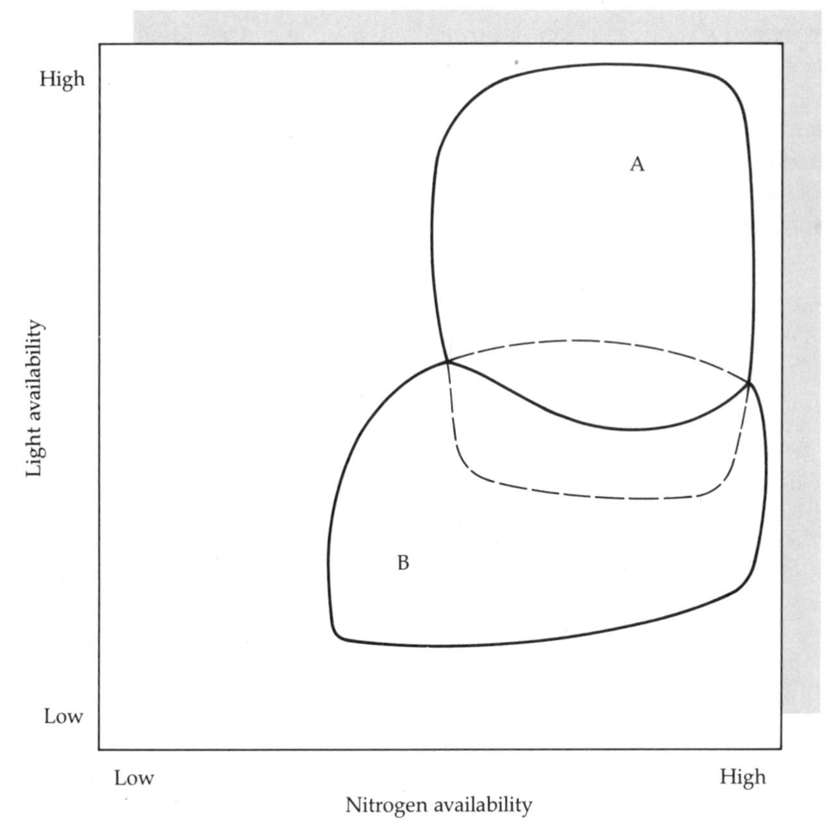

Figure 5.4

Niche relationships for two species. Fundamental niches are shown by dashed outlines, realized niches by solid lines.

absence of competition. The realized niche is that portion of the fundamental niche actually occupied in the presence of a particular set of competitors. In Figure 5.4, the realized niches of both species are considerably smaller than their fundamental niches. It must be remembered that niches are rarely equivalent to a physical location; rather, they relate to gradients of resource availability. (Much of the rich literature on niche theory relates to species–species interactions rather than to those between species and resource availability. Our focus here on ecosystem function results in this particular emphasis.)

LIMITATIONS OF THE SYSTEMS APPROACH

For all the similarities between the study of natural systems and mechanical or electrical ones, the analogy is not perfect. A major difference is that, in human-made systems, all of the subparts can be described clearly because they have been constructed to particular specifications. This is not true of ecosystems. Often the responses of individual components are only dimly known for a fairly narrow set of conditions. So systems analysis of ecosystems is not as clean or precise and is always dependent on accurate measurements and studies by field researchers from many disciplines working at many levels. In that sense, the analysis is never complete, as new information is constantly being made available. Rather, the analysis of ecosystems is more of a continuing set of approximations, hopefully becoming more complete and accurate as the information base accumulates.

REFERENCES

Hall, C. A. S., and J. W. Day. 1977. *Ecosystem Modeling in Theory and Practice*. John Wiley and Sons, New York.

Jeffers, J. N. R. 1978. *An Introduction to Systems Analysis: With Ecological Applications*. University Park Press, Baltimore.

Lovelock, J. E. 1982. Cybernetics. In *Gaia: A New Look at Life on Earth*. Oxford University Press.

Whittaker, R. H. 1975. Communities and environments. In *Communities and Ecosystems*. Macmillan, New York.

Whittaker, R. H. 1975. Community structure and composition. In *Communities and Ecosystems*. Macmillan, New York.

Part II

MECHANISMS Processes Controlling Ecosystem Structure and Function

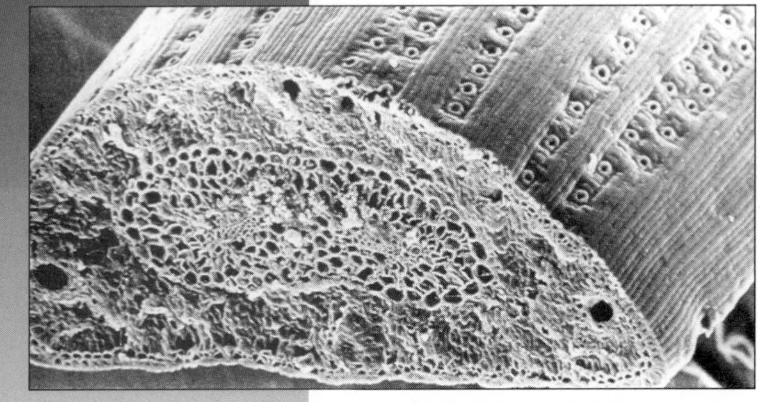

Scanning electron micrograph of a cross section of a pine leaf. Biophoto Associates/Photo Researchers

Resource use and primary production in plants are increasingly viewed as optimization problems. All plants need basically the same resources for growth (light, water, certain nutrients), and most use similar physiologic mechanisms to obtain these resources. Differences between plant species in morphology, size, and growth form represent subtle shifts in the efficiency with which different resources are acquired and used. Plants as visibly distinct as the giant redwoods of coastal California and tiny shrubs of the arctic tundra are doing basically the same things. The differences in size, shape, longevity, and length of activity during the year reflect evolutionary adaptations to the physical, chemical, and biological environment in which they grow.

However, there are a few very important differences in biochemical mechanisms of resource acquisition that separate the tremendous number of plant species in the world into a smaller number of distinct categories. One distinction is between those species that can support root symbionts, which carry out fixation of nitrogen gas (N_2) from the atmosphere, and those that cannot. A second grouping can be made by the predominant process used in carbon fixation for photosynthesis (C3, C4, and CAM; Chapter 6). These and a few other basic changes in plant physiology cause large changes in the ability of plants to acquire resources required for growth, as well as the efficiency with which they are used.

The purpose of this section is to discuss processes that control the availability of resources required for plant growth and characteristics of plant morphology and physiology that relate to the optimal acquisition and use of these resources. The presentation begins at the leaf level and then extends to the plant, canopy, and stand levels. Both the actual uptake or fixation of resources and the allocation of resources to build new tissues for future uptake are presented. These allocation patterns have important effects in reducing herbivory, affecting litterfall and litter decay rates, and altering energy, water, and nutrient dynamics of the ecosystem.

Chapter 6 Energy, Water, and Carbon Balances over Leaves

R. Krubner/H. Armstrong Roberts

INTRODUCTION

Leaves are a logical starting point for a discussion of mechanisms of energy, carbon, and water transfers in terrestrial ecosystems, as these are the tissues most directly involved. (We will use the term "leaves" in the generic sense to include needles of coniferous tress, leaf sheaths of grasses, and other photosynthetic tissues.) In all but the driest or most degraded systems, leaves intercept most sunlight before it reaches the ground. As little as 2% of photosynthetically active solar radiation may reach the soil surface in the most dense (vigorous) forests and grasslands. Radiation absorbed by foliage drives photosynthesis and CO_2 fixation.

However, we have seen in Chapter 3 that even under ideal conditions only a very small portion of solar energy is converted to chemical energy through photosynthesis. Over 95% of absorbed light energy is converted to heat, which in turn increases leaf temperature. High leaf temperatures can impair leaf function by increasing respiration rates or, at extreme levels, disrupting function. One means to reduce temperature is through evaporation of water from cell surfaces, as a large amount of energy is required to convert liquid water to gaseous (vapor) form. This water vapor within the leaf is lost to the atmosphere through the stomata (transpiration). The availability of water for transpiration is limited by soil water content and the ability of plants to extract it. Under conditions of low availability in the soil and high evaporative demand in the atmosphere, water stress occurs in plants, and stomates are closed to minimize water loss. CO_2 intake is also cut off, and photosynthesis quickly slows.

Thus the energy source for photosynthesis, sunlight, drives the fixation of CO_2 but also imposes a heat load on the leaves and a potential for transpirational water loss. Changes in the size, shape, and chemical content of leaves affect how fast photosynthesis proceeds and how quickly excess heat is lost to the surrounding air. The purpose of this chapter is to discuss this carbon–water–energy interaction, including the basic processes of energy exchange and the effects of differences in leaf morphology, chemistry, and physiology on these processes.

THE ENERGY BALANCE OF A LEAF

The energy balance of a leaf can be summarized as in Figure 6.1. Inputs include short-wave and long-wave radiation. Outputs include long-wave radiation, conduction and convection, transpiration, transmission, reflectance, and energy stored in carbon compounds (sugars) for export to the

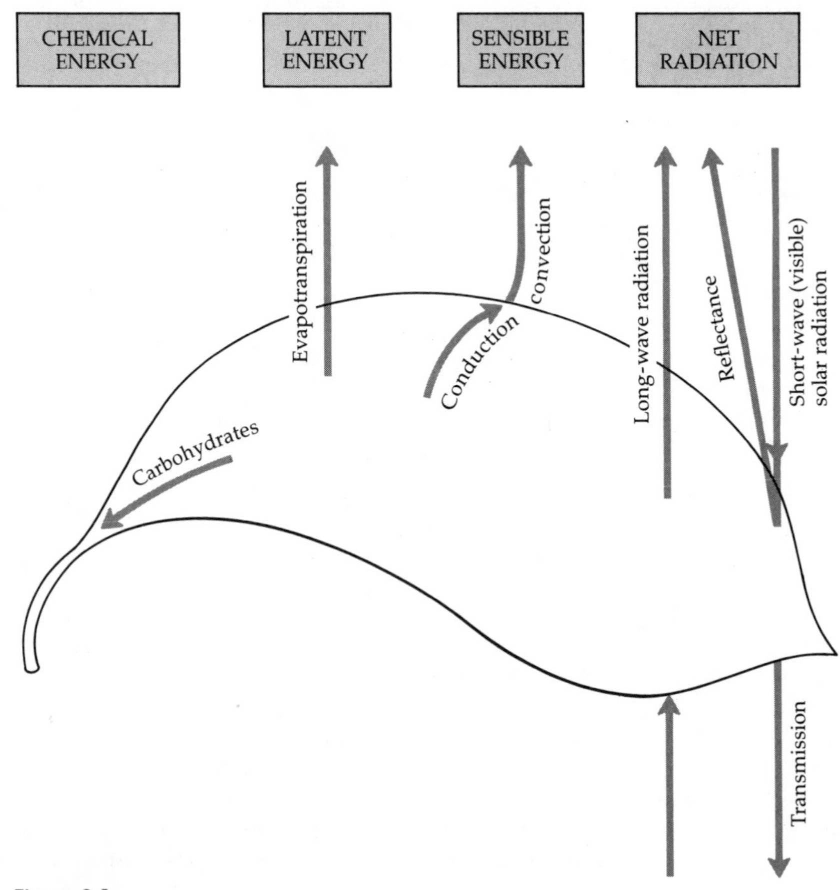

Figure 6.1
Components of the energy balance of leaves.

rest of the plant. These can be grouped into net radiation and the exchange of sensible energy (heat), latent energy (conversion of liquid water to vapor), and chemical energy (photosynthate) (Figure 6.1).

Net Radiation

The sun emits radiant energy over most of the electromagnetic spectrum. In passing through the Earth's atmosphere, some of this energy is absorbed, particularly in wavelengths associated with water vapor and major atmospheric gases (Figure 6.2). Plant leaves reflect some of this incoming radiation but absorb most. Of this energy, only that in a very narrow set of wavelengths in the visible part of the spectrum is actually converted by chlorophyll in chloroplasts into the simple sugars, which are the first products of photosynthesis. As most leaves are composed of similar compounds (chlorophyll, other pigments, water, proteins, etc.), their absorption characteristics are generally similar (Figure 6.3). Shifts in the relative abundance of different pigments can cause some plant foliage to appear more blue or more red than the normal green leaf.

Any object at a temperature above absolute zero (−273°C, or the temperature at which all molecular motion stops) radiates energy. This is the principle behind a home radiant heater or why you feel warmth from glowing coals in a fireplace even when no warmed air reaches you di-

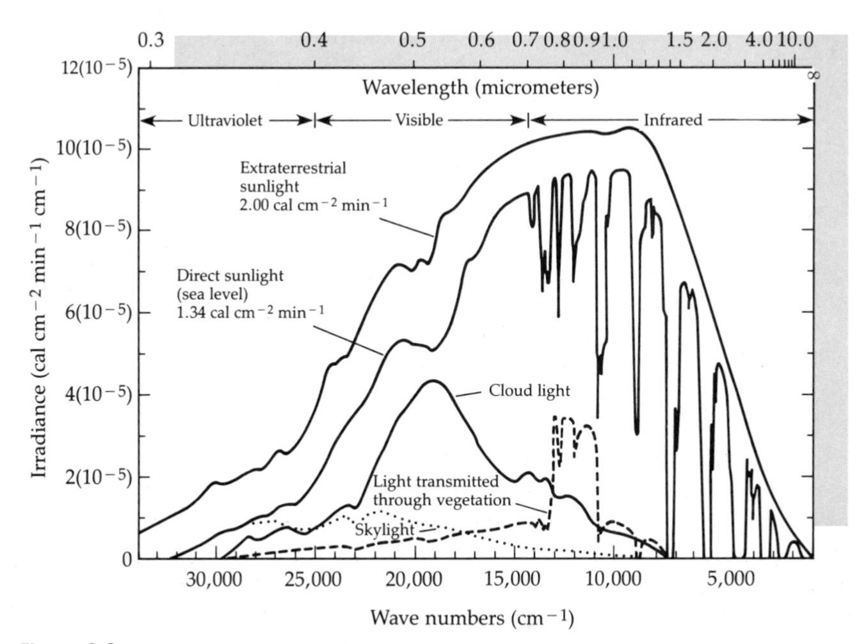

Figure 6.2
Energy in different wavelengths of light reaching the top of the Earth's atmosphere and below the atmosphere. (Spurr and Barnes 1980, after Gates 1968)

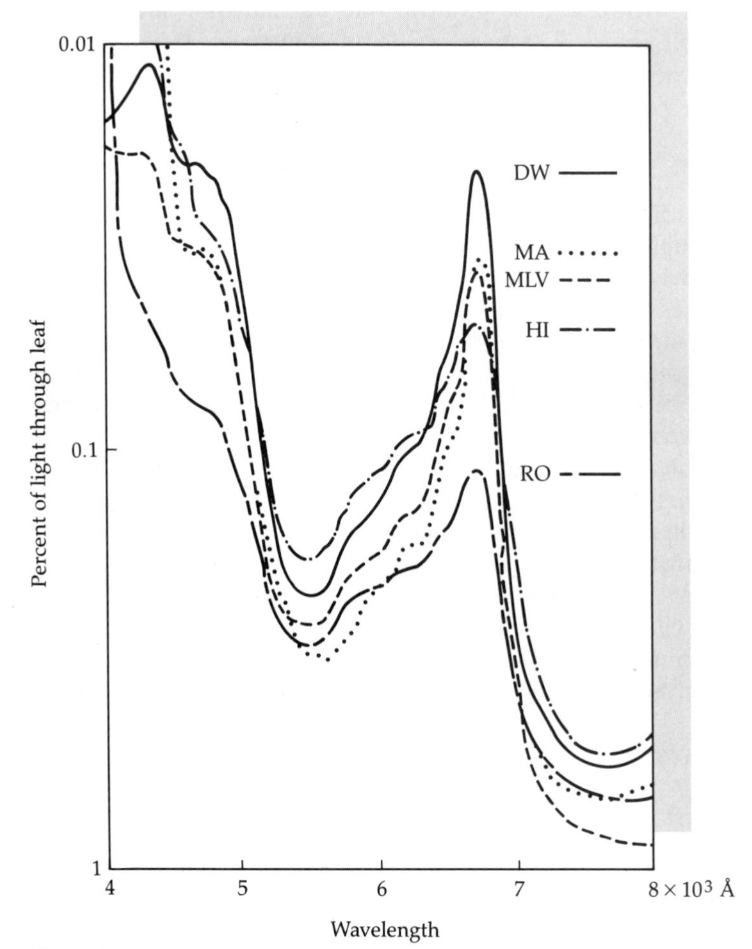

Figure 6.3
Light absorbance at different wavelengths for leaves of different species. RO = northern red oak, HI = shagbark hickory, DW = flowering dogwood, MLV = maple-leaf viburnum, MA = may apple. (Horn 1971)

rectly. At the temperatures realized by living leaves, this reradiated energy is entirely in the long-wave, or thermal, rather than the short-wave, or visible, part of the electromagnetic spectrum and increases with the fourth power of temperature expressed in degrees Kelvin (°C + 273). The fourth power on T causes radiation to increase exponentially with temperature. Actually, long-wave radiation is both an input and an output of energy for leaves, since the soil and plants surrounding the leaf are all emitting long-wave radiation, some of which reaches each leaf. Again, since most leaf surfaces are reasonably similar in the efficiency of long-wave radiation, temperature is the most important factor in determining energy losses by this mechanism.

Sensible Heat Loss: Conduction and Convection

Conduction is the transmittal of energy through a material as a function of the gradient in temperature from one side to the other and the material's ability to conduct heat (for example, Fiberglas insulation in a wall conducts more slowly than a glass windowpane).

Convection is the loss of energy from a heated surface by the warming of adjacent air, which then is removed by turbulent air flow. If the air on the upper surface of a leaf is heated, it will tend to rise above the colder air farther from the surface. This will occur in a turbulent or nonsmooth manner (Figure 6.4). At a much larger scale, this same principle causes the formation of thunderheads (cumulonimbus clouds) on hot summer afternoons. Hot pockets of air rise as a bubble through colder overlying air

Figure 6.4
Turbulent movement of heated air away from the surface of a leaf (convective heat loss). The lighter streaks are currents of warmer air escaping upward from the leaf surface. (Ray 1972, photo courtesy of David M. Gates)

layers until condensation of the moisture in the bubble occurs and a cloud forms.

Conduction and convection are closely linked processes because of the existence of a "boundary layer" of nonturbulent air that covers leaf surfaces. Think of a flat leaf surface with a wind blowing over it. Because of the friction between the leaf and the moving air, the velocity of the air decreases closer to the leaf surface (Figure 6.4). Slower air movement reduces the rate at which the warmed air near the surface is replaced by cooler air needed to increase the conduction of heat from leaf to air.

The size of the boundary layer can be greatly affected by the structure of a leaf. Three characteristics are particularly important: the size of the leaf, the degree of lobing, and the roughness of the surface. Both leaf size and lobing are important because the thickness of the boundary layer increases with increasing distance to the edge of a broad-leaved plant. Leaves with identical leaf areas can have very different mean distances to an edge, depending on the shape and degree of lobing. This is often measured as the **critical dimension** of a leaf, defined as the diameter of the largest circle that can be entirely contained within the outline of the leaf (Figure 6.5). Leaves with different critical dimensions would also have different boundary layer thicknesses.

A corollary to this is that boundary layers are negligible at the ends of broad leaves. Heat conducted to edges would therefore be lost most rapidly. Thicker leaves have greater mass for conducting heat to edges and would lose heat more rapidly. Similarly, needle-leaved plants, such as pines, spruce, and fir, tend to lose excess heat very rapidly and are almost always similar in temperature to the surrounding air.

There is a general relationship between the dryness of an environment and the occurrence of leaf hairs. These could act in two ways to increase heat loss by leaves. They could increase conductive loss to the convective layer by providing a medium for such conduction. Or they could act to increase the turbulence in air flowing near the leaf surface in much the same way as buildings redirect air movement in a city. Leaf hairs may also increase the reflectance of a leaf surface.

Smaller, thicker, and more deeply lobed leaves do in fact lose heat more rapidly and maintain lower temperatures in a given environment, and leaf morphology does change in response to environment. This can be seen in a comparison of "sun" and "shade" leaves of a given species. Deciduous, broad-leaved trees generally have larger, thinner, and less deeply lobed leaves on branches in shade than on those exposed to full sun. Figure 6.6 shows the morphologic changes that coincide with increasing thickness in leaves of American beech. Changes in the lobing patterns of oak leaves between full-sun and shaded understory conditions are shown in Figure 6.5.

There are also special adaptations for increasing heat loss. One of the best known is the "tremble" in trembling aspen and other species of the genus *Populus*. In these species the petiole stem of the leaf is flat and

(a) Leaf of a seedling (b) Leaf from a shaded branch near ground

(c) Leaves from progressively higher on the tree

Figure 6.5
The "critical dimension" (inner circle) as a fraction of total surface area (outer circle) for leaves of black oak at different heights in the canopy. (Horn 1971)

rotated 90 degrees from the plane of the leaf blade (Figure 6.7). A breeze will cause the blade to turn or rotate to present the smallest area to the wind; when the leaf has turned, the petiole is then struck, causing the leaf to turn back to its original position. This happens very rapidly, so that the leaf "trembles." This also increases turbulent airflow and decreases the boundary layer, increasing heat loss. It should also increase rates of water loss and CO_2 gain.

Chemical and Latent Energy Exchanges

Photosynthesis

Although photosynthesis never accounts for a large proportion of total energy removed from leaves, there is still a good deal of variation in the

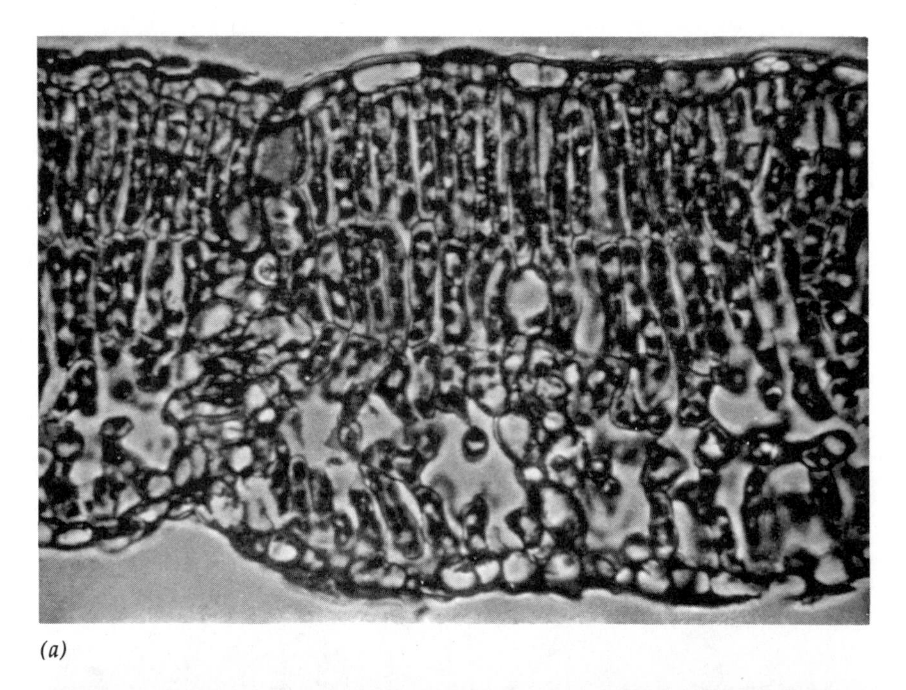

(a)

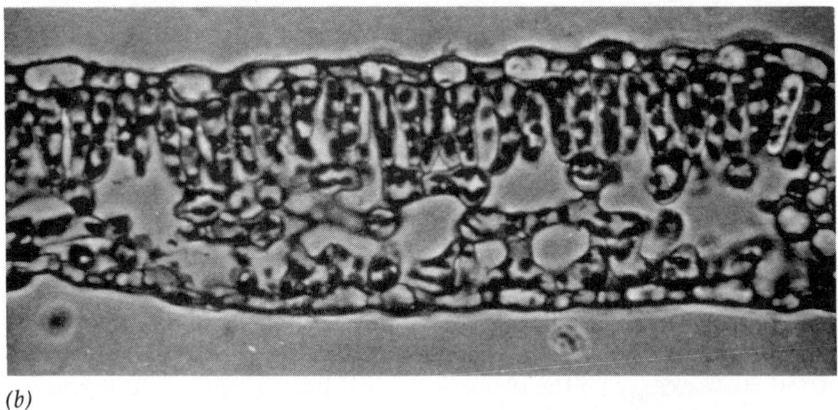

(b)

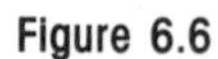

Figure 6.6
Differences in leaf morphology between *(a)* sun and *(b)* shade leaves of American beech. (Horn 1971)

rate at which carbon is fixed. For a given species and leaf type, the relationship between the rate of CO_2 fixed and light intensity is usually summarized in a "photosynthetic response curve." Sample curves are presented for two species in Figure 6.8. All curves show three characteristics.

Figure 6.7
Trembling aspen leaves actually "tremble" in light winds due to the structure of the leaf. The leaf stem or petiole is flattened at a 90-degree angle from the leaf blade. This causes the entire leaf to be turned continually, producing the trembling motion. (Doug Lee/Peter Arnold, Inc.)

First, all leaves show a "saturation" or "diminishing return" effect, such that increases in light levels provide smaller and smaller increases in photosynthesis (or none at all!). Those leaves with highest maximum rates of photosynthesis are also saturated at the highest light levels. This maximum rate of photosynthesis is called the **"light saturated" rate**.

Second, all leaves reach zero net CO_2 fixation at some light level above zero. This is called the **compensation point** and is higher for plants with the higher maximum photosynthetic rates. This results from the constant respiration of cells in leaves which, at low light levels, give off CO_2 faster than it can be fixed in photosynthesis.

Third, leaves from species adapted to growing under full-sun conditions (e.g., aspen in Figure 6.8) show higher rates of photosynthesis than those from species adapted to growing in partial shade (e.g., oak). Within a species, leaves that have developed under partial shade will have lower maximum photosynthetic rates and lower compensation points and will be saturated at lower light levels.

Both high compensation points and high maximum rates of net photosynthesis are related to leaf morphology and chlorophyll and nitrogen

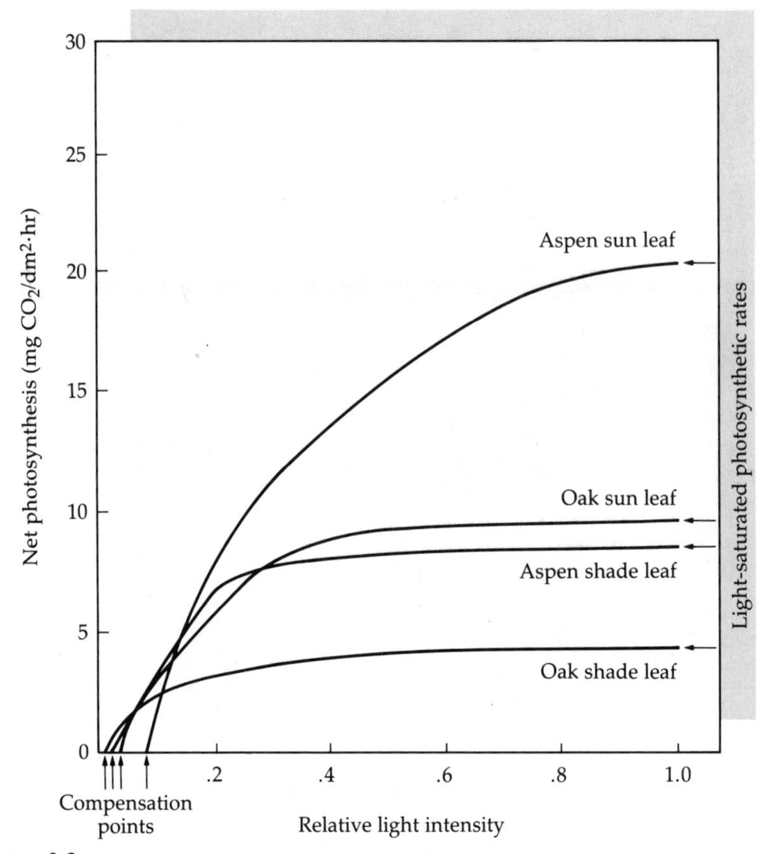

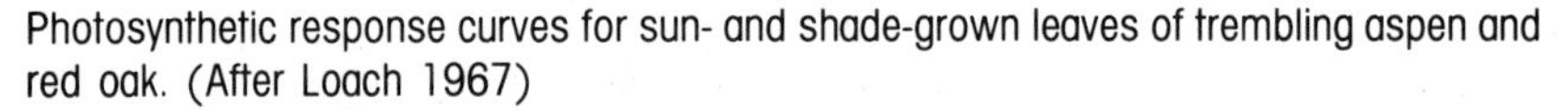

Figure 6.8
Photosynthetic response curves for sun- and shade-grown leaves of trembling aspen and red oak. (After Loach 1967)

concentrations. Look again at Figure 6.6, which shows changes in the structure of American beech leaves in sun and shade. The leaf in sun has two layers of chlorophyll-bearing palisade cells and so a greater density of chlorophyll per unit leaf surface area. There is generally a close relationship between the amount of chlorophyll and the amount of proteins and enzymes required to carry out all the biochemical reactions of photosynthesis. This higher concentration of chlorophyll and enzymes causes both the higher rate of net photosynthesis under full light (it takes more energy to saturate the photosynthetic "machinery") and the higher compensation point (it takes more light to offset the respirational costs of maintaining such a high density of actively metabolizing cells). There is a very good relationship between maximum rates of photosynthesis and leaf nitrogen concentration, which holds across many genera of plants (Figure 6.9). This suggests that differences in photosynthetic rates between leaves of most plants result from different ways of "packaging" the chlorophyll and

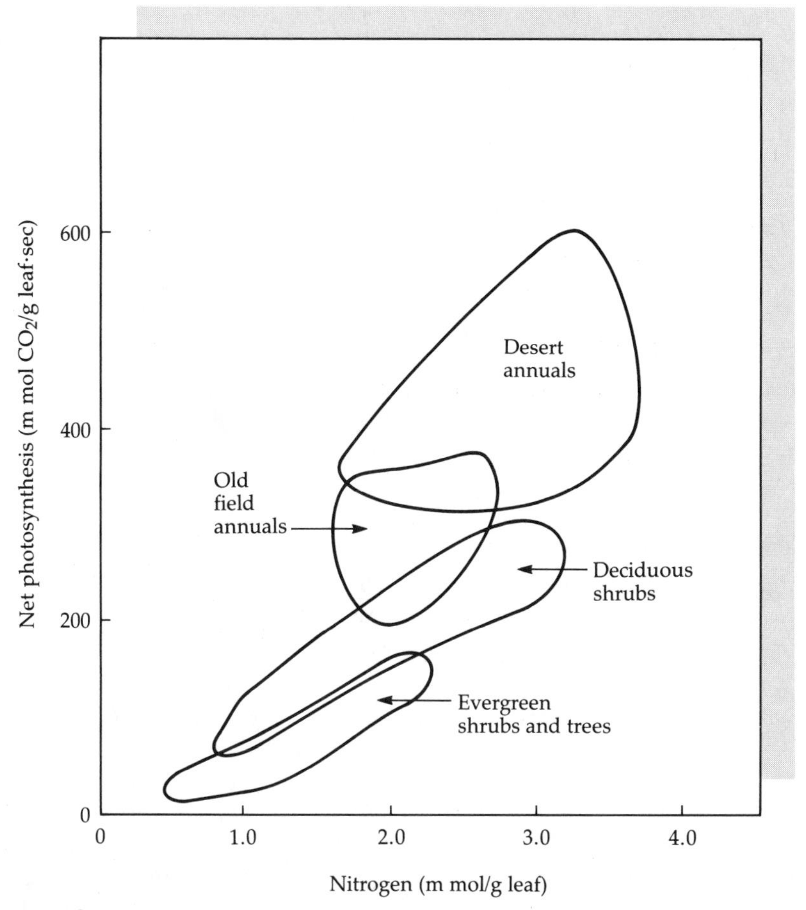

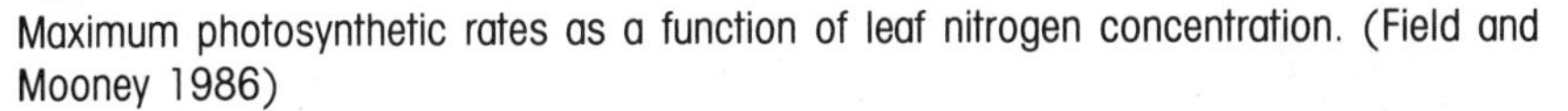

Figure 6.9
Maximum photosynthetic rates as a function of leaf nitrogen concentration. (Field and Mooney 1986)

associated enzymes (see the discussion of exceptions below). Another way of saying this is that the important variation in photosynthetic rates of plants occurs at the biochemical level rather than the species level.

Temperature also plays an important role in determining rates of net photosynthesis. The systems of enzymes active in both photosynthesis and respiration respond to changes in temperature by changing rates of function. Generally, gross photosynthesis is maximized at temperatures representative of the environment in which the plant grows, while respiration continues to increase with increasing temperature (Figure 6.10). Net photosynthesis, the difference between these two, peaks and then declines with increasing temperatures. As with light, a species' or individual plant's realized optimal temperature for photosynthesis will vary with the environment in which it is found.

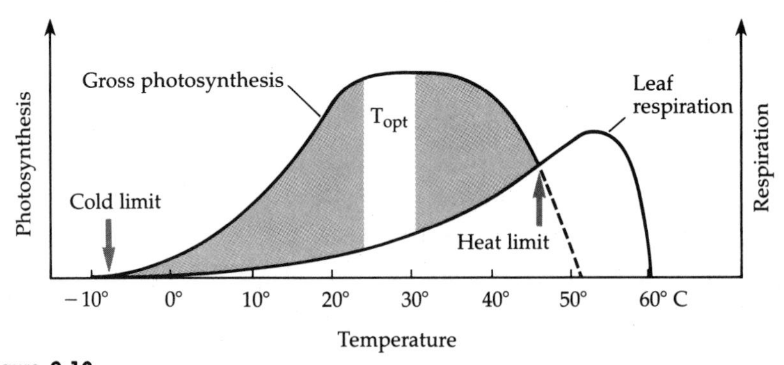

Figure 6.10

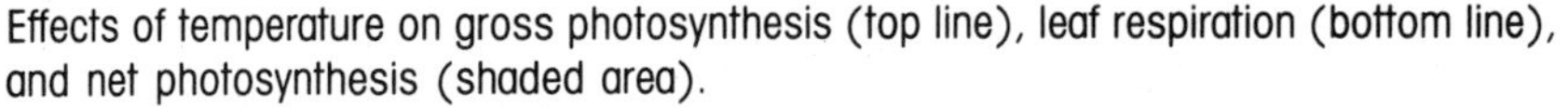
Effects of temperature on gross photosynthesis (top line), leaf respiration (bottom line), and net photosynthesis (shaded area).

Transpiration and Water Use Efficiency

Movement of CO_2 from the atmosphere and into the plant occurs through openings or pores called stomates (see Figure 8.5). This happens because the fixation of CO_2 at the cell surface within the leaf reduces the internal CO_2 concentration, causing more CO_2 to diffuse into the leaf along this concentration gradient. The same cell surfaces are constantly moist, which causes the evaporation of water into the internal leaf (intercellular) spaces. Under any conditions less than 100% humidity and full cloud cover, a gradient of water concentrations (relative humidity or vapor pressure deficit) will exist in the direction opposite to the CO_2 gradient. Thus, water loss to the atmosphere is an inevitable result of CO_2 uptake. This tradeoff will be discussed more fully in Chapter 8.

Transpiration can be an important component of the leaf energy balance. However, transpiration requires the constant movement of water to the leaf to replace that lost by evaporation. If the plant is unable to take up enough water from the soil, water stress develops, leading eventually to stomatal closure and a great reduction in both water loss and CO_2 uptake for photosynthesis.

This water cost–carbon gain process, along with the fact that a lack of available water limits plant growth over a large part of the earth, has generated interest in the concept of **water use efficiency**. This is defined as the weight of carbon gained per unit weight of water lost, or:

$$WUE = \frac{\text{C gain (g)}}{\text{H}_2\text{O lost (g)}}$$

Water use efficiency can be defined at several levels: the leaf, the whole plant, or the whole ecosystem. The concept is a classic demonstration of the idea that resources are used in exchange for one another. In this case absorbed water is required to obtain carbon.

Just as with photosynthesis, there is a tremendous amount of variation between species in the rate at which they transpire water under identical conditions. Again, species that are generally found in wetter environments have higher transpiration rates. While many measurements of transpiration rate have been made, no fundamental theory on controlling factors, equivalent to that of the relationship between nitrogen concentration and photosynthetic rate, has been generally accepted. However, there is an intriguing theory that links transpiration to leaf nitrogen content through the concepts of resource use efficiency.

Very briefly, leaves with more nitrogen per unit leaf area are also generally thicker or smaller ("sun leaves"). Maximum rates of transpiration may be closely linked to total leaf surface area, at least for species with similar leaf shape. If so, then those same leaves with higher rates of photosynthesis per unit area may also have lower rates of transpiration per unit CO_2 fixed in photosynthesis. This has been demonstrated to some extent in the field (Figure 6.11). This suggests a tradeoff between water use efficiency, which would be higher in the thicker, high-nitrogen leaves, and nitrogen use efficiency, which would be higher in the thinner, low-nitrogen leaves. This interaction suggests that there may be limitations in the extent to which the use efficiency of one resource can be increased without reducing the use efficiency of another.

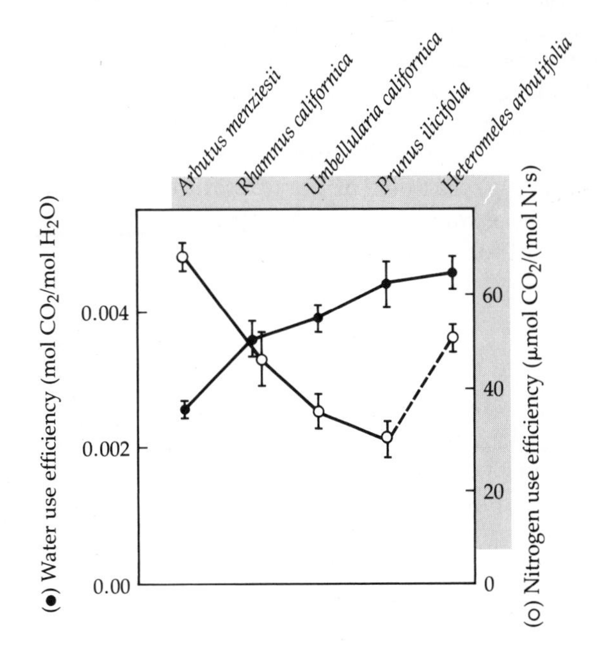

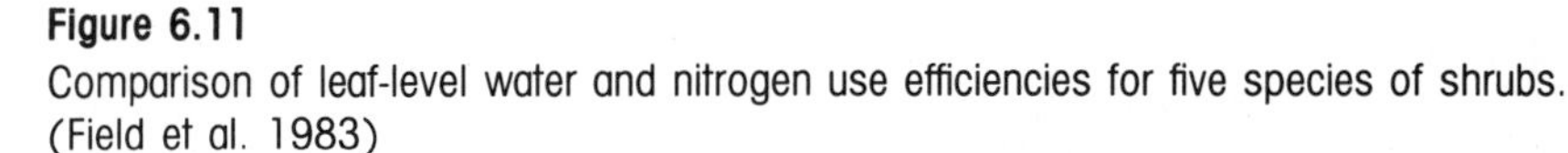

Figure 6.11

Comparison of leaf-level water and nitrogen use efficiencies for five species of shrubs. (Field et al. 1983)

SPECIAL PHYSIOLOGIC ADAPTATIONS FOR PHOTOSYNTHESIS AND WATER USE EFFICIENCY

Earlier we discussed the fact that photosynthesis is basically the same process in all plants, with differences in rates due largely to differences in structure. This is an oversimplification. There are three very different mechanisms for fixing carbon that have evolved in response to extremes of temperature and moisture. Each of these has a different leaf structure associated with it. The three processes are called C3, C4, and crassulacean acid metabolism (CAM) photosynthesis.

The C3 pathway was the first to be described and is the most widely occurring. The leaf structure associated with this processes is the one shown in Figure 6.12, with chlorophyll-bearing palisade cells arrayed more or less continuously across the top of the leaf. Carbon fixation occurs within these cells, and the first product of fixation is a three-carbon organic acid, hence the name.

In C4 plants (so named because the first organic acid formed from fixed CO_2 has four carbons rather than three), both structure and physiology are altered. Palisade cells are absent in C4 plants. Instead, the chloroplasts are located within bundle sheaths concentrated in the center of the leaves (Figure 6.12). These are surrounded by mesophyll cells in which the initial fixation of CO_2 occurs. Fixed carbon is then transported into the bundle sheath, released, and refixed into a C3 compound for use in photosynthesis. As the mesophyll cells surround the actively metabolizing photosynthetic cells, any CO_2 generated by respiration in the bundle sheath can be refixed in the mesophyll before it is lost to the atmosphere. The effect of this is to keep CO_2 concentrations within the leaf near zero. This increases the difference between atmospheric and internal leaf concentrations and causes a faster net flow of CO_2 into the plant. As a result, the carbon-fixing (carboxylating) reactions of photosynthesis can occur at a faster rate for a given rate of transpiration, so water use efficiency is increased. The cost to the plant for this improved CO_2 fixation mechanism is in the extra

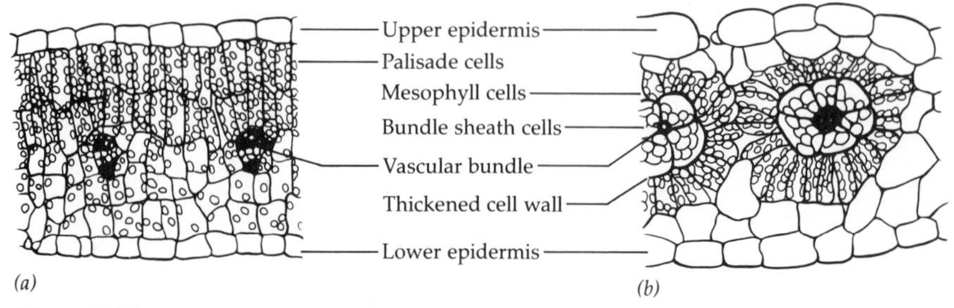

Figure 6.12
Comparison of leaf morphology in *(a)* C3 and *(b)* C4 carbon fixation systems in two species of saltbush (*Atriplex*). (Ray 1972)

energy required to operate it. The reactions require adenosine triphosphate (ATP) from the light reactions of photosynthesis and thus are most adaptive in areas where light intensity is high, water is limiting, and/or high temperatures lead to high rates of respiration.

C4 plants are most common in (but not limited to) disturbed areas in the tropics and subtropics and also in semi-arid areas where their greater water use efficiency may confer a selective advantage. Some of the most productive crop species (e.g., corn and sugarcane) are C4 grasses. It is interesting that C4 photosynthesis is not limited to one part of the plant kingdom but occurs across widely different families. Within a single genera, some species may be able to carry out C4 photosynthesis while others are not.

The CAM adaptation provides a means for taking up CO_2 at night and closing stomates during the day, reversing the pattern in C3 or C4 plants (Figure 6.13). At night, CO_2 enters the open stomates, is fixed and converted to a C4 acid, and is stored in the vacuole or large central cavity of leaf cells. During daylight hours, this acid is removed from the vacuole and transported to the site of photosynthesis, while stomates remain closed. As temperatures are lower and humidities higher at night, stomates are open when the atmospheric demand for moisture is lowest. The

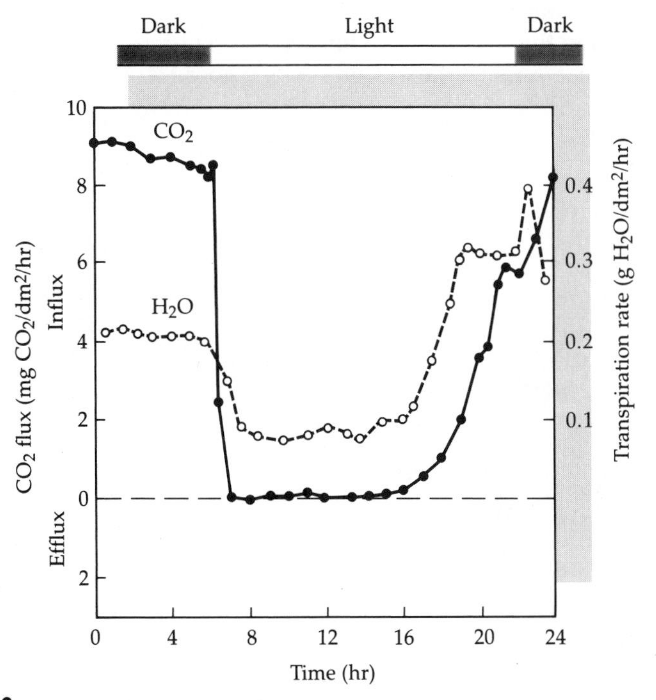

Figure 6.13
Pattern of CO_2 uptake and transpiration in a CAM plant. (Salisbury and Ross 1978, after Neales 1975)

gradient in concentration of water vapor between leaf and atmosphere is reduced, so water loss is lower. Again, water use efficiency is also higher. However, this mechanism is limited both by the amount of CO_2 that can be stored in cells in this way and by the metabolic costs of movement and storage. CAM photosynthesis is most common under extremely dry conditions. Many of the "succulent" or cactus-like desert species are CAM plants.

Thus both C4 and CAM photosynthesis are mechanisms to increase water use efficiency. The C4 pathway accomplishes this by reducing internal CO_2 concentrations in the leaf and thus increasing CO_2 diffusion from the atmosphere. The CAM plants allow stomatal opening at times when water loss to the atmosphere will be lowest.

CHANGES IN LEAF STRUCTURE AND FUNCTION BETWEEN MAJOR ECOSYSTEMS

Do changes in the size, structure, and physiology of photosynthetic tissues between major vegetation types reflect responses to changing environments? There are certainly trends that fit the patterns predicted by our discussion in this chapter. However, the diversity of leaf types within a region shows that energy balance and water use efficiency are not the only selective forces at work.

Considering the major vegetation types in Figure 2.2, there is a general decrease in leaf "critical dimension" as one goes from the tropical rainfor-

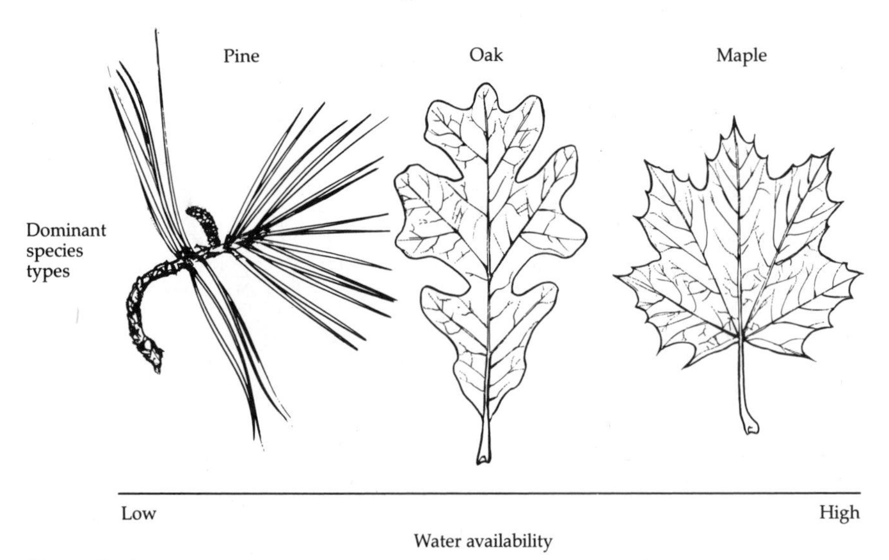

Figure 6.14
Changes in leaf shape of dominant species from moist to dry environments within the northern temperate forest region of the United States.

est through seasonal forests, grasslands, and shrublands. In the extreme desert environments, leaves disappear entirely from some plants, and instead the "stems" and "branches" become the photosynthetic tissues (for example, in cacti). A similar decrease in leaf critical dimension occurs within the temperate deciduous forest regions of North America. The stands with highest water availability tend to become dominated by maples and basswood, slightly drier sites by oak, and the driest sites by pines. Changes in leaf size and shape (sun leaves in this case) varies as expected (Figure 6.14).

The relative importance of the three different photosynthetic processes also varies as expected. All major species in moist, cool environments are C3 plants. C4 plants become more important in warm and dry to hot and moist regions, especially in disturbed environments. CAM plants increase in importance in hot and dry systems.

REFERENCES CITED

Field, C., J. Merino, and H. A. Mooney. 1983. Compromises between water-use efficiency and nitrogen-use efficiency in five species of California evergreens. *Oecologia* 60:384–389.

Field, C., and H. A. Mooney. 1986. The photosynthesis–nitrogen relationship in wild plants. In T. J. Givinish (ed.), *On the Economy of Plant Form and Function*. Cambridge University Press, New York, pp. 25–55.

Gates, D. M. 1968. Energy exchange between organism and environment. In P. Lowery (ed.), *Biometerology*. Oregon State University Press, Corvallis.

Horn, H. S. 1971. *The Adaptive Geometry of Trees*. Princeton University Press.

Larcher, W. 1975. *Physiological Plant Ecology*. Springer-Verlag, Berlin.

Loach, K. 1967. Shade tolerance in tree seedlings I. Leaf photosynthesis and respiration in plants under artificial shade. *New Phytologist* 66:607–621.

Neales, T. F. 1975. The gas exchange patterns of CAM plants. In R. Marcelle (ed.), *Environmental and Biological Control of Photosynthesis*. D. W. Junk, The Hague, pp. 299–310.

Ray, P. M. 1972. *The Living Plant*. Holt, Rinehart and Winston, New York.

Salisbury, F. B., and C. W. Ross. 1978. *Plant Physiology*. Wadsworth, Belmont, California.

Spurr, S. H., and B. V. Barnes, 1980. *Forest Ecology*. Ronald Press, New York.

ADDITIONAL REFERENCES

Gates, D. M. 1965. Heat transfer in plants. *Scientific American* 213 (Dec.):76–84.

Jackson, L. W. R. 1967. Effects of shade on leaf structure of deciduous trees. *Ecology* 48:489–499.

Knoerr, K. R., and L. W. Gay. 1965. Tree leaf energy balance. *Ecology* 46:17–24.

Vogel, S. 1968. Sun leaves and shade leaves: Differences in convective heat dissipation. *Ecology* 49:1203–1204.

Chapter 7 Canopy Structure, Light Attenuation, and Total Potential Photosynthesis

R. Krubner/H. Armstrong Roberts

INTRODUCTION

In addition to modifying leaf structure in response to differences in environment, plants can also vary the ways that leaves are arranged into canopies. In vegetation types such as forests and grasslands, the complex structure of the canopy can result in very different temperature and light conditions for leaves at different heights.

The purpose of this chapter is to describe the interactions between leaves and the light environment in these complex canopies and to use this information to discuss how estimates of total canopy photosynthesis might be made. Canopies designed for maximum photosynthesis will also be discussed and compared with those created by management techniques. Effects of water limitations on photosynthesis will not be considered here but will be taken up in the next chapter.

EFFECTS OF VEGETATION ON THE LIGHT ENVIRONMENT

Canopies create large changes in the amount of sunlight striking leaves at different heights. Walking through or under any canopy, it is clear that this is a very irregular, patchy effect. Even under the darkest canopies there are some spots, called sun flecks, where the ground receives full, direct sunlight. Describing the complete three-dimensional interaction between sunlight and canopies is extremely difficult, involving the plotting of the sun's course through the sky and the precise location of leaves.

However, if abstracted to a larger scale, this problem, like many others, becomes simpler.

Whole forest canopies can be described as uniformly dense suspensions of light-absorbing particles, much like algal cells suspended in water. Both exhibit much irregularity in the distribution of individual units, whether leaves or cells. However, the variability in leaf display in forests is on a scale apparent to the human eye, and so some effort has been expended trying to describe it. Few would think of attempting to measure the location of individual algal cells in a water column.

For whole canopies, the decrease in light intensity (light attenuation) with increasing depth can be described by the equation $IL/IO = e^{-k\,LAI(L)}$, where IL/IO is the percentage of incident light at the top of the canopy (IO) reaching depth L in the canopy, $LAI(L)$ is the cumulative leaf area (in m^2 of leaf area per m^2 of ground area, called leaf area index) from the top of the canopy to depth L, k is a stand or species-specific constant, and e is the base of natural logarithms (2.718). This relationship has the shape shown in Figure 7.1. Examples of forest, grassland, and pasture canopies, their leaf distribution, and light attenuation are shown in Figure 7.2.

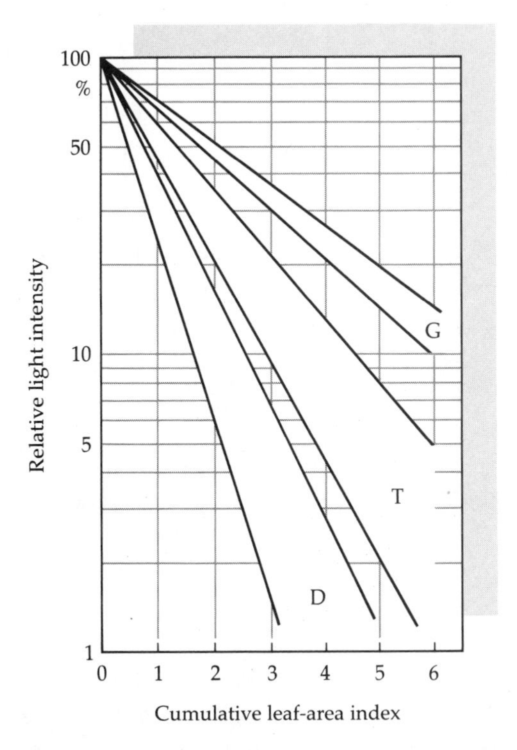

Figure 7.1
Relationship between accumulated leaf-area index and light attenuation for three types of plant canopies with different ranges of the light extinction coefficient k. G = grasses (k less than 0.5), T = range for forest canopies (k 0.5–0.65), D = short-stature broad-leaved canopies (k greater than 0.7). (After Larcher 1975)

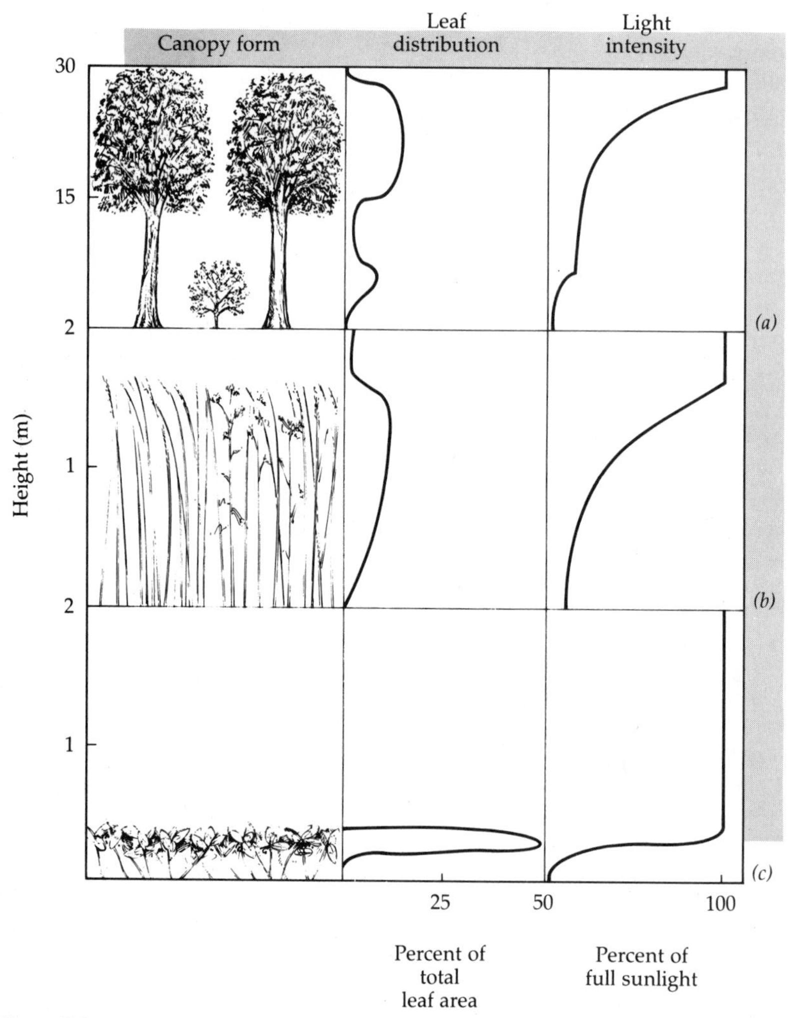

Figure 7.2
Examples of leaf distribution and light attenuation in three canopy types: *(a)* broad-leaved forest, *(b)* grassland, and *(c)* pasture dominated by broad-leaved herbs.

Different types of vegetation, as seen in Figure 7.1, can have very different k values, causing very different rates of light attenuation for the same amount of leaf area. The principal factor causing this is differences in the angle at which the foliage is displayed.

The angle at which a leaf is displayed (its "angle of inclination") changes the amount of light it absorbs in a very straightforward way. If a leaf perpendicular to the sun absorbs 1.0 cal of energy/cm^2/sec in full sunlight, the same leaf displayed at a 60-degree angle to the sun will absorb only $1.0/2 = 0.5$ cal/cm^2/sec (Figure 7.3*a*,*b*). This is accomplished, in effect, by placing the same amount of leaf area over a smaller ground area. In grasslands, the dominant growth form is the erect-leaved grasses. Thus grasslands have the lowest k values (Figure 7.2). Communities that look quite similar to grasslands, those pastures or herbaceous communities dominated by short, horizontal-leaved species, actually have the highest k values (Figure 7.2*c*).

(*a*) Leaf perpendicular to full sun

Energy
2 cal/cm^3·min

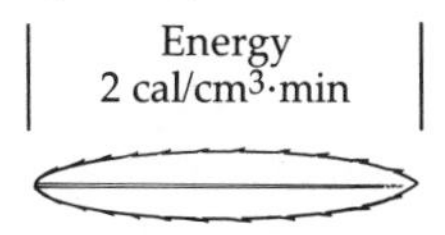

Total area of leaf 10 cm^2
Area exposed to direct sun 10 cm^2
Total energy input to leaf 20 cal/min

(*b*) Leaf at 60° angle to full sun

Energy
2 cal/cm^2·min

Total area of leaf 10 cm^2
Area exposed to direct sun 5 cm^2
Total energy input to leaf 10 cal/min

(*c*) Effect of leaf angle on photosynthetic rate

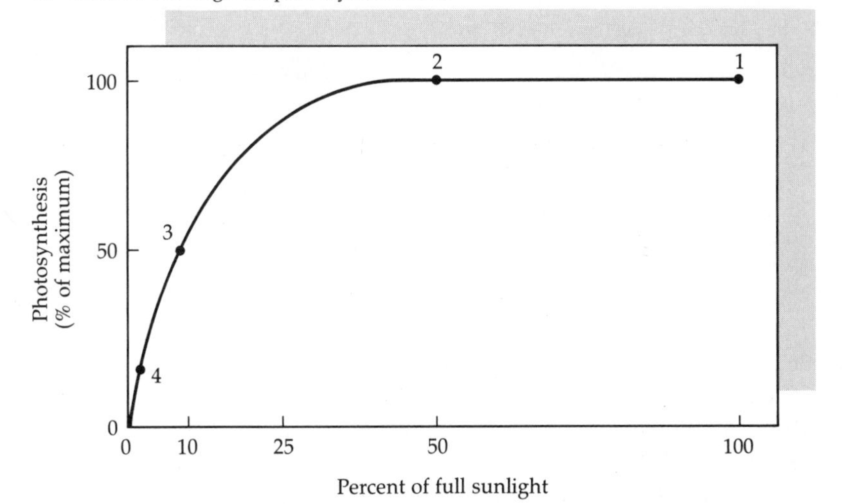

Figure 7.3
The effects of angle of leaf inclination on heat loading and photosynthesis. (*a*) A leaf perpendicular to the sun. (*b*) A leaf at a 60-degree angle from the perpendicular. (*c*) A photosynthetic response curve showing the effects of different angles of inclination.

Forest canopies are intermediate in both leaf angle and k value, but there is a tremendous amount of diversity both within and between forest tree species. This ability to vary angle of inclination may be of great value to plants in meeting the heat load/temperature/respiration problems discussed in Chapter 6 without reducing rates of photosynthesis. Assume the leaves in Figure 7.3*a* and *b* both have the same photosynthetic response curve shown in Figure 7.3*c*. Changing the angle of display from horizontal to 60 degrees reduces the heat-loading by 50% without reducing photosynthesis at all (Figure 7.3*c*, points 1, 2). This occurs because the leaves are light-saturated at 50% of full sunlight. Inclining the leaf 60 degrees in effect spreads the incident radiation over twice the leaf surface area.

The exact value of this inclination depends on the shape of the photosynthetic response curve and the leaf environment. The same degree of inclination in a leaf receiving only 10% of full sun (Figure 7.3*c*, points 3, 4) would actually reduce photosynthesis more than it would reduce heat-loading due to the steep slope in the photosynthetic response curve at this radiation level. Thus sun leaves might be expected to attain greater angles of inclination than shade leaves. This is generally true (Table 7.1), with species requiring full sun (e.g., cottonwood) developing greater variations in leaf angle than relatively shade-tolerant species such as red oak and sugar maple. In general, sun-demanding species also show greater variation in size, thickness, and shape between sun and shade leaves.

Table 7.1 Average angles of inclination for several species, arranged roughly in order from shade intolerant (top) to tolerant (bottom)*

	DEGREES FROM HORIZONTAL		Z-TEST (SIGNIFICANCE)
Species	**Sun**	**Shade**	
Cottonwood	75.7 ± 7.8 (5,5)	32.3 ± 19.8 (5,5)	.01
Plum	33.4 ± 15.5 (5,4)	16.4 ± 12.4 (5,3)	.01
Kentucky coffee tree	64.5 ± 19.2 (5,5)	10.2 ± 7.0 (5,5)	.01
Catalpa	24.2 ± 15.0 (5,5)	8.2 ± 6.4 (5,5)	.01
Redbud	35.9 ± 18.8 (4,5)	13.8 ± 14.2 (4,1)	.01
Green ash	36.8 ± 18.9 (5,4)	14.4 ± 13.8 (5,5)	.01
Red oak	10.1 ± 10.9 (_,3)	11.5 ± 8.2 (_,4)	NS
Mulberry	34.0 ± 15.4 (5,5)	10.6 ± 8.4 (5,5)	.01
Silver maple	16.9 ± 15.5 (5,5)	12.7 ± 9.8 (5,5)	NS
Silver maple	18.7 ± 12.5 (_,5)	11.1 ± 8.6 (_,5)	.01
Sugar maple	14.6 ± 10.2 (_,3)	7.8 ± 5.5 (_,4)	.01

*From McMillen and McClendon 1979.

FUNCTION OF WHOLE-CANOPY SYSTEMS

Combining the concepts discussed above, estimates of whole-canopy photosynthesis can be calculated by treating the canopy as a series of discrete layers. For each layer, measurements must be made of light penetration to that layer, the amount of leaf area by species, and each species rate of photosynthesis at the measured light intensity. Multiplying the photosynthetic rate by the leaf area and summing by species and then overall canopy layers produces the total canopy estimate.

Using these types of calculations, it is an interesting exercise to "design" plant communities for maximum photosynthesis and to compare such optimal canopies with those realized by human management or produced by competition between species in natural communities.

A canopy system "designed" for maximum photosynthesis would simply match photosynthetic response curves with declining light levels down through the canopy. As an example, Figure 7.4 describes hypothetical photosynthetic response curves for three species. The response curve for species 3 is typical of a "shade-tolerant" species, one that can survive at low light intensities but cannot grow as fast in full sun. In contrast, species 1 is "shade-intolerant," growing rapidly in full sun but unable to maintain itself in shade. Species 2 would be called "intermediate." As discussed in the previous chapter, the differences in these response curves would be strongly related to differences in leaf thickness and nitrogen content between species. We will assume that this response includes variation between sun and shade leaves within a species.

A canopy composed solely of species 3 would realize a total photosynthesis equal to the area under curve A (Figure 7.4*b*). Energy at the higher light intensities at the top of the canopy is used inefficiently. A canopy with only species 1 would do better (Figure 7.4*c*), but would fail to utilize energy present at low light intensities in the lower canopy. The optimal arrangement would put an upper layer of leaves from species 1, reducing light levels to 60% of full sun, over a middle layer of species 2, reducing light levels to 25% of full sun, over a lower layer of species 3. Photosynthesis carried out by each species is seen in Figure 7.4*d*. Such a canopy might look like that in Figure 7.5.

How do managed ecosystems compare with these idealized canopies? They are totally different. Intensive management of either forests or grasslands usually results in single-species (monospecific) stands (monocultures) of highly productive "intolerant" species such as species 1 in Figure 7.4. Two of the most widespread examples in temperate North America are loblolly pine forests and cornfields. The structure and light attenuation of two such canopies are shown in Figure 7.6.

Why this discrepancy? Three factors are involved. First, it is easier and less expensive to plant, maintain, and harvest uniform stands with mechanized equipment. Second, in managed stands, yield of harvestable material (wood, grain) is the management criterion, not total photosynthesis. Thus, if a tree produces photosynthate but not wood, it is of little eco-

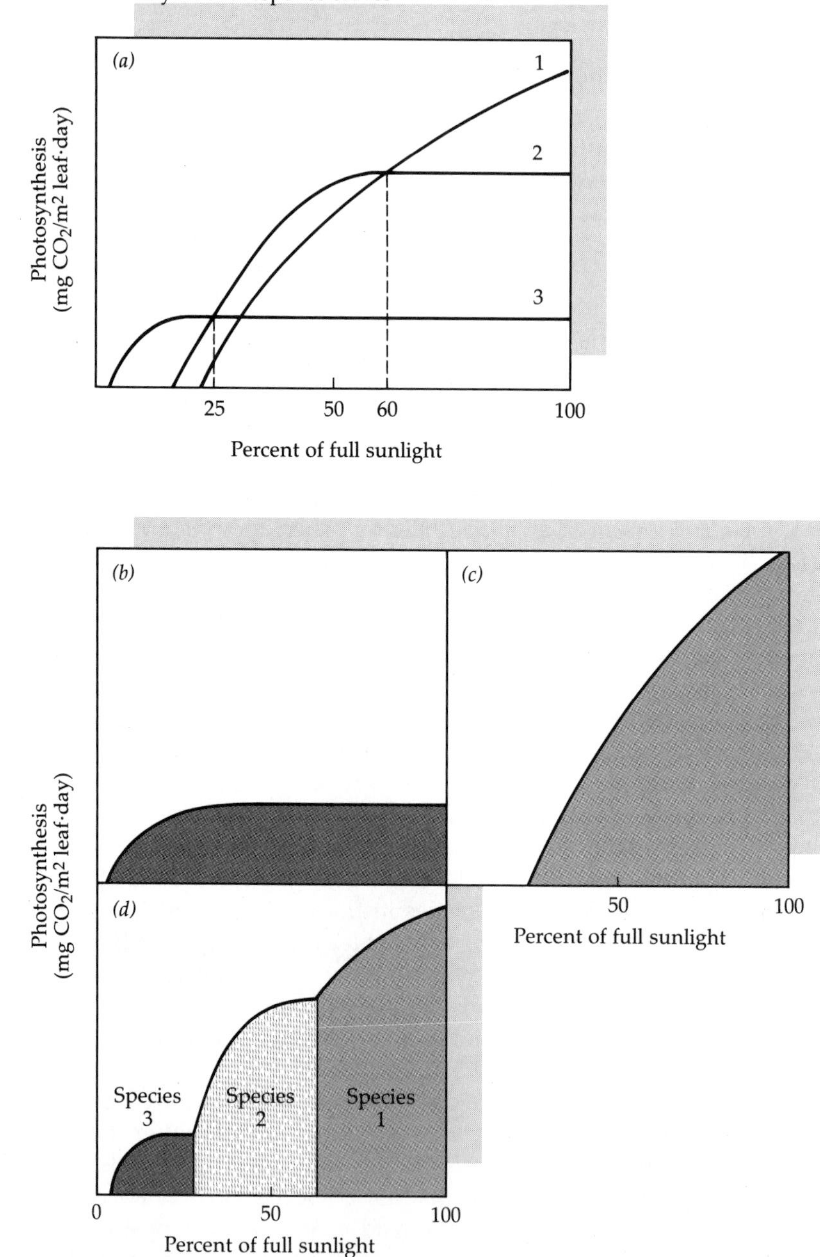

Figure 7.4
Total photosynthesis for hypothetical canopies with different species compositions. *(a)* The photosynthetic response curves of the three hypothetical species. *(b)* Total photosynthesis for a canopy composed only of species 3. *(c)* Total photosynthesis for a canopy of species 1. *(d)* Total photosynthesis for "optimal" canopy arrangement.

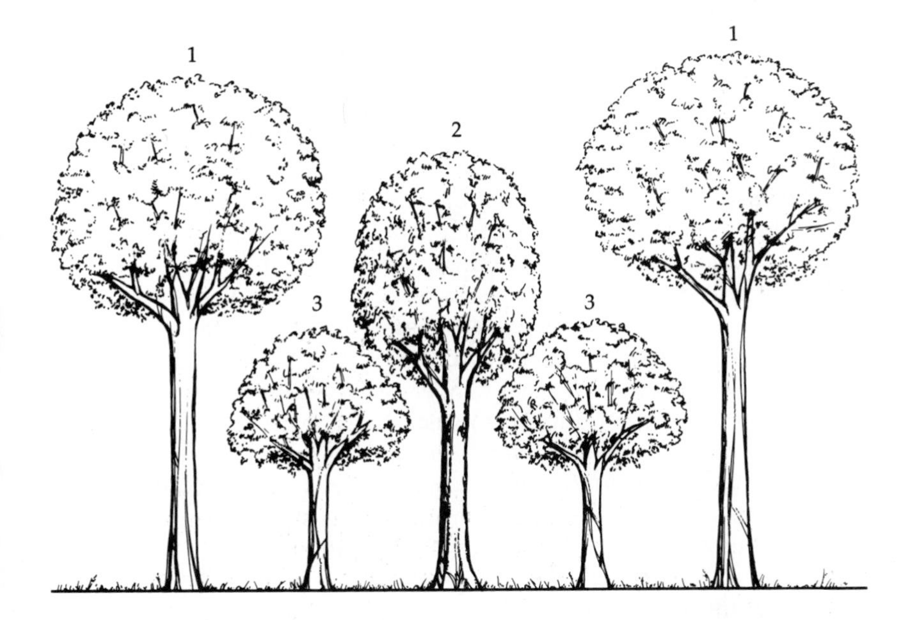

Figure 7.5
Diagram of the optimal canopy structure shown in Figure 7.4.

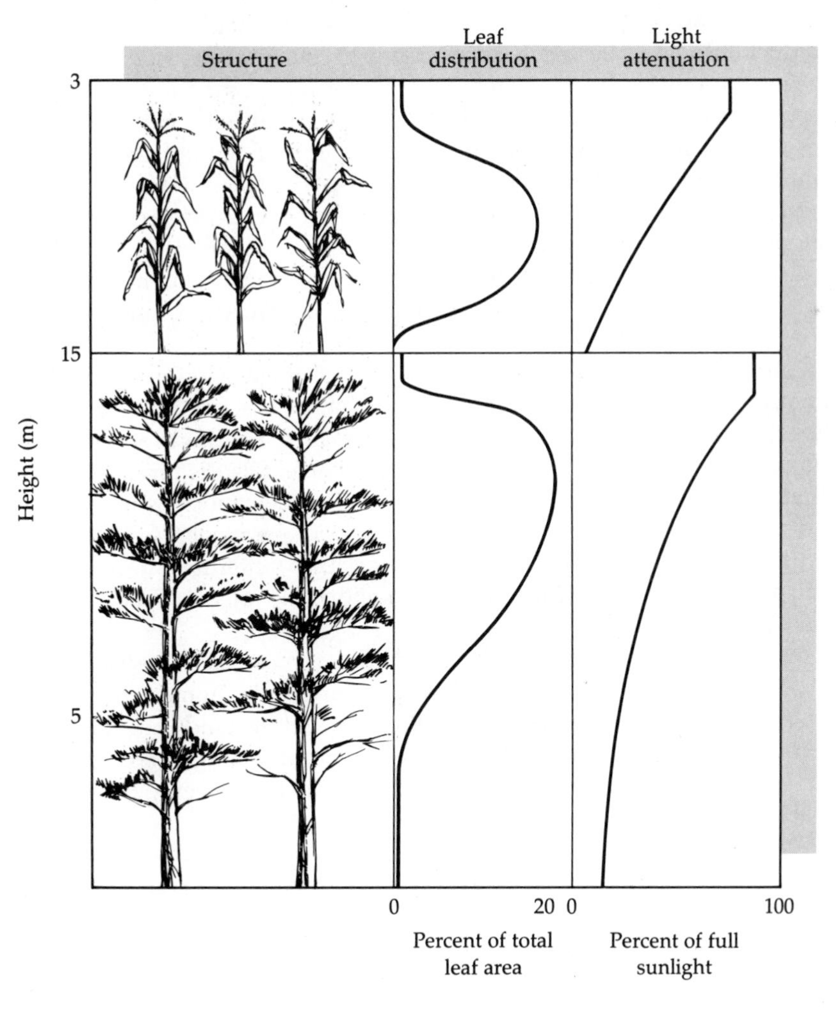

Figure 7.6
Structure, leaf distribution, and light attenuation in canopies of commercial corn and forest crops. (Tetio-Kagho and Gardner 1988, Aber unpublished)

nomic value. Slow-growing, tolerant trees, such as species 3 in Figure 7.4, tend to allocate more photosynthate to leaves, twigs, and roots. Restated, total photosynthesis is only one part of yield. The other is how that photosynthate is partitioned. Third, breeding and genetic selection of a single species may allow development of an optimal range of leaf characteristics within that species so that canopy structure and total photosynthesis may equal those of mixed stands. There is increasing interest in the use of more diverse, multispecies communities for crop production, particularly in tropical regions where traditional temperate-zone monocultural methods of agriculture have been difficult to sustain (Chapter 22).

How do canopies in natural plant communities compare with the idealized structure? This is a much more complicated question. Natural communities are complex and go through many stages of development. Disturbances such as fire and wind-throw are an integral part of community dynamics (see Part III) and alter canopy structure significantly. Should we compare our idealized canopy to old, "undisturbed" stands or to younger ones? An early hypothesis in ecosystem research held that net primary production should be greatest in old-age or mature forests.

This hypothesis can be tested in part by comparing an optimal canopy structure to different stages of development in natural communities following disturbance. An example for one type of temperate deciduous forest, the northern hardwood forests of eastern North America, is outlined in Figure 7.7. In this case, the disturbance was a clearcutting. Progressive, directional change in the structure of a community is called "succession," and Figure 7.7 represents a successional sequence. (A more detailed description of succession in northern hardwood forests is provided in Chapter 20.)

Four stages are represented. Within the first stage (age four years) total canopy leaf area is already approaching that of the mature forest, even though the height of the canopy is much smaller. Stratification of species has occurred, and the most intolerant, pin cherry, fills the top meter, with other, less tolerant species below. By year 30, the canopy is much taller and more evenly distributed. In addition, species are even further stratified by height, with pin cherry above species of intermediate tolerance, such as birch and ash, which are in turn above the tolerant species, beech and sugar maple.

By 60 years, intolerant species have died or been shaded out by the now equally tall tolerant trees of beech and sugar maple. Some intermediate trees (yellow birch) remain. By year 200+, the oldest stands begin to show a three-layer canopy of tolerant trees over tolerant saplings over extremely tolerant shrubs. Similar three-layered structures have been described for very old, undisturbed tropical rainforests.

Which of these canopies is most similar to the optimal structure? Both B and D show layering of species, but only B contains a large number of intolerants in the upper canopy. Thus B, representing a fairly early stage of succession, seems optimal for photosynthesis. This may be a general rule

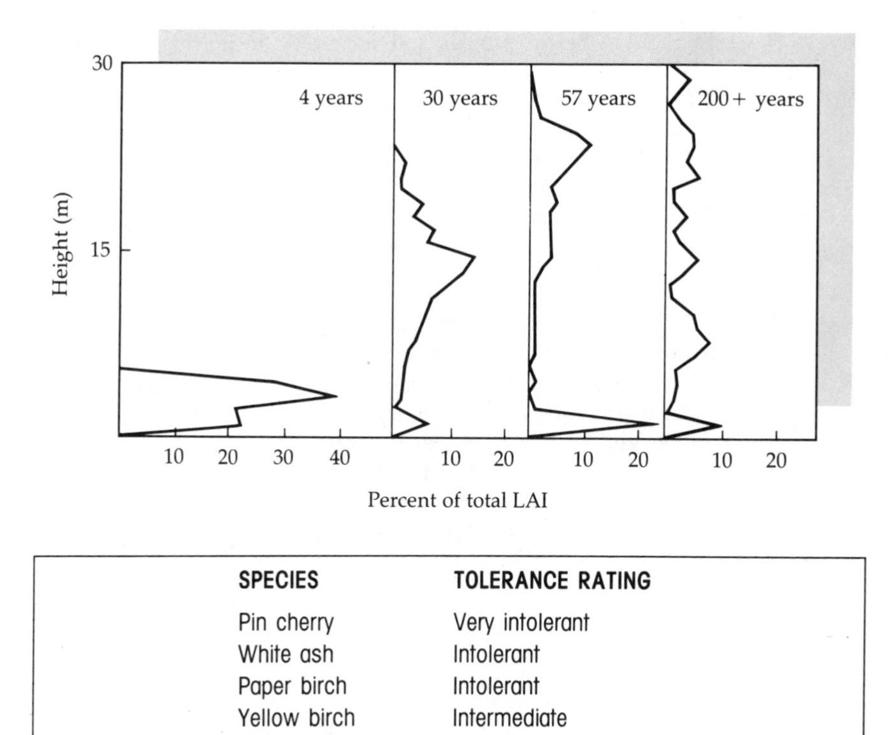

SPECIES	TOLERANCE RATING
Pin cherry	Very intolerant
White ash	Intolerant
Paper birch	Intolerant
Yellow birch	Intermediate
Sugar maple	Tolerant
Beech	Tolerant

Figure 7.7
Four stages of development in canopies of northern hardwood forests. Also shown are the shade tolerance classes of the major species present. (Aber 1979)

for forest ecosystems, that fairly recently disturbed areas with higher numbers of species arranged in height according to tolerance are the most productive, rather than older or mature forests dominated by tolerant species at all levels. Again, this is for total photosynthesis, not yield.

Why are old, mature (''climax'') communities not ''designed'' for optimal production? Very simply, intolerant species, with high compensation points, cannot grow in the shade of tolerant ones (see Figure 7.4). Intolerants grow quickly, but unless they can achieve permanently greater height than tolerants, they will eventually be replaced and disappear from the area until a new disturbance reopens the canopy. The presence of species in mature northern hardwood communities is determined more by compensation point than by maximum photosynthetic rate. In many vegetation types, natural disturbance recurs within the lifetime of the early successional species. In such cases, dominance by tolerant species may never be achieved over large areas (see Part III).

REFERENCES CITED

Aber, J. D. 1979. Foliage height profiles and succession in northern hardwood forests. *Ecology* 60:18–23.

Larcher, W. 1975. *Physiological Plant Ecology*. Springer-Verlag, Berlin.

McMillen, G. G., and J. H. McClendon. 1979. Leaf angle: An adaptive feature of sun and shade leaves. *Botanical Gazette* 140:437–442.

Tetio-Kagho, F., and F. P. Gardner. 1988. Response of maize to plant population density. I. Canopy development, light relationships and vegetative growth. *Agronomy Journal* 80:930–935.

ADDITIONAL REFERENCES

Ford, E. P., and P. J. Newbould. 1971. The leaf canopy of a coppiced deciduous woodland. *Journal of Ecology* 59:843–862.

Hedman, C. W., and D. Binkley, 1988. Canopy profiles of some piedmont hardwood forests. *Canadian Journal of Forest Research* 18:1090–1093.

Kira, T., K. Shinozaki, and K. Hozumi. 1969. Structure of forest canopies as related to their primary productivity. *Plant and Cell Physiology* 10:129–142.

Miller, P. C. 1967. Leaf orientation and energy exchange in quaking aspen and Gambrell's oak in central Colorado. *Oecologia Plantarum* 2:241–270.

Monsi, M., Z. Uchijima, and T. Oikawa. 1973. Structure of foliage canopies and photosynthesis. *Annual Review of Ecology and Systematics* 4:301–327.

Chapter 8 Ecosystem Water Balances and Realized Photosynthesis

H. Armstrong Roberts

INTRODUCTION

In the previous chapter, whole-canopy photosynthesis was described only as a function of light levels and species-specific photosynthetic response curves. However, water loss from leaves through evaporation from internal leaf surfaces (transpiration) is an inevitable consequence of photosynthesis. This water must be replaced by uptake from the soil. In natural ecosystems, the availability of water for transpiration is frequently less than the total amount needed to maintain photosynthesis at optimum rates. Under such conditions water stress develops in plants, and one of several mechanisms comes into play to either reduce the movement of water through the leaf or maintain this flow by physiologic modifications to reduce water stress in leaf cells.

The purpose of this chapter is to discuss water as a resource for plant growth. Environmental factors determining its availability will be discussed along with the process by which water moves from the soil, through the plant, and into the atmosphere. Finally, plant responses to moisture stress and effects of suboptimal moisture availability on photosynthesis and canopy development will be examined.

WATER AS A RESOURCE

Water is required in much greater quantities than are nutrients per unit of biomass produced in all terrestrial ecosystems. This is because nutrients allocated to the production of a given tissue tend to remain in that tissue until it is shed as litter. In contrast, water is continuously given up to the environment through transpiration. Nutrients in foliage may even be retranslocated back into perennial parts of long-lived plants for reuse in the following growing season. Up to 40% of annual plant nutrient requirements can be met by this mechanism (Chapter 11). Only very specialized plants, such as desert succulents, can store a significant amount of water relative to daily transpirational demands. In the massive coniferous forests of the Pacific Northwest of North America, perhaps one-half day's worth of water for transpiration can be stored in the woody tissues on tree stems.

In most terrestrial ecosystems, the only major storage for water is in the soil. Thus the total amount and seasonal pattern of water availability is determined by two nonbiological (abiotic) factors—precipitation and soil water holding capacity (determined by soil volume and texture).

THE CONCEPT OF WATER POTENTIAL

Soil, plant, and atmosphere form a tightly coupled system through which transpiration occurs. Water moves through this system in response to purely physical forces, although the rate of flow can be controlled through physiologic modifications in the leaf.

The concept of "water potential" is used to quantify the forces causing the movement of water through plants. Technically, water potential is defined as the free energy of water, or its capacity to do work. In some cases, it can also be thought of as the effective concentration of water in a solution or of water vapor in the atmosphere. Distilled water in liquid form at 20°C and atmospheric pressure is defined as having **zero water potential**. Three types of water potential are important in transpiration: osmotic potential, physical potential, and matric potential.

Liquid water with any solutes added (e.g., ions such as the sodium and chloride in dissolved salt or small organic molecules such as simple sugars) will have a negative water potential caused by the reduction of the free energy of the water present due to interaction with the solutes. This is called **osmotic potential**. The classic example is a beaker of distilled water divided into two sections by a semipermeable membrane (Figure 8.1). If a solute such as salt (NaCl) is introduced into one side of the beaker and the membrane allows water to pass, but not the ions of sodium and chloride, then the negative water potential in the side with the salt will draw water through the membrane. This will continue until the positive physical pressure against the membrane, evident in the expansion of the membrane, counters the osmotic potential. This simple example is actually quite similar to the process by which turgor is maintained in leaves, as discussed below.

The **physical water potential** of the atmosphere rapidly becomes very negative as relative humidity drops below 100% (Figure 8.2), since the effective concentration of water in the atmosphere is very low relative to that of liquid water. Atmospheric water potentials can reach much more negative values than are found either in plants or in most soils. The more

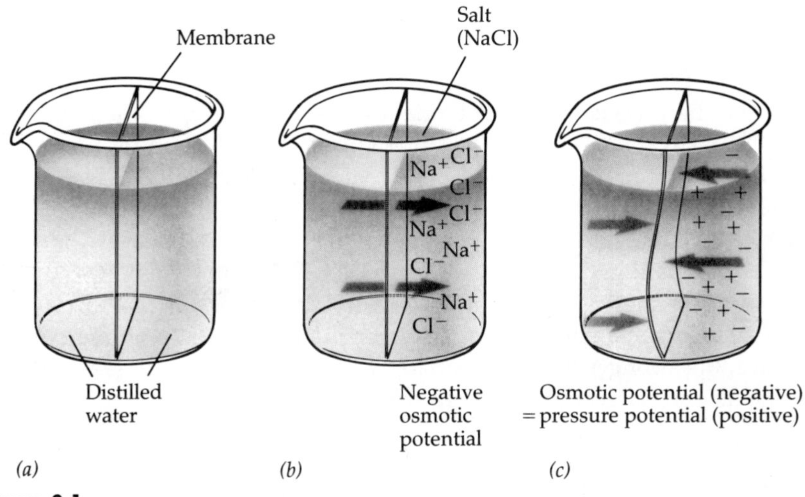

Figure 8.1

An example of osmotic potential. *(a)* The beaker contains distilled water and is separated into two compartments by a semi-permeable membrane. *(b)* Sodium chloride is added to the right compartment and dissociates into ions that create a negative osmotic potential, pulling water across the membrane. *(c)* Water moves across the membrane until the increase in pressure in the right compartment equals the osmotic potential.

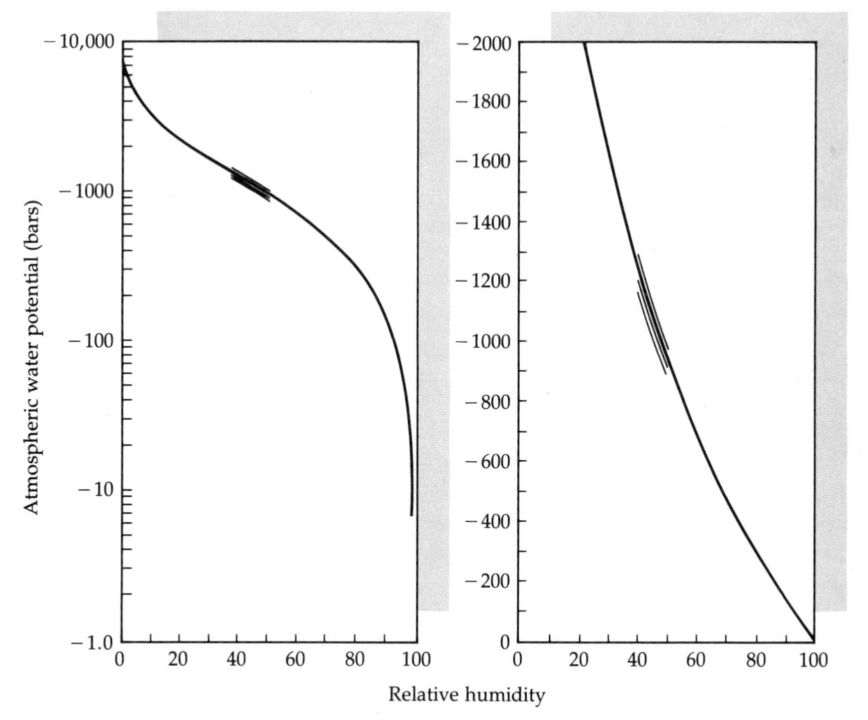

Figure 8.2

Atmospheric water potential as a function of relative humidity. The four parallel lines show the relatively small differences caused by changes in temperature over a range from 0 to 30°C. (Salisbury and Ross 1969)

negative the physical water potential of the atmosphere is, the greater its capacity to evaporate water and drive transpiration. An atmosphere with low relative humidity will place a much greater transpirational demand on a plant than will a very humid one.

Matric potential expresses the tendency for water to adhere to surfaces and is important in determining the rate at which water can be withdrawn from soils. Water is held in soils as very thin films on the surfaces of mineral or organic particles. The thinner the film of water on the particle surface, the stronger the attraction between the particle and the outermost molecules of water and the more difficult it is to remove (Figure 8.3).

SOIL TEXTURE AND WATER AVAILABILITY

The total amount of water a soil can hold per unit soil volume is a function of the surface area of all the particles within that volume and the amount of air space present between these particles. A soil dominated by clay

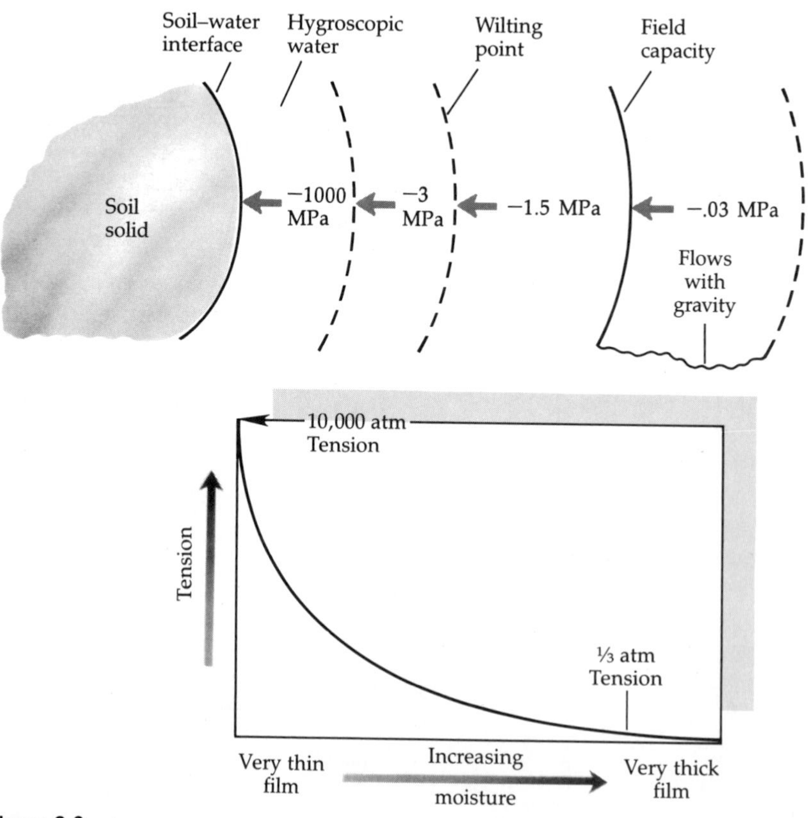

Figure 8.3

The relationship between thickness of water films on soil particles and the tension with which water is held. (Buckman and Brady 1969)

particles (<0.005 mm in diameter) would normally hold more water than a soil dominated by silt particles (0.005 to 0.074 mm) or by sand particles (0.074 to 2.0 mm). These kinds of relationships break down if a soil is severely compacted, as occurs under some types of intensive, mechanized management. In a heavily compacted clay soil, much of the particle surface area would be compressed against other particle surfaces, leaving little space between for the storage of water.

Figure 8.4 shows the interaction between water content of soils, water availability, soil matric water potential, and soil texture. Soil water is separated into three categories. Many plants cannot extract water from soil with matric potentials more negative than about −1.5 megapascals (MPa). This point has been termed the **permanent wilting point** and determines the position of the line separating available water (AW) from unavailable water (UW). Soils with significant water content may be completely "dry" as far as plants are concerned! As discussed below, the −1.5 MPa cutoff does not apply to native plants adapted to very dry conditions, which can extract water at much more negative matric potentials.

Soil matric potentials less negative than about −0.01 MPa are too weak to retain water against the force of gravity. Water will drain freely under these conditions until this matric potential is reached. Soil water content at this point is called the **field capacity**. The amount of plant-extractable water that a soil can hold (AW) is the difference between field capacity and the permanent wilting point, or between −0.01 and −1.5 MPa. In the

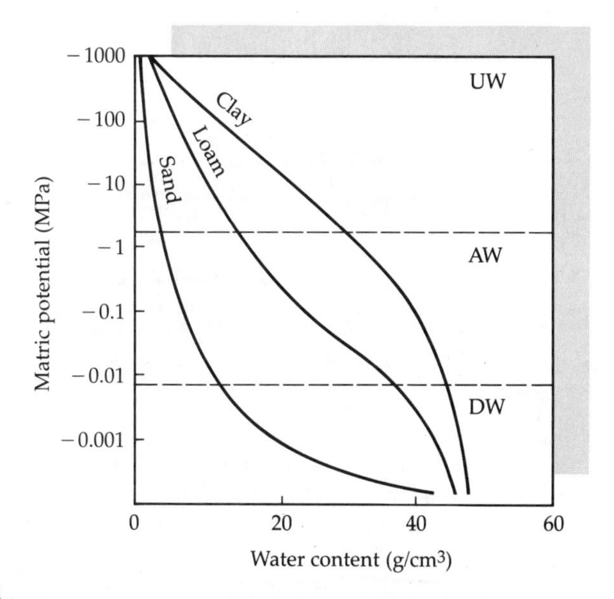

Figure 8.4
Water content and soil matric potential for soils with different textures. UW = water unavailable to plants, AW = plant-available water, DW = drainage water. (After Waring and Schlesinger 1985, Ulrich et al. 1981)

example in Figure 8.4, the loam soil (loam is a mixture of sand, silt, and clay plus organic matter) has the greatest amount of available water (about 22 g/cm^3), followed by the clay soil (about 18 g/cm^3) and the sandy soil (about 9 g/cm^3). Note that the clay soil contains more water at −1.5 MPa (permanent wilting point) than the sandy soil does at −0.01 MPa (field capacity). The frequency with which the "available water" is replenished depends on the frequency and quantity of precipitation.

THE SOIL–PLANT–ATMOSPHERE CONTINUUM

Water moves from soil to plant to atmosphere along gradients from high (less negative) to low (more negative) water potential. How fast the water moves is a function of the quantitative difference between the potentials in different parts of this soil–plant–atmosphere continuum (sometimes abbreviated SPAC) or of the steepness of the water potential gradient and the resistance to movement between two potentials.

The plant will conduct water from the soil to the atmosphere only if the atmosphere has a more negative water potential than the soil. This is the case under most circumstances. Figure 8.2 shows that at anything less than complete saturation, atmospheric water potentials become very negative compared to the −0.01 to −2.0 MPa range usually encountered in soils. Extended periods of high humidity can be important in some ecosystems. The coastal redwood forests of California and Oregon attain high growth rates and tremendous size despite relatively low rainfall. This is thought to be at least partly due to frequent fogs, which increase atmospheric water potential (make it less negative) and reduce water loss by plants. This in turn reduces water uptake from soils and slows the rate of soil drying.

The plant is the vehicle for water movement from soil to atmosphere. Water evaporates from the moist surfaces of cells within the leaf. This film of water is continuously replenished by water movement up through the plant in small tubes called the xylem or vessel elements, which serve the purpose of water transport. The evaporation of water at the cell surfaces creates a tension (the opposite of pressure; a negative water potential) on the water column in the xylem cells, which pulls the water up through the plant, much as a rope pulls. This tension is conducted out through the roots and, through intimate contact between roots and soil, to the water adhering to soil particles.

This moving water column must be continuous. Any air gaps in the system will relieve the tension and stop the water movement. Transpiration from very dry soil can be limited by the lack of continuous water films in the soil. Frost damage to trees can also disrupt water flow if the water frozen in xylem elements breaks cell walls and creates air gaps in the water-conducting system. Resistance to flow within the plant results mainly from friction between water and the walls of the xylem elements through which it passes and from the force of gravity pulling the water back down.

PHYSIOLOGIC CONTROL OVER WATER LOSS

Thus far, the plant has been described passively, much like a pipe. Plants do exert considerable control over water transport and transpiration. When water movement from soil to root to leaf is too slow to keep up with evaporation within the leaf or movement from the leaf to the atmosphere, leaves can wilt and be permanently damaged or killed. Control over this is exercised at the stomata, the leaf pores through which gases are exchanged with the atmosphere. Under conditions of high water stress, the stomata are closed to severely reduce the further evaporation of water from internal plant surfaces.

To examine the operation of this system for control over water loss, we can use the example of a shrub growing in full sun in a relatively dry soil (Figure 8.5). Because the stomata are closed at night, the water potential is

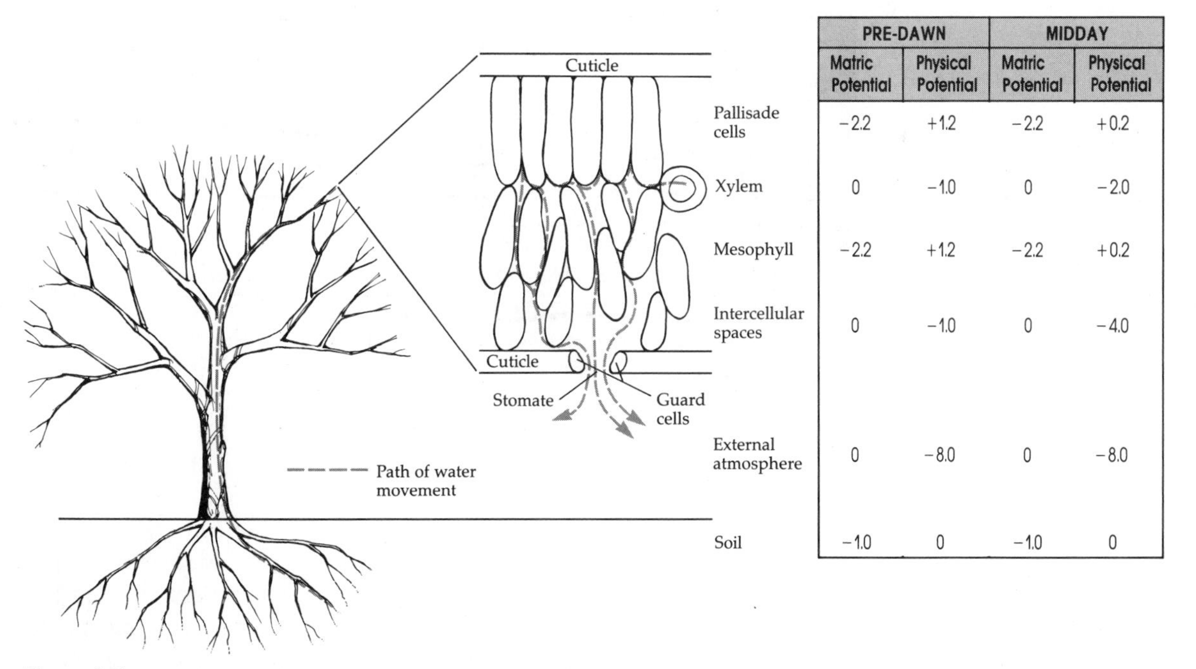

	PRE-DAWN		MIDDAY	
	Matric Potential	Physical Potential	Matric Potential	Physical Potential
Pallisade cells	−2.2	+1.2	−2.2	+0.2
Xylem	0	−1.0	0	−2.0
Mesophyll	−2.2	+1.2	−2.2	+0.2
Intercellular spaces	0	−1.0	0	−4.0
External atmosphere	0	−8.0	0	−8.0
Soil	−1.0	0	−1.0	0

Figure 8.5
Example of changes in the components of water potential in different parts of the soil–water–atmosphere continuum from pre-dawn to midday (all values in MPa (megapascals)).

relatively constant throughout the plant just before dawn and is close to the soil value of −1.0 MPa. (One of the most common ways to measure soil water potential is to actually measure the tension on the water column in xylem of low-growing twigs just before dawn.) If the atmosphere has a relative humidity of 50%, then it will have a water potential (physical) of −80.0 MPa. As the sun rises and the stomata on the plant leaves open, a steep water potential gradient exists between the intercellular air spaces in the leaf and the external atmosphere, and transpiration will begin.

As the relative humidity in the air spaces within the leaf drops, its water potential becomes more negative, and a gradient is established between the air spaces and the liquid films of water on the plant cell surfaces. Evaporation from the cell surfaces lowers the potential there, and this lower potential is transmitted throughout the water column in the plant. By midmorning, a gradient of water potential exists from the atmosphere to the plant and into the soil immediately surrounding the roots (Figure 8.5). Water moves from soil to atmosphere in response to this gradient. The rate of flow is a function of the steepness of the gradient and the resistance to flow through the system. If the rate of water movement is not sufficient to keep up with transpirational water losses, the water potential of the water within the leaf xylem elements and on the evaporating cell surfaces decreases and greater tension is exerted on the entire water column, increasing the rate of water flow to the leaf.

So far, all of these potentials have been physical, a pulling of water in response to tension on the water column. This tension is also exerted across the membranes of the cells in the leaf. This tension would lead to the loss of water from the cell if it were not countered by a more negative potential within the leaf—an osmotic potential created by the concentration of solutes within the cell. In general, the osmotic potential is greater than the tension on the water column passing along the cell surfaces, creating a physical pressure against the cell walls. It is this pressure that keeps leaves turgid and prevents wilting. If the water potential on the water column at the cell surfaces becomes more negative than the osmotic potential in the cells, then water is lost across the cell membrane, turgor is lost, and wilting occurs.

In Figure 8.5, if the cells have an osmotic potential of −2.2 MPa, the water will flow from the exterior water film into the cell until a positive pressure is created against the cell wall, in this case +0.2 MPa, which equalizes total water potentials inside and outside the cell wall.

The −2.0 MPa at the cell surface results from a balance between the rate of water movement up through the plant and the rate of evaporation and diffusion out through the stomata. Imagine now that the atmosphere becomes progressively drier during the day. The steepness of the water potential gradient between air and leaf is increased and the rate of flow out of the leaf increases. The tension on the water on the cell surfaces increases and may drop below −2.2 MPa. If this happens, water will again start to move out of the cells, and wilting will occur. Over longer time periods, reductions in soil water content could also reduce the gradient between plant and soil and slow water movement to the plant. This could also result in lower water potentials at the cell surfaces and loss of water from cells.

PLANT RESPONSE TO WATER STRESS

Plants have both a short- and a long-term response to prevent wilting of leaves under conditions of water stress. In the short-term response, the stomata can close. This reduces diffusion out of the leaf. Thus, the humidity in the air within the leaf increases, and evaporation from cell surfaces declines markedly. At the same time, however, the flow of CO_2 into the plant is also reduced and photosynthesis is suppressed.

The long-term response is to modify the osmotic potential of the leaf cells. Increasing the concentration of sugars, cations, and other low-molecular-weight solutes in cells creates a more negative osmotic potential. This can both increase the difference between water potentials of leaf and soil and slightly decrease the gradient between leaf and atmosphere, helping to bring the rates of water movement to and from the leaf back into balance. However, the increasing solute concentrations can interfere with other cell functions. Thus, longer periods of photosynthesis are made possible by stomata remaining open longer, but this comes at the cost of less efficient cell operation.

These two mechanisms interact in a plant's response to increasing water stress, as shown by a study of the water relations of four species on north- and south-facing mountain slopes in northern Idaho. This is a region of winter snows and relatively dry summers, so the soil is moist at the beginning of the growing season and becomes progressively drier from spring through fall.

This study showed first that, during the course of a given day, the water potential of the xylem in the leaf never became significantly more negative than the osmotic potential of the leaf cell sap (Figure 8.6*a*). During the course of a day, the atmosphere pulled water from the leaf, and the water potential of the water within the xylem elements in the leaves and at the cell surfaces increased. As this approached the osmotic potential, stomatal closure occurred to disallow wilting. From 10 A.M. to about 6 P.M., stomata would keep opening and closing to keep this xylem very near the osmotic potential in the cells.

As the summer progressed, the soil became drier and the water stress on the plants increased. Stomata would therefore be closed for a greater portion of each day and photosynthesis further reduced. In response to increased stress, many plants decrease (make more negative) the osmotic potential of the cell sap. Progressive decreases in osmotic potential occurred for most of the species in this study. Different species also showed very different osmotic potentials at the same time of year (Figure 8.6*b*). It appeared that the ability to create more negative osmotic potentials in

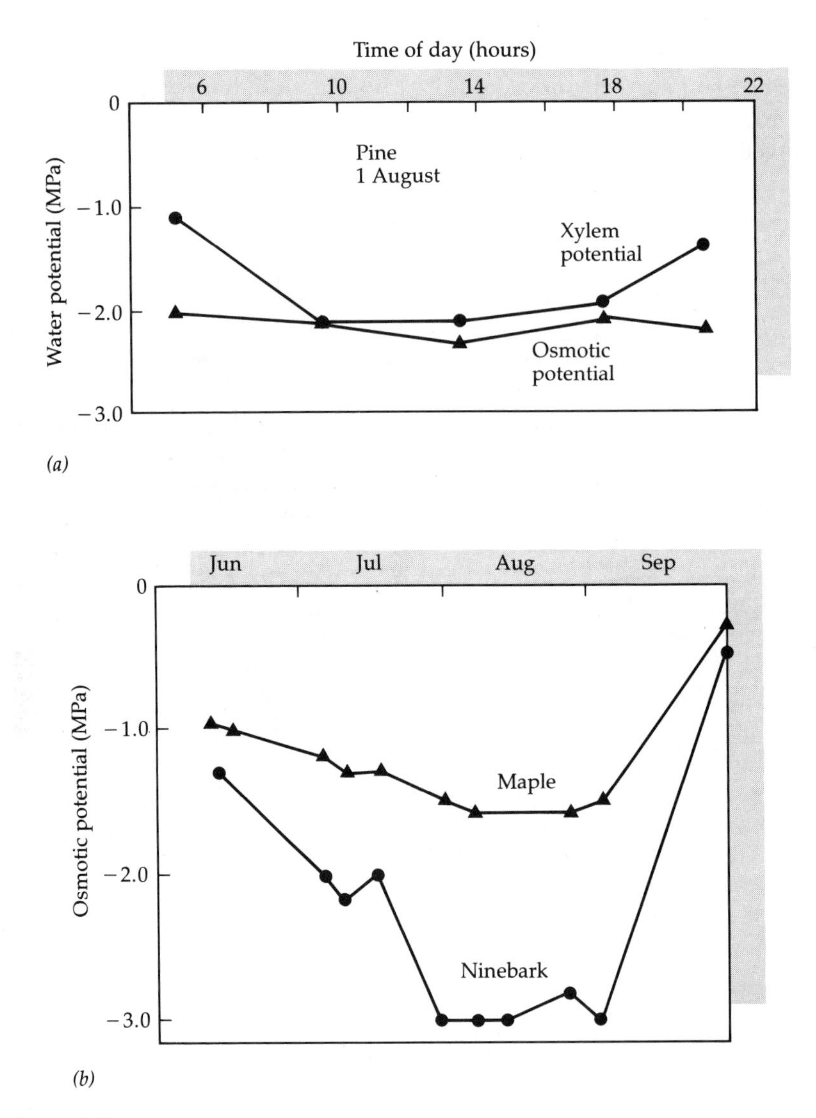

Figure 8.6
Changes in leaf xylem potential and osmotic potential. *(a)* Changes in xylem and osmotic potential of pine over the course of one day. *(b)* Changes in osmotic potential for two species during the growing season. (Cline and Campbell 1976)

leaves conferred an adaptive advantage and a greater chance of survival on the drier slopes.

If altering osmotic potentials and thereby tolerating more negative xylem potentials (greater tension on the water column and cell surfaces) is an important mechanism for adapting to dry sites, then measured xylem potentials in different ecosystems should be quite different. Drier sites such as deserts should have more negative xylem potentials.

This hypothesis has been tested using an elegantly simple device known as a Schollander pressure bomb (Figure 8.7). The device measures leaf xylem water potential, the tension being exerted on the water column coming up through the plant. The inventor and several co-workers used this device to measure midday xylem potentials across a transect in California running from coastal saline wetlands through dry shrublands (chaparral), up the coastal mountain range where water stress declines with elevation, and over the mountain into the rain-shadow deserts of that region. Measurements from freshwater pond plants were included to yield the widest possible range of water availabilities. Figure 8.8 summarizes the findings. Xylem water potential does indeed become much more negative in the drier habitats.

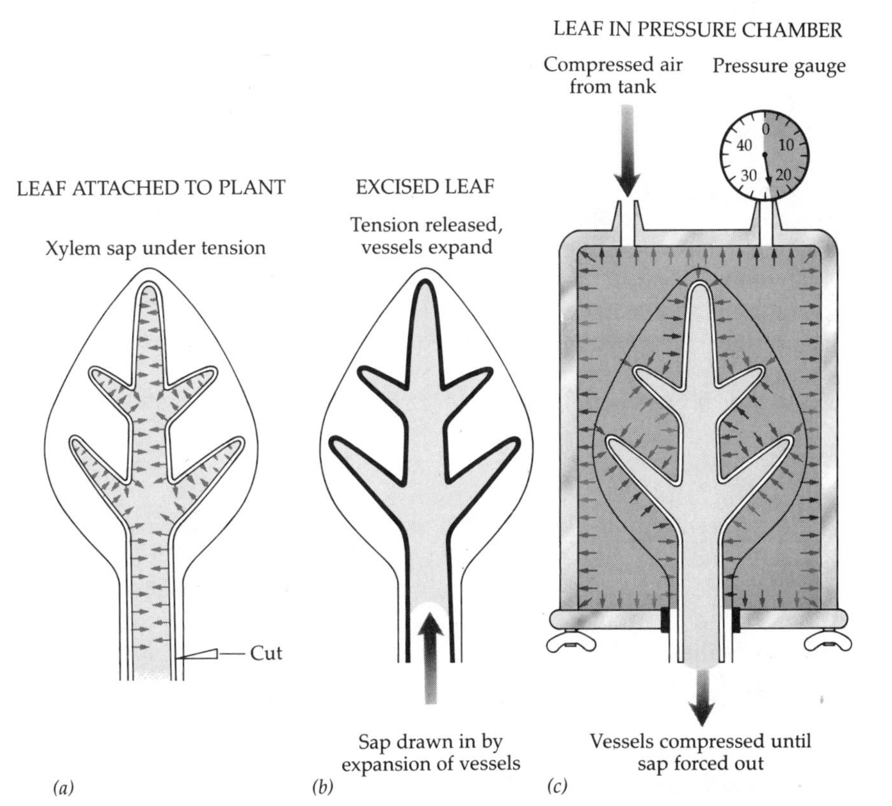

Figure 8.7
Operation of the Scholander pressure bomb for measuring tension on the water column in the xylem of plant twigs (xylem water potential). *(a)* Tension on the water column in the xylem results in shrinkage of the xylem elements. *(b)* Cutting the twig releases this tension and the xylem elements expand, drawing water away from the cut edge. *(c)* In the pressure bomb, pressure is exerted on the twig until the water in the xylem returns to the cut edge. It is assumed that this pressure is equal to the tension on the water column before the twig was cut. (Ray 1972)

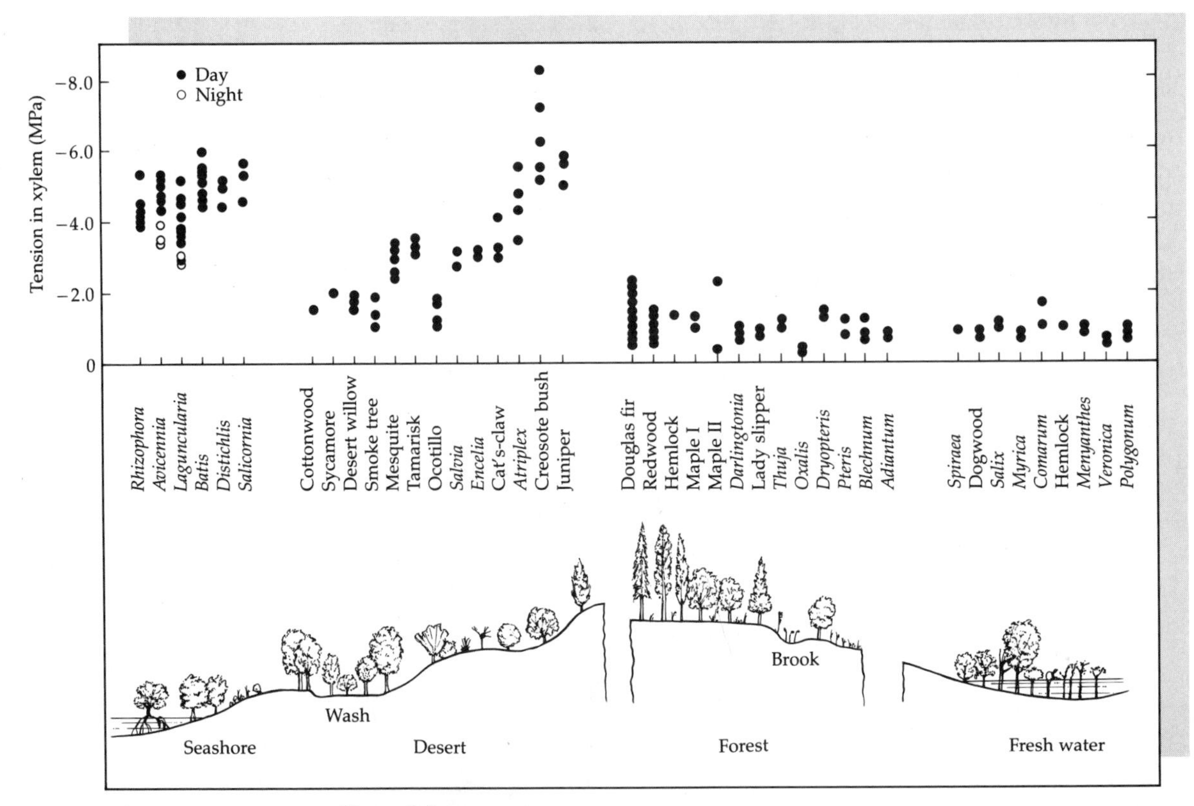

Figure 8.8

Range of midday xylem water potentials for plants in several habitats. (Scholander et al. 1965)

It is interesting that plants rooted in ocean saltwater had intermediate potentials, more negative than freshwater plants. Although these plants are surrounded by water, the salt concentration in the water creates an osmotic potential of about −3.0 MPa. This is functionally equivalent to a matric potential of the same magnitude in soils, so this habitat appears fairly "dry" to these rooted, emergent plants.

Seasonal Water Balances and Effects on Canopy Structure and Total Annual Photosynthesis

There is an additional but rather drastic response plants can make to severe drought. They can shed their leaves through a process called **senescence**, becoming partially dormant until favorable conditions recur. This is common in the seasonal tropical forests where senescence and reactivation are in response to moisture conditions rather than seasonal temperature shifts as in most temperate seasonal forests. Senescence is also common in grasslands, shrublands, and deserts where the timing and length of the "growing season" may depend on when and how much precipitation occurs. The annual "blooming of the desert" is an ephemeral and highly erratic example.

An even longer-term response seen in plant communities is the modulation of total leaf area produced during each growing season in response to average, long-term climatic conditions. A striking example of this can be seen along a transect similar to the one described above for California, but this time running from coastal Oregon to the deserts in the eastern part of that state. Topography plays a large role in causing very different climatic conditions along this gradient. Rainfall is abundant near the coast and at higher elevations in the mountains. Rain-shadow effects cause increasingly dry conditions to the east. Summer drought plays a major role in limiting plant growth. Potential evapotranspiration (PET; Chapter 2) increases from west to east, while precipitation decreases. In response to this, the total leaf surface area present in the canopies of the vegetation declines remarkably (Figure 8.9).

Chronic water stress also alters the structure and total plant biomass of a system. In the prairie–forest border region of Wisconsin, changes in soil

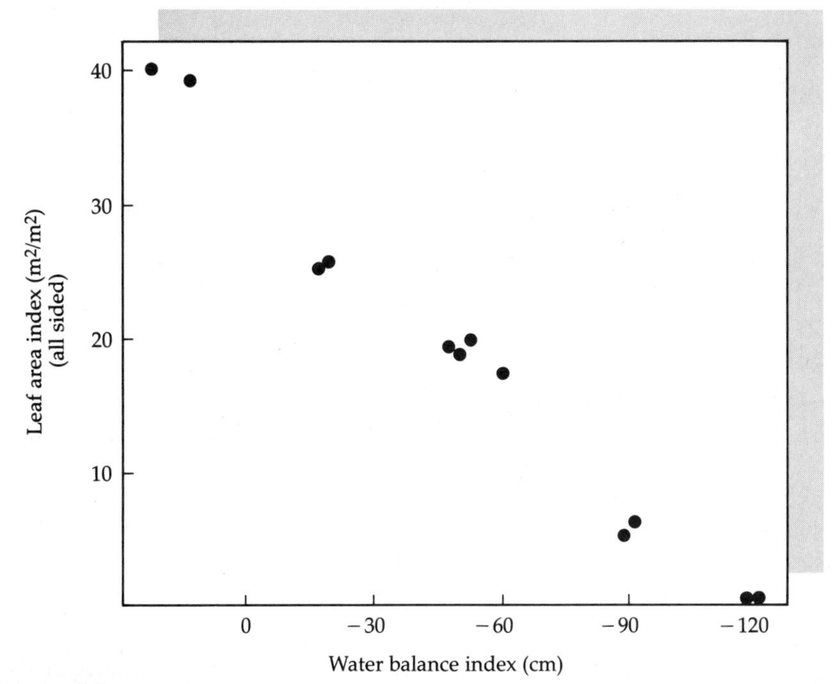

Figure 8.9
Changes in forest leaf biomass in relation to an index of annual water stress, along a gradient in water availability in Oregon. (After Grier and Running 1977, Gholz 1982)

texture play an important role in determining annual water availability. As soils grade from silt clay loams (fine-textured) to sandy loams (coarse-textured), both mean leaf height and canopy layering change continuously, along with changes in species composition (Figure 8.10).

To understand the interaction between water availability and ecosystem function, we can examine seasonal patterns of temperature, precipitation, soil water availability, leaf area display, and net photosynthesis for five types of temperate zone ecosystems (Figure 8.11). These represent four of the systems of the temperate zone in Figure 2.2 and a true desert.

In the driest type, the desert (Figure 8.11*a*), total production is very low and occurs mostly in a single flux. In this example, this occurs following winter rains, which, in the absence of transpiration and with reduced evaporation, partially recharge the soil with water. Increased soil moisture causes dormant seeds to germinate and grow. Deciduous shrubs also refoliate, and "evergreen" succulents—cacti—become active. The duration and intensity of this growth flush depends directly on the amount of water available and so is quite variable from one year to the next. Photosynthesis is always less than the potential determined by temperature. Water is always limiting, although somewhat less so during this post-winter period. When the soil water is exhausted, the nonsucculents die or resume dormancy.

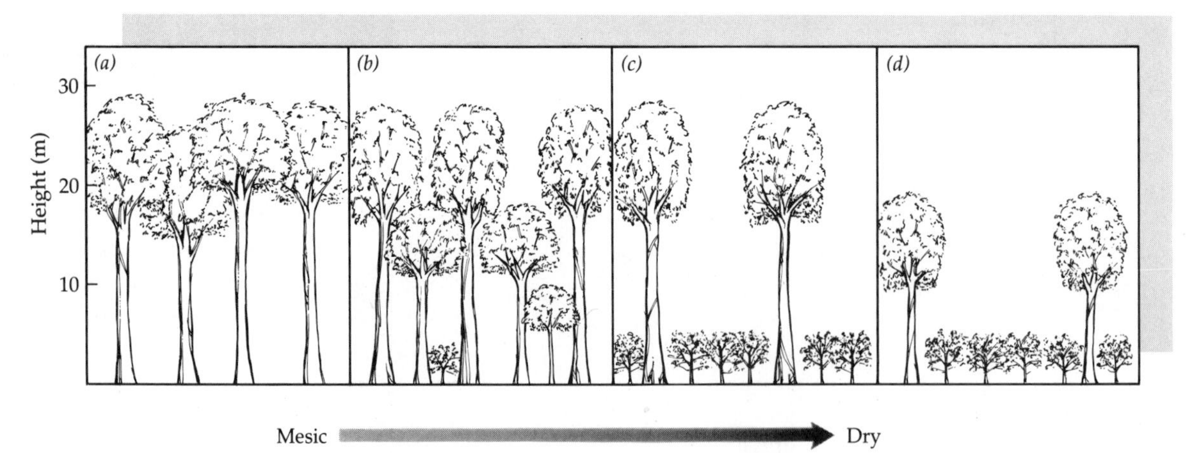

Figure 8.10

Changes in forest canopy structure with changes in soil texture in forests of the forest–prairie border region in Wisconsin. *(a)* Fine-texture soils have closed upper canopies and little understory. As soils become progressively drier, canopy structure grades to *(b)* evenly distributed, *(c)* open upper canopy with significant shrub layer, and *(d)* open upper canopy of reduced height, with significant shrub layer. Sites drier than those covered in this study would support prairie systems. (After Aber et al. 1982)

As precipitation increases, ecosystems become dominated by grasses and shrubs. The relative mixture of these two may depend on the timing of the precipitation. Figure 8.11*b* shows the pattern for a shrub (chaparral) ecosystem in coastal Southern California. In this case, precipitation occurs mainly in the winter. Many of the shrub species are evergreens, which maintain leaf area year-round and so can use this winter-available water. There is also an additional deciduous component of shrubs, grasses, and annual plants, which are also active at this time, becoming dormant as water availability declines. In contrast to the desert, the evergreen shrubs in the chaparral continue to function throughout the year, making use of rare summer rains, deep water tables, or the scarce water reserves in the soil. Photosynthesis is rarely, if at all, temperature-limited in these systems.

In contrast, grasslands (Figure 8.11*c*) tend to occur in areas of moderate summer rainfall and cold winter temperatures. Here, soil water increases during winter even though precipitation is light because of low temperatures, which rule out plant function. As temperatures increase in the spring, growth and photosynthesis begin and may be more limited by temperature than by water, depending on the amount of winter precipitation. Summer rains tend to recharge the soil but are generally insufficient to eliminate water stress. Depending on the amount of rainfall and the length of the frost-free season, senescence and the end of the growing season may result either from drought or from cold. In the example in Figure 8.11*c*, drought is the cause.

The seasonal evergreen forests such as those in the interior regions of the Pacific Northwest are similar in climatic pattern to the chaparral but cooler and with greater precipitation, which occurs mainly as snowfall (Figure 8.11*d*). Cooler temperatures reduce evapotranspiration in winter, and photosynthesis, when it occurs, is mostly temperature-limited. High temperatures and low rainfall cause water stress in summer, and photosynthesis is mainly water-limited. Unlike the previous three systems, leaf area of the vegetation does not vary significantly between seasons; there is no large-scale canopy senescence. Conditions are never so extreme as to favor the development of either a summer-dormant or winter-dormant vegetation.

In the temperate deciduous forest, on the other hand, total and synchronous canopy senescence is one of the most conspicuous processes. This type tends to occur in regions of generally high precipitation and great extremes of temperature (Figure 8.11*e*). Limitations on photosynthesis by water stress are less common and generally less severe than in the seasonal evergreen forest. Low temperatures inhibit leaf-out in the spring and cause senescence in fall, so that leaves are displayed only during the frost-free season. This compares to the seasonal evergreen forest where up to 70% of the total photosynthesis can occur outside the normal "growing season."

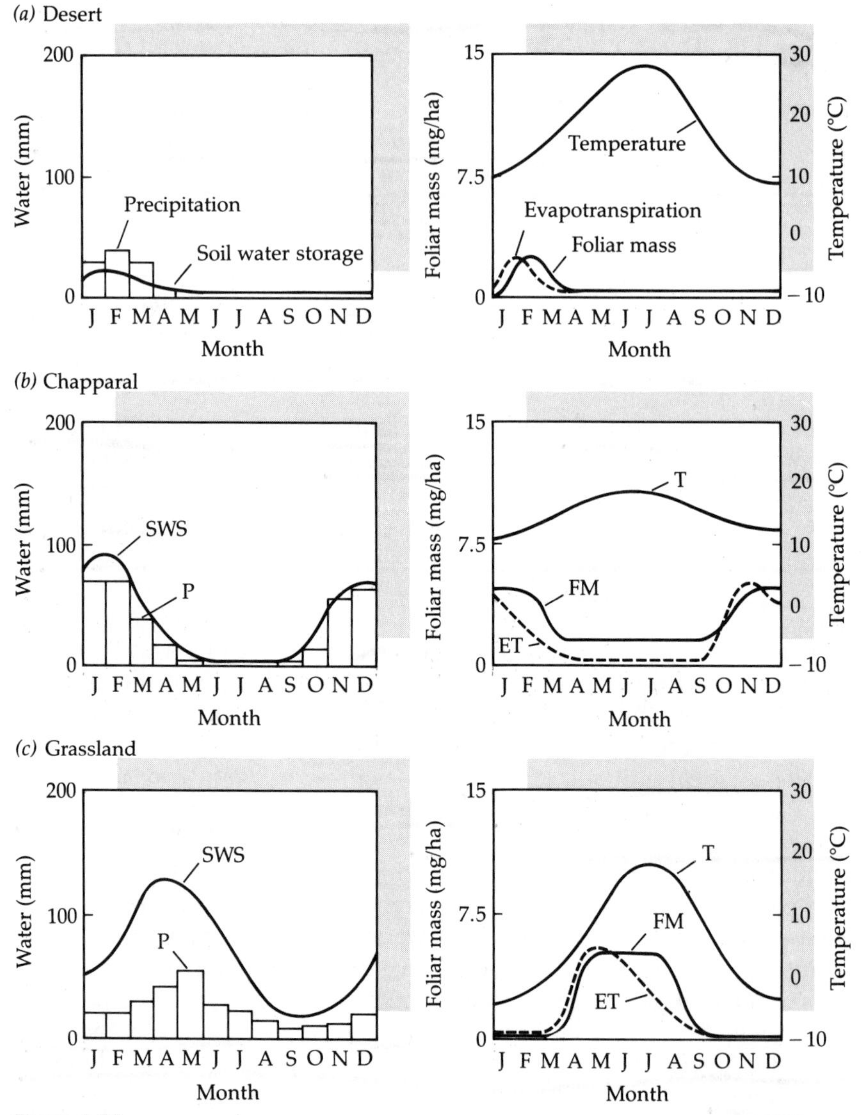

Figure 8.11

Seasonal patterns of temperature, rainfall, soil water storage, foliar biomass, and evapotranspiration for five idealized temperate-zone ecosystems. Potential net photosynthesis is proportional to the temperature curve. Realized photosynthesis is proportional to the evapotranspiration curve. (Emmingham and Waring 1977, Walter 1979, Parton et al. 1981, Running and Coughland 1988, Vorosmarty et al. 1990)

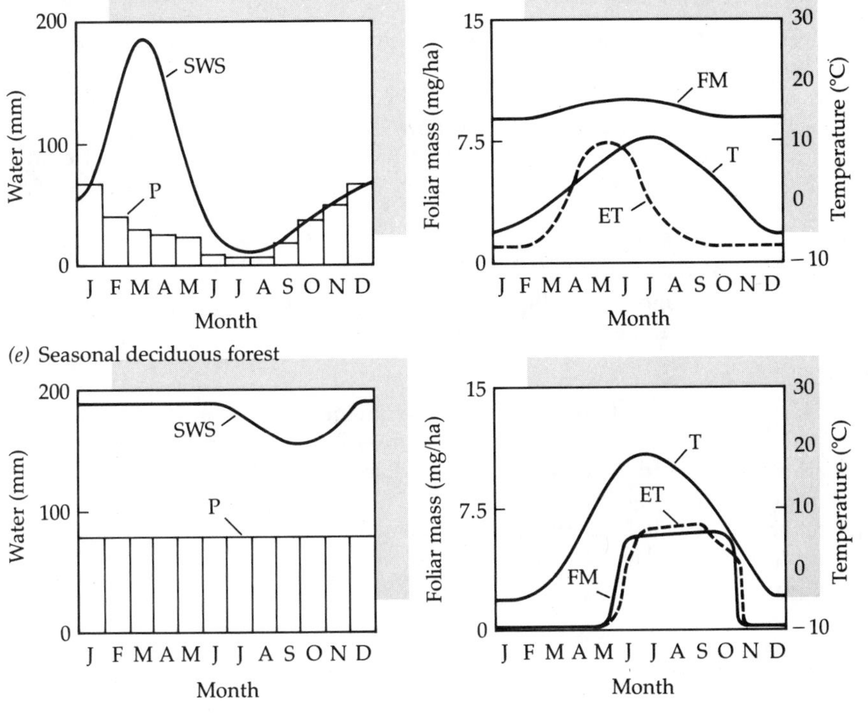

Figure 8.11 ***continued***

REFERENCES CITED

Aber, J. D., J. Pastor, and J. M. Melillo. 1982. Changes in forest canopy structure along a site quality gradient in southern Wisconsin. *American Midland Naturalist* 108:256–265.

Buckman, H. O., and N. C. Brady. 1969. *The Nature and Properties of Soils.* Macmillan, New York.

Cline, R. G., and G. S. Campbell. 1976. Seasonal and diurnal water relations of selected forest species. *Ecology* 57:367–373.

Emmingham, W. H., and R. H. Waring. 1977. An index of photosynthesis for comparing forest sites in western Oregon. *Canadian Journal of Forest Research* 7:165–174.

Gholz, H. L. 1982. Environmental limits on aboveground net primary production, leaf area, and biomass in vegetation zones of the Pacific Northwest. *Ecology* 63:469–481.

Grier, C. C., and S. W. Running. 1977. Leaf area of mature northwestern coniferous forests: Relation to site water balance. *Ecology* 58:893–899.

Parton, W. J., W. K. Lauenroth, and F. M. Smith. 1981. Water loss from a shortgrass steppe. *Agricultural Meteorology* 24:97–109.

Ray, P. M.. 1972. *The Living Plant.* Holt, Rinehart and Winston, New York.

Running, S. W., and J. C. Coughlan. 1988. A general model of forest ecosystem processes for regional applications. I. Hydrologic balance, canopy gas exchange and primary production processes. *Ecological Modeling* 42:125–154.

Salisbury, F. B., and C. Ross. 1969. *Plant Physiology*. Wadsworth, Belmont, California.

Scholander, P. F., et al. 1965. Sap pressure in vascular plants. *Science* 148: 339–346.

Ulrich, B., et al. 1981. Soil Processes. In D. E. Reichle (ed.), *Dynamic Properties of Forest Ecosystems*. Cambridge University Press, London.

Vorosmarty, C. J., et al. 1990. Continental scale models of water balance and fluvial transport: An application to South America. *Global Biogeochemical Cycles* 3:241–265.

Walter, H. 1979. *Vegetation of the Earth and Ecological Systems of the Geo-Biosphere*. Springer-Verlag, New York.

Waring, R. H., and W. H. Schlesinger. 1985. *Forest Ecosystems: Concepts and Management*. Academic Press, New York.

ADDITIONAL REFERENCES

Hinkley, T. M., J. P. Lassoie, and S. W. Running. 1978. Temporal and spatial variations in the water status of forest trees. *Forest Science Monograph* 20.

Waring, R. H., and B. D. Cleary. 1967. Plant moisture stress: Evaluation by pressure bomb. *Science* 155:1248–1254.

Whittaker, R. H. 1975. Communities and environments. In *Communities and Ecosystems*. Macmillan, New York.

Chapter 9 Soil Chemistry and Nutrient Availability

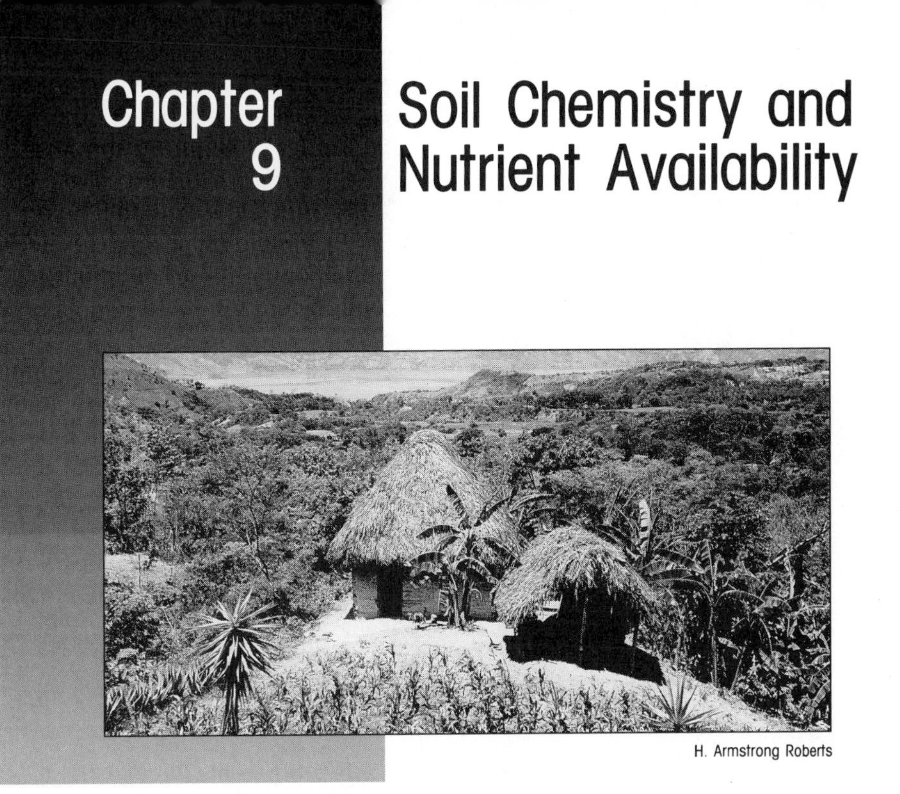

H. Armstrong Roberts

INTRODUCTION

Canopies are visually and structurally complex, yet they are much simpler than soils. Canopy interactions are dominated by the energy, water, and gaseous balances of leaves. Physically, canopies are dominated by plants alone. In contrast, nutrient and water uptake by plants are only two of several important processes occurring in soils. Microbial decomposition, water storage and leaching, and the physical and chemical reactions of weathering, cation exchange, and sorption all go on simultaneously. These processes all occur at such a microscopic scale and are so strongly interrelated that it is nearly impossible to separate them. Because of this, soils are more difficult to study in terms of the rates at which basic processes occur, and our understanding of them is less complete.

The purpose of this chapter is to introduce the chemical elements essential for plant growth and the chemical reactions in soils that affect their availability to plants. This will be followed, in the next two chapters, by a discussion of how plants can alter the soil environment to increase nutrient availability and how nutrients and carbon taken up by plants are allocated to different types of tissues (roots, stems, leaves) in order to optimize resource availability and growth rate.

THE ESSENTIAL NUTRIENTS

Fifteen elements are generally considered to be required for plant function (Table 9.1). These are separated into the "macronutrients" and "micronutrients," according to the amounts generally found in plant tissues. With

Table 9.1 Elements known to be required for plant growth*

ELEMENT	CHEMICAL SYMBOL
Macronutrients	
Carbon	C
Hydrogen	H
Oxygen	O
Nitrogen	N
Potassium	K
Calcium	Ca
Magnesium	Mg
Phosphorus	P
Sulfur	S
Micronutrients	
Chlorine	Cl
Iron	Fe
Boron	B
Manganese	Mn
Zinc	Zn
Copper	Cu
Molybdenum	Mo

*From Salisbury and Ross 1978.

one exception, the macronutrients are major components of important structural and metabolic molecules such as carbohydrates, proteins, chlorophyll, deoxyribonucleic acid (DNA), ribonucleic acid (RNA), sugar phosphates, and phospholipids. The exception is potassium, which is used in stomatal control, for charge balance during ion movements across membranes, and as a coenzyme in many important biochemical reactions (Table 9.2). The micronutrients are generally used either as structural components of less common molecules or as coenzymes—ions needed to catalyze specific reactions.

Elements required by animals are generally the same as for plants, with some notable exceptions. One is sodium. Mammals require relatively large amounts of sodium for proper function of the nervous system. Selective use of sodium-rich plants (for plants do take up elements other than those required for growth) and even the ingestion of sodium-rich soils (salt-licks) are two common animal responses to a lack of this element in plants.

The smaller requirement for micronutrients in plants does not necessarily mean they are less important or exert less control on ecosystem function. However, this is usually the case. Only those systems growing on unusual geologic formations, very old and fully weathered soils, or on soils subject to extreme human disturbance are likely to show limitations

Table 9.2 Macronutrients, their biochemical uses, form of uptake by plants, typical concentrations in leaves and wood, and whether they have been shown to be limiting to plant growth and mobile within plants*

NAME	USES	TAKEN UP AS	CONCENTRATION IN PLANTS		MOBILITY	LIMITING?
Carbon (C), hydrogen (H), oxygen (O)	Carbohydrates and derivatives, basic building blocks for nearly all plant products	CO_2 H_2O	90–98%		Variable	As seen before
Nitrogen (N)	Amino acids, proteins, enzymes, nucleic acids, chlorophyll	NO_3^-, NH_4^+	1–4%	0.1–0.3%	High	Yes
Phosphorus (P)	Sugar phosphates (ATP, ADP), nucleic acids, phospholipids	$H_2PO_4^-$	1/10 of N		High	Yes
Potassium (K)	Not structural, enzyme co-factor catalyzes protein formation; stomata; charge balance across membranes	K^+	1%	0.1%	Very high	Rarely
Sulfur (S)	Amino acids, proteins, enzymes	SO_4^-	0.2%	0.02%	Low	Very rarely
Magnesium (Mg)	Chlorophyll, enzyme co-factor	Mg^{2+}	0.2%	0.02%	Very low	No (but see Chapter 23)
Calcium (Ca)	Crucial to membrane function; binds wood fibers together	Ca^{2+}	0.8%	0.2%	Very low	No (but see Chapter 23)

After Salisbury and Ross 1978.
*Mobile elements are those that can be retranslocated by plants before leaf senescence. Elements such as sulfur and magnesium are somewhat mobile in plants, but are rarely retranslocated due to excess availability in soils.

by low or unbalanced availability of micronutrients. In keeping with our emphasis on factors controlling ecosystem function, the rest of this presentation on nutrient availability and uptake will deal with the macronutrients.

Carbon, hydrogen, and oxygen form the majority of plant biomass—up to 96% for some types of tissues. These are derived from CO_2 and water and are made available to the plant as simple sugars through photosynthesis. The remaining six macronutrients exist in a variety of states in soils, and their availability to plants is affected by several important and different processes. The summary figure at the end of Chapter 1 (reproduced here as Figure 9.1) lumped these into four states: organic, exchangeable, sorbed, and locked-in primary and secondary minerals. The latter three pools are affected by the interacting soil chemical processes of cation (and anion) exchange, sorption, and weathering, respectively.

While these states and classes of reactions represent a grand oversimplification of soil chemical processes, they provide a basis for discussing the differences in the chemical factors affecting the availability of the soil-derived macronutrients (nitrogen, phosphorus, sulfur, calcium, magnesium, and potassium). These chemical processes are the subject of this chapter. The effect of biological processes on availability will be discussed in Chapters 10, 12, and 13.

WEATHERING OF PRIMARY MINERALS

Weathering is the process by which newly created or newly exposed geologic substrates are converted into soils. It involves the physical and chemical alteration of the geologic substrate underlying an ecosystem. Weathering is initiated by the retreat of a glacier, the creation of newly hardened lava flows by volcanic activity, or by any other process exposing geologic material that previously has been protected from chemical dissolution and physical wear.

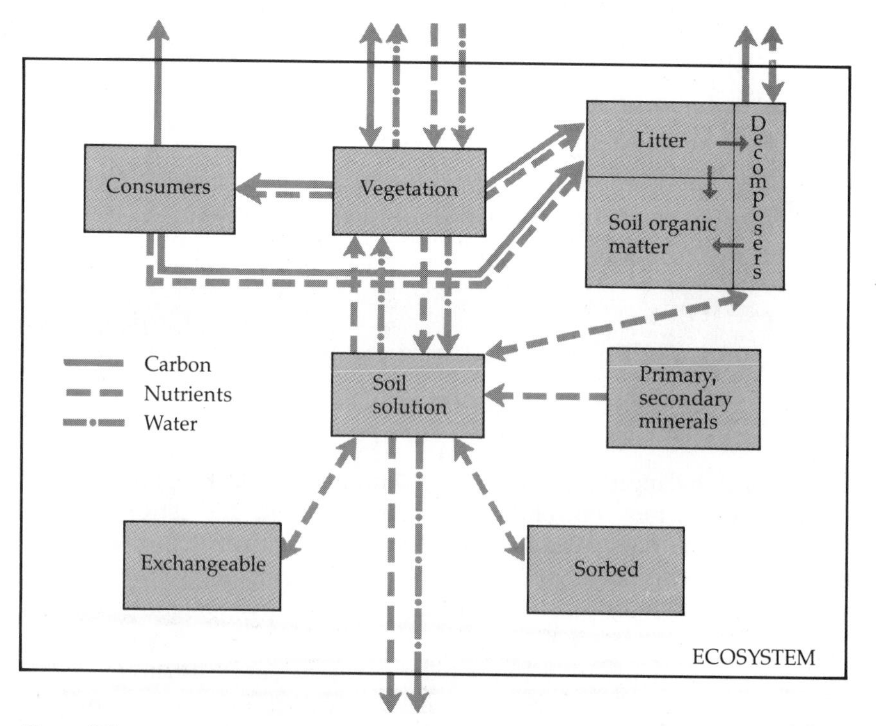

Figure 9.1
A summary diagram of nutrient cycling in terrestrial ecosystems.

Three aspects of the larger study of weathering rates and reactions are important in the function of ecosystems: (1) the rate at which the geologic material in the system weathers, (2) the nutrient elements released during weathering, and (3) the type of secondary minerals formed. The first two factors determine the effect of weathering directly on pools of nutrients available to plants in the system. The third affects the ion exchange and nutrient sorption potentials of derived soils.

There are three major types of rocks. **Sedimentary rocks** are formed by deposition of mineral particles eroded from the landscape and deposited in lakes, seas, and oceans. As the sediments into which these particles settle get deeper, the pressure produced by the weight of overlying material causes the sediments to solidify into rock. The characteristics of these rocks, and the types of soils derived from them by weathering, are largely determined by the type of material deposited. For example, sandstones formed by the compression of sand particles generally weather back into sandy soils, releasing few nutrients in the process. In contrast, limestone is formed by the sedimentation of the remains of aquatic organisms that concentrate calcium and magnesium during life in shells and other structures. Weathering of this type of rock is rapid and releases large amounts of calcium and magnesium.

Igneous rocks are formed by the cooling of volcanic flows (magma) either at or beneath the surface of the Earth. The properties of these rocks depend on the rate and temperature conditions at which they formed. Igneous rocks consist mainly of iron and aluminum silicates with lesser concentrations of calcium, potassium, magnesium, and sodium, reflecting the relative abundance of these elements in the Earth's crust (Table 9.3).

Table 9.3 Percent elemental composition of the Earth's crust*

ELEMENT	WEIGHT (%)
Oxygen (O)	45.2
Silicon (Si)	27.2
Aluminum (Al)	8.0
Iron (Fe)	5.8
Calcium (Ca)	5.1
Magnesium (Mg)	2.8
Sodium (Na)	2.3
Potassium (K)	1.7
Titanium (Ti)	0.9
All other elements	1.0
Total	100.0

*From Flint and Skinner 1974.

Metamorphic rocks are either sedimentary or igneous rocks that have been altered by the pressure and heat generated by overlying rocks but that have not returned to the melted or magma state. These rocks are modified more in appearance and large-scale structural features than in the mineralogic characteristics affecting weathering.

Figure 9.2 shows the relative content of different minerals in different types of igneous rocks, as well as the relative nutrient contents and rates of weathering. The darker, more easily weathered, and more nutrient-rich rock types are generally formed by rapid cooling after movement to the Earth's surface. Few rock types consist entirely of one mineral. In igneous rocks of mixed mineralogy, the more easily weathered minerals (e.g., olivene and pyroxene families) will weather first, leaving a residue of the more slowly weathered minerals, plus secondary minerals (discussed below).

Resistance to weathering in igneous rocks is a function of the degree to which individual silicate units have condensed or polymerized to form chain, sheet, or three-dimensional crystal structures (Figure 9.3). Again,

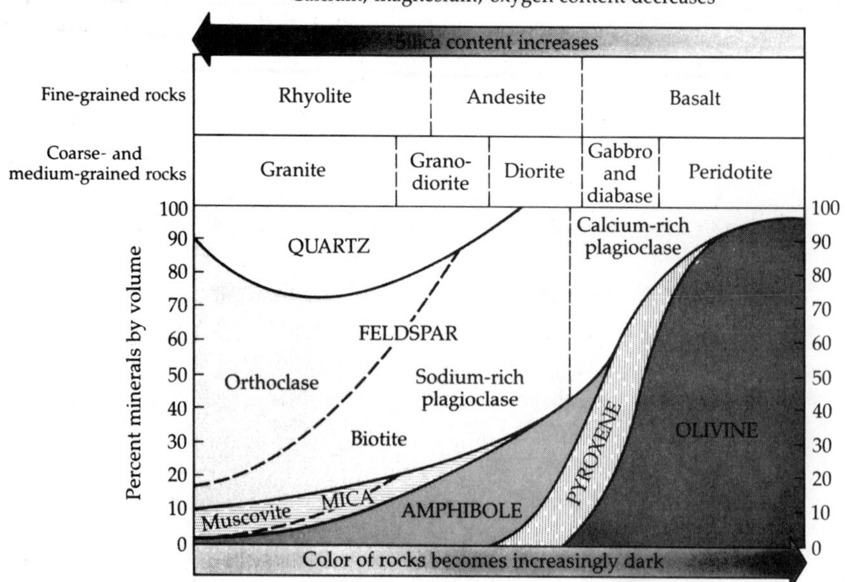

Figure 9.2

Types of rocks and the minerals they contain as a function of the conditions under which they formed. The content of plant nutrient elements in the rocks and the rate at which they weather also change along this gradient. (After Flint and Skinner 1974)

	Arrangement of silica tetrahedra	Formula of the complex anions	Typical mineral
Isolated tetrahedra		SiO_4	The olivine family
Isolated polymerized groups	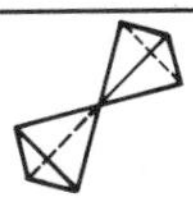	$(Si_2O_7)^{-6}$	Lawsonite
	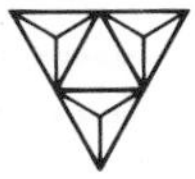	$(Si_3O_9)^{-6}$	Wollastonite
	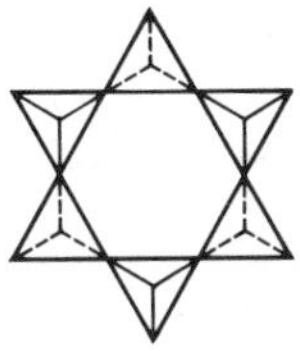	$(Si_6O_{18})^{-12}$	Beryl
Continuous chains	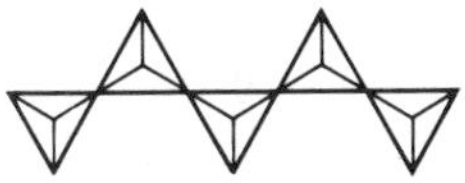	$(SiO_3)_n^{-2}$	The pyroxene family
	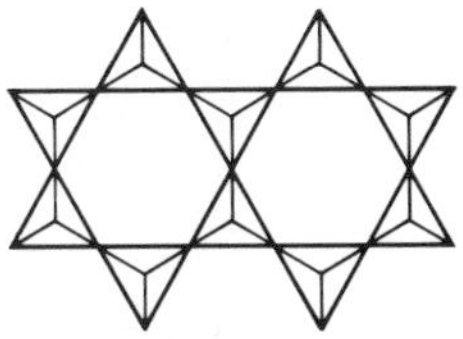	$(Si_4O_{11})_n^{-6}$	The amphibole family
Continuous sheets	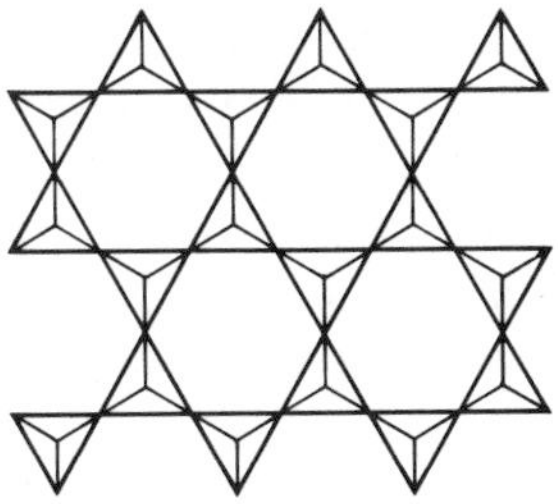	$(Si_4O_{10})_n^{-4}$	The mica family
Three-dimensional networks	Too complex to be shown by a simple two-dimensional drawing	(SiO_2)	Quartz

Resistance to weathering increases

Figure 9.3
Changes in the degree of polymerization of the silica tetrahedra in different minerals. Resistance to weathering increases with increasing polymerization. (Flint and Skinner 1974)

the more gradual the cooling of magma, or the lower the temperature under which the rock solidifies, the greater the degree of silicon polymerization and the slower the weathering rate. Quartz forms during the relatively slow cooling that occurs in magmas held well beneath the Earth's surface.

The content of elements other than silicon and oxygen is also affected by conditions during rock formation. The harder rocks tend to have greater substitution of aluminum for silicon and lower contents of calcium and magnesium as "bridging" ions between the crystal structures. Thus young soils derived from volcanic flows that solidified on the Earth's surface are often rich in nutrient cations (Ca, Mg, K) as well as phosphorus.

The macronutrients sulfur and phosphorus are present in trace amounts in many geologic formations, usually in minerals that are easily weathered. Nitrogen is absent from all major rock types and is not made available through weathering.

Weathering rates are increased by high temperatures and high (but not saturated) soil water content. Thus similar rock types will be altered more quickly in tropical rainforest conditions than in either cool or dry systems (Figure 9.4). This has the positive effect of liberating the nutrients held in fresh rocks more quickly under tropical conditions and also the negative effect of more rapid depletion of weatherable minerals under these same conditions.

FORMATION OF SECONDARY MINERALS

In igneous rocks, weathering does not occur through complete dissolution of the primary minerals. Rather, secondary minerals are formed by the alterations that occur during the weathering reactions. There are four important classes of secondary minerals that can be produced by the weathering of igneous rocks and one additional class formed from more easily weathered minerals under arid conditions (Table 9.4). The type of secondary mineral formed is strongly affected by environmental conditions and also has important effects on nutrient and water retention characteristics of soils.

Under cool, acidic conditions, aluminum oxides are more soluble than silicon oxides. So secondary minerals formed under these conditions tend to be enriched in silicon at the expense of aluminum. Under the most extreme conditions, quartz (sand) can be the principal weathering product. Under warm and less acidic conditions, silicon is weathered selectively, leaving a soil enriched in aluminum (and iron). When this process goes to completion, the residual material is gibbsite (oxides of aluminum) or hematite (oxides of iron). Quartz and these oxides represent extremes of the weathering spectrum and require thousands of years to become dominant soil minerals. Most soils contain a mixture of both fully and only partially weathered minerals.

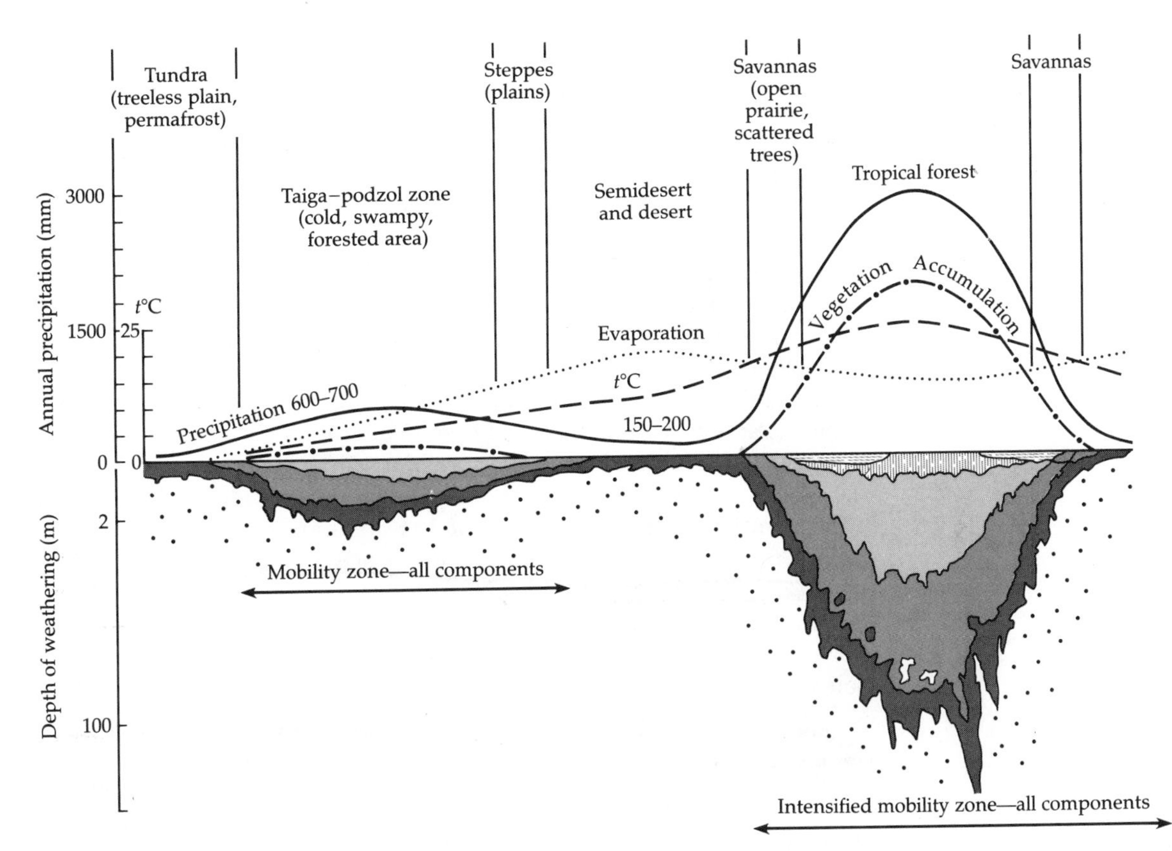

Figure 9.4
A schematic view of the effects of climate on the depth of soil weathered. The warm and wet conditions of the tropical forest result in the most deeply weathered soils. This is reduced by drier conditions in the semi-arid regions and by both reduced precipitation and colder temperatures in the boreal forest and tundra zones. (Strakhov 1967)

Table 9.4 The five major classes of secondary products of weathering reactions and conditions favoring their formation

SECONDARY MINERAL TYPE	CONDITIONS FAVORING FORMATION
Quartz	Cool, acid, humid
2 : 1 layer clays (e.g., montmorillonite)	Cool, neutral, semihumid
1 : 1 layer clays (e.g., kaolinite)	Warm, humid
Iron and aluminum oxides (e.g., gibbsite, hematite)	Excessively warm and humid
Evaporites (e.g., calcite)	Warm, dry, alkaline

The two classes of secondary minerals found in less fully weathered soils are the 1:1 and 2:1 layered clays (Table 9.4). These are often placed in a weathering sequence, with the 2:1 layer clays (montmorillonite, illite) considered more easily weathered and precursors to the formation of 1:1 layer clays (kaolinite). However, under moderately cool and dry conditions, 2:1 clays make up a considerable portion of the clay fraction even in old soils. Conversely, primary minerals can weather directly to kaolinite under favorable conditions, skipping the 2:1 layer stage.

While moisture and temperature are important in determining weathering rates and the relative solubility of aluminum and silicon, other characteristics of the weathering environment can also be important in determining the type of secondary mineral formed. For example, sandstone substrates, rich in silica and poor in aluminum, can weather to form silicon oxides even under tropical conditions. This occurs because reduced aluminum concentrations in the substrate fail to provide enough soluble aluminum to cause the formation of 1:1 clays or aluminum oxides. The lack of nutrient cations in sandstone also leads to low soil pH and, again, increased mobility of aluminum. As another example, montmorillonite (a 2:1 layer clay) is more stable at near-neutral pH than under acid conditions, and its formation may depend on the presence of potassium in the soil solution.

Evaporites are formed in arid regions where water leaching through the soil is minimal. In these areas, easily weathered minerals such as calcium carbonate are dissolved in the upper soil horizons during brief wet periods and deposited in lower horizons as percolating water either evaporates or is taken up by plants. Desert subsoils often contain a hard, cement-like layer of deposited calcium carbonate or "caliche."

Sedimentary rocks are described both by their mineralogic content and the size of the particles of which the sediment was composed (Table 9.5). Sedimentary rocks high in aluminosilicates will weather chemically in the same way as described above, but the particle size distribution in the sediment will partially determine the particle size distribution (sand, silt, clay) in the resulting soil.

In contrast, limestone weathers to calcium and carbonate rapidly and completely in humid climates, leaving no secondary minerals. In very dry climates, calcium carbonate is one of several minerals, called evaporites, that are dissolved in the upper soil horizons by infrequent rainfall events and deposited in lower horizons as soil water is taken up by plants or evaporated. Halite (sodium chloride) is another evaporite.

WEATHERING AND SOIL TYPES

It is not by coincidence that this discussion of weathering has run somewhat parallel to the discussion of soil-forming factors in Chapter 2. The type of weathering regime active in a soil, and the secondary products formed, are important determinants of soil type. Important secondary products of weathering can be placed on a temperature–precipitation grid

Table 9.5 Classes of sedimentary rocks and the particle sizes from which they are formed*

NAME OF PARTICLE	RANGE LIMITS OF DIAMETER		NAME OF LOOSE SEDIMENT	NAME OF CONSOLIDATED ROCK
	mm	Inches (approx.)		
Boulder	>256	>10	Gravel	Conglomerate and sedimentary breccia
Cobble	64 to 256	2.5 to 10	Gravel	
Pebble	2 to 64	0.09 to 2.5	Gravel	
Sand	1/16 to 2	0.0025 to 0.09	Sand	Sandstone
Silt	1/256 to 1/16	0.00015 to 0.0025	Silt	Siltstone
Clay†	<1/256	<0.00015	Clay	Claystone, mudstone, and shale

*From Flint and Skinner 1974.
†"Clay," used in this context, refers to a particle size. The term should not be confused with clay minerals, which are definite mineral species. Many geologists prefer to use the term "clay-sized particle" to avoid confusion.

similar to the one used in Chapter 2 (Figure 9.5*a*). The tropical rainforest environment produces kaolinite or, in extreme cases, gibbsite and hematite clays associated with lateritic soils. Subtropical forest soils are rich in the 1:1 clays. The northern temperate and boreal rainforest environments produce quartz, associated with the E horizon of Spodosols. The intermediate environments of the temperate forests produce a mixture of 1:1 and 2:1 clays, which can be leached to the B horizon (lessivage) to create Alfisols. Drier and more neutral soil conditions associated with melanization of grassland soils favor 2:1 clays and only slight horizon development (Mollisols). The driest conditions lead to the formation of evaporites, which are characteristic of Aridisols. A distribution of major soil types can be overlain on the distribution of major types of secondary minerals (Figure 9.5*b*; the correlation between soil, climate, and vegetation is also presented in Chapter 2).

SECONDARY MINERALS AND SOIL STRUCTURE

The different types of secondary minerals formed by weathering confer different physical properties, which affect water retention and soil structure. The deposition of evaporites can create a cement-like layer in the soil, which restricts water movement down through the profile and further inhibits soil development. The 2:1 clays differ significantly from other products because of their tendency to swell upon wetting and to shrink upon drying. This can cause reduced infiltration of water during periods of wet soil conditions and large-scale cracking of surface soils during dry periods (Figure 9.6). Otherwise, the accumulation of clay-sized particles generally increases water retention capacity (Chapter 8).

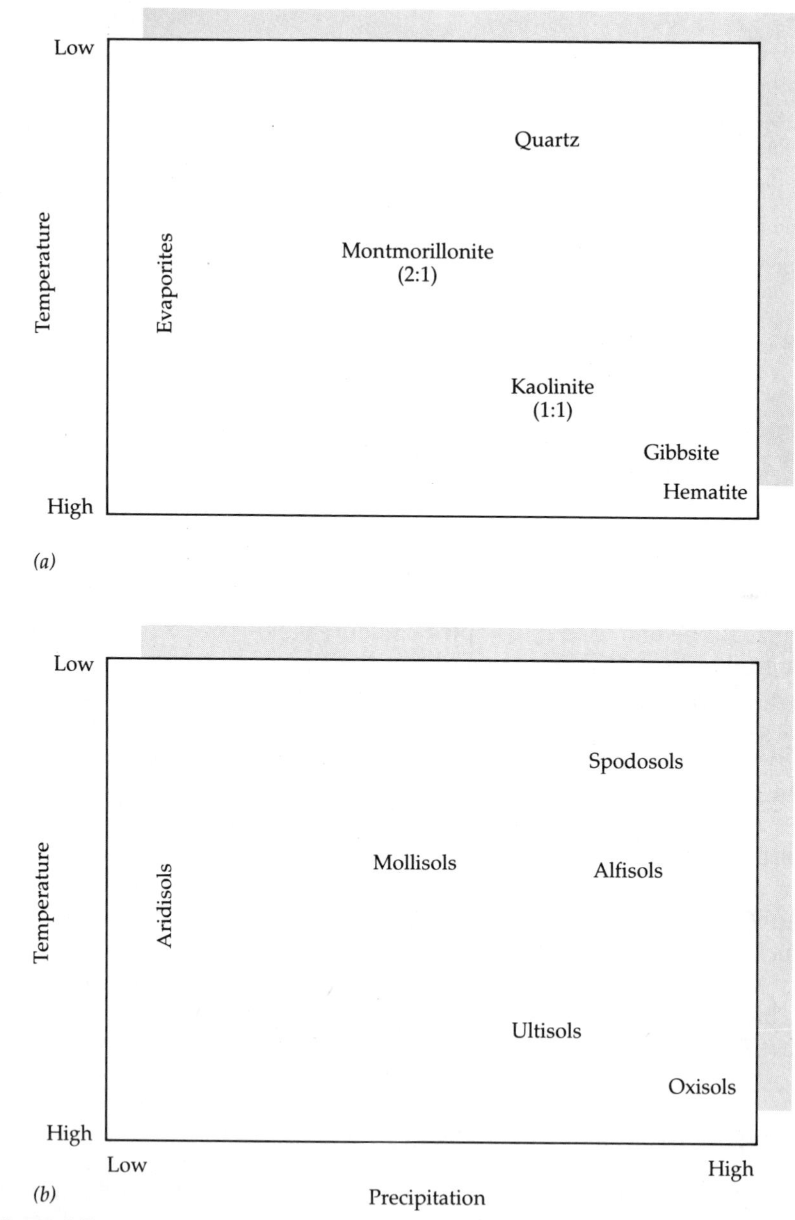

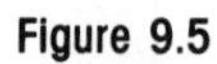

Figure 9.5
Effects of climatic conditions on *(a)* the type of secondary minerals formed by the weathering process and *(b)* the distribution of major soil types.

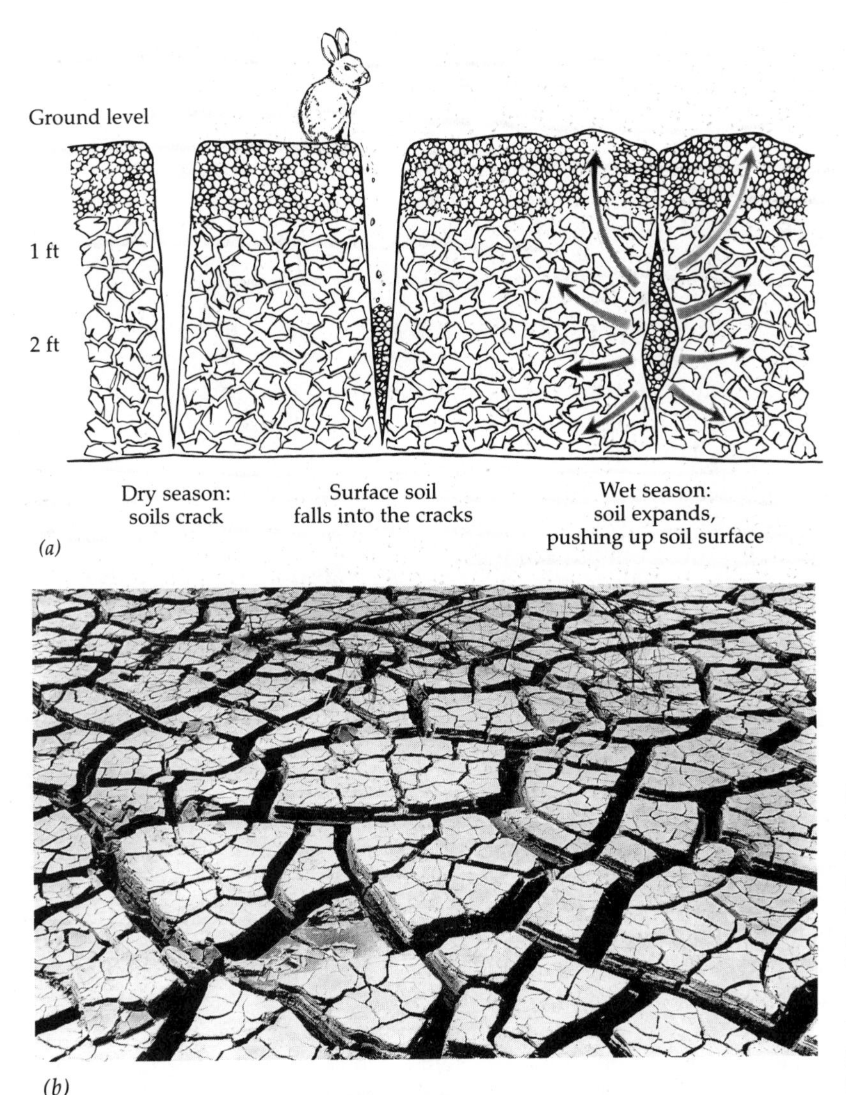

Figure 9.6
The effect of extreme drying on the structure of a soil rich in 2:1 layer clays. *(a)* Cross section showing shrinkage and expansion with drying and wetting (Buol et al. 1980). *(b)* Surface cracking during a dry period. (H. Armstrong Roberts)

WEATHERING PRODUCTS, ION EXCHANGE CAPACITIES, AND BASE SATURATION

Exchangeable nutrients (Figure 9.1) are the most readily available for plant uptake. They are held in soils by the simple attraction of oppositely charged particles and are in constant interchange with the soil solution. The types of secondary minerals produced by weathering reactions play a large role in determining the ion exchange capacities of soils.

A soil's **ion exchange capacity** is defined as the total ionic charge equivalent of positive (cation) or negative (anion) ions that can be held in the soil. This can be thought of as the total number of negatively and positively charged sites on particles within the soil body.

In most temperate-zone soils, cation exchange predominates over anion exchange because of the prevalence of negatively charged particles, or colloids, in soils. These negatively charged sites arise from the dissociation of hydrogen in carboxyl groups in soil organic matter and from exposed edges of clay particles. The 2:1 clay types have much greater cation exchange capacities per unit weight than 1:1 clay types. This is because their structure and element content facilitate what are called isomorphic substitutions, or the substitution within the clay lattice of one ion for another of similar size. For example, the substitution of a divalent magnesium ion (Mg^{2+}) for trivalent aluminum (Al^{3+}) creates a negative charge on the lattice. The role of organic matter in cation exchange is pH-dependent. Under low pH conditions, the reversible dissociation reaction of the hydrogen ion on the carboxyl group (Figure 9.7) is pushed to the left, and hydrogen ions are held more tightly and are more difficult to displace for measurement as cation exchange.

There are far fewer positively charged sites in temperate soils. For this reason, anions such as nitrate (NO_3^-) and sulfate (SO_4^{2-}) are not retained on exchange sites in soils but tend to leach away quickly if not taken up by plants. The total amount of negatively charged exchange sites, or the total **cation exchange capacity**, is a basic measure of soil quality and increases with increasing clay and organic matter content.

Exchange sites are occupied by hydrated cations (cations surrounded by water molecules but still bearing a positive charge). These include the macronutrients calcium, potassium, magnesium, and nitrogen (as ammonium, NH_4^+), nonnutrient cations such as aluminum, and hydrogen ions. The percentage of the sites occupied by ions other than hydrogen (or hydrogen plus aluminum) is called the **percent base saturation**. Acidic (low pH) soils, those with relatively high exchangeable hydrogen content, also have low percent base saturation.

The relative abundance of different ions on exchange sites is a function of their concentration in the soil solution and the relative affinity of each for the sites. In general, the physically smaller the ion (when hydrated) and the greater the positive charge on it, the more tightly it is held. Thus

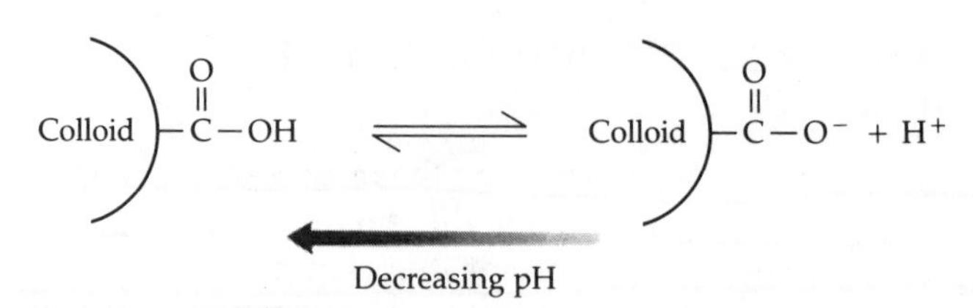

Figure 9.7
The effect of soil pH on the cation exchange capacity of organic matter. High hydrogen ion concentration in the soil solution decreases the dissociation of protons from the carboxyl group, decreasing the number of negatively charged sites.

hydrogen ions are harder to displace than ammonium ions (same charge, different size; the ratio of charge to ion size is called its "charge density"). The lyotropic series places the major cations in order in terms of decreasing affinity for cation exchange sites:

$$Al^{3+} > H^+ > Ca^{2+} > Mg^{2+} > K^+ = NH_4^+ > Na^+$$

However, high concentration in the soil solution can overcome these differences in affinity. Percent base saturation and the relative concentrations of different elements on exchange sites thus reflect the net effect of several processes that release cations and hydrogen ions into the soil solution (see discussion below and Figure 9.8).

Weathering reactions are an important source of cations. For example, soils developing on limestone receive a continuous input of calcium by weathering. This increases the calcium concentration in the soil solution so that the differences in concentration outweigh the tendencies for hydrogen to displace calcium. Instead, calcium tends to displace hydrogen, base saturation goes up, and the soil becomes less acidic. This is precisely what happens when limestone is applied to an acid soil. The calcium displaces the hydrogen, which then tends to recombine with the carbonate to form bicarbonate or carbonic acid, which may then be leached through the soil and out of the system (Figure 9.8*a*). Many weathering reactions of iron and aluminum silicate rocks also release cations (including iron and aluminum in ionic form) and consume hydrogen ions.

Soils can be acidified by both physical/chemical and biological processes. Rainfall is somewhat acidic (pH 5.0–5.5) even in relatively unpolluted areas, in part because of the formation and dissolution of carbonic acid by reactions between water vapor and CO_2 in the atmosphere. In industrialized regions, precipitation pH can be well below 4.0 for individual storm events because of the formation of nitric and sulfuric acids in the atmosphere due to reactions between water and oxides of sulfur and nitrogen released by the combustion of fossil fuels ("acid rain"). As acidified precipitation percolates through the soil, hydrogen ions may displace cations from exchange sites, reducing base saturation (Figure 9.8*b*; see more detailed discussion in Chapter 23).

Biological acidification of soils results both from respiration by roots and soil microbes and from imbalances in anion/cation uptake by plants. Soil acidification due to hydrogen ion release by plant roots is greatest when nitrogen is taken up as ammonium. Charge balance on the root is then maintained by "pumping" a hydrogen ion out into the soil (Figure 9.8*c*; see also Chapter 10). Soil pH can also be reduced by the release of CO_2 into the soil by respiration. This combines with water to produce carbonic acid, which then dissociates to produce hydrogen ion and bicarbonate (Figure 9.8*c*). The extent to which this dissociation occurs depends on the pH of the soil and is minimal below pH 5.0.

Measured soil pH represents a balance between the processes of acidification due to the acidity of rainfall and the effects of biological activity

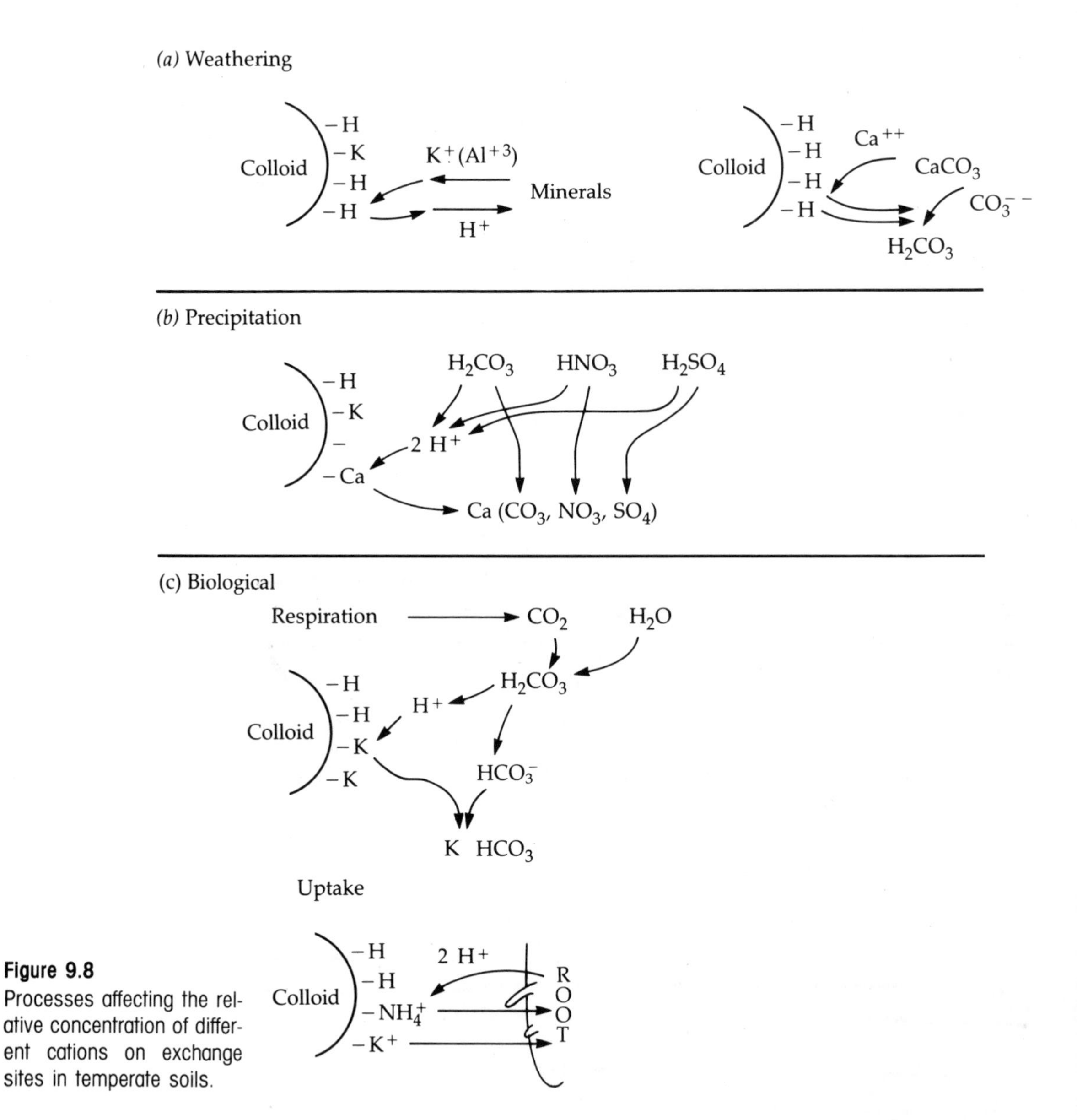

Figure 9.8
Processes affecting the relative concentration of different cations on exchange sites in temperate soils.

and the neutralizing effects of weathering. Organic horizons (the O horizon described in Chapter 2) are generally more acidic than mineral horizons (the E and B horizons) due to higher biological activity and the low levels of weatherable minerals present. However, when weathering rates are low in the mineral soil, pH values may be similar between horizons.

We will return to the processes controlling soil acidity in Chapter 23 on the acid rain phenomenon.

In tropical zones, where 1:1 clays such as kaolinite and the oxides of aluminum and iron predominate in the clay fraction of soils, the ion exchange process is very different. Substitutions within the crystal structure are rare. Instead, the clay particles have a surface sorption capacity that will attract either hydrogen ions (H^+) or hydroxyl ions (OH^-), depending on the relative abundance of these two in the soil solution (Figure 9.9). Thus, the net charge on the particle will be determined by soil pH. In acid soils, which are the rule, there will be greater adsorption of hydrogen ions, resulting in a net positive charge, and anion, rather than cation, exchange capacity. This "variable charge" system is somewhat similar to that described for organic matter above.

Although tropical soils have measurable anion retention capacities, these capacities are much lower than the cation exchange capacities of the 2:1 clays. This combines with the extreme leaching of tropical soils by excessive rainfall to produce soils with very low exchangeable cation or anion content.

SORPTION

Sorption is a generalized term for the movement of ions onto the surfaces of soil particles. It implies a stronger chemical reaction than the charge-based attractions discussed under ion exchange. Sorption is an important process affecting the availability and movement of sulfur and phosphorus, as both sulfate (SO_4^{2-}) and phosphate (PO_4^{3-}) can be strongly retained by this mechanism. Although there are still some uncertainties as to the actual reactions involved, the sorption potential for these anions depends on the types of secondary minerals developed in soils and the amount of organic matter present.

Sorption potential is described as the quantity of an element that can be removed from the soil solution and retained by the soil as a function of its

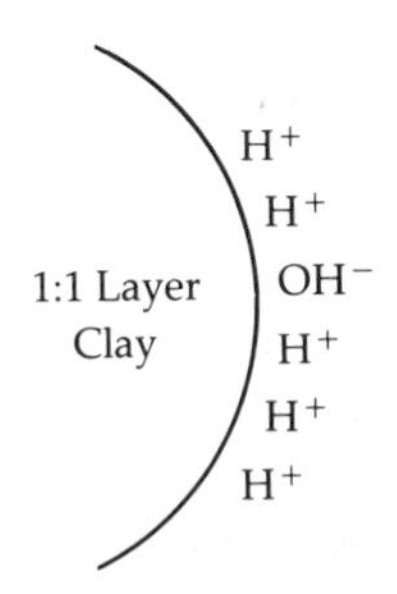

Figure 9.9
Adsorption of hydrogen and hydroxyl ions onto the surface of 1 : 1 clays and the resulting effect on cation and anion exchange capacity. (After Uehara and Gillman 1981)

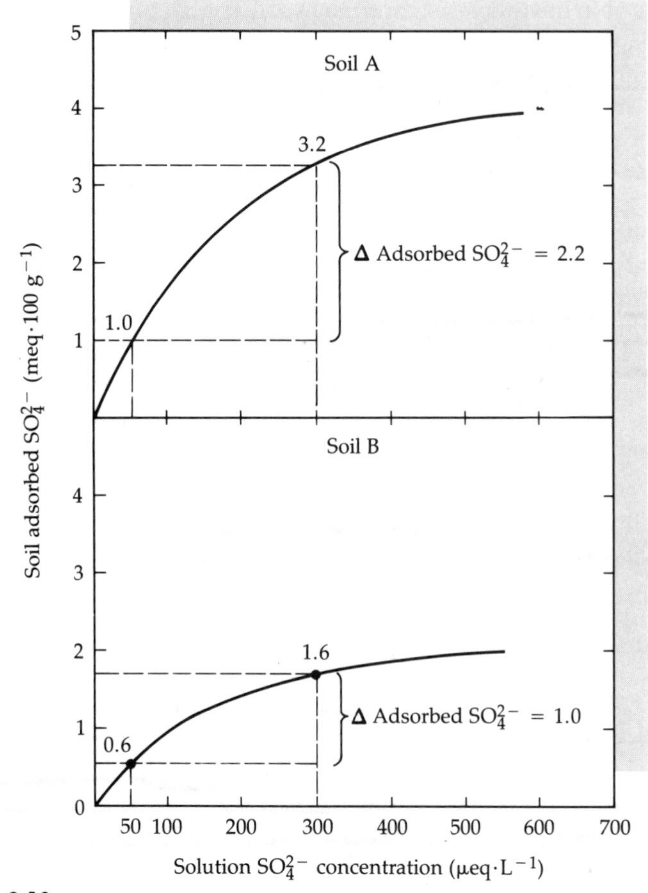

Figure 9.10

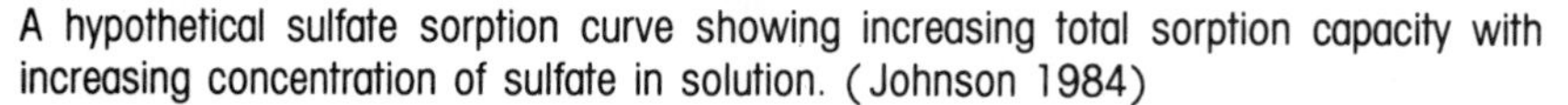
A hypothetical sulfate sorption curve showing increasing total sorption capacity with increasing concentration of sulfate in solution. (Johnson 1984)

concentration in the soil solution (Figure 9.10). In field soils, a soil horizon that has a lower total quantity of sorbed sulfate or phosphate than the sorption isotherms predict for the concentration of that element in the soil solution will remove those ions from the solution, reducing their availability to plants. This reaction can be reversed (desorption) by a lowering of soil solution concentrations, which will cause previously sorbed sulfate or phosphate to be released again to the soil solution. Biological activity in the soil can play an important role in increasing the availability of sorbed nutrients (see Chapter 10).

The sorption potential of a soil horizon is increased by the presence of hydrous oxides of iron and aluminum and decreased by the presence of organic matter. This explains the general occurrence of suboptimal phosphorus availability in tropical soils. Sorption can also be high in the B horizons of Spodosols, where the iron and aluminum oxides leached from the A horizon are deposited, unless these horizons are also enriched in organic matter.

Sorption of anions in soils dominated by 1:1 layer clays can alter the ion exchange properties. The sorption reaction converts the positively charged anion retention site to a negatively charged cation retention site (Figure 9.11). The same holds for phosphate, and large phosphate additions have occasionally been prescribed to increase the cation retention capacity of tropical soils.

The different processes described here as the chemical controls over nutrient availability operate at very different time scales. Ion exchange and sorption/desorption reactions occur very rapidly in response to very short-term changes in soil solution concentrations. However, these reactions are constrained by more slowly varying soil characteristics of cation exchange capacity and anion sorption potential, which result from the processes of weathering and soil development. For example, phosphate ion concentration in the soil solution comes into rapid equilibrium (within minutes to hours) with the sorption potential determined by current soil conditions (Figure 9.10). By contrast, changes in soil development and weathering rates over geologic time (thousands of years) determine both the inputs of phosphorus from weathering of primary minerals and the potential for sorption and removal by secondary weathering products (Figure 9.12).

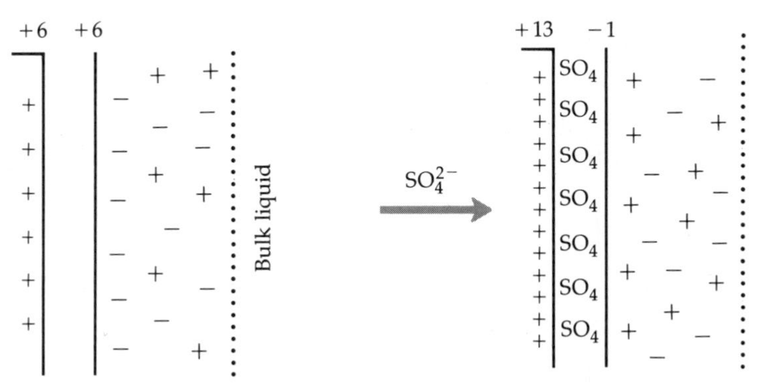

Figure 9.11
The conversion of positively charged clay surface to negatively charged clay surface by the adsorption of sulfate. (Uehara and Gillman 1981)

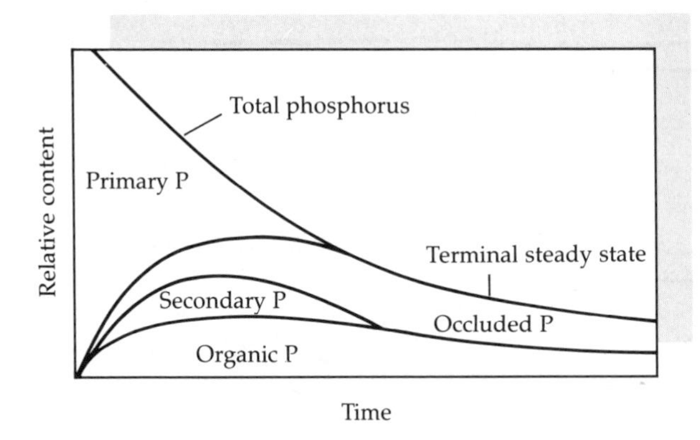

Figure 9.12
The generalized effects of long-term weathering and soil development on the distribution and availability of phosphorus. Newly exposed geologic substrate is relatively rich in easily weatherable minerals, which release phosphorus. This release leads to accumulation in both organic and readily soluble forms (secondary P—for example, calcium phosphate). As primary minerals disappear and secondary minerals capable of sorbing phosphate accumulate, an increasing proportion of the phosphorus remaining in the system is held in unavailable (occluded) forms. Availability of phosphorus to plants peaks relatively early in this sequence and declines thereafter. (After Walker and Syers 1976)

REFERENCES CITED

Buol, S. W., F. D. Hole, and R. J. McCracken. 1980. *Soil Genesis and Classification.* Iowa State University Press, Ames.

Flint, R. F., and B. J. Skinner. 1974. *Physical Geology.* John Wiley and Sons, New York.

Johnson, D. W. 1984. Sulfur cycling in forests. *Biogeochemistry* 1:29–44.

Salisbury, F. B., and C. W. Ross. 1978. *Plant Physiology.* Wadsworth, Belmont, California.

Strakhov, 1967. *Principles of Lithogenesis.* Oliver and Boyd, London.

Uehara, G., and G. Gillman. 1981. *The Minerology, Chemistry and Physics of Tropical Soils with Variable Charge Clays.* Westview Press, Boulder.

Walker, T. W., and J. K. Syers. 1976. The fate of phosphorus during pedogenesis. *Geoderma* 5:1–19.

ADDITIONAL REFERENCES

Drever, J. I. (ed.). 1985. *The Chemistry of Weathering.* B. D. Reidel, Dordrecht, Netherlands.

Epstein, E. 1972. *Mineral Nutrition of Plants: Principles and Perspectives.* John Wiley and Sons, New York.

Henderson, G. S., et al. 1978. Nutrient budgets of Appalachian and Cascade region watersheds: A comparison. *Forest Science* 24:385–397.

Jenny, H. 1980. *The Soil Resource.* Springer-Verlag, New York.

Schlesinger, W. H. 1985. The formation of caliche in soils of the Mojave Desert, California. *Geochimica Cosmochimica Acta* 49:57–66.

Tiessen, H., J. W. B. Stewart, and C. V. Cole. 1984. Pathways of phosphorus transformations in soils of differing pedogenesis. *Soil Science Society of America Journal* 48:853–858.

Chapter 10 Biological Modification of Nutrient Availability

H. Armstrong Roberts

INTRODUCTION

Soil chemistry plays a large role in determining the availability of nutrients in ecosystems. Biology plays an important role as well. Acting within the constraints set by the basic chemical processes of weathering, sorption, and ion exchange, plants and associated soil organisms can cause significant changes in the relative availabilities of essential nutrients. The processes involved generally require energy and carbon for both structure and respiration, and this establishes tradeoffs in the use of carbon from photosynthesis to increase the uptake of potentially limiting nutrients. The processes can operate directly by increasing availability or uptake from existing soil nutrient pools or indirectly by altering the soil environment.

The purpose of this chapter is to present traditional, chemical methods of assessing nutrient availability to plants, as well as the biological processes that can be employed to increase nutrient availability or uptake. At the beginning of the last chapter we stressed that much of what goes on in soils is not clearly understood. This is particularly true for the types of interactions discussed in this chapter. While the processes presented have been shown to occur, neither their quantitative importance, nor the carbon costs involved, have been determined. This is currently an area of active research.

MEASURES OF NUTRIENT AVAILABILITY

Static Measures of the Available Pool

Most methods for assessing nutrient availability come from agricultural research and are attempts to mimic chemically the ability of plants to extract nutrients from soils. Each method includes, to a different extent, nutrients in the several soil chemical pools shown in Figure 9.1.

One frequently used measurement is that of total exchangeable nutrient content. This is accomplished by saturating a soil sample with a high-concentration ionic solution, displacing all cations and anions from exchange sites. Potassium chloride may be used for nitrate and ammonium extraction and ammonium acetate for other cations. This assumes that only exchangeable nutrients and those already in the soil solution are available to plants and that measuring availability at one point in time is a good indicator of availability over the whole year.

For elements present in excess of plant demand, there may be a fairly good relationship between soil exchangeable pools and plant uptake. High cation retention capacities in temperate soils cause cations present in excess of plant demand to be retained on soil colloids. However, exchangeable amounts of either limiting nutrients or of nutrients that can be either leached from soils or sorbed onto clays (including the often limiting nutrients nitrogen and phosphorus) are generally very low, much less than one year's plant requirement for uptake. In general, both nutrient uptake and plant productivity are not related to exchangeable N, P, or S. There are two major reasons for this.

First, nutrient availability is a rate phenomenon. Annual uptake is not so much a function of concentration at any one time as it is a rate at which nutrients enter the available pool by organic matter decomposition, weathering, rainfall, etc. In fact, it is a general principle that nutrients that are limiting to production or other ecosystem functions will be present in the smallest amounts in the soil solution due to high demand and rapid uptake by plants and microbes (i.e., they have very high turnover rates).

Second, the biological activity of plants and microorganisms can significantly alter availability. There are several mechanisms employed, which involve either the separate function of plants and microbes or their joint function as symbionts. These will be discussed under "biological modification" below.

Measures Based on Rate of Mineralization from Organic Matter

An improvement on the one-time measurements of nutrient availability is incubation techniques that isolate soil cores for a period of time and measure the rate of release from organic matter. This approach is particularly valuable for elements with cycles dominated by the biological decomposition of organic matter (e.g., nitrogen, sulfur). Incubation techniques involve isolating a soil core either in the laboratory or in the field, allowing it to incubate for some length of time, and then measuring the accumulation of the mineral forms of the nutrient (e.g., ammonium and nitrate). The stable isotope ^{15}N has also been used to trace the dynamics of nitrogen metabolism during incubation.

Measuring phosphorus availability is altogether more difficult. In Chapter 9 we mentioned the tendency for phosphate (PO_4^{3-}) to be sorbed by soils with high concentrations of iron and aluminum. These reactions make it impossible to use the same incubation techniques described for nitrogen. Phosphorus mineralized in the incubating soils could be sorbed quickly and not accumulate in ionic form. The radioisotope ^{32}P has been used with some success, but most researchers use one of several chemical extraction techniques. Each technique extracts a different proportion of the total precipitated phosphate pool (Table 10.1). The best technique would be the one most similar to the ability of plants and microbes to solubilize and take up P. However, that ability is quite variable. Estimating P availability and cycling rate remains a challenge.

Measurement of sulfur offers some of the same problems as measurement of phosphorus. Decomposition of organic matter is a principle source of sulfate (SO_4^{2-}) for plant uptake, but potential sorption clouds its measurement by incubation techniques in some soil types. In general, sulfur is rarely a limiting nutrient in natural ecosystems. It has received more attention in relation to acid rain effects (Chapter 23) than as an important nutrient limiting plant growth.

Table 10.1 Different methods of extracting phosphorus from soils and the different form of phosphorus removed by each*

EXTRACTING SOLUTION	PHOSPHATE FRACTION EXTRACTED
Water	P in soil solution
Dilute acid fluoride ($HCl + NH_4F$)	Easily soluble forms: calcium phosphates plus some iron and aluminum phosphates; considered plant-available
Dilute acid ($HCl + H_2SO_4$)	A stronger extraction for soils with high P-sorption potentials; considered plant-available
Sodium bicarbonate ($NAHCO_3$)	For extraction of P from neutral to alkaline soils
Perchloric acid ($HClO_4$)	Total phosphorus content of soil

*From Olson and Dean 1965.

Biological Modification of Nutrient Availability

There is a potential problem in looking at nutrient availability in isolated soil cores devoid of plant activity. Plants can do much to alter the chemistry of soils. Even more can be accomplished in symbiotic relationships with microorganisms.

We will discuss three types of processes by which biological activity can increase nutrient availability: (1) increasing the root mass and altering root physiology; (2) altering the environment surrounding the soil by exuding simple organic compounds; and (3) direct symbiosis. The thrust of this discussion is to demonstrate that nutrient availability is not a constant to the plant but rather that increases in the availability of nutrients are frequently possible by the use of extra carbon for their acquisition. In other words, more of a nutrient can be obtained with a greater allocation of photosynthate. This creates a tradeoff between carbon and nutrient availability within the plant.

Increased Root Mass The presence of nutrients on exchange sites or in the soil solution does not ensure uptake by plants. The ion must also move to the surface of the root. The rate of nutrient uptake is frequently limited by this rate of movement.

Movement occurs either because the ion moves with the flow of soil water to the root for transpiration or because of a gradient in concentration within the solution caused by nutrient uptake. In temperate-zone soils, movement with water is generally limited to anions, such as nitrate, which are not retained by soils either by ion exchange sites or sorption. Mobility of cations in soil is reduced by retention on exchange sites. Phosphate mobility is very low in soils with high phosphate sorption potentials. The same processes that reduce nutrient losses to groundwater also cause resistance to the movement of ions to the root.

Figure 10.1 shows a new root growing in soil surrounded by available nutrients. Initially, the nutrient concentration is constant throughout the soil (t0). Uptake by the root will begin to deplete nutrients from the soil solution (t1). This causes a gradient in nutrient concentration to develop within the soil, and nutrients will then move toward the root. As uptake continues, a localized zone of nutrient depletion will develop around the root. This depletion zone will expand away from the root (t2 – t4) until the gradient in concentration between root and soil at some distance is not enough to overcome the resistance to movement of the ions. Further uptake would then depend on replacement of nutrients by processes of organic matter decomposition and weathering in the same immediate area around the root (this zone of root influence within the soil is termed the **rhizosphere**).

Depletion zones are of very different sizes for different nutrients. In temperate soils dominated by cation exchange, the resistance to movement of ammonium may be 100 times that for nitrate. The depletion zone

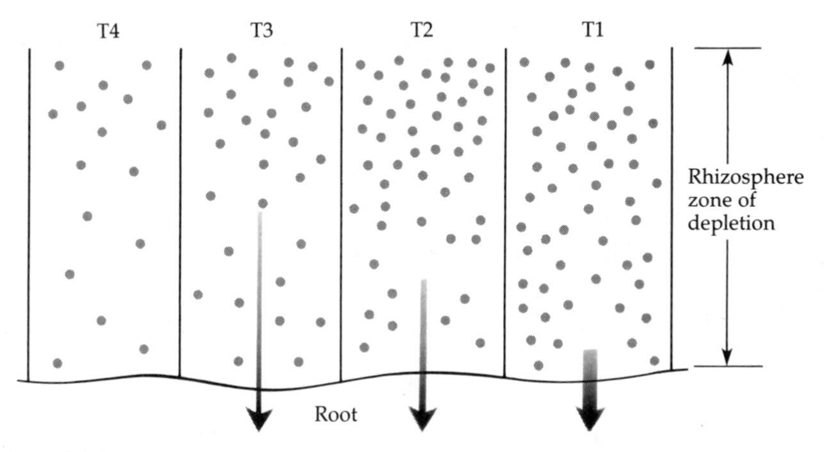

Figure 10.1
Development of ion depletion zones in soils as a result of nutrient uptake. As uptake occurs, a concentration gradient develops in the soil, resulting in movement of ions to the root surface. As uptake reduces nutrient concentration in the developing depletion zones, the concentration gradient to the root declines and the uptake rate is reduced. Eventually the gradient is insufficient to overcome resistance to movement in the soils, and uptake stops.

for nitrate will be correspondingly higher. Depletion zones for phosphate are generally much smaller even than for ammonium because of sorption potentials.

In the context of the whole soil body then, not all nutrients that are chemically exchangeable need be "available" to plants. Those outside the rhizospheres of all plant roots or beyond the zones of depletion will not be taken up. Under these conditions a simple way to increase effective nutrient availability to the plant is to increase the root mass or surface area (Figure 10.2).

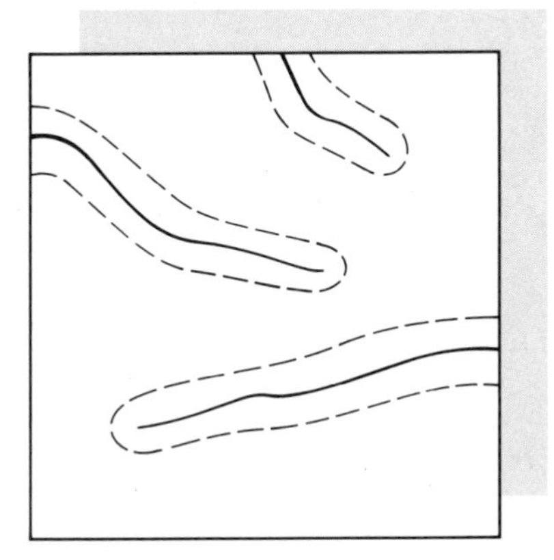

Roots
Depletion zone around roots

Figure 10.2
The effect of increasing root density on the total fraction of soil within the depletion zone of roots.

It follows from this that plant communities on nutrient-poor sites often have greater root masses. In nitrogen-limited systems, those communities growing on soils where nitrate is the dominant form of available N will have lesser root masses than similar communities where ammonium predominates (Figure 10.3). It has been hypothesized that root density should be lowest, but rooting depth highest, in ecosystems limited mainly by water availability and that root density should increase in systems limited by nitrate, ammonium, and phosphate, respectively.

Increasing nutrient uptake by increasing rooting density assumes that there is some excess pool of available nutrients in the soil. It does not increase the rate at which nutrients are transformed from unavailable to available forms. A densely rooted soil ensures complete uptake of all available nutrients by the plant, if required. As an example, continued uptake of nitrogen cannot exceed the rate at which it is released from organic matter (Chapters 12, 13) for any length of time. Increasing root mass only ensures that all mineralized nitrogen is indeed taken up and not

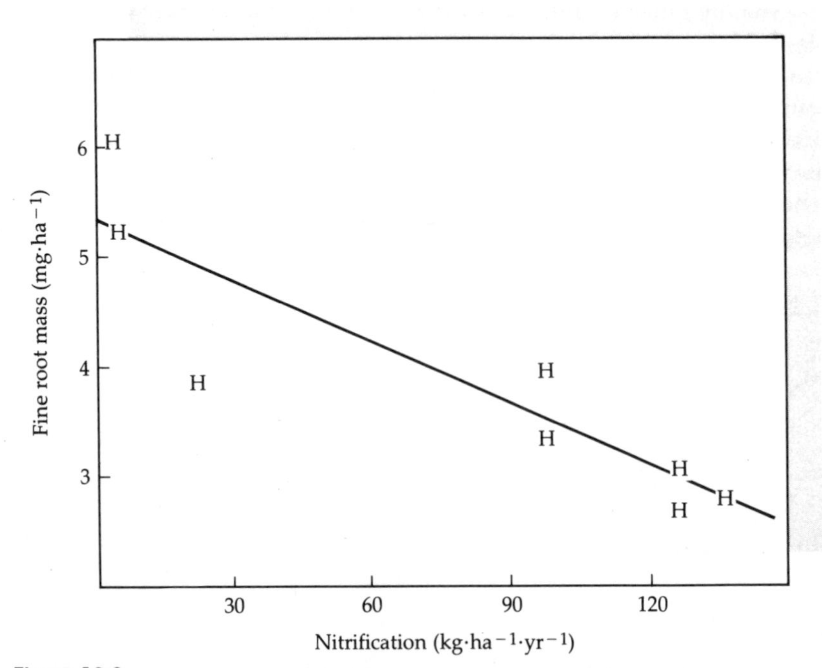

Figure 10.3
The relationship between annual availability of nitrate (nitrification) and total fine root biomass for several temperate deciduous forests. (Aber et al. 1985)

lost to groundwater leaching. Beyond this, there must be some direct effect of roots on soil processes to foster an increase in mineralization or weathering or desorption of nutrients for root mass to exert an effect on the rate of nutrient availability.

There are three broad categories of processes that are active in plants and that increase the rate at which nutrients become available in the rhizosphere. The first is that plants can themselves produce enzymes that speed decomposition or even chemical weathering of nutrients. Many plant roots exude enzymes that catalyze the release of phosphate from organic matter (phosphatase enzymes). There are also suggestions that cellulase enzymes (for decomposing cellulose, the primary constituent of plant cell walls; see Chapter 12) may be released, perhaps creating a role for the plant in the decomposition of organic matter. The direct extraction of nutrients, particularly potassium, from fresh minerals has been shown for several species, although the mechanism is not clear.

Second, plants can affect decomposition or soil chemistry indirectly by exuding from the root simple organic compounds that stimulate microbial activity and increase organic matter decay. Much of the organic matter in soils is of very poor quality and yields little or no net energy gain to microbes (Chapters 12, 13). The presence of more easily decomposed carbohydrates from root exudation may provide an energy subsidy, which increases the rate of organic matter decay and nutrient release. Organic acids, amino acids, and other growth-enhancing compounds are present in the exudate as well, creating a rich "soup" of substrates for microbial growth (Table 10.2). This interaction can be thought of as a loose symbiosis with the plant supplying energy and even some nitrogen in amino acids to the microbes to promote more rapid release of nutrients from organic matter.

Plants can also alter the pH of the rhizosphere as a byproduct of nutrient uptake and charge balance limitations. Roots cannot accumulate a significant net electrical charge, either positive or negative. Thus the total net negative charge of the important anions (NO_3^-, SO_4^{2-}, PO_4^{3-}) taken up must be equal to the total positive charge of cations (NH_4^+, CA^{2+}, Mg^{2+}, K^+), or the difference must be made up by movement of H^+ and OH^- between root and soil.

In natural ecosystems, the form of nitrogen taken up is an important consideration. Uptake as nitrate (anion) can be balanced by passive uptake of excess cations (Figure 10.4*a*). However, where ammonium is the dominant form of N taken up, this causes an imbalance that cannot be made up by increased uptake of the anions phosphate and sulfate. The charge balance is maintained by "pumping" hydrogen ions (H^+) out of the root and into the soil (Figure 10.4*b*). This lowers soil pH and can have an effect on solubility of nutrients, particularly phosphorus, and the displacement of cations from exchange sites.

The third method by which plants can affect nutrient availability is the most complex and effective, by symbiotic relationships with soil microor-

Table 10.2 Organic compounds detected in root exudates from three tree species*†

EXUDATES	TREE SPECIES		
	Betula alleghaniensis	*Fagus grandifolia*	*Acer saccharum*
Carbohydrates			
Glucose	12.36 ± 5.21‡	6.32 ± 3.36	0.21§
Fructose	16.59 ± 2.54‡	11.08 ± 6.40	0.83 ± 0.62
Ribose	3.35 ± 0.74	ND	ND
Sucrose	21.17§	ND	13.14 ± 5.26
Amino acids/amides			
Alanine	1.42§	0.04§	1.21 ± 0.47
Arginine	0.60§	0.12 ± 0.08	ND
Asparagine	0.31 ± 0.13	0.07 ± 0.03	ND
Aspartic acid	0.52 ± 0.28	0.11§	ND
Cystine	0.02 ± 0.01‡	0.02§	0.03§
Glutamine	0.78 ± 0.68	0.13§	0.75§
Glycine	0.45§	0.06 ± 0.02	0.05 ± 0.01
Homoserine	0.15 ± 0.01‡	0.23 ± 0.06	0.22 ± 0.03‡
Leucine/Isoleucine	ND	ND	ND
Lysine	1.07 ± 0.57	0.59 ± 0.16	0.18 ± 0.14‡
Methionine	0.08 ± 0.03‡	0.01§	ND
Phenylalanine	ND	ND	0.13 ± 0.08‡
Proline	1.69§	2.36 ± 0.64	ND
Serine	1.19§	0.51§	0.32§
Threonine	0.03 ± 0.003	0.08 ± 0.03	0.29 ± 0.24
Tyrosine	ND	ND	0.03§
Valine	0.27§	0.07 ± 0.05	0.37§
Organic acids			
Acetic	15.38 ± 6.99	5.92§	21.42 ± 10.35‡
Aconitic	ND	17.47§	ND
Citric	10.06§	ND	10.93 ± 4.29‡
Fumaric	20.02§	5.62 ± 5.54‡	ND
Malic	20.74 ± 8.86‡	ND	ND
Malonic	ND	ND	1.85§
Oxalic	10.28 ± 2.26	24.68 ± 0.10‡	ND
Succinic	4.03§	3.74 ± 1.98	ND
Total organic material	142.56	79.23	50.75

*Values are in units of 0.1 microgram released per milligram of root over 14 days; ND = not detected.
†From Smith 1976.
‡Detected in exudate from two trees.
§Detected in exudate from one tree.

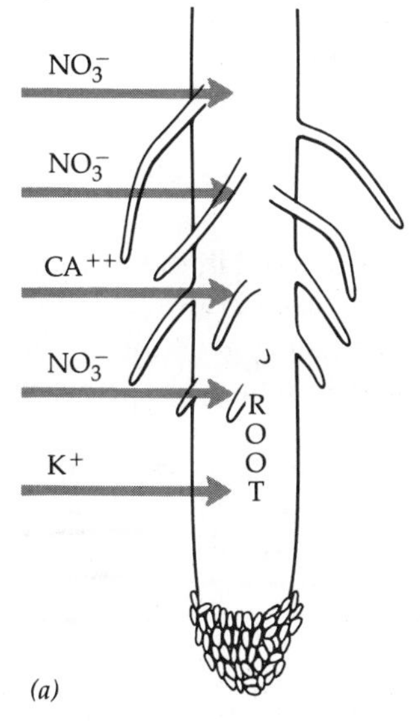

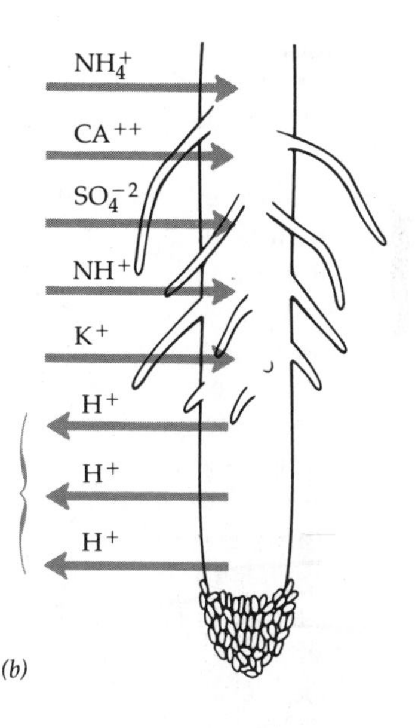

Figure 10.4
The effects of nitrate versus ammonium uptake on the movement of hydrogen ions from root to soil.

ganisms. The most important of these in natural ecosystems are with mycorrhizal fungi and species that fix atmospheric nitrogen.

Mycorrhizal (*myco* = fungus, *rhizo* = root) associations are nearly ubiquitous in natural ecosystems. While the species and even the morphology of the symbiosis vary, nearly all plants in nearly all terrestrial ecosystems have mycorrhizal symbionts. There are two major groups of mycorrhizal fungi: ectomycorrhizal and endo- or vesicular–arbuscular-mycorrhizal (VA-mycorrhizal). The ectomycorrhizal fungi form large mats called Hartig nets, which sheath the infected root tips. The sheath may account for 40% of the total weight of the root plus sheath. From this net, hyphae extend both into the intercellular spaces within the root and out into the soil. Thus, the fungi form a bridge between regions of soil nutrient availability and the plant root.

The VA mycorrhizae contain a more direct connection between plant and soil. There is no sheath or net in this case. Rather, the hyphae actually penetrate the cells of the root and are in direct contact with cytoplasm. They then also penetrate the root surface and extend into the soil. Fungal weight in the VA mycorrhizae is generally less than 15% of the total weight.

In both cases, mycorrhizal symbioses are viewed primarily as a means of increasing the surface area of the nutrient-absorbing network. In terms of carbon-nutrient tradeoffs, they do this at a lower carbon cost than root production because the thinner hyphae require less carbon per unit length than roots. An increased affinity for nutrients, especially phosphorus, may also increase the efficiency of nutrient removal within the depletion zone. The effectiveness of VA mycorrhizae on increasing P uptake and plant growth can be seen in Figure 10.5.

There are important differences between the two fungal groups. VA-mycorrhizal fungi are thought to be obligate symbionts, meaning that they cannot grow actively without the carbon subsidy obtained in the relationship with the plant. Ectomycorrhizal fungi generally can survive and grow as free-living decomposers in the soil. VA forms are more generally implicated in increased uptake of phosphate and nitrate and ecto- forms in increased uptake of ammonium.

As discussed above, increasing uptake of already available nutrients is of limited value compared with actually increasing availability. There are some intriguing indications that mycorrhizae may be capable of increasing availabilities by increasing desorption of phosphate or decomposing organic forms of nitrogen and phosphorus. If these processes are found to be more effective with mycorrhizae than for plants growing alone, then the symbiosis can actually increase the rate of nutrient availability to the plant. The use of plant-derived photosynthate to speed decomposition would also blur the distinction between plants and decomposers.

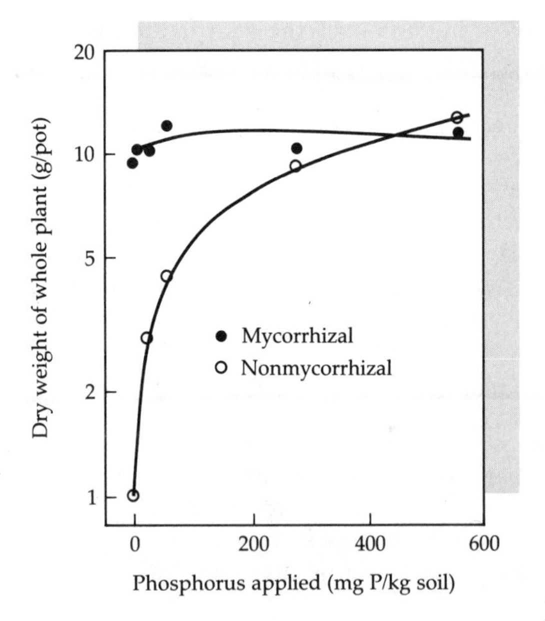

Figure 10.5
The effect of VA mycorrhizal infection on the growth of Brazilian sour orange with different levels of phosphorus addition. The ability of VA mycorrhizae to increase phosphorus availability is shown by the high growth rate at low levels of added P and by the lack of response to P fertilization. (Abbot and Robson 1985)

Symbiotic nitrogen fixation is far less common than mycorrhizal associations in natural ecosystems but has received much attention because of its economic importance. Nitrogen is an important limiting nutrient in both natural and managed ecosystems. Producing nitrogen fertilizers requires lots of fossil fuel energy and is expensive. Increasing our ability to use solar energy through photosynthesis to provide fixed nitrogen from the atmosphere can increase yields in both agriculture and forestry and reduce costs.

The most widely known N-fixing symbiosis is between leguminous plants and bacteria of the genus *Rhizobium*. This is the combination involved in growth of soybeans, green beans, peas, etc. The legumes are a very large plant family, especially in the tropics, where both herbaceous and tree species are found in abundance. In temperate zones, legumes are less common and are generally herbaceous, with only a few tree species.

The legume–*Rhizobium* symbiosis functions through the "infection" of roots of host plants by the bacteria. The infection stimulates the construction of a specialized organ called a root nodule (Figure 10.6). The specialized environment in the nodule shields the microbial symbionts from soil conditions (acidity, oxygen) that inhibit fixation.

Other plant–microbe symbioses also produce nodule-like systems and the capacity to fix atmospheric nitrogen. At least 20 genera of angiosperms in seven families interact with species of actinomycete fungi (Table 10.3).

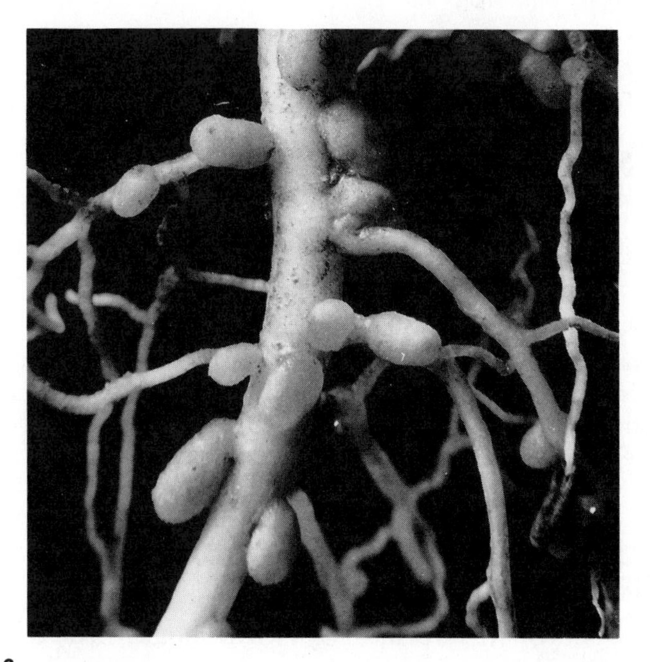

Figure 10.6

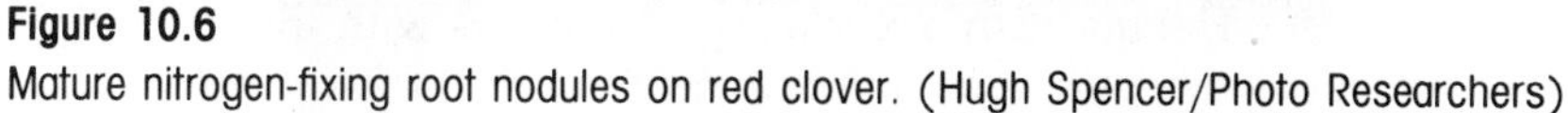

Mature nitrogen-fixing root nodules on red clover. (Hugh Spencer/Photo Researchers)

Table 10.3 Genera of plants in which some species form symbiotic nitrogen-fixing relationships with actinomycete fungi*

GENUS	NUMBER OF SPECIES RECOGNIZED IN GENUS	PRESENT DISTRIBUTION	NUMBER OF SPECIES RECORDED BEARING NODULES
Casuarina	45	Australia, tropical Asia, Pacific islands	24
Myrica	35	Many tropical, subtropical, and temperate regions	26
Alnus	*35*	*Europe, Siberia, N. America, China, Japan, Andes*	*39*
Dryas	4	Arctic, mountains of north temperate zone	3
Cercocarpus	20	West and southwest U.S., Mexico	4
Purshia	2	N. America	2
Chamaebatia	2	California	1
Cowania	5	Southwest U.S., Mexico	1
Rubus	250	Many regions, especially north temperate	2
Coriara	15	Mediterranean, Japan, China, New Zealand, Chile, Mexico	16
Colletia	17	Temperate and subtropical America	3
Discaria	10	S. America, New Zealand, Australia	5
Trevoa	6	Andes	2
Talguenea	1	Chile	1
Kentrothamnus	2	Bolivia, Argentina	1
Ceanothus	55	N. America	31
Elaeagnus	45	Asia, Europe, N. America	28
Hippophaë	3	Europe, temperate Asia to Kamchatka, Japan, Himalaya	1
Shepherdia	3	N. America	2
Datisca	2	Mediterranean to Himalaya and central Asia, southwest U.S.	2

*From Bond 1983.

Most of these are woody shrubs or trees, frequently of wetland habitats. Use of these species in mixed plantations with trees or even in crop-rotation forestry (similar to corn–soybean rotations in agriculture) is currently being tested as a means of increasing site fertility and yield.

LIMITATIONS ON THE ADAPTIVE ADVANTAGE OF INCREASING NUTRIENT AVAILABILITY

Several processes by which plants may increase nutrient availability have been presented. The question arises as to why plants with these abilities do not simply overcome any nutrient deficiencies so as to always be growing at the maximum potential rates allowed by light, temperature, and water. Why don't all plants possess N-fixation symbioses and high-phosphatase activities? Using the arguments of natural selection and evolution, why have species that possess those abilities not outcompeted and replaced those that do not?

The simplest answer is that each mechanism has a carbon cost. The extra nutrient uptake occurs at the expense of higher root/symbiont tissue construction costs and higher respiration. For a plant to realize an adaptive advantage for supporting root symbionts, the increase in photosynthesis due to the added nutrient availability must be greater than the carbon "invested" in supporting the symbiosis. Such will not be the case in environments that are already rich in nutrients.

It is generally recognized that the degree of subsidy to root symbionts varies with site quality or nutrient availability. Fertilizing a soybean field leads to a reduction in N fixation. Forests on richer sites may have less complete infection of roots by mycorrhizal fungi. Natural ecosystems dominated by symbiotic N-fixers tend to be those with high availabilities of light and water and relatively low availabilities of N. These include certain frequently disturbed or burned areas in the temperate zone and some tropical forests on poor, fully weathered soils.

In sum, we can begin to think of plants in ecosystems not as maximizing nutrient uptake but rather optimizing the allocation of carbon for tissue growth (leaves, roots, etc.) and support of physiologic mechanisms or symbionts to maintain a relatively constant and proportional acquisition of all the resources (light, water, nutrients) required for growth. The actual measurement of the carbon cost/nutrient gain ratio associated with each of these mechanisms for increasing availability remains an intriguing problem.

REFERENCES CITED

Abbott, L. K., and A. D. Robson. 1985. The effect of VA mycorrhizae on plant growth. In Powell, C. L., and D. J. Bagyaraj, *VA Mycorrhiza*. CRC Press, Boca Raton, Florida.

Aber, J. D., et al. 1985. Fine root turnover in forest ecosystems in relation to quantity and form of nitrogen availability: A comparison of two methods. *Oecologia* 66:317–321.

Alexander, M. 1977. *Introduction to Soil Microbiology*. John Wiley and Sons, New York.

Bond, G. 1983. Taxonomy and distribution of non-legume nitrogen-fixing systems. In: J. C. Gordon and C. T. Wheeler (eds.), *Biological Nitrogen Fixation in Forest Ecosystems: Foundation and Applications.* Martinus Nijhoff/Dr. W. Junk Publishers, The Hague.

Olson, S. R., and L. A. Dean. 1965. Phosphorus. In Black, C. A. (ed.), *Methods of Soil Analysis*. American Society of Agronomy, Madison.

Smith, W. H. 1976. Character and significance of forest tree root exudates. *Ecology* 57:324–331.

ADDITIONAL REFERENCES

Binkley, D., and D. Richter. 1987. Hydrogen ion budgets and nutrient cycles of forest ecosystems. *Advances in Ecological Research* 16:1–51.

Dodd, J. C., et al. 1987. Phosphatase activity associated with the roots and the rhizosphere of plants infected with vesicular–arbuscular mycorrhizal fungi. *New Phytologist* 107:163–172.

Fogel, R. 1980. Mycorrhizae and nutrient cycling in natural forest ecosystems. *New Phytologist* 86:199–212.

Goldstein, A. H., D. A. Baertlein, and R. G. McDaniel. 1988. Phosphate starvation inducible metabolism in *Lycopersicon esculentum*. I. Excretion of acid phosphatase by tomato plants and suspension-cultured cells. *Plant Physiology* 87:711–715.

Harrison, A. F. 1987. Mineralization of organic phosphorus in relation to soil factors determined using ^{32}P labelling. In Rowland, A.P. (ed.), *Chemical Analysis in Environmental Research.* Institute of Terrestrial Ecology, Abbots Ripton, England.

Nye, P. H., and P. B. Tinker. 1977. *Solute Movement in the Soil–Root System.* University of California Press, Berkeley.

Pastor, J., et al. 1984. Above-ground production and N and P cycling along a nitrogen mineralization gradient on Blackhawk Island, Wisconsin. *Ecology* 65:256–268.

Van Veen, J. A., et al. 1987. Turnover of carbon, nitrogen and phosphorus through the microbial biomass in soil incubated with ^{14}C, ^{15}N and ^{32}P labelled bacterial cells. *Soil Biology and Biochemistry* 19:559–565.

Chapter 11 Resource Allocation and Net Primary Productivity

R. Krubner/H. Armstrong Roberts

INTRODUCTION

The previous chapters in this section have discussed the chemical, physical, and biological factors affecting the availability of resources (water, light, nutrients) required by plants. The physiologic potential for, and selective advantage of, applying one resource to increase the availability or uptake of another has been presented as a key to understanding limitations on productivity. Resource use efficiency is another concept that has surfaced several times.

The purpose of this chapter is to discuss the interactions between resource availability, total net primary production, the allocation of both net primary production and nutrients to different tissue types, and the effects of allocation and other physiologic processes on nutrient use efficiency.

It should be clear from the outset that the ecosystem level patterns of allocation and resource use efficiency derive from the process of natural selection acting on plant species. In general, high rates of primary production increase a species' competitive ability. Faster-growing plants can grow taller and shade out slower-growing ones. They may also have more resources available for the production of seeds. However, plants genetically "programmed" for fast growth or large size may do very poorly in dry or nutrient-poor systems. For example, the tree growth form is not adaptive in semiarid climates because water availability is too low and photosynthesis too restricted to support the high respirational costs of maintaining a tall stem. Natural selection favors species that have resource requirements, growth forms, and life history characteristics that match a site's resource availabilities and seasonal changes in climate.

RESOURCE LIMITATIONS ON PRODUCTION — A SIMPLIFIED VIEW

In 1840, Justus von Liebig first presented the theory that plant growth is limited by a single factor "presented to it in minimum quantities." Stated with true 19th-century certitude, this became known as Liebig's Law of the Minimum. It has served as a springboard for discussions of the nature of resource limitations on growth.

What is meant by "minimum quantities"? This term is not meant in the absolute sense. Phosphorus can be provided in much smaller quantities than nitrogen, and still nitrogen could be limiting due to a higher demand for nitrogen in plant growth. We have already seen that water is required in much larger quantities than any nutrient. Thus the expression of "minimum quantities" must be made relative to plant requirements.

A simplified view of annual resource availabilities to plants can be constructed as in (Figure 11.1*a*. This can be matched with the relative demand for resources of a given species or plant community, in this case to produce a unit of biomass (Figure 11.1*b*). The nutrient ratios represent the weighted mean concentration of that nutrient in net primary production. The water demand is the total evapotranspiration of the system for a year divided by net primary productivity (NPP). By comparing these two sets of ratios, it quickly becomes apparent that nitrogen is going to be the limiting factor in this case. Assuming that all of the 80 kg/ha of available N is taken up, the ratio of 0.020 kg N taken up per kg NPP yeilds a total NPP of 4000 kg/ha/yr. Uptake of other resources would be in the ratios shown in Figure 11.1*a*.

Productivity in this system could be increased by fertilization with nitrogen. Adding 100 kg N/ha/yr would increase net primary productivity until the associated increase in demand for another resource exceeded availability. By comparing the ratios of requirement to availability in Figure 11.1*a* and *b*, we can see that large increases in the availability of nitrogen would result in a phosphorus limitation.

Another way to realize increased production on this site would be to plant a species that was more efficient in the use of N and had a lower demand per unit of plant production, as shown in Figure 11.1*c*. This may come at the expense of water use efficiency (see Figure 6.11). A species represented by the resource use ratios in Figure 11.1*c*, growing on a site with availabilities as in Figure 11.1*b*, would be limited by the availability of water.

This analysis treats nutrient availability as a constant with no consideration of the effect of the mechanisms described in the last chapter for increasing availability by spending carbon. It also ignores potential mechanisms for altering nutrient use efficiency (the weighted mean concentra-

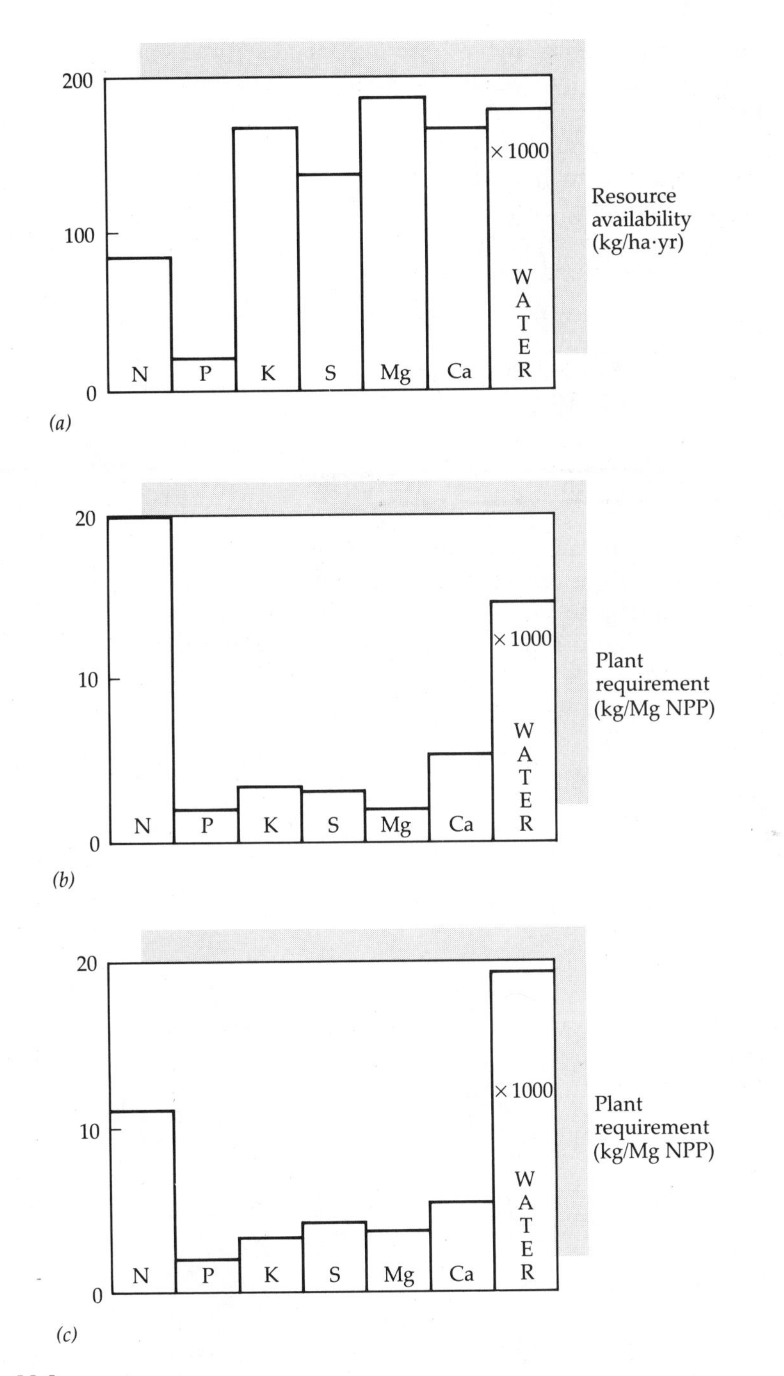

Figure 11.1
A simplified view of the ratios determining nutrient limitations on net primary productivity. *(a)* Relative resource availability. *(b)* Resources required to produce a unit of biomass. *(c)* Resources required by a species that is more efficient in the use of nitrogen than the species in part *b*.

tion of nutrients in net primary production) and the effects that the accumulation of large amounts of biomass can have on net primary production by increasing plant respiration. In addition, changes in climatic conditions (percent cloud cover, relative humidity, precipitation, temperature) from year to year at a given site can alter water use efficiency (the water requirement bar) significantly from one year to the next. The actual interactions between resource availability and production are considerably more complex than in this simple example.

RESOURCE POOLS IN PLANTS AND THEIR ALLOCATION

Immediately following uptake or fixation, nutrients and carbon are present in the plant in mobile forms. These resources are allocated to meet the demands for respiration of existing tissues and construction of new foliage and fine roots and the structural tissues in stems, which support roots and foliage and allow movement of material between them. Resources must also be allocated for reproduction and protection against herbivory (Figure 11.2). How are these allocation patterns within a plant determined?

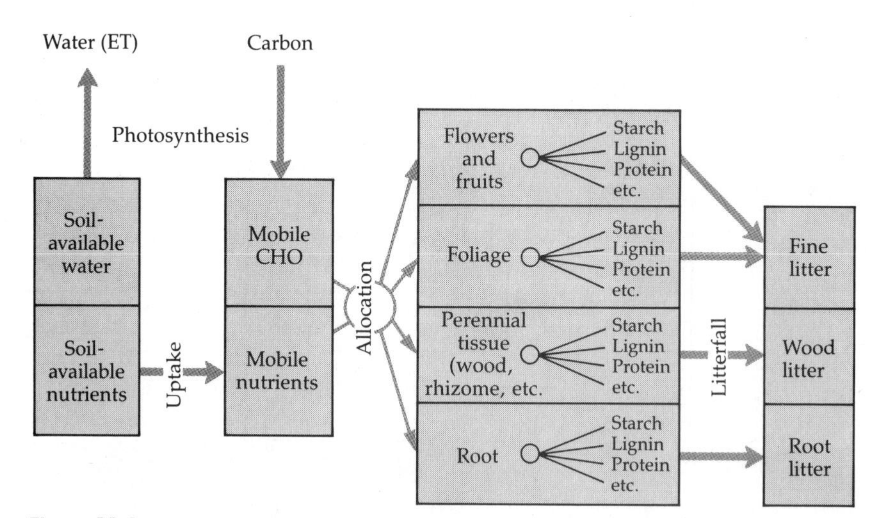

Figure 11.2

Schematic diagram of pathways for carbon and nutrient allocation from internal mobile pools in plants. The relative importance of each path depends on the growth form and tissue chemistry of the species involved as well as on the environmental conditions of the site.

General Framework: Source–Sink Relationships

Patterns of allocation vary with species and stage of growth and are anything but random. The current working hypothesis is that the direction and rate of flow of carbon and nutrients through plants are controlled by the relative demands for resources exerted by different tissues at different times. Areas where demand for carbon is greater than for photosynthesis (such as newly growing leaves) are termed carbon "**sinks**." Mature leaves that are actively photosynthesizing but not growing would be carbon "**sources**." The stronger the sink strength, the greater the allocation to that tissue.

Seasonal patterns of allocation in an annual grass plant and a perennial tree illustrate differences in the timing and intensity of sink strength (Figure 11.3). In the newly germinated annual grass, early products of photosynthesis may go initially to roots as the plant establishes an equilibrium between the fixation of carbon and energy from the atmosphere and the uptake of water and nutrients from the soil. The growing top may then become a center for faster growth, creating a greater sink and moving carbon up from established leaves. After the vegetative portion of the plant is fully grown, allocation of carbon to flowers and then to fruits begins. In annual plants, a large fraction of the annual uptake of carbon and nutrients is directed to seed production, which becomes a very strong sink at the end of the growing season.

In perennial plants such as trees, carbohydrates (starches) stored throughout the plant and nutrients stored in buds, twigs, and branches are mobilized in spring to produce new leaves as well as some new top and stem growth. The leaves quickly become net sources of carbon, sending carbohydrates back down the plant. Different parts of the tree act as important sinks at different parts of the growing season. Top growth is usually accomplished first, followed by stem expansion (by radial increment). In many temperate- and boreal-zone species, these stages are both completed by the middle of the growing season. Carbon fixed after this period goes either to roots or to storage as starch, which is deposited throughout the stems, twigs, and roots. It is a general characteristic of perennial plants that a significant amount of carbohydrate is always held in reserve to buffer catastrophic losses of roots or foliage by insect attack, fire, wind, and so forth. This is in strong contrast to annual plants, in which all available carbohydrate reserves are directed to the maturing fruit.

Throughout plants, the availability of nutrients, particularaly nitrogen, affects the distribution of allocated carbon between proteins and structural carbon compounds such as cellulose and lignin. Lignin is a large, amorphous, and complex polyphenolic compound, which confers both "woodiness" to stems and branches of woody plants and a measure of resistance to leaf consumption by herbivores (see Chapter 15). The nutrient content of different tissues and their palatability to consumers may

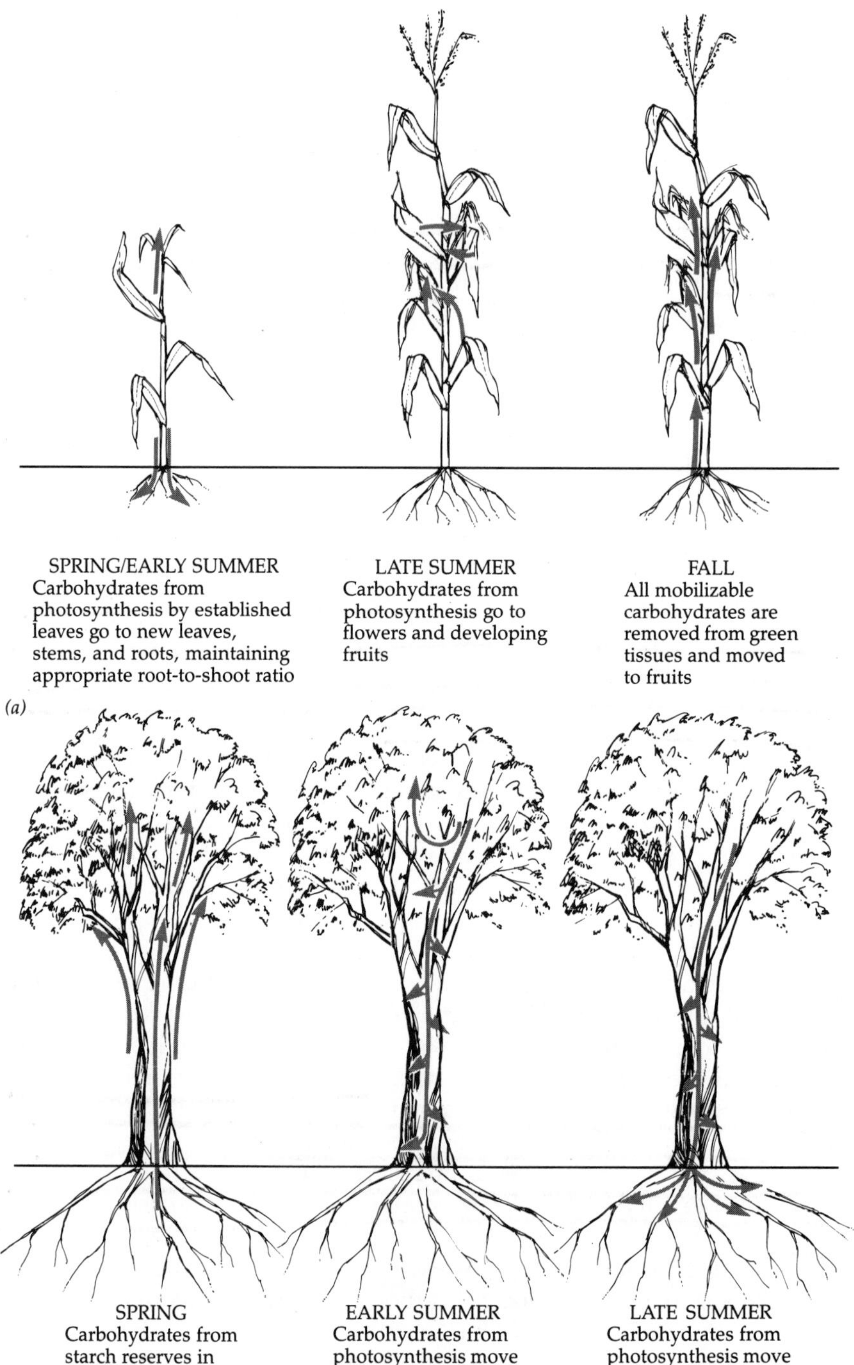

Figure 11.3
Seasonal changes in the allocation of carbon in *(a)* an annual grass and *(b)* a tree.

be affected by the relative availabilities of carbon and nutrients in the plant mobile pools.

If rapid growth rates of tissues result in the creation of strong carbon sinks, what then controls the stimulation of tops, roots, fruits, or flowers to initiate rapid growth? A proximate or partial answer is that plant hormones (a hormone is a "catalytic" molecule that, when present in small quantities, can cause large changes in plant metabolism) exert control over allocation patterns and that these are produced in response to different environmental conditions in patterns controlled by the genetics of the species. Ultimate control resides in the genetic makeup of the plant, which in turn is the result of natural selection of individuals within a species for genetically "programmed" allocation patterns that best fit the environment in which they grow.

Respiration—the Variable Cost of Maintaining Biomass

Respiration can create large carbon sinks within plants. In general, it is thought that respirational demands in plants are met before allocation is made to growth of new tissues. The amount of respirational demand for carbon changes with the accumulation of biomass resulting from plant growth.

This effect can be seen in a single forest stand through time, for example in a plantation of slash pine growing in Florida (Figure 11.4). Needle mass in these plantations rises rapidly following establishment, reaching a maximum in year seven, when crown closure occurs. The lowered needle mass past year seven may result from reduced nutrient availability.

Net photosynthesis for the whole stand (Figure 11.4*b*) would follow the trend in needle mass. Right after planting, respirational demands of stems and branches are low, so that net primary production is high, peaking at about the time that adjacent tree crowns approach each other and the canopy becomes closed. As live biomass accumulates, nonleaf respiration increases, causing reductions in net primary production.

The distribution of this biomass between harvestable biomass, belowground biomass, and litter production also changes with time. Litter production is initially very low, as both foliar and branch mass in the evergreen canopy are increasing, and needle retention is high. As the canopy closes, both needle and branch litterfall increase. Harvesting occurs before tree death can become a major component of litter production. Allocation to root growth is high initially and declines after the root network has been established. Rates of accumulation of harvestable yield are highest near the mid-point of the growth cycle.

Why are trees in plantations not planted more densely so that crown closure and the initiation of maximum biomass increment occur sooner? Respiration would be higher in a denser stand, due to a larger respiring

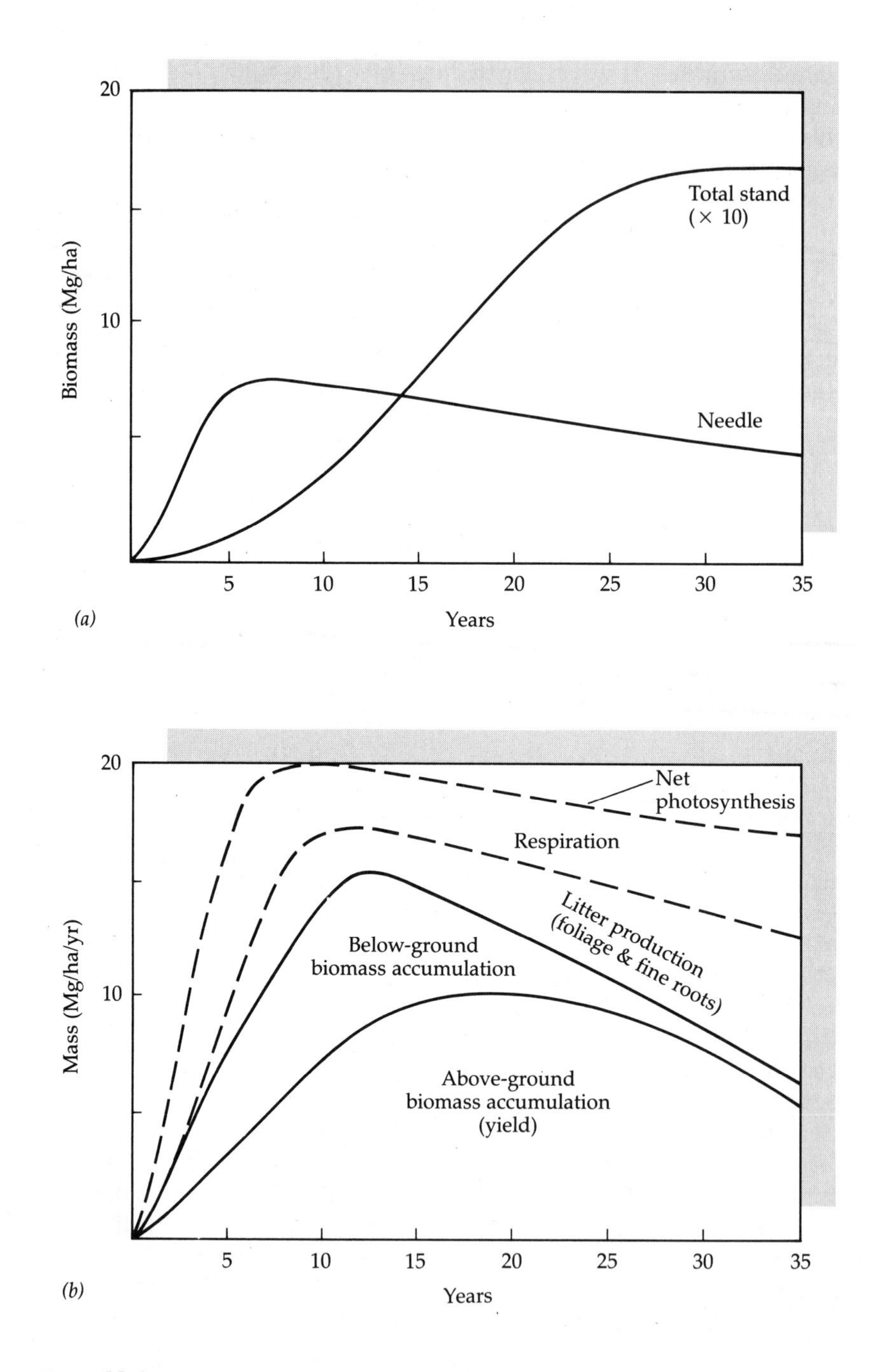

Figure 11.4

Changes in carbon, allocation with time in a slash pine plantation. *(a)* Accumulation of needle and total tree biomass. *(b)* Net photosynthesis and carbon allocation. (After Gholz and Fisher 1982)

surface area on the greater number of stems. Litterfall would be higher as well, as some trees would be outcompeted and die. Photosynthate in stems that die before the harvest is lost from the yield. An additional consideration is that individual stems get larger in more widely spaced plantations, as the accumulated biomass is distributed over fewer trees. Larger stems can be used for solid wood products as well as for pulp for papermaking and may be more valuable than smaller trees.

Allocation of Net Primary Production — Interactions with Nutrient Use Efficiency

Nutrient use efficiency was defined as the amount of biomass produced per unit of nutrient taken up. Productivity and uptake are linked through the concentration (as fraction, not percentage) of nutrients in the biomass produced. Uptake is the same as the sum of production in each type of tissue produced times the concentration in that tissue:

$$\text{nutrient use efficiency} = \frac{\text{NPP}}{\text{uptake}} = \frac{\Rightarrow \sum_{i=1}^{t} \text{prod}_i}{\Rightarrow \sum_{i=1}^{t} (\text{prod}_i \times \text{conc}_i)}$$

where t is the number of tissue types measured (e.g., leaves, roots, wood, bark, branches, etc.). However, the actual nutrient requirement for producing a tissue is not its concentration while alive but the concentration it carries with it at senescence and litterfall. Foliage can lose up to 75% of its nitrogen content between the green, midsummer condition and the time it is shed at the end of the growing season. Retranslocation is an invisible process that accompanies the very visible changes of color in foliage during senescence. The "missing" nutrients can be lost by either of two mechanisms: They may be leached from the foliage before senescence or they may be retranslocated back into the perennial part of the plant as part of the process of senescence.

The magnitude of nutrient decline in foliage and the relative importance of leaching and retranslocation are very different for different nutrients. Table 11.1 shows how the decline in foliage content of several macronutrients is partitioned between leaching and retranslocation for a mature northern hardwood forest. Nitrogen, phosphorus, and potassium content of the canopy decreased substantially during senescence. Magnesium content declined to a lesser extent and calcium content actually increased. Both nitrogen and phosphorus were retranslocated strongly, while losses of potassium and magnesium were mostly as throughfall. These results are representative of many temperate- and boreal-zone systems in that calcium and magnesium are generally considered to be less mobile than nitrogen, phosphorus, and potassium. Of the latter three,

Table 11.1 Decline in foliar content of several macronutrients during senescence, partitioned between leaching losses (throughfall) and retranslocation in northern hardwood forests of two different ages

	5-YEAR-OLD FOREST					55-YEAR-OLD FOREST				
	N	P	K	Ca	Mg	N	P	K	Ca	Mg
Foliar contents before senescence (F_1)	63	4.4	37.6	20	7.2	71	5.6	28	20	4.9
Foliar contents after senescence (F_2)	26	1.6	19.5	24	5.5	33	2.1	14	25	4.1
Foliar leaching during senescence (L)	1.4	0.2	9.4	2	0.6	2	0.1	11	2	0.6
Resorption (F_1-F_2-L)	35.6	2.6	8.7	−6	1.1	36	3.4	3	−7	0.2

nitrogen is often removed from leaves in the largest amounts by retranslocation, with lesser removals of phosphorus. Potassium losses are generally through leaching. This ion is not fixed in organic molecules in the leaf, so it is very mobile and easily leached.

Nutrients retranslocated from leaves are stored in adjacent woody tissues (in trees and shrubs) or in underground perennial organs (grasses and forbs) and are available for reuse by the plant in the next growth cycle. Part of next season's nutrient requirement for leaf production can be met from this internal plant pool. This is an effective mechanism both for reducing the total amount of nutrient required for growth in a given year and for "buffering" seasonal and annual variations in soil nutrient availability.

Retranslocation reduces nutrient concentrations in leaves and therefore increases nutrient use efficiency. If this is of value on nutrient-poor sites, then we might expect some correlation between nutrient availability from soils and the concentration of nutrients in litterfall.

Using the total amount of nitrogen cycling through above-ground litter production as an index of nitrogen availability, there is indeed a very strong relationship between availability and nitrogen concentration in above-ground litter (Figure 11.5*a*). For phosphorus (Figure 11.5*b*), the same pattern emerges only for tropical forests where litterfall concentrations are generally lower than for other forest and shrubland types at a given rate of phosphorus cycling. Both these findings suggest that a relative deficiency of N or P availablity (the latter particularly apparent in the tropical sites) results in increased retranslocation or at least retranslocation to a lower concentration in foliage. In contrast, the relationship

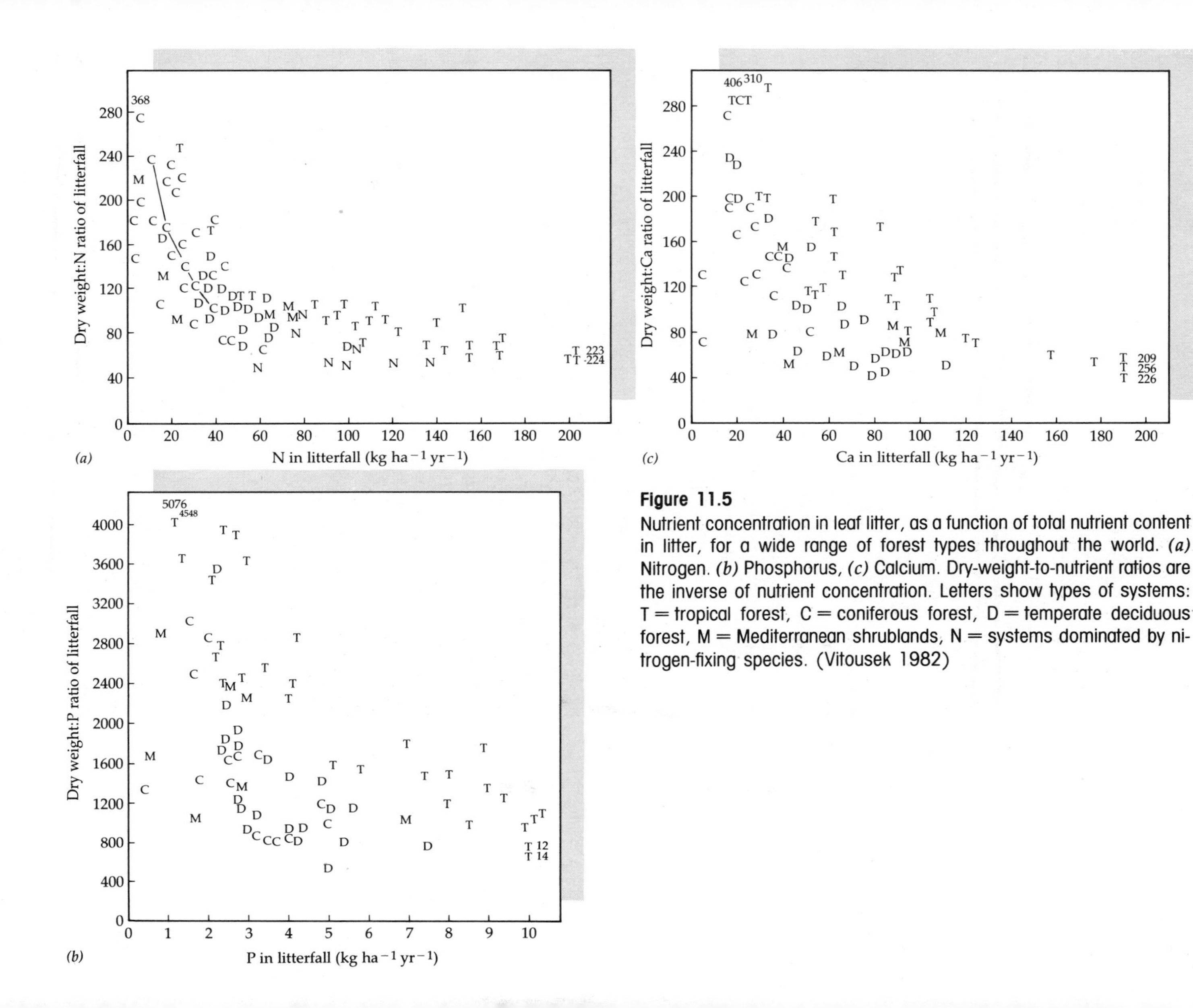

Figure 11.5

Nutrient concentration in leaf litter, as a function of total nutrient content in litter, for a wide range of forest types throughout the world. *(a)* Nitrogen. *(b)* Phosphorus, *(c)* Calcium. Dry-weight-to-nutrient ratios are the inverse of nutrient concentration. Letters show types of systems: T = tropical forest, C = coniferous forest, D = temperate deciduous forest, M = Mediterranean shrublands, N = systems dominated by nitrogen-fixing species. (Vitousek 1982)

between calcium concentration and total calcium in leaf litterfall is much less pronounced (Figure 11.5*c*).

These patterns could arise in either of two ways: (1) species that create low-nutrient-content litter could have a competitive advantage and replace species with high-content litter on poor sites, or (2) foliage of individuals of the same species could produce leaf litter with lower nutrient content on poorer sites. In the first alternative, species replace one another along a gradient from rich to poor sites. In the second, a species responds to lower nutrient availability by altering litter nutrient contents.

It is clear that species do replace each other along resource availability gradients. This could contribute to the large changes in retranslocation measured in a series of boreal forests in Alaska (Figure 11.6*b*). It is also clear from fertilizer experiments that nutrient contents of both green and senescent foliage can be altered significantly in single-species stands by increasing nutrient availability and that retranslocation is generally reduced following such additions (see Figure 11.6*a*). However, results from studies of retranslocation by a single species along measured gradients of nutrient availability show that while there are significant differences in retranslocation between species, there is relatively little difference within a species between sites (Figure 11.6*c,d*).

These results suggest both genetic (species-specific) and environmental (nutrient-deficiency) controls over retranslocation but that genetic control —differences between species—may be the most important in natural systems.

There is still much to be learned about retranslocation. An important unknown is whether or not fine roots also undergo retranslocation during senescence. If they do, the total size of this internal plant nutrient pool could be much larger than that supplied by leaves alone.

Even after retranslocation, roots, leaves, and stems have very different nutrient concentrations. In general, fine roots (those less than about 3 mm in diameter, which are most active in nutrient uptake) have the highest concentration, followed by foliage, followed by stems. Woody stems are particularly low in nutrient content, especially in conifer species. The allocation of carbon to these different tissues thus carries different nutrient costs and creates different nutrient use efficiencies.

A tradeoff is set up by the conflicting needs to conserve nutrients on poor sites and to produce enough of the high-nutrient-content foliage and fine roots required for photosynthesis and nutrient uptake. Increasing nutrient use efficiency by allocating more carbon to wood instead of leaves may not prove to be very adaptive if this seriously reduces leaf area and total photosynthetic capacity. Another approach is to maintain foliage (or roots) for more than a single year. If foliage is retained for two years, then a significant increase in carbon gain may be realized per unit of nutrient put into leaf tissue. Indeed, the "evergreen" habit is considered a primary adaptation to low-nutrient sites in the temperate forest region, where the benefits of increased nutrient use efficiency offset the respira-

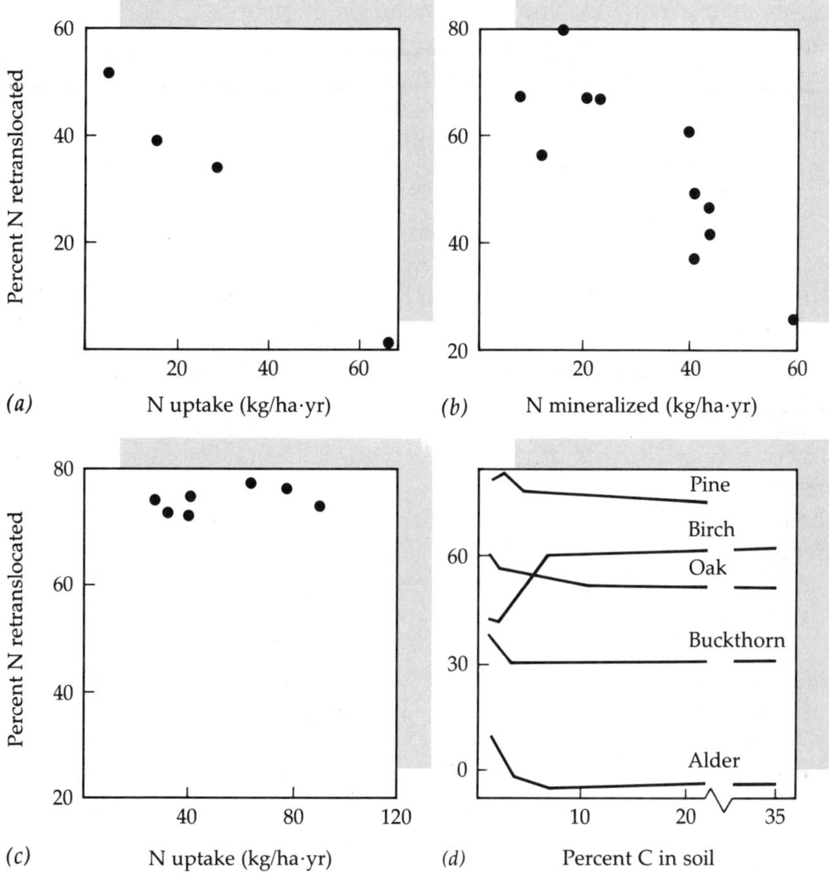

Figure 11.6
Nitrogen retranslocation from senescing foliage over four different nitrogen-availability gradients: *(a)* fertilized Douglas fir (Turner 1977), *(b)* taiga forests (Flanagan and Van Cleve 1983), *(c)* sugar maple stands (Lennon et al. 1985), *(d)* several forests in Poland (Zimka and Stachurski 1976; percent carbon in soil assumed related to nitrogen mineralization).

tional costs of maintaining foliage through the winter. These same allocation characteristics partially explain the preference of forest managers for evergreen species. Even on rich sites, planted pines and other conifers will allocate less to foliage and more to wood production than deciduous species, which need to replace the entire leaf canopy each year.

Some evergreen species can respond to particularly low site quality by increasing needle retention time. For example, black spruce can retain needles for ten years and more on the poorest sites.

Does the same relationship between site quality and retention time apply to fine roots? Do fine roots live longer on poor sites? This is a harder question to answer. In general, plants maintain higher fine root biomass on poorer sites as a mechanism for more complete uptake of nutrients. However, there is also evidence that, once the higher biomass is produced, it turns over more slowly on poor sites. Changes in the type of mycorrhizae associated with fine roots appear to respond to nutrient availability and possibly affect root longevity.

Fine roots remain one of the most difficult and important aspects of ecosystems to study. With some effort, estimates of the changes in biomass through time can be obtained. But, unlike foliage, which turns color and falls at senescence, fine roots are difficult to separate into live and dead tissues. Most approaches to measuring the growth or death of fine roots modify the soil environment in some way and may invalidate results. Yet obtaining good estimates of fine root production is crucial for calculating both parts of the nutrient use efficiency equation, total NPP and total nutrient uptake. This is why the analysis in Figure 11.5 was limited to foliage. Relatively few estimates of production and nutrient requirement of fine roots exist. Because of this there are disturbingly few measurements of total NPP in terrestrial ecosystems. Most productivity values published apply to above-ground tissues only.

Two divergent views of fine root production have emerged from the research done to date. The first concludes that fine root turnover rate is not greately affected by site quality and that low root biomass and small seasonal changes in that biomass on rich sites (Figure 11.7*a*) indicate lower total root production. This means a higher percentage of total productivity allocated to wood and foliage (Figure 11.7*b*). This idea matches well with the idea of lower allocation of carbohydrate to roots on rich sites where less may need to be "invested" to obtain adequate nutrient supplies. A problem with this is that fine roots are higher in nutrient content than senescent foliage and growing more roots actually decreases nutrient use efficiency and increases total calculated nitrogen uptake. However, if fine roots do retranslocate nutrients before senescence, then this problem may be answered in part.

The second view holds that fine root turnover is analogous to the retention time of foliage in evergreens—that richer sites will lead to tissues with higher nutrient content, different mycorrhizal associations, and shorter life spans. In this view, low and constant fine root biomass on rich sites is consistent with rapid and constant production and mortality of fine roots. Using a nitrogen budgeting approach, one comparison across sites of different nitrogen availability concluded that fine root turnover rate does increase with site quality (Figure 11.8) and that this relationship counterbalances the lower fine root biomass on rich sites to cause carbon allocation to different tissues to be relatively constant across the gradient (Figure 11.9).

The question of allocation to roots has also been addressed using soil carbon balances. In mature ecosystems, the carbon balance over soils is

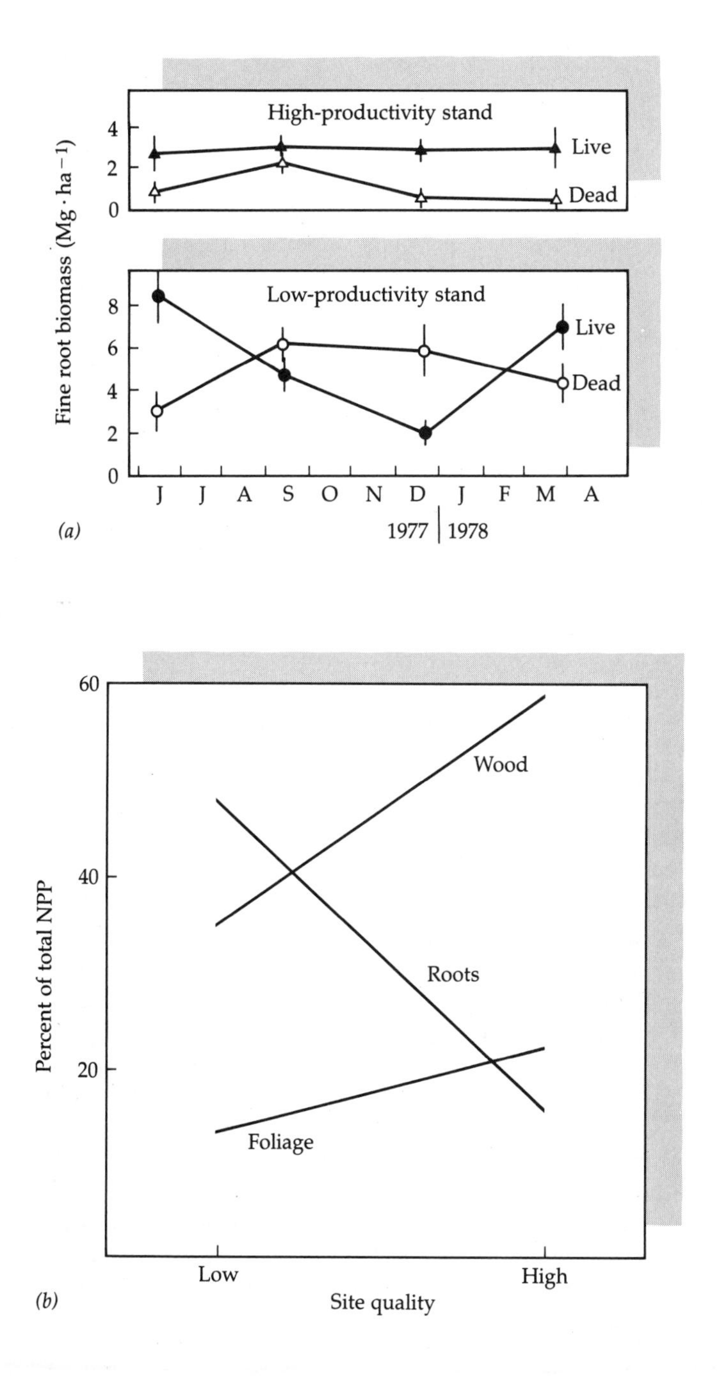

Figure 11.7
Carbon allocation among foliage, wood, and fine roots in Douglas fir on rich and poor sites. *(a)* Measured changes in fine root biomass. *(b)* Estimated changes in allocation of production. (Keyes and Grier 1981)

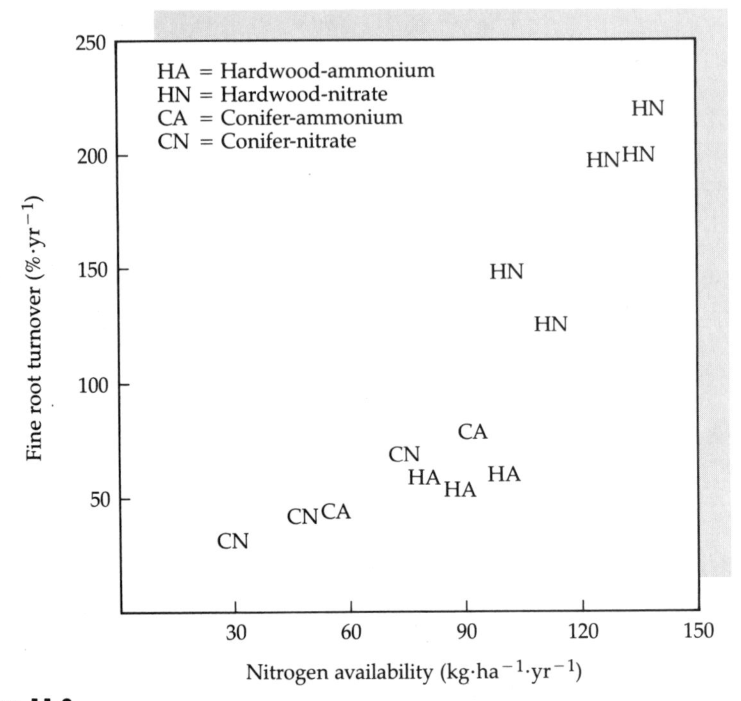

Figure 11.8
Estimated rate of fine root turnover in several forests as a function of nitrogen availability. (Aber et al. 1985)

close to zero, and the CO_2 released from decay of above-ground litter is nearly equal to the carbon deposited in that litter. Under these conditions, subtracting carbon inputs in above-ground litter from total CO_2 release from soils provides an estimate of the carbon allocation by plants to below-ground tissues, for both growth and respiration. A survey of data from many systems suggests that a relatively constant fraction of fixed carbon is allocated below ground over a very wide range of above-ground litter inputs (Figure 11.10). Only on sites with very low above-ground litter production does this analysis suggest significant increases in percentage allocation of carbon to roots. Both the nitrogen and carbon budgeting approaches to estimating fine root production represent uses of ecosystem-level measurements to place limits on processes that are difficult to measure directly.

The idea of faster root turnover on rich sites is consistent with an emerging, generalized theory of plant function along resource availability gradients. In Chapter 6 we presented data showing that maximum rates of net photosynthesis increased with increasing leaf nitrogen content and that the same quantitative relationship held for all C3 species. We also showed that respiration increased in leaves with high nitrogen content.

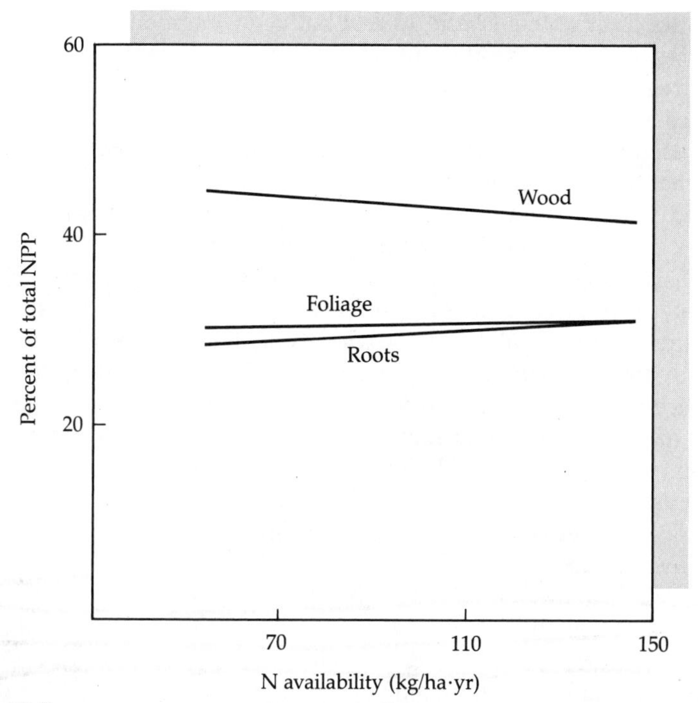

Figure 11.9
Estimated patterns of carbon allocation in oak stands as a function of nitrogen availability. (From data of Nadelhoffer et al. 1985, Pastor et al. 1984, and unpublished data)

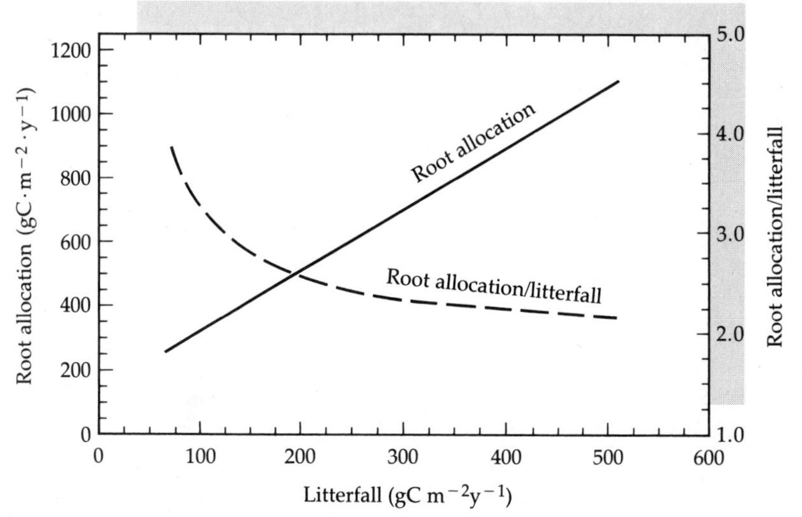

Figure 11.10
Summary relationship between above-ground litterfall and total carbon allocation to roots for both growth and respiration. The relationship was derived from many published studies. (Raisch and Nadelhoffer 1989)

We will see in Chapters 12 and 15 that both consumption of live leaves by herbivores and decomposition of litter by microbes are also increased by high nitrogen concentrations.

Leaves with high photosynthetic rates can "repay" the cost of the carbon used to construct them more rapidly than less active leaves, providing that the environment is otherwise favorable. In environments with restricted light or moisture availability, high nitrogen foliage would be unable to realize this fast "repayment" (Figure 11.11). If N-rich foliage is retained for longer periods to reach a positive carbon balance, then it becomes more susceptible to herbivore attack and may be consumed before the positive total carbon balance is achieved. In response to both the unfavorable environment and the need for longer foliage retention and hence greater protection from herbivores, leaves of poor-site species tend to have lower concentrations of nitrogen and higher concentrations of carbon-rich compounds like lignin, which reduce their palatability to herbivores.

The linkage between site quality, nutrient use efficiency, and overall metabolic rate can be summarized as follows (Figure 11.12). Rich environments provide the potential for high rates of photosynthesis, which can only be realized with high concentrations of nitrogen and other nutrients in foliage. Natural selection and competition then favor the occurrence of plants with high foliar nitrogen levels on these rich sites. The higher

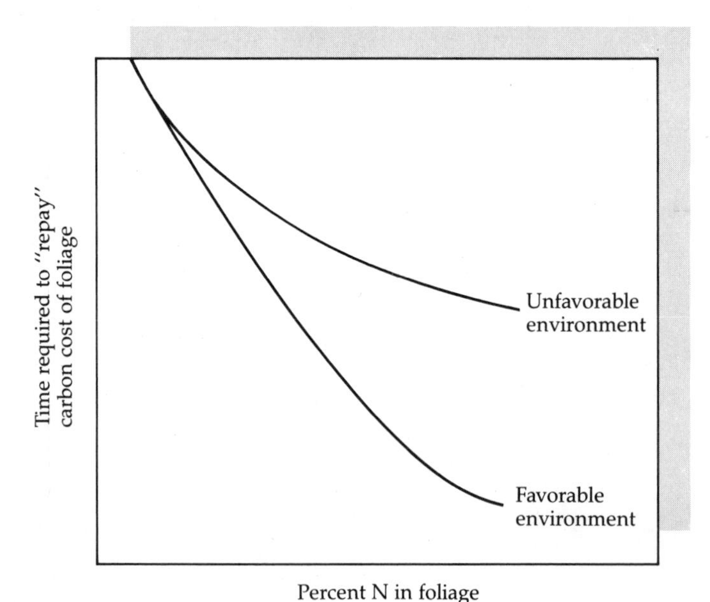

Figure 11.11

Estimated differences in time required to "repay" the carbon investment in leaves, as a function of nitrogen concentration in leaves, for generally favorable and unfavorable growing environments. (After Mooney and Gulmon 1982)

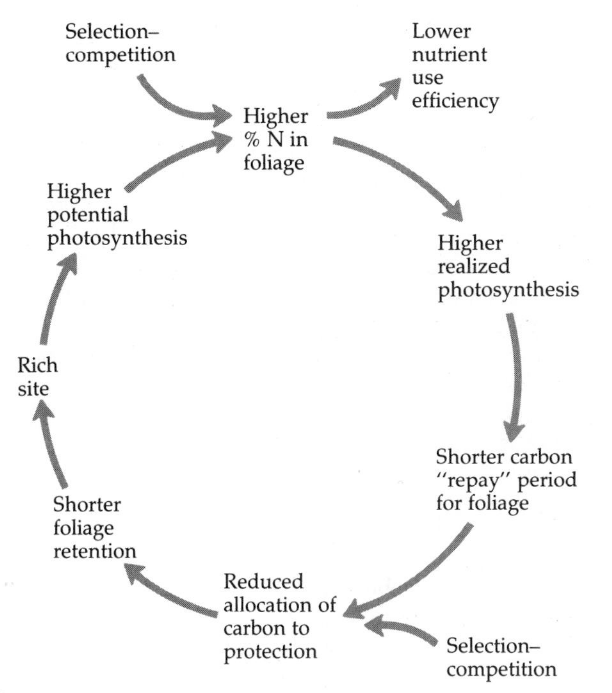

Figure 11.12
Potential interactions between site quality, foliar nitrogen content, and leaf longevity. (After Mooney and Gulmon 1982)

photosynthetic rates mean a shorter period is required to repay the carbon costs of building leaves, leading to a selective advantage in reducing allocation to compounds designed to protect leaves from herbivores (Chapter 15). This, in combination with the fast "repay" period, leads to shorter leaf retention times. These interactions would all be reversed on low-quality sites, leading to longer foliage retention (evergreens) and high nutrient use efficiency. To summarize, rich sites lead, in general, to high rates of metabolic activity and reduced retention time or longevity for leaves and perhaps also for roots. In the next chapter we will see that tissues with high nitrogen content and low content of generalized herbivore inhibitors (such as lignin) also decay and release nutrients more rapidly, completing the positive feedback loop (Figure 11.12).

REFERENCES CITED

Aber, J. D., et al. 1985. Fine root turnover in forest ecosystems in relation to quantity and form of nitrogen availability: A comparison of two methods. *Oecologia* 66:317–321.

Chapin, F. S., P. M. Vitousek, and K. Van Cleve. 1986. The nature of nutrient limitation in plant communities. *The American Naturalist* 127:48–58.

Flanagan, P. W., and K. Van Cleve. 1983. Nutrient cycling in relation to decomposition and organic matter quality in taiga ecosystems. *Canadian Journal of Forest Research* 13:795–817.

Gholz, H. L., and R. F. Fisher. 1982. Organic matter production and distribution in slash pine (*Pinus elliotti*) plantations. *Ecology* 63:1827–1839.

Keyes, M. R., and C. C. Grier. 1981. Above- and below-ground net production in 40 year old Douglas-fir stands on low and high productivity sites. *Canadian Journal of Forest Research* 11:599–605.

Lennon, J. M., J. D. Aber, and J. M. Melillo. 1985. Primary production and nitrogen allocation of field grown sugar maple in relation to nitrogen availability. *Biogeochemistry* 1:135–154.

Mooney, H. A., and S. L. Gulmon. 1982. Constraints on leaf structure and function in relation to herbivory. *BioScience* 32:198–206.

Nadelhoffer, K. J., J. D. Aber, and J. M. Melillo. 1985. Fine root production in relation to total net primary production along a nitrogen availability gradient in temperate forests: A new hypothesis. *Ecology* 66:1377–1390.

Pastor, J., et al. 1984. Above-ground production and N and P cycling along a nitrogen mineralization gradient on Blackhawk Island, Wisconsin. *Ecology* 65:256–268.

Raisch, J. W., and K. J. Nadelhoffer. 1989. Below ground carbon allocation in forest ecosystems: global trends. *Ecology* 70:1346–1354.

Ryan, D. F., and F. H. Bormann. 1982. Nutrient resorption in northern hardwood forests. *BioScience* 32:29–32.

Turner, J. 1977. Effects of nitrogen availability on nitrogen cycling in a Douglas-fir stand. *Forest Science* 23:307–316.

Vitousek, P. M. 1982. Nutrient cycling and nutrient use efficiency. *American Naturalist* 119:553–572.

Zimka, J. R., and A. Stachurski. 1976. Vegetation as a modifier of carbon and nitrogen transfer to soil in various types of forest ecosystems. *Ekologia Polska* 24:493–514.

ADDITIONAL REFERENCES

Chapin, F. S. 1980. The mineral nutrition of wild plants. *Annual Review of Ecology and Systematics* 11:233–260.

Nambiar, E. K. S. 1987. Do nutrients retranslocate from fine roots? *Canadian Journal of Forest Research* 17:913–918.

Vogt, K. A., et al. 1983. Organic matter and nutrient dynamics in forest floors of young and mature *Abies amabilis* stands in Western Washington, as suggested by fine-root input. *Ecological Monographs* 53:139–157.

Chapter 12 Litter Decomposition and Nutrient Balances

H. Armstrong Roberts

INTRODUCTION

Carbon allocation and nutrient use efficiency are important processes in overall plant function. They also control the quantity and biochemical content, or "quality," of dead organic matter (litter) produced. Litters of different "quality" are decomposed at different rates. For example, fine roots generally decay more slowly than do leaves, so root:shoot production ratios affect not only plant function but also decomposition rates and the development of soil organic matter. These, in turn, affect site quality and future root:shoot ratios.

Why is decomposition important? Litter materials contain nutrients. The quantity of essential nutrients entering an ecosystem each year is generally low compared to the amount cycling within that system. Plant production depends on the recycling of nutrients within the system, and recycling depends on the decomposition of organic matter and release of the nutrients it contains. The long-term stabilization of organic matter through the formation of humus (discussed in Chapter 13) also has important implications for soil structure, water-holding capacity, and ion exchange.

The purpose of this chapter is to discuss litter as a resource for microbial growth and to introduce the major chemical components of litter that determine the "quality" of this resource and control rates of weight loss and nutrient balances during decomposition. We will also discuss the relative concentrations of these different constituents in different types of litter and the resulting differences in decomposition between them.

LITTER AS A RESOURCE FOR MICROBIAL GROWTH

Why is it that leaves that fall to the ground in a forest can disappear in a year or two while a fallen log may last a century or more? Why do certain types of leaves disappear more rapidly than others, even on the same site? Often the answer lies in differences in the "quality" of the different types of litter relative to the energy and carbon and nutrient requirements of the microbes involved in the decay process.

Litter materials are a resource for microbial growth, and both the rates of litter decay and the nutrient balances during the decay process reflect the interaction between microbial requirements and relative resource availability in litter. For example, decay rates of materials that are rich in high-energy-yielding compounds (high carbon quality) may be reduced by a low concentration of nitrogen. Low nitrogen availability reduces the rate of protein synthesis by microbes, reducing both population levels and the rate at which carbon is processed. Low nitrogen levels in litter may result in the removal of mineral nitrogen from the soil by microbes, leading to competition between microbes and plants for available nitrogen. Limitations on microbial growth due to other nutrients, particularly phosphorus, can also occur. More generally, the concepts of resource limitation and growth rates, discussed relative to primary production in Chapter 11, apply to microbial growth as well.

Three general characteristics determine the quality of litter materials relative to microbial decay: (1) the types of chemical bonds present and the amount of energy released by their decay, (2) the size and three-dimensional complexity of the molecules in which these bonds are found, and (3) the nutrient content. The types of carbon bonds present and the energy they yield constitute the "carbon quality" of the material. Nutrient content and the ease with which those nutrients can be made available constitute the "nutrient quality."

Decomposition occurs through the production of enzymes by microbes, which break the chemical bonds formed during the construction of plant tissues. Enzymes are expensive to produce in terms of carbon and nitrogen. These costs must be more than offset by energy and/or nutrient gain for the microbial community to grow through the decomposition of a given material.

This energy return is greatest for small, high-energy-content molecules such as simple sugars (see discussion below), which have saturated carbon–carbon bonds and which can be taken up into microbial cells and metabolized internally. However, these constitute only a small fraction of plant litter. The majority of plant residues consist of large polymers (many small molecules connected into long chains) formed from simpler compounds. These larger molecules cannot be taken into microbial cells directly. Instead, the required enzymes must be produced within the microbial cells and then released into the environment. Not only are these enzymes expensive, but once in the soil they might not reach the intended substrate and not affect any decay at all. Consequently, the decay of these larger molecules is less efficient, even if the substrate itself is of high quality.

The three-dimensional structure of large molecules may offer an additional barrier to decay. External decay enzymes cannot be effective unless they encounter an appropriate three-dimensional bond structure. Compounds that do not have a consistent structure, or in which a complex three-dimensional structure impedes enzyme access to reaction sites (see the discussion of lignin and proteins below), will decay more slowly than the carbon quality of the material would suggest.

Nutrient incorporation or release during decomposition depends on the ratio of nutrient-to-energy content in the material. Microbes growing rapidly on high-carbon-quality substrates will create large nutrient demands for the synthesis of new biomass. If these demands are not met, microbial activity and decomposition will be inhibited. Decomposition of low-carbon-quality organic matter will not support net growth in microbial populations and so will result in a net release of nutrients stored in the decomposing substrate.

BIOCHEMICAL CONSTITUENTS OF LITTER AND THEIR RATES OF DECAY

Glucose (Figure 12.1*a*) and other simple sugars (or carbohydrates) are among the first products of photosynthesis and are very high-quality substrates for decomposition. The molecules are small and the chemical bonds in them are energy-rich, yielding much more energy than is required to create the enzymes necessary to initiate the chemical breakdown reaction. They also can be taken into microbial cells and metabolized internally.

Simple sugars not immediately required for respiration and growth by plants may be stored as starch, a carbohydrate polymer formed by bonding of the 1 and 4 position carbons of adjacent molecules (Figure 12.1*b*). Decomposition of starch is somewhat slower than that of simple sugars because the longer molecule must be severed before the sugar units can be metabolized. Still starch is a rapidly decomposed, high-energy-yielding substrate. Both sugars and starch are present in small amounts in plant litter material (Table 12.1), as they tend to be used for respiration before senescence is complete.

Cellulose is also a polymer of simple sugars linked by bonds between the 1 and 4 carbons of adjacent molecules. However, the three-dimensional structure of the bond is slightly different (Figure 12.1*c*). This slight difference allows a totally different function. Cellulose is the main component of primary cell walls in plants. Carbohydrates converted to cellulose cannot be remobilized for respiration or growth by the plant. Respiration of this carbon must be accomplished by microbial decay. Individual polymers of cellulose, consisting of from 2000 to 15,000 sugar units, are

(a) Simple sugars

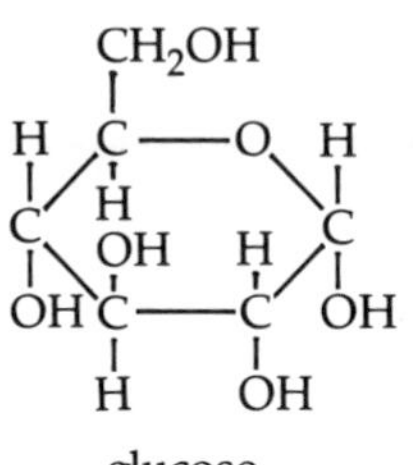

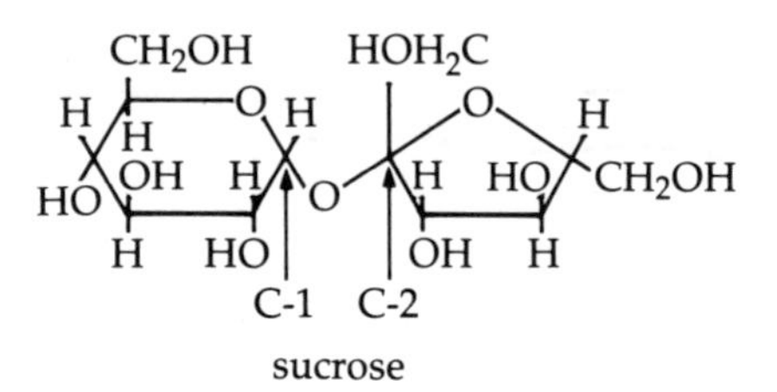

(b) Starch

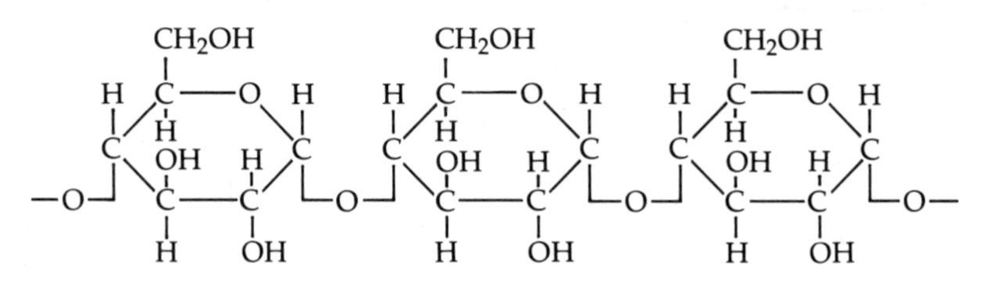

(c) Cellulose

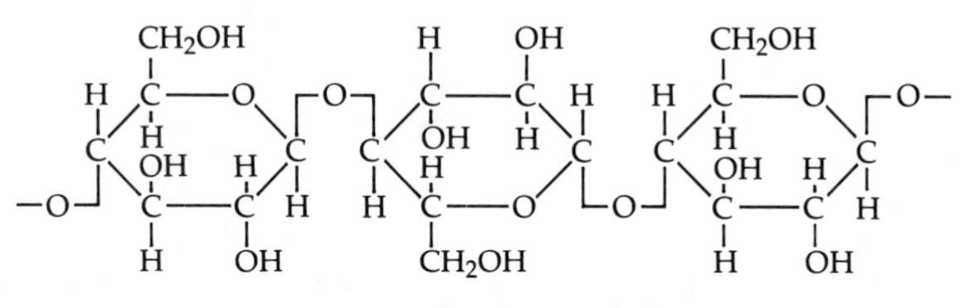

structure with β-(1 → 4) linkages

(d) Hemicellulose

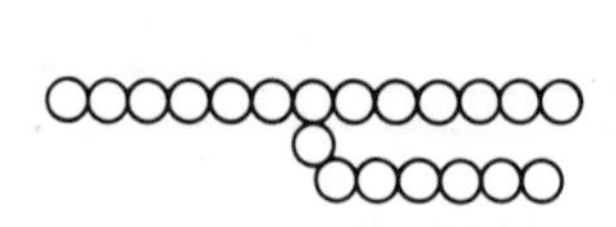

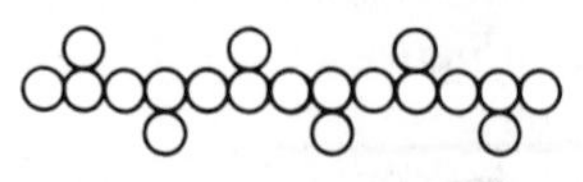

Figure 12.1
Chemical structures of several important types of carbohydrates in plants.

Table 12.1 Concentrations of major carbon compounds in different plant materials*

	SUGARS AND STARCH (%)	OTHER SOLUBLES (%)	CELLULOSE (%)	LIGNIN (%)
Woody plants				
Foliage				
Sugar maple	7.2	37.6	43.1	12.1
Red oak	7.3	25.1	47.4	20.2
White pine	5.7	27.1	44.7	22.5
Fine roots				(Suberin)
Sugar maple	3.9	14.6	47.7	33.8
White pine	5.2	20.0	49.5	25.3
Wood				
Red maple	1.1	5.9	80.5	12.5
Hemlock bark	4.1	16.7	40.3	38.9
Herbaceous plants				
Foliage and stems				
Salt marsh grass				
Tall-form, live		34.4	52.5	13.1
Tall-form, dead		28.9	57.7	14.4
Tall-form, stems		30.3	56.0	13.7
Ryegrass stems				3–9
Leaves				2–6
Timothy stems				5–9
Leaves				3–6
Roots				
Salt marsh grass		36.2	41.6	12.2
Mixed pasture grasses		20	58	22

*Data from McClaugherty et al. 1985, Larsson and Steen 1988, Morrison 1980, Hodson et al. 1984.

further entwined into larger strands or fibers, which are laid down in roughly parallel form to create the cell walls (Figure 12.2).

Cellulose is probably the most common molecule in the plant component of terrestrial ecosystems. It is also the source of fiber for paper and paper products, which are nearly pure cellulose. Cellulose is of moderate quality as a decay substrate. External enzymes are required to cleave the large polymers into simple sugars, which can be taken up and metabolized. Two types of enzyme systems are known: One cleaves bonds within the molecule to create two shorter molecules, and the other separates individual sugar units from the ends of polymers.

Mixed in with cellulose and serving the same basic function are the hemicelluloses. These are also polymers but consist of several different basic sugar units combined into both straight and branched chains (Figure

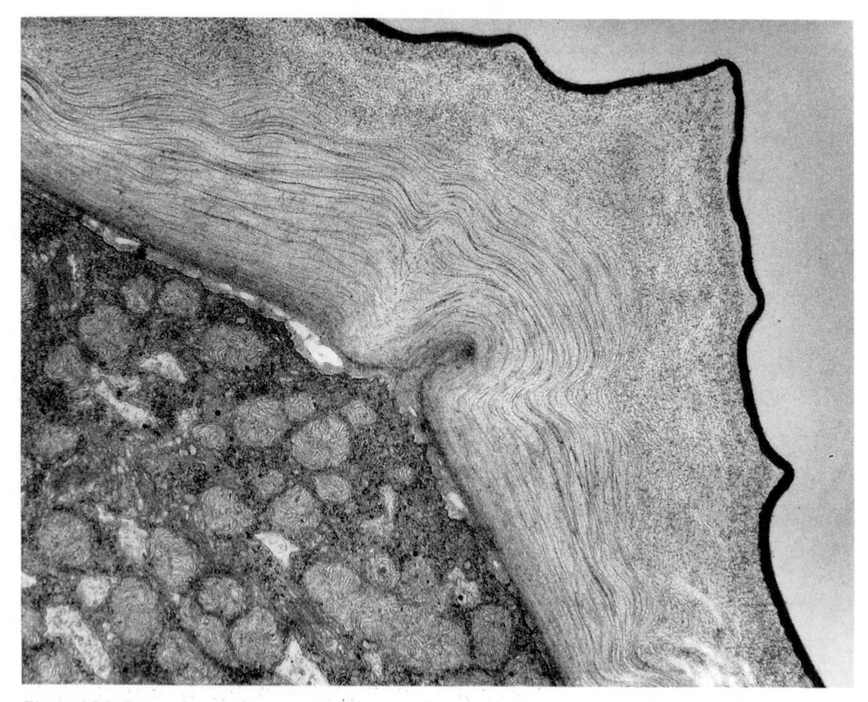

Figure 12.2
Arrangement of cellulose fibrils in primary cell wall. (Biophoto Associates/Photo Researchers)

12.1*d*). The branched structure helps bind the long, straight cellulose fibers together in the cell wall. There is little difference between cellulose and the hemicelluloses in rate of decay or energy yield.

A number of compounds containing unsaturated carbon–carbon bonds are very important in plant function and for decomposition. Particularly important are those based on the six-carbon phenolic ring. Two classes of such compounds are generally recognized: smaller phenol polymers (polyphenols) made from several phenolic acids (Figure 12.3*a*) and often called "tannins" and the larger, amorphous, and very complex compounds collectively called "lignin" (Figure 12.3*b*).

The **tannins** are so named because they have historically been extracted from plant tissues rich in these substances (e.g., the bark of some tree species) and used to "tan" leathers for shoemaking. The tanning process causes the combining (or condensation) of polyphenols with proteins present in animal hides to increase the strength and durability of the leather (and to decrease its decomposability). In plants, they are thought to be primarily a defense mechanism against animal consumption and attacks by pathogenic fungi and bacteria. Many polyphenols are easily extracted from leaf tissues, indicating their potential mobility within plants.

(*a*) Common phenolic acids

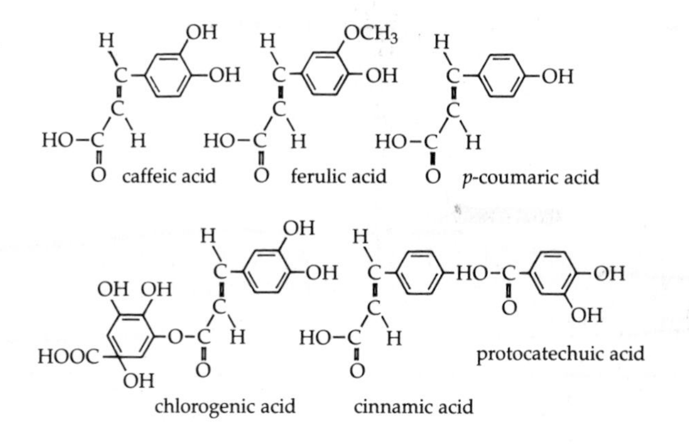

(*b*) Proposed subunit of a lignin molecule

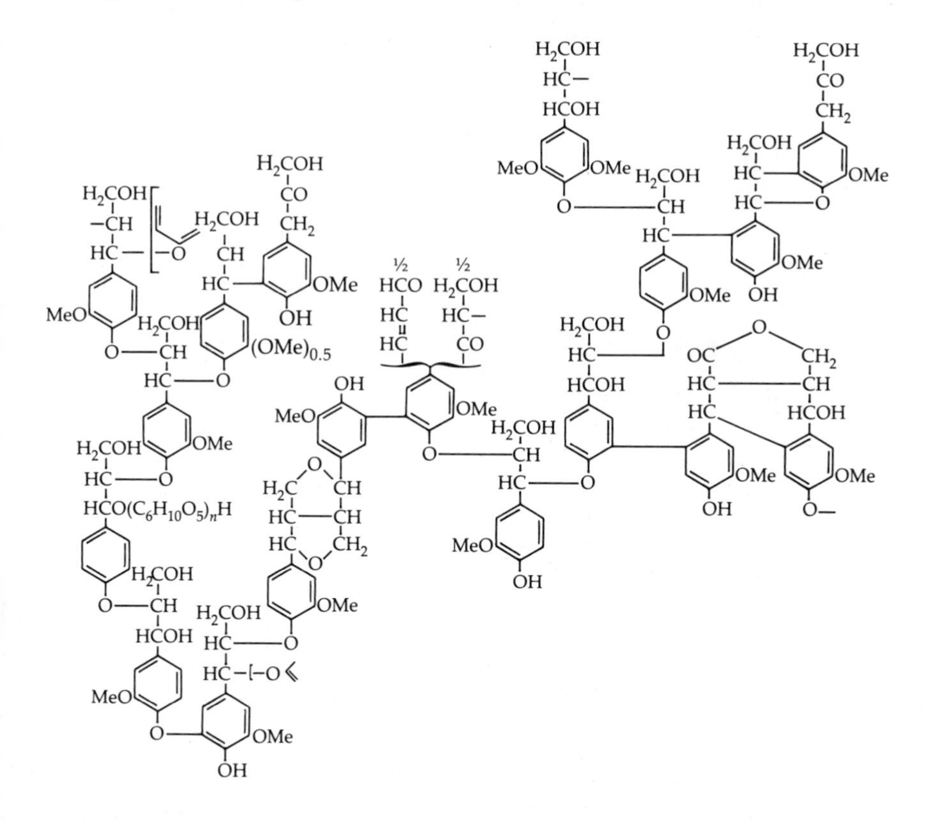

Figure 12.3
Chemical structure of some phenolic compounds found in plants.

Although the phenolic ring yields less energy than saturated carbon bonds, the smaller polyphenols can be metabolized. The rate in nature is hard to determine, however, because of the mobility of the molecules and their capacity to condense with proteins or other polyphenols to form lignins. Thus polyphenols may "disappear" from decomposing litter but may not be decomposed. They may be present as "new" lignin or polyphenol–protein complexes.

The much larger **lignin** molecules are among the most complex and variable in nature. There is no precise chemical description of "lignin" as this term actually applies to a class of compounds with variable structure. In fact, there is no precise way of chemically separating all of the molecules of lignin from plant material when performing tests of plant chemistry. Rather, a series of "proximate" analyses are performed. Figure 12.4 lists a series of treatments that divide the components of plants into different chemical fractions that have proven useful in explaining differences in decay rates. When we say a tissue is 20% lignin we are actually saying that 20% of the tissue remains after the series of treatments shown in Figure 12.4. There are different sets of proximate analyses, and each gives a different set of results.

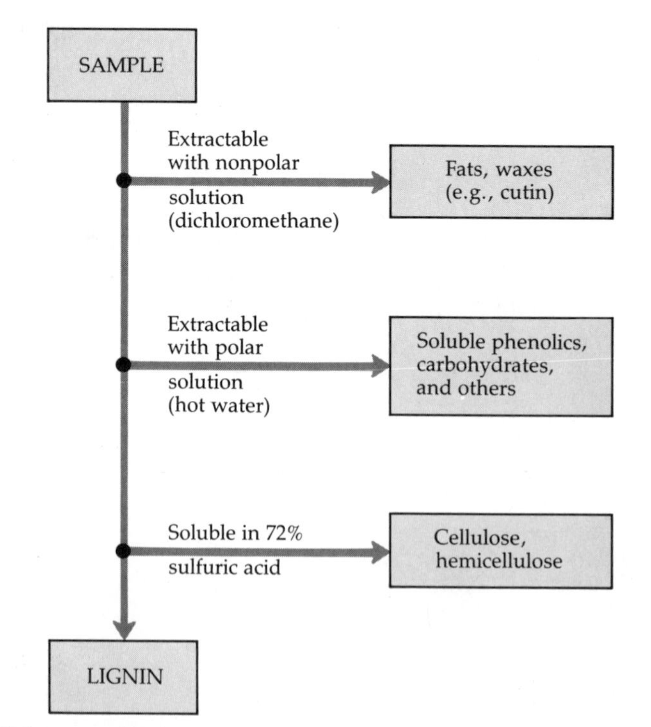

Figure 12.4
Flow diagram of a series of treatments in the "proximate analysis" of plant tissues.

It is possible to synthesize specific types of lignin-like compounds in the laboratory or, by chemically degrading a lignin molecule into smaller units, to get some idea of the structure of these subunits. One example of part of a lignin molecule is presented in Figure 12.3*b*. Basic characteristics of lignin are the density of phenolic rings and the frequency of side chains, both of which reduce decomposition rates significantly. In addition, these very large molecules can be intricately folded into complex three-dimensional structures that effectively "shield" much of the internal portions of the molecule from enzymatic attack.

Lignin is one of the slowest of the common plant components to decay. It yields almost no net energy gain to microbes during decomposition because of the large amounts of energy required to initiate its decomposition. There is evidence that energy derived from the decay of higher-quality substrates may be required to decompose lignin. For example, very complex enzyme systems that simultaneously degrade both lignin and cellulose-like carbohydrate polymers recently have been identified. In addition, laboratory studies indicate that the release of carbon as CO_2 from lignin can be increased by the addition of high-quality carbon substrates (sugars).

Yet lignin is second only to cellulose in quantitative importance in most plant tissues (Table 12.1). Lignin is what makes wood "woody." It is encrusted on and around cellulose in cell walls to provide rigidity and strength. Paper is easier to bend or tear than thin slices of wood largely because the lignin has been removed in the papermaking process. Lignin can be as concentrated in leaves as it is in wood. Within these tissues, lignin may actually "shield" some of the cellulose present from microbial processing, causing some of the cellulose to decay more slowly than would be expected by its biochemical quality.

An intermediate set of compounds exists between the carbohydrates and phenolic compounds of plants. The "hydrocarbons" or fats and waxes are partially saturated long-chain carbon compounds in which hydrogen ions have largely displaced the hydroxyl (OH) group. This makes them more impervious to water (hydrophobic), which, along with their size, makes them intermediate between carbohydrates and phenolics in decay rate.

The most common compound in this category is *cutin* (Figure 12.5), which is present in the cuticle, or outer, waxy layer in plant leaves. Again, no strict definition of this substance is known, as it is an amorphous combination of constituents into variable structures. Its major role is to reduce evaporation through and to protect leaf surfaces.

Suberin is an interesting class of molecules composed of roughly half hydrocarbons and half phenolics. This makes it both hydrophobic and of low energy quality, and consequently it is one of the slowest classes of compounds to decompose. Suberin is found in fine roots largely in place of lignin. When fresh, white roots become darker in color and more rigid, they are called "suberized," as this change is due to the deposition of

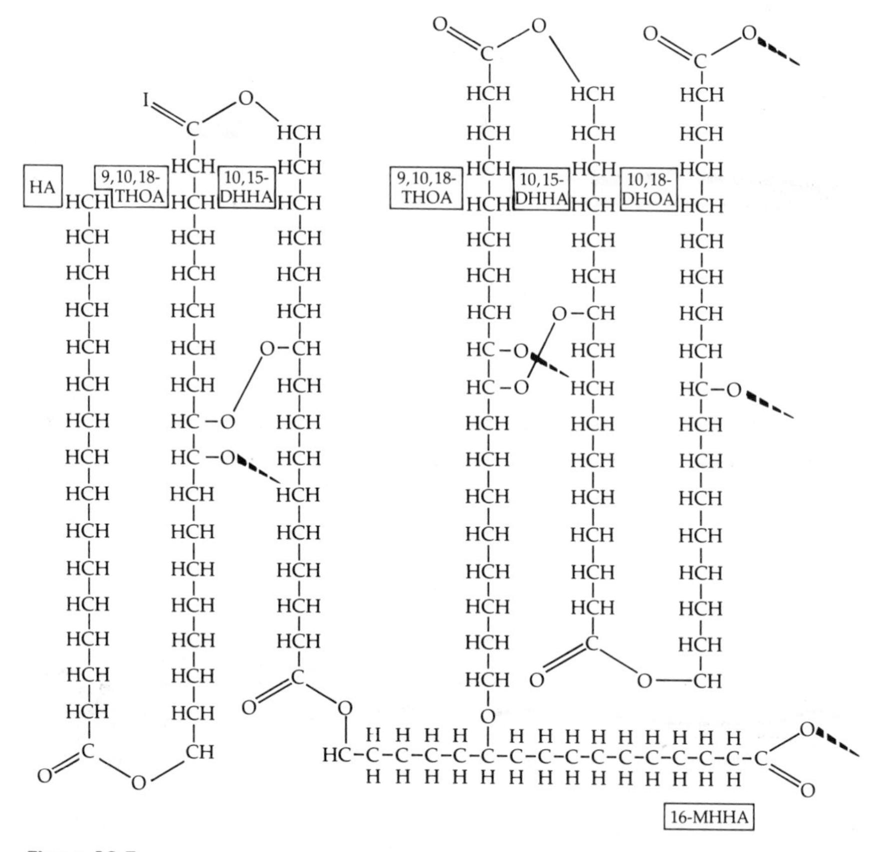

Figure 12.5

Chemical structure of cutin. HA = hexadecanoic acid, MHHA = monohydroxyhexadecanoic acid, DHHA = dihydroxyhexadecanoic acid, DHOA = dihydroxyoctadecanoic acid, THOA = trihydroxyoctadecanoic acid.

suberin within the cell walls. Because suberin is of less economic importance than lignin, even less is known about its structure and the compounds from which it is formed.

Proteins are another important constituent of litter, both because of their carbon quality and because they contain nitrogen. Nitrogen is present in plants mainly as structural or functional (enzymatic) proteins and the amino acids from which they are constructed (Figure 12.6). The simple amino acids can either be metabolized or used directly in construction of new proteins within the microbial cell. Proteins are polymers of amino acids and can contain more than 10,000 amino acid units. The largest proteins can become resistant to decay by the complexity of their three-dimensional structure, even though the individual units are very energy-rich. In addition, proteins are often condensed with, and deactivated by, polyphenols and lignin. These ligno–protein complexes are very

Basic structure of amino acids

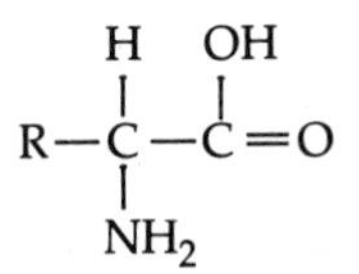

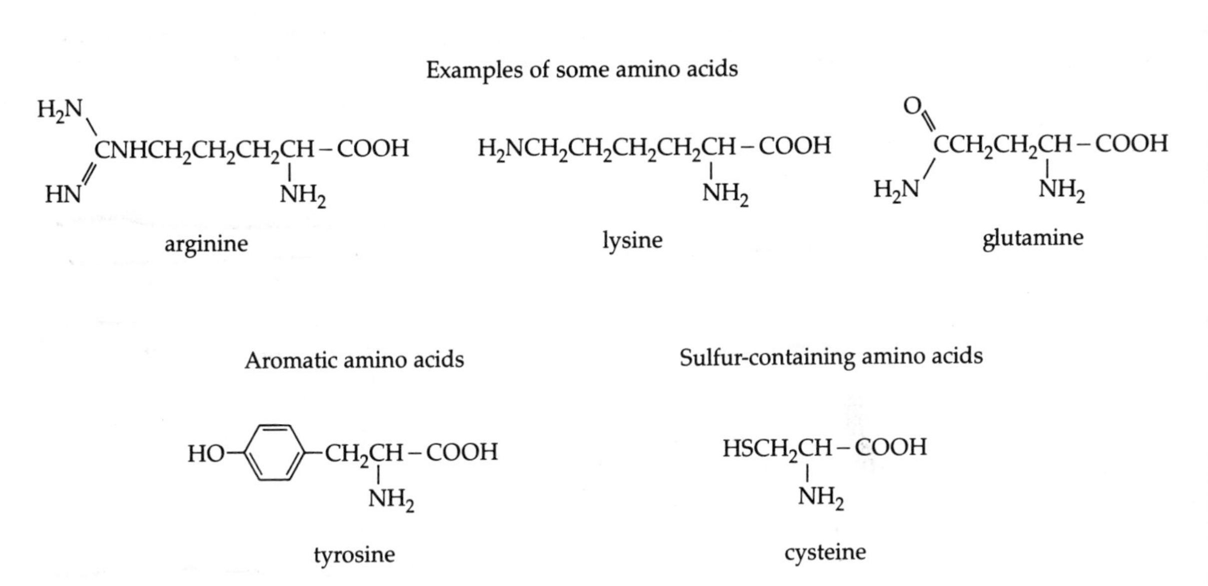

Figure 12.6
Basic structure of amino acids and examples of some common amino acids.

resistant to decay, even though they have a high nitrogen content. These condensation reactions may involve the very enzymes released by microbes to effect decomposition! In this way, the polyphenolic content of litter may have an additional negative effect on decay rates.

DECOMPOSITION RATES OF DIFFERENT LITTER CONSTITUENTS

Decomposition rates for litter have often been described by a constant percentage weight loss per unit time. This yields a curvilinear pattern for percent of original weight remaining, which can be fit to an exponential equation of the form:

$$\% \text{ original remaining} = e^{-kt}$$

where t = time and k = litter-specific constant. This equation should look familiar. It is the same one used to describe light extinction in forest canopies as a function of cumulative leaf area index (Chapter 7).

The concept of a constant decomposition factor, k, for a given litter type is an integral part of the standard approach to litter decay but is actually only indirectly related to the actual decay process. The constant value k assumes that litter decay rates do not change through time. It also implies that all substrates within a litter type are processed simultaneously. This is not the case. Three phenomena demonstrate some of the actual complexity of the decomposition process.

First, significant weight loss can occur by the purely physical process of leaching by precipitation as it washes over the litter. Low-molecular-weight sugars, polyphenols, and amino acids can be lost in this way. The total weight loss by leaching will depend on the amount of water-soluble material in the litter and the amount of water passing over it.

Second, the constituents that are not water-soluble do not all decompose simultaneously. The similar carbon compounds, which can be degraded rapidly and yield the most net energy, are attacked preferentially. So when sugars and free cellulose are available, enzymes for the degradation of lignin tend not to be produced. Lignin decomposition will not begin until more easily decomposed materials are nearly exhausted. In fact, many "lignin-like" substances (ones that appear in the lignin fraction of proximate analyses; Figure 12.4) are produced as byproducts of the decay process. This can cause actual increases in the total amount of lignin in a decaying material.

Third, lignin, and probably suberin in roots, are so intimately entwined with the cellulose strands in most materials that they effectively "shield" much of the cellulose from microbial attack. After the nonshielded cellulose has been decomposed, lignin must be degraded to allow microbial access to the remaining cellulose.

A summary of the effect of these processes is reflected in the rates of disappearance of different classes of compounds in Figure 12.7. The simple extractable compounds are both leached and decomposed very quickly to low levels. The values never reach zero, however, as small amounts are either stabilized by physical shielding by inorganic or organic components or because a small amount is always present within the populations of microbes intimately associated with the decomposing litter. Simple sugars also decay quickly. Cellulose decays first at an intermediate rate until all the non-lignin-shielded cellulose is gone. Up to this point, lignin decay is negligible, and formation of lignin-like compounds by the decay process may cause actual increases in total lignin content. Beyond this point, lignin and cellulose are degraded simultaneously.

For all of this complexity, the exponential decay function remains a statistically useful method for summarizing mass loss during the early stages of litter decay and the one most frequently encountered in the literature. It is important, however, to realize that this model does not describe the timing or rate of decomposition of any one fraction within the material, nor does it apply over the complete decomposition sequence.

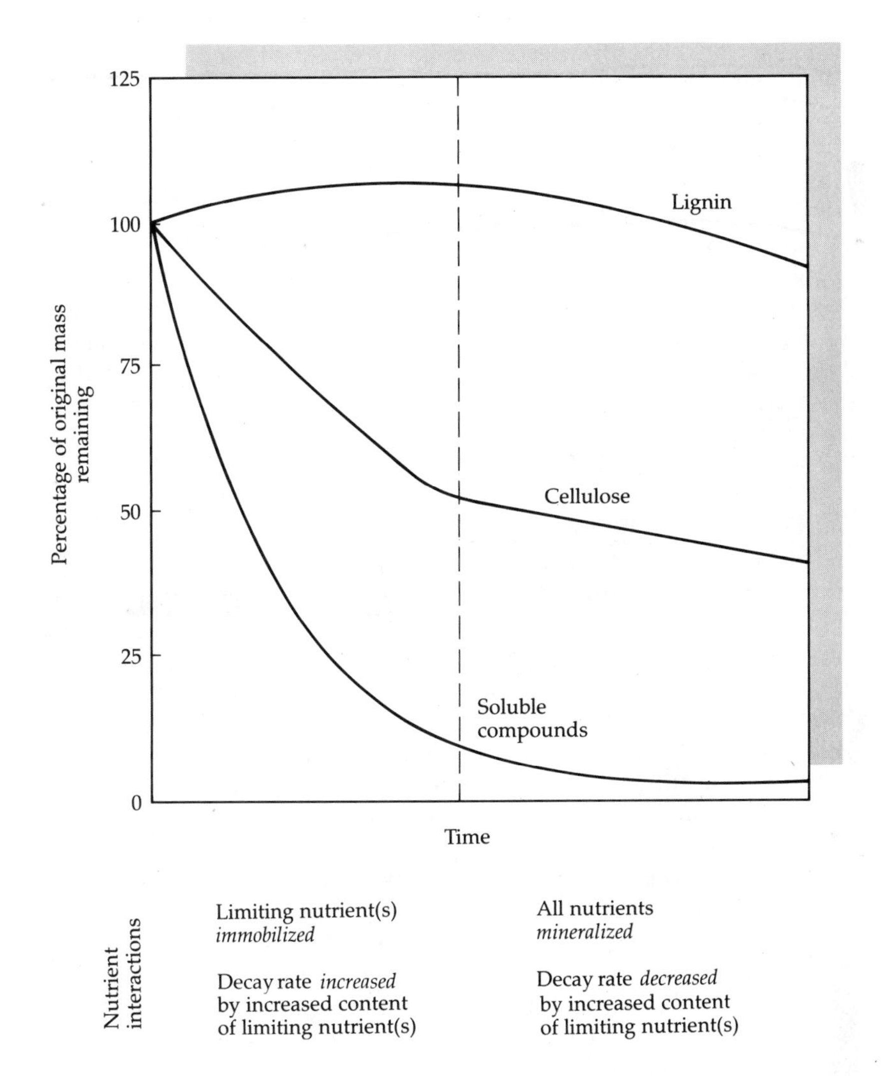

Figure 12.7
Generalized pattern of mass loss by carbon fraction during decomposition and associated interactions with nutrient availability.

NUTRIENT DYNAMICS DURING LITTER DECOMPOSITION

The rate of weight loss from decomposing litter is an indication of the amount of energy and carbon available to microbes for growth. As discussed at the beginning of this chapter, microbes also require nutrients in given concentrations to build cells. For microbial growth to proceed at the rate allowable by available energy and carbon, nutrients must also be available from either the decomposing substrate or from the surrounding soil solution. Suboptimal nutrient availability to microbes will reduce decomposition rates of the litter and also lead to removal of available nutrients from the soil solution. Conversely, if nutrients are present in litter in excess of microbial requirements (microbial activity is limited by carbon quality rather than nutrient availability), then those nutrients will be released as decomposition proceeds.

Figure 12.8 shows weight loss and nutrient dynamics accompanying decomposition of scots pine needle litter in a forest in Sweden. While the needles are of only moderate carbon quality and decay relatively slowly, they are also very low in content of several nutrients. Microbial growth on these needles creates demands for nitrogen, phosphorus, and sulfur, which are greater than the supply within the litter material. In Figure 12.8, the net increase in the total content of these nutrients within the decomposing litter results from microbial uptake from outside the leaf material and incorporation into the litter.

This net increase in the absolute amount (not just the concentration) of nutrients in decomposing litter is called **net immobilization**. It actually decreases the amount of nutrients available for plant growth. Plants and microbes growing in the same soil can be competing for available nutrients. This is why materials like fresh sawdust and wood chips can seriously reduce plant growth when used in gardens and planting mixtures; they are rich in carbon relative to the low concentration of nutrients.

The rate of nutrient immobilization in the scots pine litter is highest in the earliest stages of decay, when the most easily decomposed compounds are being degraded and microbial growth is most rapid. After about a year and a half in the field, the concentrations of nitrogen and phosphorus in the partially decayed litter have increased substantially, and the decomposition rate has decreased. As the carbon and energy yield from decomposition declines, so does the demand for nutrients. Nutrient release from the decaying material, and also from dead and now decaying microbial tissues, is greater than continuing microbial demand, and the excess nutrients are released in mineral form (e.g., ammonium, sulfate). This nutrient release is called **net mineralization** and results in a net reduction in the absolute amount of nutrients in the litter and increased availability to plants.

Where do the immobilized nutrients come from? Two sources are most likely. For tissues decaying on or above the soil surface, throughfall is the most likely source. For tissues decaying below ground, microbes have access to nutrient mineralized from older litter and soil organic matter. By taking up these nutrients, microbes compete with plants for what might be an important limiting resource. A third potential source for nitrogen is fixation by free-living microbes living within litter. However, free-living nitrogen fixation (as opposed to symbiotic N fixation; see Chapter 10) is generally very low in most terrestrial ecosystems.

The patterns discussed here for nutrient immobilization and minerali-

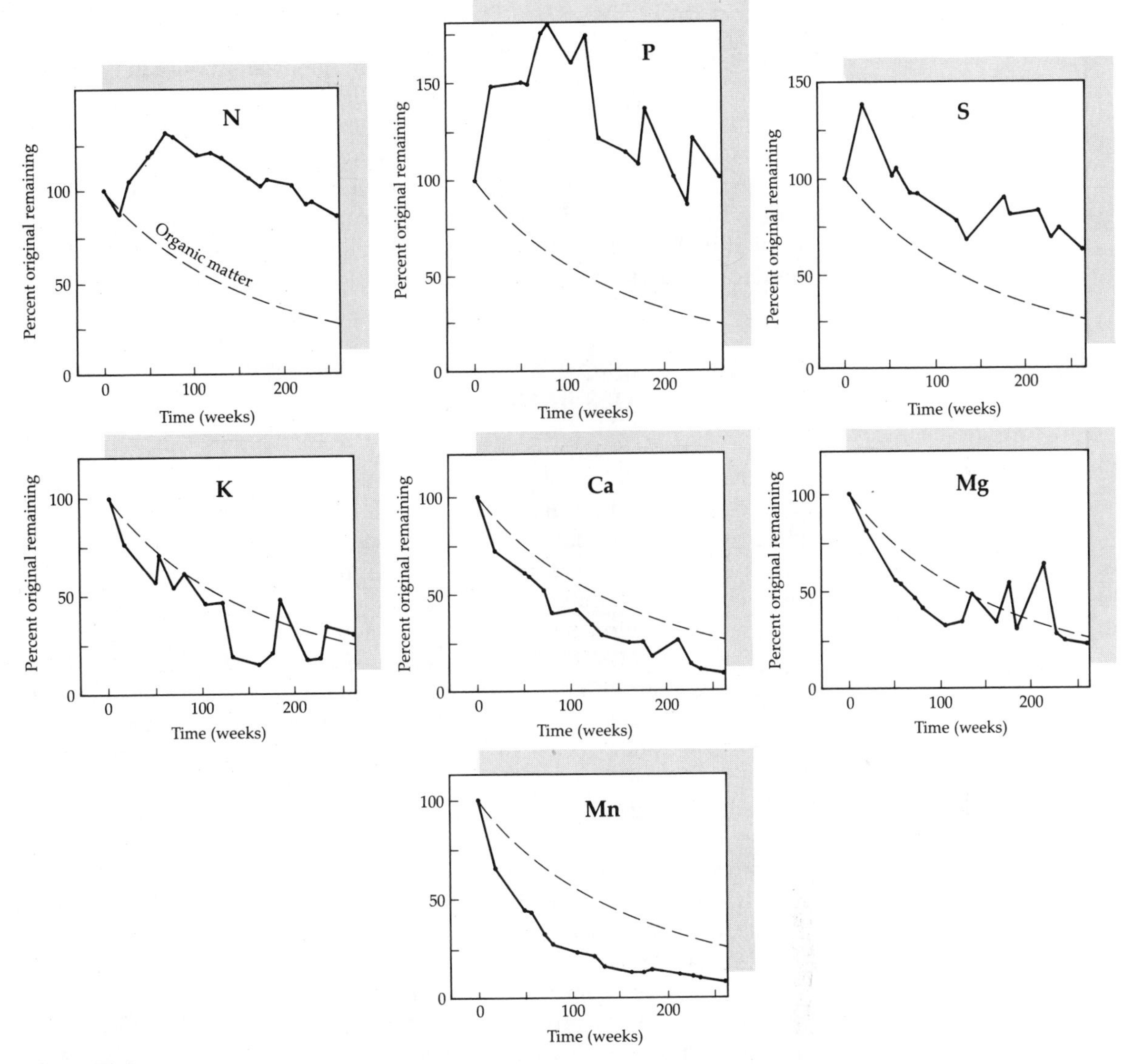

Figure 12.8
Rates of nutrient immobilization and mineralization from decomposing Scots pine litter over a five-year period. (Staff and Berg 1982)

zation are fairly typical. Nitrogen and phosphorus tend to be the two nutrients most likely to be immobilized. However, other elements may be immobilized in materials where they are present in low concentrations. The important principle is the supply of the element from decaying litter compared with the nutrient demands of the growing microbes.

When does nutrient mineralization begin? According to our discussion of microbial resource requirements, it should begin when carbon quality declines to the point that nutrient availability is no longer limiting to microbial activity. There is evidence that mineralization of N in N-limited ecosystems begins at the same time that lignin begins to be decomposed. Both the beginning of nitrogen mineralization and the initiation of lignin decay would indicate exhaustion of relatively high-quality carbon compounds. In Figure 12.7, decay is separated into an initial period in which immobilization of limiting nutrient(s) occurs and the decay process is nutrient limited and a later period when all nutrients are mineralized and the decay process is limited by the carbon quality of the material.

EFFECT OF CARBON AND NUTRIENT QUALITY ON DECAY RATES IN NATURE

Given this understanding of the role of carbon and nutrient quality in litter decay, are there generalized, quantitative relationships that can be derived for predicting decay dynamics for all ecosystems? Forest foliage litters are the only plant tissue for which enough data are available to make meaningful comparisons. In short-duration studies, decay rates have been related to the initial ratio of lignin to nitrogen in the material (Figure 12.9*a*). In long-term studies, the role of nutrient content is much less important, and rates covering the entire period of rapid weight loss can be predicted by carbon quality alone (Figure 12.9*b*).

Tissues of similar quality will decay at different rates under different climatic conditions. The two lines in Figure 12.9*a* represent two study sites, one in New Hampshire, one farther south in North Carolina. It is apparent that litter with similar initial "quality" as expressed by the lignin : nitrogen ratio will decompose faster at the North Carolina site. This is because of the longer growing seasons at this site.

By considering both the initial quality of the material and the climate of an area, as characterized by actual evapotranspiration (AET; Chapter 2), it is possible to predict decomposition rates for different litter types under different climatic regimes (Figure 12.10). This ability to predict the rate of a central ecosystem process when only a few important variables are known is important to the understanding and prediction of ecosystem function at the global level (see Chapter 24).

DECOMPOSITION OF WOOD

Decomposition of woody stems in forest ecosystems varies significantly from the simple models presented above. Wood decays much more slowly than would be predicted by its carbon chemistry alone, due mainly to the presence of chemical inhibitors in heartwood and to the time required for microbes to fully colonize the large mass of material. For example, the heartwood of temperate-zone conifers can contain any of several carbon-

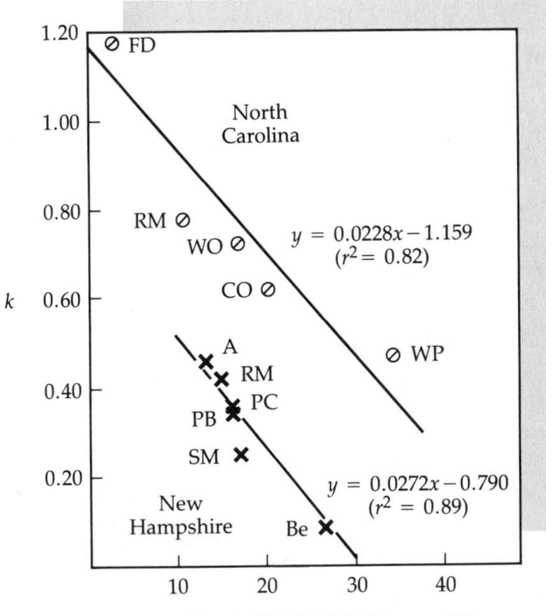

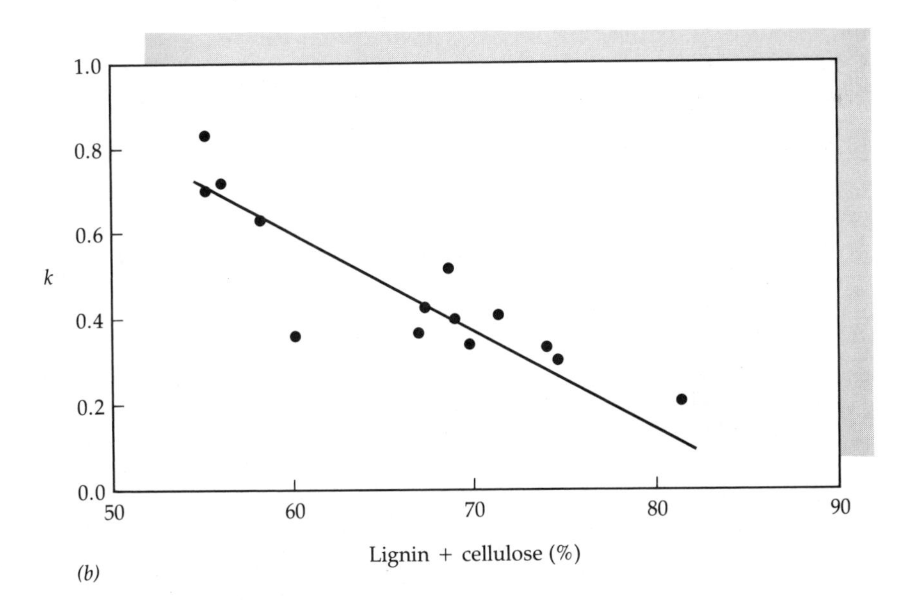

Figure 12.9
Decomposition constant k for foliar litter in relation to initial tissue chemistry. *(a)* First-year decomposition as a function of the lignin-to-nitrogen ratio (Melillo et al. 1982; includes data from Cromack 1973). *(b)* Long-term decay as a function of initial lignin + cellulose concentration (Aber et al., in press).

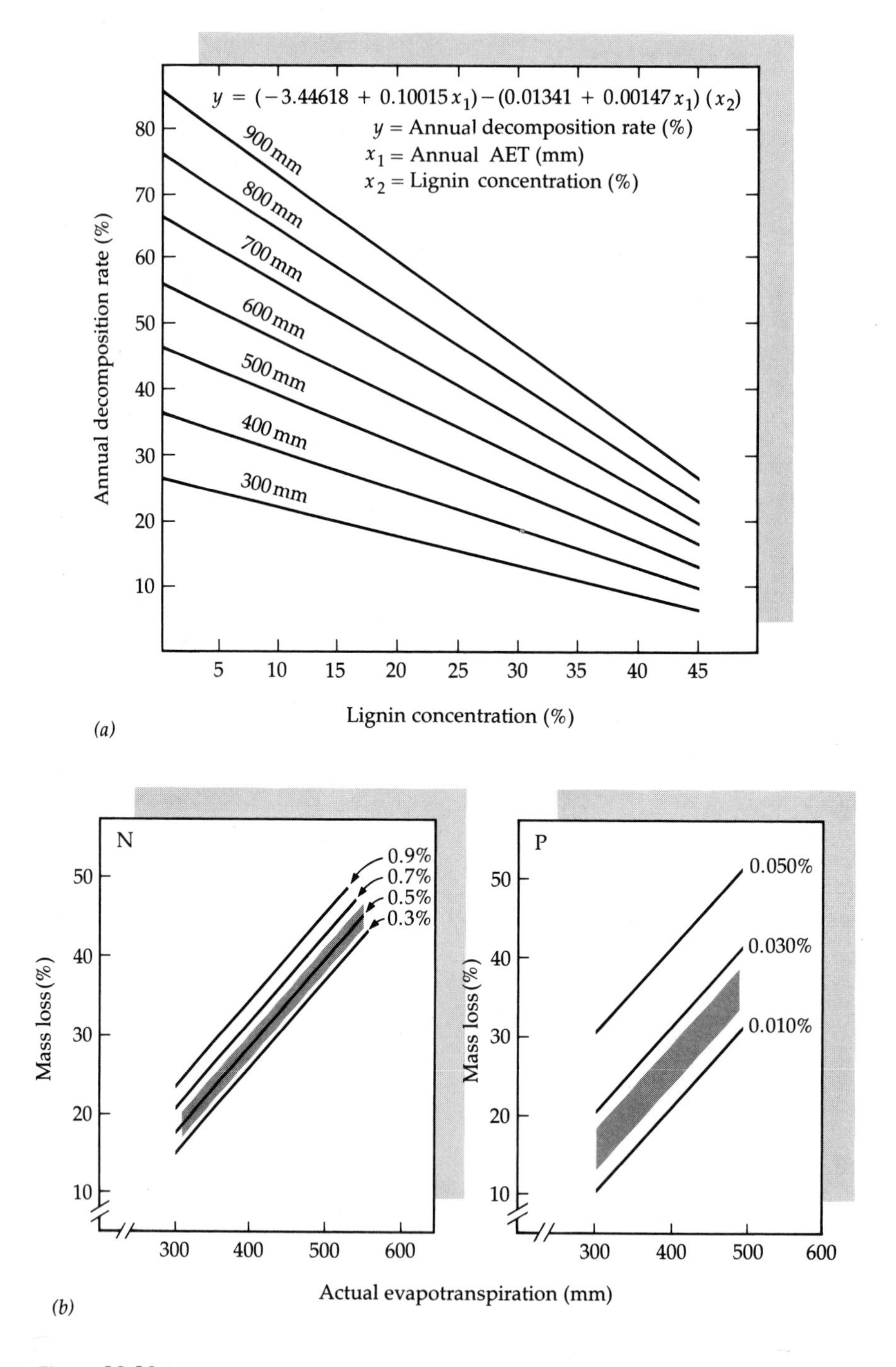

Figure 12.10

Summary relationships of first-year decay rates as a function of litter quality and climate: *(a)* using lignin as quality parameter and several litter types (Meentemeyer 1978), and *(b)* using nitrogen and phosphorus as quality parameters in Scots pine needles for several sites in Scandinavia. (Meentemeyer and Berg 1986)

based decay inhibitors. During the tree's lifetime, these protect the structural integrity of the stem. After tree death, they impede decomposition.

Unlike foliage and roots, which offer a large amount of surface area to decomposers relative to their mass, large tree boles may require many years for complete microbial colonization. Figure 12.11 is an example of the effect of stem diameter on the percentage of stem wood colonized with time. Decay will be delayed for decades at the center of large boles.

Another aspect of stem size is its effect on immbolization rates. Wood is very low in nutrient content, yet the interior of a decaying stem is largely isolated from sources of nutrients in either throughfall or soils. Nutrient

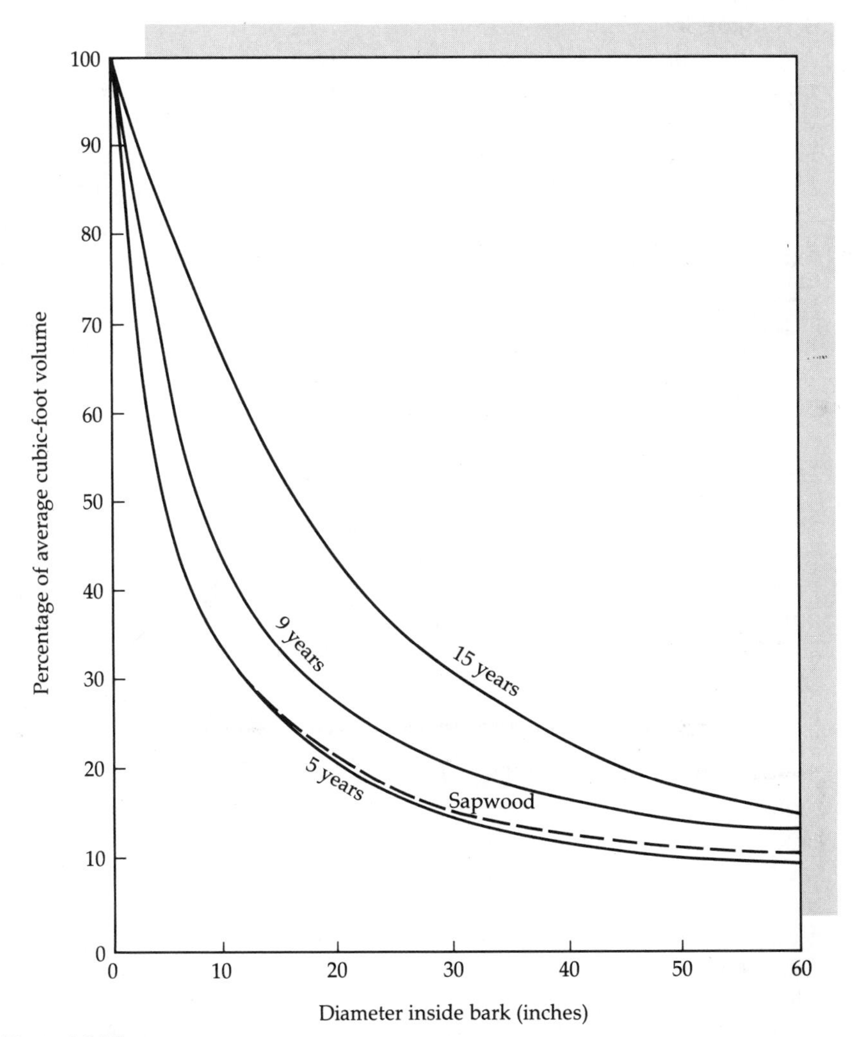

Figure 12.11
Percent colonization of Douglas fir logs by decomposer organisms as a function of time and the diameter of the log. (Harmon et al. 1986)

additions to decaying wood have shown significant increases in decomposition rates.

EFFECTS OF SOIL ANIMALS ON DECOMPOSITION

Most of the previous discussion has pertained to forest ecosystems, as this is where most of the studies have been carried out. There are indications that things may not be quite so neat in the drier ecosystems—grasslands, shrublands, and deserts. Particularly in deserts, much of the small amount of litter produced may be removed from the surface by animals and stored in burrows and tunnels. This greatly complicates the study of the fate and decomposition of litter!

In general, the role in decomposition played by larger animals, including organisms as common as earthworms, is relatively unknown, even in the better-studied humid regions. It is clear that they can move or ingest large amounts of material. Charles Darwin, in an early and classic study of earthworms in field soils, calculated that earthworms deposited more than 24 tons of castings per year on a hectare of English meadow. There is a general relationship between the mass of earthworms present and the rate at which surface litter "disappears." However, disappearance is not decomposition. A leaf moved from the surface to an earthworm burrow has not changed biochemically, but it is in a very different environment. Whether it then decays differently is unknown.

PRODUCTION OF SOIL ORGANIC MATTER

We have not yet described the end point of the litter decay process. Litter passes through periods of nutrient immobilization and mineralization and periods of more and less rapid mass loss, but neither mass loss nor nutrient mineralization proceeds to completion. When litter reaches the stage of very slow weight loss and becomes embedded in the surrounding soil matrix, it becomes the conceptual province of soil organic matter research. The boundary between litter and soil organic matter has not been clearly defined, either operationally or intellectually.

One emerging characteristic of litter decay is that the process begins with a wide variety of materials of very different chemical quality and produces a much more homogeneous material as a product. For example, several different analyses suggest that the end product of litter decay, regardless of initial chemical quality, will have a lignin : cellulose ratio (as measured by the proximate analyses in Figure 12.4) of about 1 : 1. Whether this ratio has real meaning in terms of the fundamental energetics of decomposition or is only a useful empirical result is unknown. The nutrient content of the organic matter produced by this "decay filter" will vary considerably depending on the initial content of the litter and the availability of nutrients in the soil solution.

The most recent studies of organic matter dynamics in terrestrial ecosystems have begun to treat both litter and soil organic matter as part of a single decay continuum. However, traditional methods used to study soil organic matter are very different from those applied to litter, as we will see in the next chapter.

REFERENCES CITED

Aber, J. D., J. M. Melillo, and C. A. McClaugherty. Predicting long-term patterns of mass loss, nitrogen dynamics and soil organic matter formation from initial litter chemistry in forest ecosystems. *Canadian Journal of Botany* (in press).

Cromack, K. 1973. Litter production and litter decomposition in a mixed hardwood watershed and in a white pine watershed at Coweeta Hydrologic Station, North Carolina. Ph.D. dissertation, University of Georgia.

Harmon, M. E., et al. 1986. Ecology of coarse woody debris in temperate ecosystems. *Advances in Ecological Research* 15:133–302.

Hodson, R. E., R. R. Christian, and A. E. Maccubbin. 1984. Lignocellulose and lignin in the salt marsh grass *Spartina alterniflora:* Initial concentration and short-term post-depositional changes in detrital matter. *Marine Biology* 81:1–7.

Larsson, K., and E. Steen. 1988. Changes in mass and chemical composition of grass roots during decomposition. *Grass and Forage Science* 43:173–177.

McClaugherty, C. A., et al. 1985. Forest litter decomposition in relation to soil nitrogen dynamics and litter quality. *Ecology* 66:266–275.

Meentemeyer, V., and B. Berg. 1986. Regional variation in rate of mass loss of scots pine needle litter in Swedish pine forests as influenced by climate and litter quality. *Scandinavian Journal of Forest Research* 1:167–180.

Morrison, I. M. 1980. Changes in the lignin and hemicellulose concentration of ten varieties of temperate grasses with increasing maturity. *Grass and Forage Science* 35:287–293.

Staaf, H., and B. Berg. 1982. Accumulation and release of plant nutrients in decomposing scots pine needle litter. Long-term decomposition in a scots pine forest. II. *Canadian Journal of Botany* 60:1561–1568.

ADDITIONAL REFERENCES

Alexander, M. 1977. *Introduction to Soil Microbiology*. John Wiley and Sons, New York.

Berg, B. 1986. Nutrient release from litter and humus in coniferous forest soils—a mini review. *Scandinavian Journal of Forest Research* 1:359–369.

Berg, B., and C. A. McClaugherty. 1989. Nitrogen and phosphorus release from decomposing litter in relation to the disappearance of lignin. *Canadian Journal of Botany* 67:1148–1156.

Darwin, C. R. 1882. *The Formation of Vegetable Mould through the Action of Worms, with Observations on the Habits.* D. Appleton, New York.

Dighton, J., E. D. Thomas, and P. M. Latter. 1987. Interactions between tree roots, mycorrhizas, a saprotrophic fungus and the decomposition of organic substrates in a microcosm. *Biology and Fertility of Soils* 4:145–150.

Fogel, R., and K. Cromack. 1977. Effects of habitat and substrate quality on Douglas fir litter decomposition in western Oregon. *Canadian Journal of Botany* 55:1632–1640.
Freudenberg, K., and A. C. Neish. 1968. *Constitution and Biosynthesis of Lignin.* Springer-Verlag, Berlin.
Hunt, H. W., et al. 1988. Nitrogen limitation of production and decomposition in prairie, mountain meadow and pine forest. *Ecology* 69:1009–1016.
McClaugherty, C. A., J. D. Aber, and J. M. Melillo. 1984. Decomposition dynamics of fine roots in forested ecosystems. *Oikos* 42:378–386.
Meentemeyer, V. 1978. Macroclimate and lignin control of decomposition. *Ecology* 59:465–472.
Melillo, J. M., J. D. Aber, and J. F. Muratore. 1982. Nitrogen and lignin control of hardwood leaf litter decomposition dynamics. *Ecology* 63:621–626.
Melillo, J. M., et al. 1989. Carbon and nitrogen dynamics along the decay continuum: Plant litter to soil organic matter. In Clarholm, M., and L. Bergstrom (eds.), *Ecology of Arable Land.* Kluwer Academic Publishers, Dordrecht, The Netherlands.
Staff, H. 1987. Foliage litter turnover and earthworm populations in three beech forests of contrasting soil and vegetation types. *Oecologia* 72:58–64.
Swift, M. J., O. W. Heal, and J. M. Anderson. 1979. *Decomposition in Terrestrial Ecosystems.* University of California Press, Berkeley.
Whitford, W. G., et al. 1981. Exceptions to the AET model: Desert and clear cut forests. *Ecology* 62:275–277.

Chapter 13 Origin and Decomposition of Soil Organic Matter

H. Armstrong Roberts

INTRODUCTION

Our discussion of litter decomposition in the previous chapter was limited to the first few years after senescence. Most of the examples were from forests and for above-ground material, where both the quantity of litter production and its decomposition are most easily measured. Relatively less is known about the rates of decay of roots. There is also a gap in our understanding of the processes by which older litter materials change from the recognizable, distinct original tissue to the dark, homogeneous-looking organic material called humus. The relative contribution of different types of litter to this humus pool are unknown. Even the structure and biochemistry of humus can only be guessed at. Yet humus is the source of much of the mineralized nutrients available for plant growth in most natural systems. It also represents the largest pool of stored carbon in many systems. Humus is a fundamental component of all ecosystems, yet its complexity and variation have limited the success of attempts to decipher its structure, function, and chemistry.

The purpose of this chapter is to discuss briefly the production, stabilization, and decomposition of soil organic matter, as well as its role in nutrient dynamics and nutrient availability to plants. Again we will see that important processes in soils are often less well known than important processes in plants.

Table 13.1 Proximate analyses of typical plant materials and humus*†

COMPONENT	IN PLANT TISSUE (%)	IN SOIL ORGANIC MATTER (%)
Cellulose	20–50	2–10
Hemicellulose	10–30	0– 2
Lignin	10–30	35–50
"Protein"	1–15	28–35
Fats, waxes, etc.	1– 8	1– 8

*The "protein" fraction includes a large fraction of undetermined nitrogen-containing compounds.
†From Foth and Turk 1972.

WHAT IS HUMUS?

There is general agreement that **humus** is a series of high-molecular-weight polymers with a high content of phenolic rings and quite variable side chains. In comparison with plant materials, it is very high in nitrogen and large polyphenolic molecules that are analyzed as "lignin" by the series of proximate analyses listed in Figure 12.4 and low in cellulose and hemicellulose (Table 13.1). Only part of that nitrogen content can be identified with a certain type of compound. As much as 40% of the nitrogen in soil organic matter is neither in protein nor amino acid form. It is unclear how this nitrogen is bound into humus, but a significant fraction may be present as chitin (Figure 13.1), a product of insect and fungal metabolism.

As with litter, the classes of compounds contained in soil organic matter are determined by proximate analysis, specifically by their solubility in different acidic and alkaline solutions. The most common methods of separation are outlined in Figure 13.2. First, the soil is placed in alkali. Organic matter not removed into solution by this treatment is called **humin** and may be the fraction most tightly bound to particles of mineral material in the soil (see the later discussion of stabilization). Most charac-

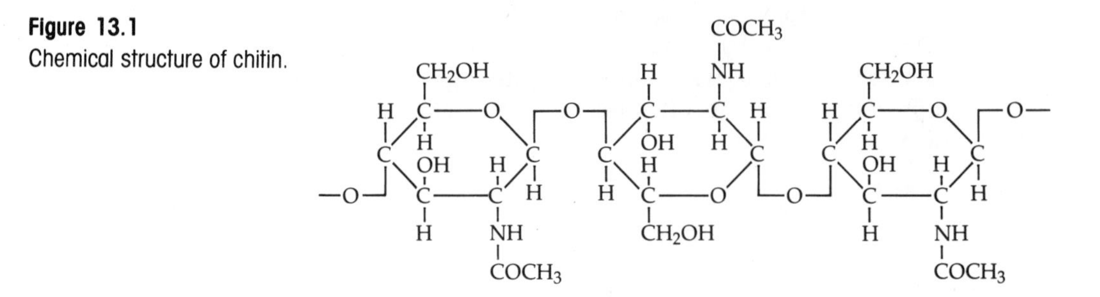

Figure 13.1
Chemical structure of chitin.

SOIL
Not soluble in alkali
Soluble in alkali
HUMIN
Precipitated by acid
Not precipitated by acid
HUMIC ACIDS
FULVIC ACIDS
Coagulated in presence of electrolyte
Not coagulated
GRAY HUMIC ACIDS
BROWN HUMIC ACIDS

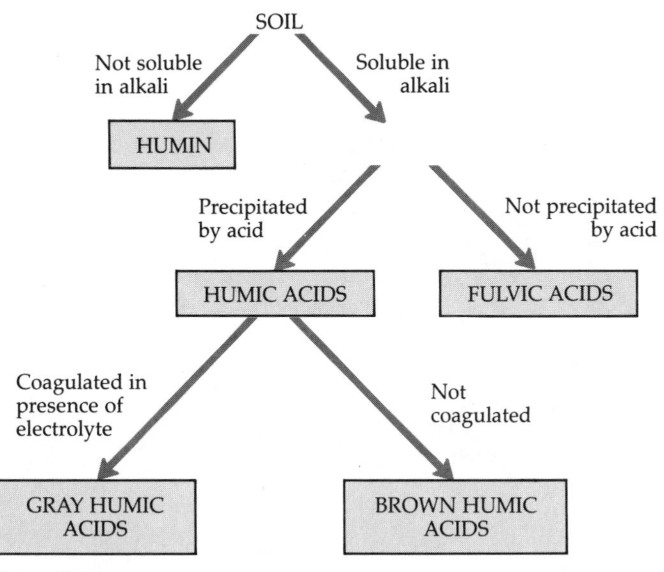

Figure 13.2
Chemical methods for separation of the different fractions of soil organic matter. (After Stevenson 1985)

terizations of soil organic matter have been done on the compounds that are soluble in alkali. When the extraction is acidified, humic acids are those that precipitate out and can be removed by filtration; the remaining compounds are termed **fulvic acids.** These fractions can be further subdivided by separations in alcohols and electrolytic solutions.

These different fractions vary considerably in the weight of the molecules extracted and in their carbon and oxygen content (Figure 13.3), and

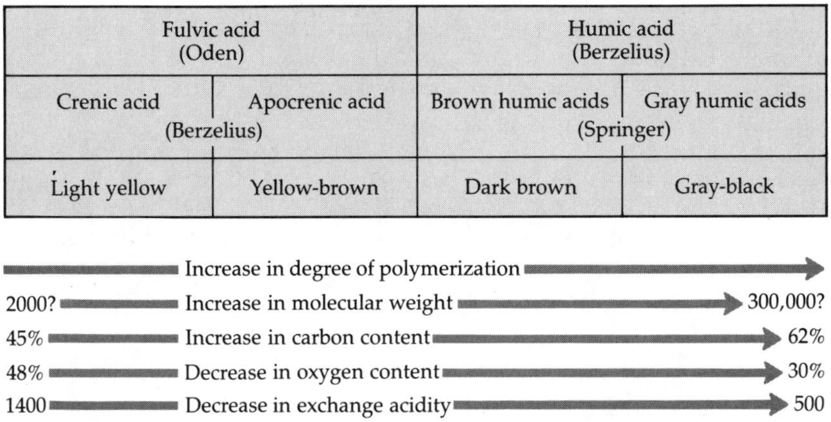

Fulvic acid (Oden)
Humic acid (Berzelius)
Crenic acid
Apocrenic acid
(Berzelius)
Brown humic acids
Gray humic acids
(Springer)
Light yellow
Yellow-brown
Dark brown
Gray-black
Increase in degree of polymerization
2000? Increase in molecular weight 300,000?
45% Increase in carbon content 62%
48% Decrease in oxygen content 30%
1400 Decrease in exchange acidity 500

Figure 13.3
Chemical characteristics of different fractions of soil organic matter. (Stevenson 1985)

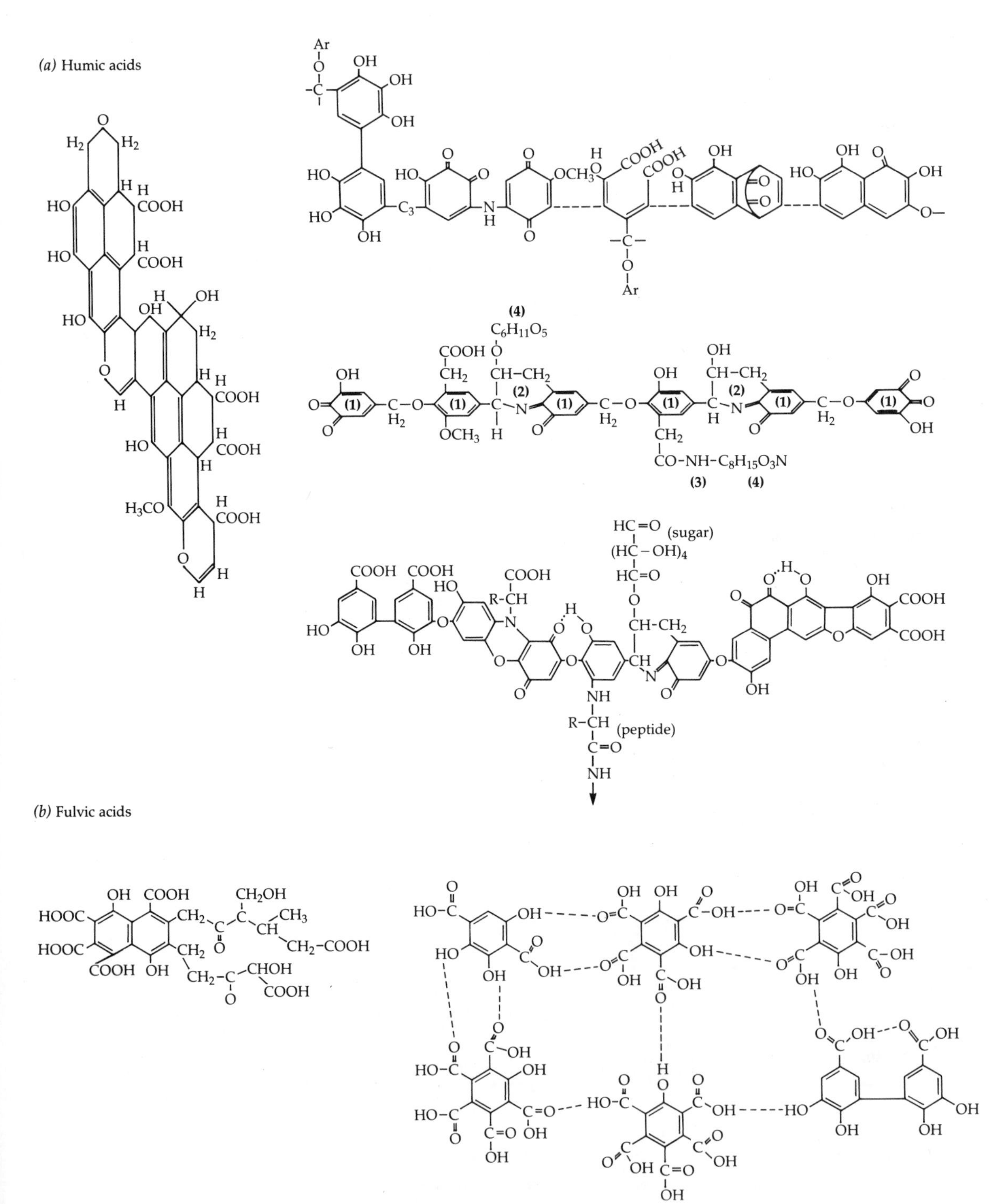

Figure 13.4

Proposed chemical structures of humic and fulvic acids. (Cited in Stevenson 1985)

there are no distinct boundaries between the classes. Rather, we can picture the major component of soil organic matter as a sequence of compounds varying in the degree of polymerization (the size of molecules) and in elemental content.

Using several traditional and modern analytic techniques, different researchers have developed different model compounds, which may represent the basic structure of humic and fulvic acids (Figure 13.4). These all show highly phenolic structures but vary considerably in the type and number of side chains and in whether the nitrogen is present in the ring structures themselves or as components of side chains. Some researchers do not include nitrogen in the structure at all, assuming that the nitrogen is present as proteins condensed with, or associated with, the humic and fulvic acids.

FORMATION OF HUMUS

There is still considerable uncertainty regarding the types of compounds from which humus is formed and the reactions involved. The range of working hypotheses can be summarized as in Figure 13.5.

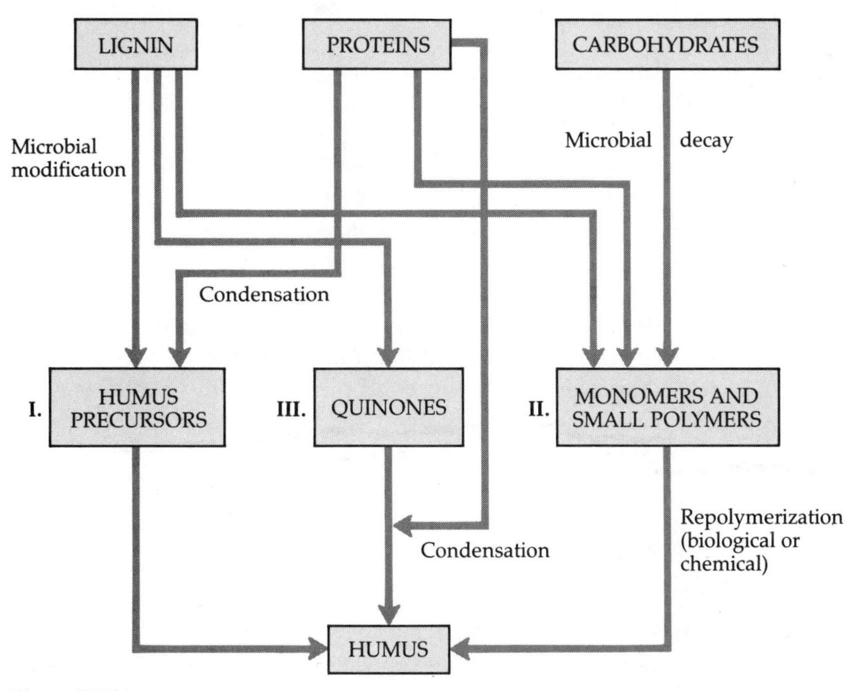

Figure 13.5

Summary diagram of hypothesized pathways for humus formation. (After Schnitzer 1978; Melillo et al. 1989)

The first, and perhaps the oldest, hypothesis is that humus is formed by the modification of existing plant residues. In this view, lignin is a crucial precursor to the formation of humus, and humus is formed by the continual modification of the initial lignin molecules by microbes. This modification includes condensation into larger and larger molecules, with nitrogen added to the molecule by condensation of lignin with proteins.

The second hypothesis is that microbes break all large litter molecules into smaller molecules, which then repolymerize to form the high-molecular-weight humus. There are three alternatives for the repolymerization. One is that microbes produce low-molecular-weight polyphenols and amino acids, which are exuded into the soil and oxidized and polymerized. The second is that the precursors of humus are produced in microbial cells and released upon the death and breakup of those cells. The third holds that the high-molecular-weight substances are actually synthesized in the microbial cell and released by cell death.

The third hypothesis is intermediate between the first two. It holds that phenolic substances originating either directly from plant litter or from microbial synthesis are converted first to quinones (Figure 13.5), which then polymerize either with nitrogen-containing compounds or other carbon compounds.

There is general agreement that all three of the proposed mechanisms may, in fact, be operating simultaneously. However, two lines of evidence suggest that the second pathway may not be the dominant one in most systems.

First, the second pathway predicts that humic acids are the first substances formed and that fulvic acids arise from the decomposition of humic acids. The first and third mechanisms would predict that fulvic acids are formed first, with humic acids resulting from continued condensation of the lower-molecular-weight fulvic acids. Radiocarbon dating indicates that the carbon in humic acids is generally older than in fulvic acids, suggesting that humic acids are formed from fulvics, rather than the other way around.

Second, a ten-year experiment involving the addition of different kinds of plant litter to a sandy soil showed that the accumulation of soil organic matter increased with higher lignin concentration in the added material (Figure 13.6). This indicates that lignin is the major precursor to soil organic matter, supporting hypotheses 1 and 3.

DECOMPOSITION AND STABILIZATION OF HUMUS

Humus decomposes very slowly under field conditions. Yet it is generally present in such large amounts that it represents a significant portion of the carbon and nutrient release from soils. There are at least two reasons why humus decays slowly.

First, the carbon compounds present are of low biochemical quality. We

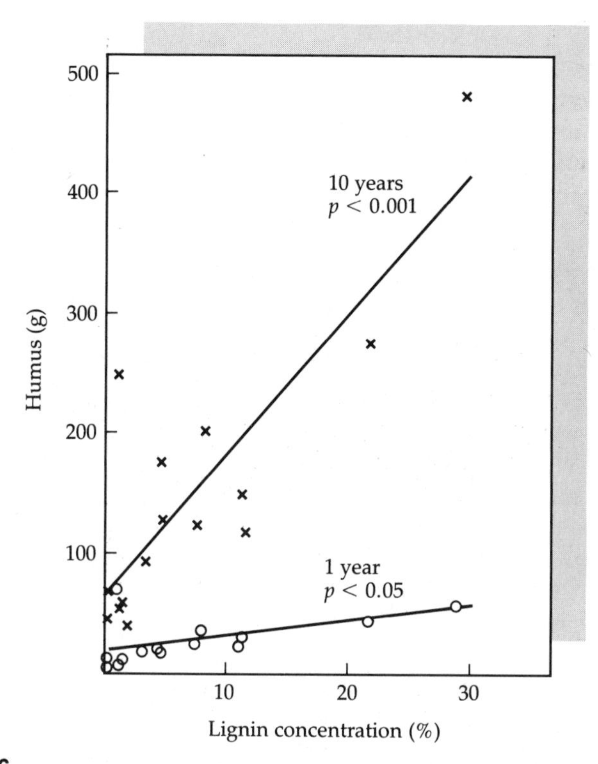

Figure 13.6
Effects of 1-year and 10-year additions of plant materials with different concentrations of lignin on the accumulation of soil organic matter in a sandy soil. (de Haan 1977)

saw in the last chapter that litter with large amounts of lignin decayed at slower rates. As decomposition continues, the concentration of lignin and lignin-like secondary products increases, reducing the energy yield from the material. While this is a continuous process, there may be a ratio of cellulose to lignin in organic matter at which decay becomes significantly reduced.

Second, organic matter tends to form colloids with mineral particles in soils. Figure 13.7 is a schematic of the ways in which humus molecules and clay particles can be bound together by metal cations, water, sugars, and other substances. This sort of interaction tends to disrupt the alignment of degrading enzymes with the organic humus molecules and reduces the effectiveness of those enzymes. In fact, enzymes produced by microbes for the purpose of decomposing organic matter can themselves become deactivated, or fixed, by humus molecules and clay particles.

Organic matter combined to different degrees with mineral particles can be separated by a process called **density fractionation**. This process relies on the very different bulk densities (or weight per unit volume) of organic and mineral materials. Water has a bulk density of 1 g/cm^3.

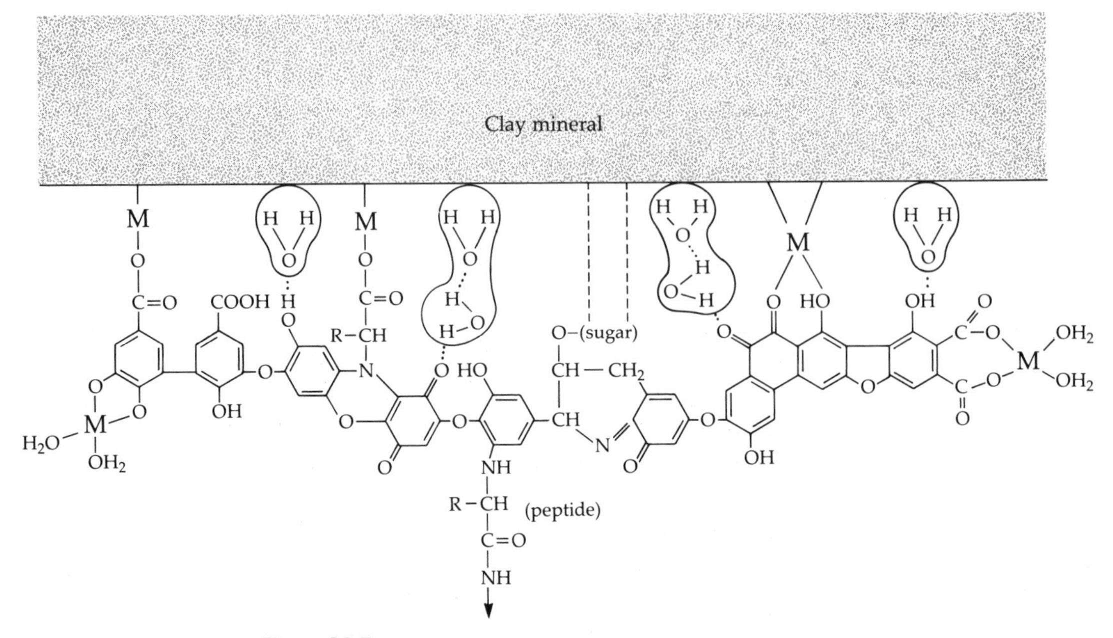

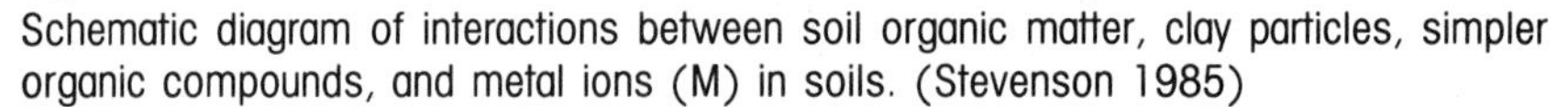

Figure 13.7
Schematic diagram of interactions between soil organic matter, clay particles, simpler organic compounds, and metal ions (M) in soils. (Stevenson 1985)

Organic matter in soils has a much lower bulk density because of its loose, porous structure. Mineral particles in soils have bulk densities of around 2.2. A soil colloid consisting of both organic and mineral materials combined will have a bulk density somewhere between the two, depending on the relative amount of each in the colloid. "Heavy" soil fractions would have higher bulk densities and more mineral matter. "Light" soil fractions would have low bulk densities and more organic matter. The two soil fractions can be separated by suspending a soil sample in a solution of known bulk density; 1.65 is often used. The light soil fraction will float in this solution; the heavy fraction will sink.

Separation of light and heavy soil organic matter has been carried out in several Pacific Northwest forests dominated by either conifers or alder. The light and heavy fractions from these soils had very different carbon-to-nitrogen ratios and nitrogen release rates (Table 13.2). In general, the light fraction consisted of still-recognizable plant litter with small amounts of mineral material encrusted on it, while the heavy fraction was a combination of mineral particles and more decomposed organic fragments. The light fraction had a lower nitrogen concentration (higher C:N ratio) and much lower rates of nitrogen mineralization as measured by a laboratory incubation method. The light fraction averaged about 25% of the total soil organic matter through the soil profile but was a higher percentage of the total near the soil surface and also changed in total mass by as much as

Table 13.2 Carbon-to-nitrogen ratios and a relative index of nitrogen mineralization in light and heavy fractions of forest soils dominated by conifers or alder in the Pacific Northwest*†

	ALDER		CONIFER	
	Light Fraction	Heavy Fraction	Light Fraction	Heavy Fraction
C:N ratio	24.8	16.7	47.5	23.1
N mineralized (%)	0.71	2.34	0.39	2.91

*Values are means for forest type.
†From Sollins et al. 1984.

50% from one season to the next, while the total mass of heavy-fraction material remained relatively constant.

The same separation methods were also used on a series of soils of different ages. The sampled soils had developed on mudflows caused by volcanic activity on the flanks of Mount Shasta in California. In these soils, the total amount of heavy-fraction carbon increased continually with age, while the total amount of light-fraction carbon tended to level off after about 600 years of soil development.

These characteristics all suggest that light-fraction carbon is younger and less completely decomposed and that organic matter becomes increasingly associated with mineral particles as decomposition proceeds. They may also suggest that once organic matter is heavily shielded by mineral particles, it is more difficult to decompose and will remain in the ecosystem for a longer time.

The amount of organic matter that can be stabilized in this way increases as soils become finer-textured, as indicated by increasing nitrogen and phosphorus contents of soils (Figure 13.8). Smaller soil particles, like clays, bind more effectively with organic matter and provide more particle surface area for organic–mineral interactions to occur. This form of stabilization is thought to be particularly important in tropical systems in which decomposition of unshielded organic matter is generally very rapid. Physical shielding may be a major reason why tropical soils contain at least as much organic matter as temperate-zone soils with similar vegetation.

DECOMPOSITION AND NUTRIENT RELEASE FROM HUMUS

Because so little is known about the nature of humus, it is very difficult to predict what its rate of decomposition is going to be in any given system without measuring it directly. This is in sharp contrast to the prediction of

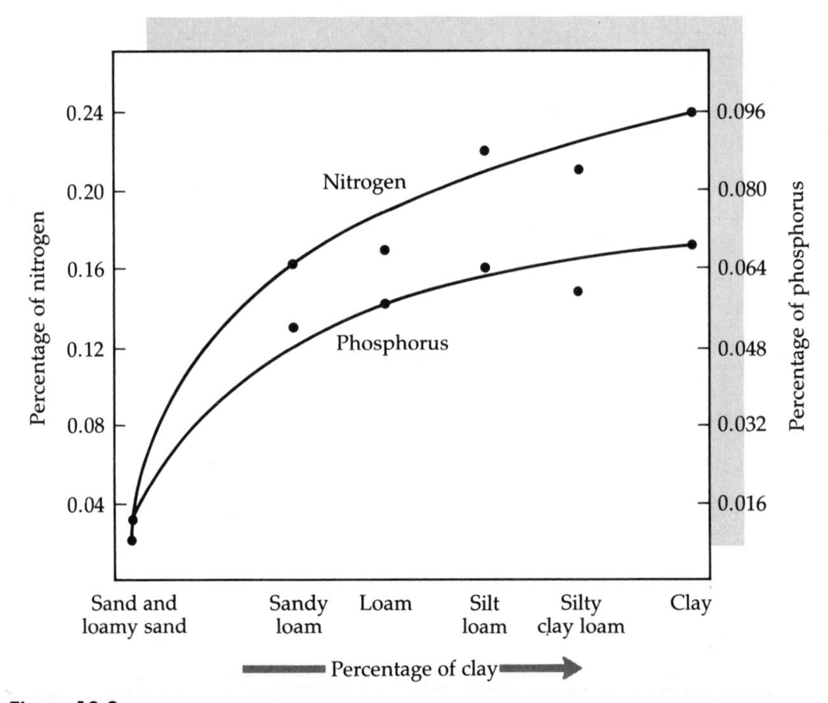

Figure 13.8
Differences in nitrogen and phosphorus concentrations in soils as a function of texture for several soils in New York state. (Foth and Turk 1972)

decay rates for litter (Chapter 12), which can be done with a good deal of accuracy given the initial composition of the material.

Even direct measurements of CO_2 flux from soils are difficult to relate to rates of humus decay. If CO_2 loss is measured from the top of the soil, then loss from relatively fresh decomposing litter is also included. If the fresher material can be separated, as in the case of an unmixed forest floor (mor type), then CO_2 loss from the remaining soil column still includes root respiration (see discussion on CO_2 balances in Chapter 3).

One set of measurements has been made for several taiga-zone forests in Alaska using simple nylon mesh bag enclosures more commonly employed in studying decomposition of fresh litter. In stands dominated by four different species (birch, aspen, white spruce, and black spruce), forest floor material was placed in the mesh bags and the loss of organic matter content was measured as weight loss through time. There was a distinct difference in decay rates between stands, with the birch and aspen stands having turnover rates of about 3% per year, and the spruce stand forest floors decaying at about 1% per year (Table 13.3A).

It may be easier, in some soils, to measure organic matter turnover rates by measuring nitrogen rather than carbon mineralization. Field soil incubation techniques have been devised that measure the accumulation (net

Table 13.3 Measured turnover rates of carbon in several forest types in the taiga zone of Alaska* and nitrogen in several forests in Wisconsin†

	TURNOVER RATE (% PER YEAR)
A. Taiga forests	
Dominant species	
Aspen	3.1
Birch	3.0
White spruce	1.0
Black spruce	1.3
B. Wisconsin forests	
Dominant species	
Red pine	1.8
White pine	4.9
Sugar maple	3.5
Red oak	7.9
Black oak	2.8
White oak	4.8

*From Flanagan and Van Cleve 1983.
†From Nadelhoffer et al. 1983.

mineralization) of ammonium and nitrate in isolated soil cores. Assuming that the ratio of carbon to nitrogen in the soil organic matter does not change during the incubation period, the percentage of soil nitrogen mineralized during a year is equivalent to the percentage of soil carbon mineralized.

Through the use of this technique, a wide range of organic matter turnover rates have been measured in a series of temperate forest ecosystems with similar soil structure and climatic regime but different dominant tree species and past history of disturbance (Table 13.3B). In well-mixed soils, including some of those in Table 13.3B, this method may measure N turnover for all soil organic matter, as fresh litter is rapidly mixed with older litter and the two cannot be separated. This, along with climatic differences, may partially explain the differences in turnover rates between forest soils in Alaska and Wisconsin.

Another approach to measuring the rate of humus decay is to measure how rapidly soil organic matter content declines following the initiation of plowing and farming. Agriculture affects soil organic matter content by reducing the annual input of litter as compared with natural systems and also because the physical effects of plowing increase decay rates. These effects include alteration of soil structure, increased aeration, and mixing of organic and mineral horizons.

Losses of soil organic matter have often been measured as associated losses in nitrogen content. There is a distinct exponential decrease in soil nitrogen beginning with the initiation of cultivation. Figure 13.9 shows changes in nitrogen content of four different prairie soils over the first few decades of plowing and planting. These changes represent losses of just over 1% per year. The carbon lost from soils in association with this nitrogen represents a potentially important source of CO_2 to the atmosphere (Chapter 24).

A very different view of the turnover rate of humus can be obtained from radiocarbon dating of extracted soil organic matter. Table 13.4 gives data on the mean age of carbon in the different fractions of organic matter commonly separated from soils. A soil with a turnover rate of 4%, typical of the values obtained with the CO_2 and nitrogen balance methods presented above, would have a mean residence time of 25 years. Yet even the youngest soil fraction in Table 13.4, the fulvic acids, ranges from 495 to 630 years in average age. Humic acids were much older.

What this means is that a certain portion of soil organic matter is very inert, decaying at a rate of less than 0.2% to 0.1% per year. Since this is a mean age for this class of compounds, some of the carbon atoms have

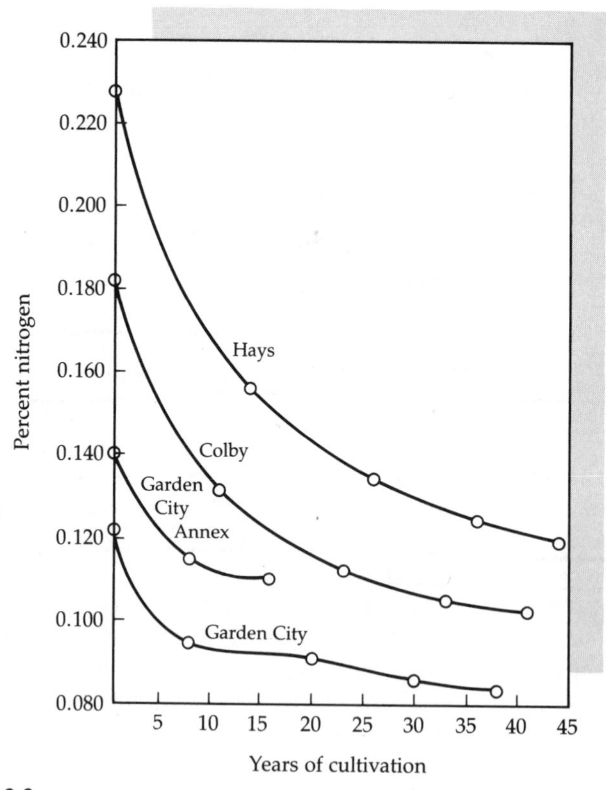

Figure 13.9
Changes in nitrogen storage in several prairie soil types following the initiation of plowing and cropping. (Cited in Allison 1973)

Table 13.4 Radiocarbon ages of soil organic matter fractions in two northern prairie soils

SOIL ORGANIC MATTER FRACTION	MEAN AGE (YR)
Campbell et al. 1967	
Calcium-humates	1400
"Mobile" humic acids	780
Fulvic acids	550
Stevenson 1985	
Humin	1240
Humic acids	1308
Fulvic acids	630

undoubtedly been in place for much longer than 1000 years. The radiocarbon ages also mean that the CO_2 and nitrogen methods described above are undoubtedly measuring relatively fresh material as well as mineralization from humus.

One way of visualizing this wide range in the dynamics of different types of soil organic matter is as a series of different fractions with different turnover rates (Figure 13.10; this is actually the outline of a computer model of soil organic matter dynamics; see modeling discussion in Chapter 21). The distribution of litter into the faster-turnover compartments is determined by the initial chemical quality (lignin content) of the plant material. Transfers to the very long-term pools are determined by soil texture, which affects the degree to which organic matter can be stabilized by association with mineral soil particles. The response of a soil to disturbance, such as plowing, will depend to a great degree on the distribution of organic matter between these different pools.

To summarize, soils contain a very wide range of organic materials—combined to different degrees with mineral material—which turn over at very different rates. The net turnover of CO_2 and nitrogen for the whole soil body represents the combined and averaged effects of all of these different fractions and their different decay rates. Much of our inability to predict the turnover of soil organic matter without measuring it directly results from an inability to accurately quantify the amount of soil organic matter in these different fractions and also perhaps to understand how disturbing soil structure alters these rates.

Humus represents a large store of essential nutrients. As with other parts of the humus story, it is not exactly clear in what form these nutrients occur. Nitrogen content was discussed earlier. Sulfur is also bound in humus with the amino acid or protein component and also as ester sulfate. Metal cations can be complexed by the humus molecule or can act as bridging ions between humus and clay. Whatever the form, as

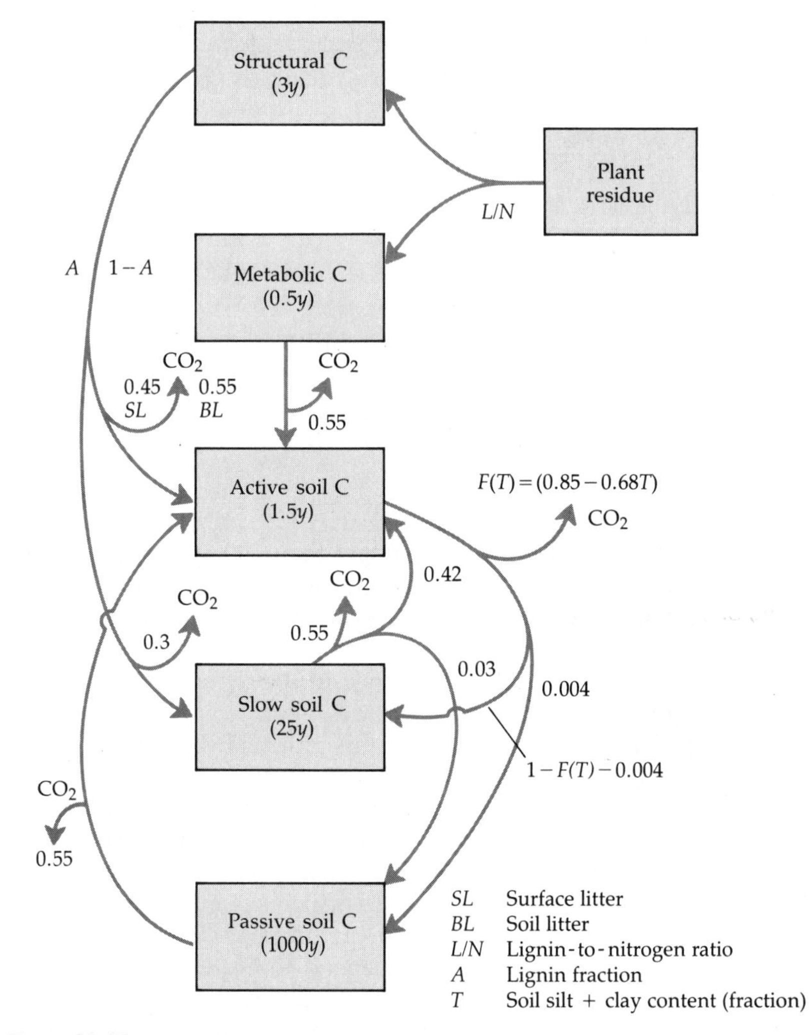

Figure 13.10

Compartmentalization of soil organic matter in a computer model (CENTURY) of soil organic matter dynamics in undisturbed and plowed prairie ecosystems. Values in parentheses give the residence time for each compartment (in years). Initial distribution between structural (cell wall) and metabolic (cell contents) compartments is determined by the lignin-to-nitrogen ratio of the material. Further divisions between CO_2 respired and transfer to longer-term organic compartments depend on soil texture, origin of material, and lignin content. This model has successfully predicted patterns of organic matter and nutrient loss from plowed prairie soils similar to those shown in Figure 13.10. (Parton et al. 1988)

the organic matter is decayed, these nutrients are mineralized and made available to plants and microbes.

In an ecosystem context, a large pool of nutrients that becomes available very slowly has the effect of moderating annual fluctuations in nutrient availability. Field measurements in several forest ecosystems have shown that nitrogen mineralization is quite constant year to year, despite significant changes in precipitation and temperature. Humus can be thought of as a slow-release fertilizer that acts to preserve site quality or maintain a given level of nutrition. Thus a forest or grassland severely disturbed by fire or insects will be "buffered" against drastic nutrient losses. Plant growth can begin again immediately after disturbance because a large nutrient reserve is already in place. In fact, as we will see in Part III, such disturbances often increase the rate of nutrient release form this long-term storage pool, which may increase the rate of ecosystem recovery.

REFERENCES CITED

Allison, F. E. 1973. *Soil Organic Matter and its Role in Crop Production.* Elsevier Scientific Publishing Company, Amsterdam.

de Haan, S. 1977. Humus, its formation, its relation with the mineral part of the soil and its significance for soil productivity. In *Organic Matter Studies*, vol. 1. International Atomic Energy Agency, Vienna.

Flanagan, P. W., and K. Van Cleve. 1983. Nutrient cycling in relation to decomposition and organic matter quality in taiga ecosystems. *Canadian Journal of Forest Research* 13:795–817.

Foth, H. D., and L. M. Turk. 1972. *Fundamentals of Soil Science.* John Wiley and Sons, New York.

Melillo, J. M., et al. 1989. Carbon and nitrogen dynamics along the decay continuum: Plant litter to soil organic matter. In Clarholm, M., and L. Bergstrom (eds.), *Ecology of Arable Land.* Kluwer Academic Publishers, Dordrecht, The Netherlands.

Nadelhoffer, K. J., J. D. Aber, and J. M. Melillo. 1983. Leaf litter production and soil organic matter dynamics along a nitrogen-availability gradient in Southern Wisconsin (U.S.A.). *Canadian Journal of Forest Research* 13:12–21.

Parton, W. J., J. W. B. Stewart, and C. V. Cole. 1988. Dynamics of C, N, P and S in grassland soils: A model. *Biogeochemistry* 5:109–132.

Schnitzer, M. 1978. Humic substances: Chemistry and reactions. In Schnitzer, M., and S. U. Khan (eds.), *Soil Organic Matter.* Elsevier Scientific Publishing Company, Amsterdam.

Sollins, P., G. Spycher, and C. A. Glassman. 1984. Net nitrogen mineralization from light- and heavy-fraction forest soil organic matter. *Soil Biology and Biochemistry* 16:31–37.

Stevenson, F. J. 1985. Geochemistry of soil humic substances. In McKnight, D. M. (ed.), *Humic Substances in Soil, Sediment and Water: Geochemistry, Isolation and Characterization.* John Wiley and Sons, New York.

ADDITIONAL REFERENCES

Alexander, M. *Introduction to Soil Microbiology*, 2nd ed. 1977. John Wiley and Sons, New York.

Campbell, C. A., et al. 1967. Factors affecting the accuracy of the carbon-dating method in soil humus studies. *Soil Science* 104:81–85.

Chapter 14 Plant–Soil Interactions: Summary Effects on Nutrient Cycles

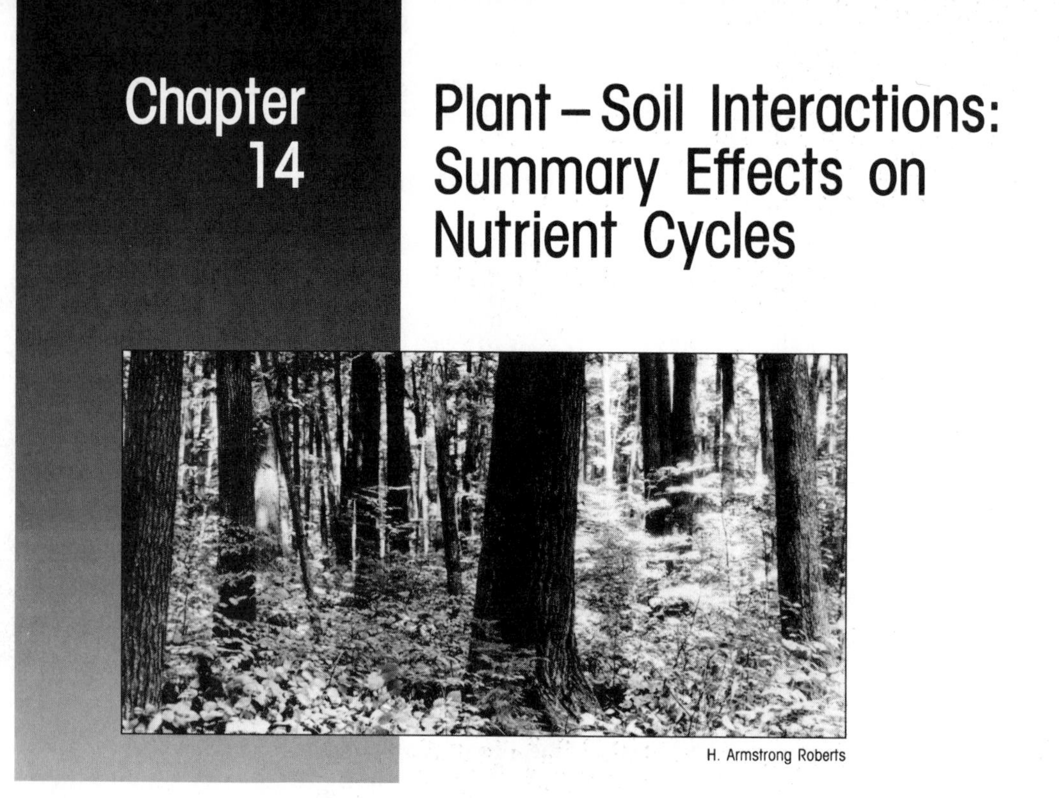

H. Armstrong Roberts

INTRODUCTION

The last eight chapters have dealt with interactions between physical climate, the chemistry of soils, and the biological processes of production and decomposition. Much of terrestrial ecosystem research has dealt with the combined effects of these interactions on the movement of nutrients through ecosystems.

The purpose of this chapter is to synthesize much of the information already presented into a comparative discussion of nutrient cycles in terrestrial ecosystems. Three approaches will be taken. First, we will return to the generalized diagram of nutrient cycling developed in Chapter 1 and present side-by-side comparisons of the relative importance of each major process for each macronutrient. We will then use the movement of water through ecosystems as a means of summarizing where in ecosystems the different processes occur and how those processes might vary under different conditions. We will conclude with some examples of the extent to which different species of plants can modify the distribution and cycling of nutrients within ecosystems.

COMPARISONS OF GENERALIZED NUTRIENT CYCLES

The cycles of the macronutrients are very different. We have seen that carbon resides in the system only as organic matter and that the atmosphere is the source of CO_2 for plant uptake, through the leaves, during photosynthesis. At the other extreme, potassium is taken up by the root and is never combined in any organic form. It moves through the plant in ionic form and is easily leached from plant surfaces. While these cycles are very different, they can be presented as part of a larger, generalized nutrient cycle similar to the one developed in Chapter 1 (Figure 14.1). The relative importance of each process and compartment for each nutrient is presented graphically in Figure 14.2 by the relative size of lines surrounding boxes or marking transfers between components.

Carbon (Figure 14.2*a*) Carbon available for plant uptake is in the atmosphere as CO_2. It is fixed by photosynthesis and is then respired or becomes part of the plant biomass. In plants it is present in the wide variety of compounds described in Chapter 12. Some of these (simple sugars) move readily through the plant and are respired. Some are converted to structural compounds which can no longer be translocated (lignin, cellulose). The immobile compounds eventually become litter either through seasonal senescence, as with leaves, or through the death of whole plants. The litter is decomposed by microbes (Chapters 12 and 13) and is respired to CO_2 or transformed to more stable organic complexes and eventually humus.

Only small amounts of organic carbon enter into the soil solution and are transported to lower soil horizons or groundwater in most systems. Large amounts of carbon may be present in systems underlain by carbon-

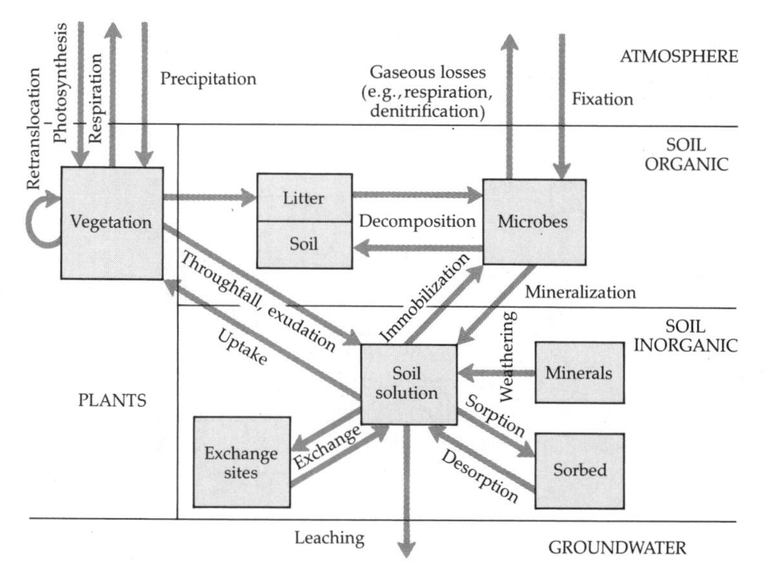

Figure 14.1
A generalized nutrient cycle.

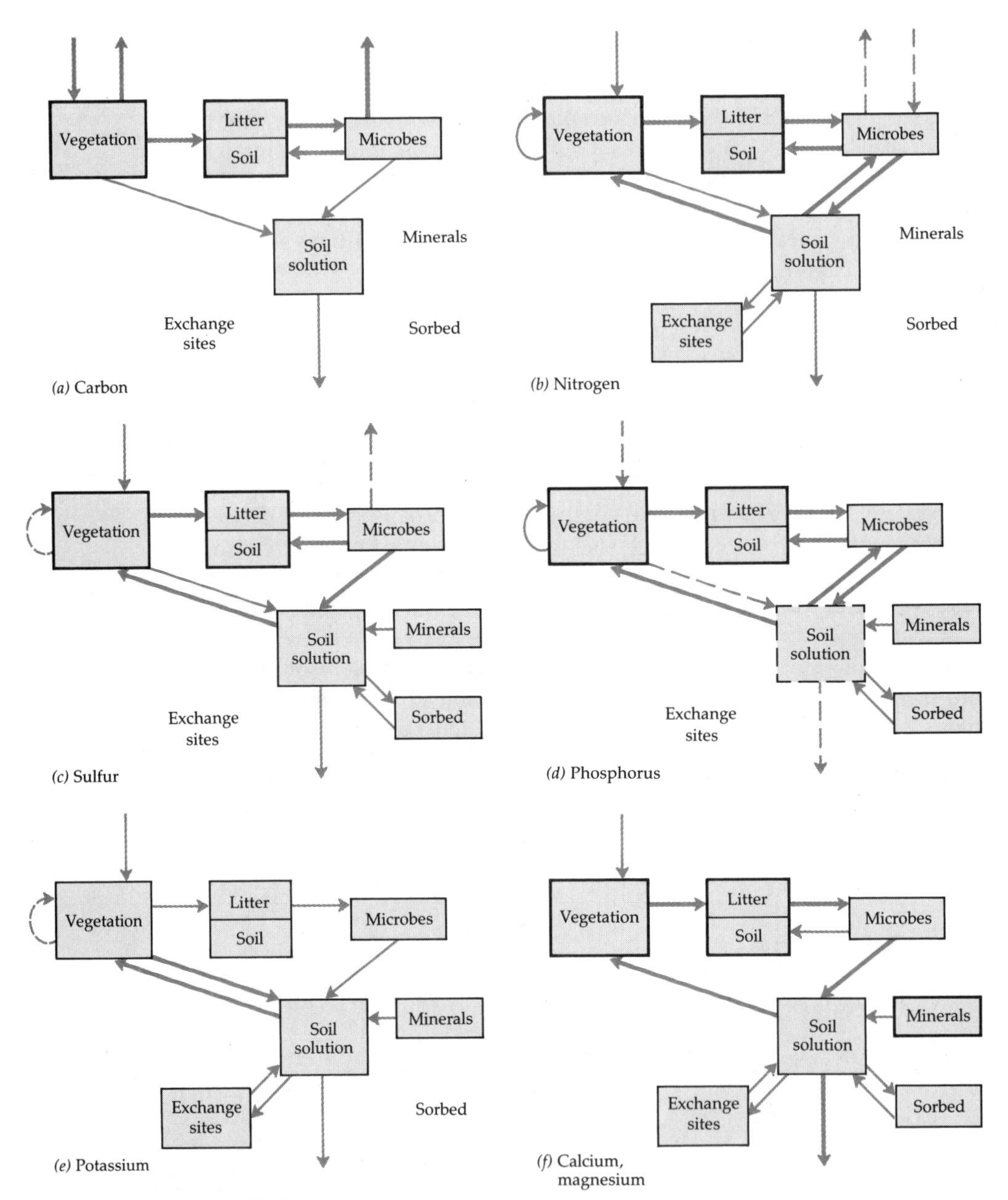

Figure 14.2
Comparative cycles for carbon and major macronutrients. The widths of the lines around compartments indicate the relative amounts stored, and the widths of arrows indicate the relative amounts transferred by those pathways.

ate rocks, such as limestone. Weathering release of carbonates has a significant effect on soil chemistry and pH, but this carbon moves through the ecosystem in a very different way from carbon fixed in organic matter. (This pathway is omitted in Figure 14.2*a*.)

The organic carbon cycle is dominated by photosynthesis and respiration. It is also an "open cycle" because large inputs and outputs occur each year relative to the amount transferred internally. Litterfall and decomposition are the major internal transfers, and internal storage is in plant and soil organic matter.

Nitrogen (Figure 14.2*b*) The nitrogen cycle is similar to that for carbon in some respects but also differs in significant ways. While the atmosphere is 79% N_2, the substantial energy investment required to convert this to a usable form limits the amount of nitrogen fixed into organic matter. While we have seen that several types of plants have symbiotic relationships with microorganisms that allow nitrogen fixation to occur, in most natural ecosystems enough N is available from internal recycling to cause N fixation to be a disadvantageous way of gaining N. Thus, N fixation is relatively low in most, but not all, natural ecosystems. Both N fixation and the complementary process of denitrification (the loss of N_2 or N_2O gas to the atmosphere) tend to occur at low and roughly equal rates in most terrestrial ecosystems.

Nitrogen is also present as ammonium (NH_4^+) and nitrate (NO_3^-) in the atmosphere and is deposited in precipitation. These inputs are usually greater than inputs from fixation and are increasing in many areas due to human activities (Chapter 23). Still, both inputs combined are generally much less than the amount of N available internally through the mineralization of nitrogen in soil organic matter. The nitrogen cycle is relatively "closed" in most systems.

The internal cycle is dominated by uptake of NH_4^+ and NO_3^- from the soil solution and exchange sites, incorporation into plant tissues, deposition in litter, and decomposition and eventual release to the soil solution. This cycle differs from that for carbon in several ways.

First, significant amounts of nitrogen can be withdrawn from the leaves of plants as they senesce (retranslocation). Second, decomposition can result in either the release of N from organic matter (mineralization) or the incorporation of available N into the litter (immobilization). Thus, microbes, using the litter as an energy source, can actually compete with plants for available N. Third, there are two forms of N that plants can take up, ammonium (NH_4^+) and nitrate (NO_3^-). Ammonium is the first inorganic product of mineralization. Under certain conditions, this is oxidized by soil microbes to nitrate (this process is called nitrification). Fourth, nitrogen is not a significant component of primary or secondary minerals. The nitrogen cycle is similar to the carbon cycle in that it is dominated by internal organic matter transfers, with major storage occurring in the plants and soil organic matter.

The occurrence of nitrification has been of special interest in ecosystem studies for some time. In Chapter 4 we saw that different forest ecosystems responded very differently, in terms of nitrate leaching losses, to clearcutting. Systems in which nitrification did not occur could not lose nitrate to streams after disturbance. Understanding the factors that control or suppress nitrification became a high priority.

An early hypothesis was that plants inhibited nitrification by exuding chemicals that interfered with the process into the soil. This is called **allelopathy,** the production and exudation of compounds harmful to other species or their function. This explanation has been refuted in most, but not all, cases.

Current theory holds that nitrification is controlled by a combination of soil pH and ammonium availability. A broad-scale comparison of nitrification rates in natural ecosystems shows that nitrification tends to be high in soils above pH 5 and very low or nonexistent in those below pH 3.5 (see the example for temperate-zone forests in Figure 14.3). However, nitrification can occur at very high rates in acid soils (even pH 4 and lower) when ammonium is produced by mineralization in excess of plant demand (see Chapter 23). It can also be low in high-pH soils if the ammo-

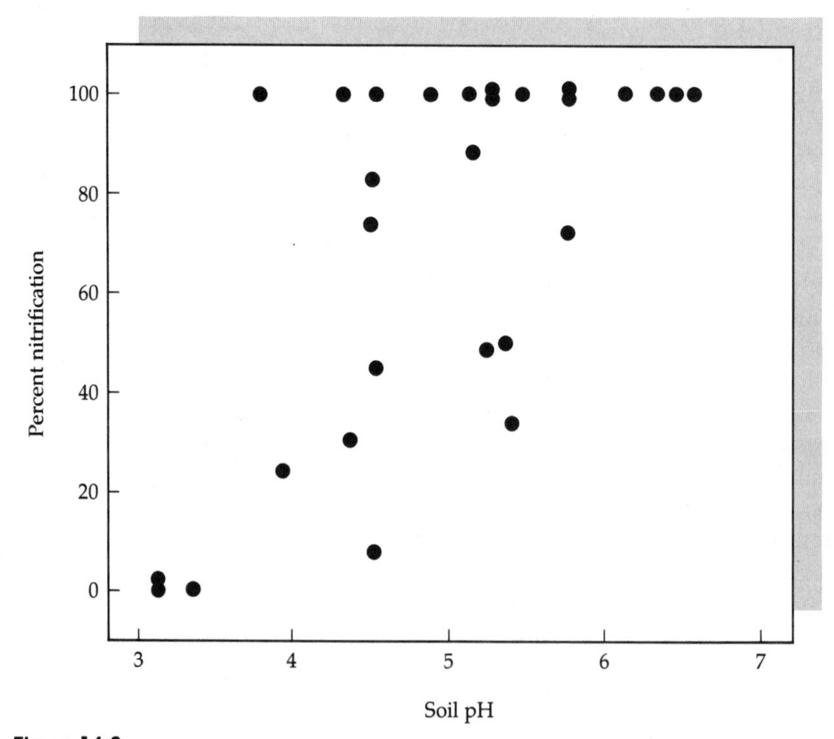

Figure 14.3

Nitrification as a fraction of total annual N mineralization in relation to soil pH, for several temperate forest soils. (Data from Pastor et al. 1984 and Nadelhoffer et al. 1983)

nium supply is low and competition for ammonium between nitrifying microbes and mycorrhizal fungi is intense.

Sulfur (Figure 14.2*c*) The sulfur cycle is similar to that for nitrogen in many ways. Most sulfur cycling occurs through plant uptake of sulfate, deposition in litterfall, and mineralization from soil organic matter. Some important differences are that retranslocation and immobilization are less important because S is generally less limiting in most terrestrial ecosystems. Sulfur is also present in certain primary minerals, can be supplied in part by weathering, and can also be sorbed strongly in soils with high iron and aluminum content. Gaseous exchanges with the atmosphere do occur, but they are generally small for non-wetland systems compared with precipitation inputs. Sulfuric acid is a major component of the acidity in "acid rain," so sulfate input rates are higher in and near heavily industrialized regions than in remote areas. As inputs increase, leaching losses may also rise, unless soil sulfate sorption capacity is high. The leaching of sulfate from soils is a major contributor to soil acidification in regions heavily affected by acid rain (Chapter 23).

Phosphorus (Figure 14.2*d*) The internal organic cycle of phosphorus is much like the nitrogen cycle. The vegetation and soil organic pools are major reservoirs of P, with uptake, litterfall, retranslocation, and decomposition as major transfers. Net immobilization into litter does occur in some systems, causing plants and microbes to compete for this element.

There are two important differences between the cycles of N and P. The first is that atmospheric inputs of P in precipitation are much smaller and gaseous exchanges are negligible. Instead, new inputs of phosphorus into terrestrial ecosystems come mainly from weathering of minerals. This will be small relative to the annual internal cycle and will vary considerably depending on the type of minerals present in the system and the degree to which they have already been weathered. In general, phosphorus inputs decline as the soil profile develops over geologic time.

The second difference is that certain soils, due to the presence of "free" iron and aluminum as a result of the weathering process, will have high sorption potentials for phosphorus. These can severely limit the availability of phosphorus for plant growth or increase the energetic cost of phosphorus uptake, especially in tropical soils. These competing demands for phosphorus cause leaching losses of this element to be minimal from nearly all terrestrial ecosystems, which is one reason why primary production in lakes and streams is often phosphorus-limited. P sorption potential tends to increase with time. Trends in phosphorus availability over geologic time were summarized in Chapter 9 (see Figure 9.13).

Potassium (Figure 14.2*e*) The cycle for this metal cation has little in common with those of C, N, S, or P. Potassium is not bound into any known organic compound but moves through the plant in ionic form.

Since it is never bound in organic form, it is susceptible to leaching by rainfall as it washes over plant surfaces. In fact, throughfall and stemflow, the two avenues for water percolation through plant canopies, play a large role in the potassium cycle. Both throughfall (water reaching the soil surface after passing over canopy leaves) and stemflow (water running down the stem) are usually greatly enriched in K as compared to rainfall and together return more K to soils than litterfall in many systems. Retranslocation has been reported for K but is difficult to detect accurately because of the large leaching losses from leaves in throughfall. The K in litter is also only loosely held, and it leaches quickly into the soil solution. Net immobilization is not important and microbial demand is small. In the soil, K is held only on exchange sites and is not involved in reactions to insoluble products.

Thus the potassium cycle is very simple. Inputs are from precipitation and weathering. Plant uptake occurs from exchange surfaces and the soil solution. Leaching occurs from plant surfaces and litter, replenishing the exchange sites. Potassium present in excess of plant demand and exchange site capacity is leached to groundwater.

Calcium (Figure 14.2*f*) The calcium cycle is similar to that for potassium in that organic cycling, weathering, and cation exchange are important processes. It differs in that calcium is included in stable organic compounds. Some of these compounds do not have a direct physiologic use but represent accumulated waste products, such as calcium oxalate in leaves. This may be no less crucial a function than calcium's role in membrane function, and its accumulation in leaves before senescence may have important effects on soil chemistry (see the later discussion of "cation pump" species).

Throughfall and stemflow are less important in the calcium cycle than litterfall and decomposition. Immobilization is rarely, if ever, reported. Mineralized Ca, along with inputs from weathering (which can vary widely from limestone to nonlimestone substrates), is either taken up, held on exchange sites, or leached to groundwater. In deserts, where leaching is minimal, calcium can accumulate as calcium carbonate (or caliche) in lower soil horizons.

Magnesium The magnesium is very similar to the calcium cycle, with the exception that it is generally present in lower concentrations in biological materials.

CHANGES IN SOLUTION CHEMISTRY IN ECOSYSTEMS

A second way of summarizing and comparing nutrient cycles is to follow changes in nutrient concentrations in water as it passes through an ecosystem. The major types of solutions generally sampled include: (1) precipitation (collected above the canopy), (2) throughfall (collected at the soil surface away from stems), (3) stemflow (collected as it runs down stem surfaces), and (4) leachates from below one or more soil horizons.

Table 14.1 compares concentrations of macronutrients in solutions collected at several points in this sequence for forests under two very different environmental regimes. The H. J. Andrews Forest (HJA) is an old-growth Douglas fir stand in the Cascade Range in Oregon. This stand is in an area largely unaffected by air pollution and is growing in soil derived from relatively young volcanic substrate (andesite). The other two stands are dominated by spruce (SSP) and beech (SB) growing on soils derived from sandstone in the Solling District of central Germany. This area experiences very high loading levels of several air pollutants, including gaseous, dissolved, and particulate sources of sulfur and nitrogen.

Although the absolute amounts vary widely between nutrients and between sites, there is a general pattern to the increases and decreases in nutrient flux between the different layers within the stands. Fluxes (in kg/ha/yr) generally increase between precipitation and throughfall plus

Table 14.1 Comparison of solution chemistry in an old-growth Douglas fir stand in Oregon (HJA)* with beech (SB) and spruce (SSP) stands in the Solling District of Germany†‡

	PRECIPITATION	THROUGHFALL+ STEMFLOW	LEACHATE FROM	
			Litter	Root Zone
Nitrogen				
HJA	2.0	3.4	4.7	1.5
SSP	22.6	28.3	76.3	14.9
SB	22.6	23.8	72.8	6.0
Sulfur				
HJA				
SSP	24.1	80.0	82.0	32.4
SB	24.1	47.6	50.8	25.6
Phosphorus				
HJA	0.3	1.2	1.8	0.7
SSP	0.8	0.7	5.1	.01
SB	0.8	0.3	4.6	0.1
Potassium				
HJA	0.9	13.8	25.5	9.5
SSP	4.1	20.6	33.4	2.1
SB	4.1	15.2	26.7	2.9
Calcium				
HJA	3.6	7.2	23.0	123.1
SSP	12.8	41.3	44.3	13.5
SB	12.8	26.7	33.9	12.7

*From Sollins et al. 1980.
†From Heinrich and Mayer 1977.
‡All values are given in kg/ha/yr.

stemflow, increase still further between throughfall and leachate below the litter or humus layer, and then decrease substantially as they pass through the mineral soil.

Differences in pollution loading in these two different locations can be clearly seen in the nitrogen precipitation input values. Nitrogen compounds are an important component of several forms of air pollution, and these can be washed from the air by precipitation and deposited on ecosystems. The N input to the Solling stands is ten times that in Oregon. Inputs of sulfur are also much higher, while inputs of P, K, and Ca are increased 2.5- to 4-fold.

Throughfall plus stemflow is further enriched in nitrogen. This can occur by leaching of internal nutrients from plants or by washing off of particulate pollutants that have been filtered from the air by the foliage. It is difficult to distinguish between leaching and washoff, but the difference is important. Nutrients leached from the plant represent an internal cycle rather than an additional input to the system. Washoff means that the pollution loading to the system is higher than measured in precipitation alone. The potential effects of washoff, and of different species in increasing washoff, can be seen in the throughfall plus stemflow data for sulfur in the two Solling stands. This flux is much higher than precipitation in both stands but is also 32.4 kg/ha/yr higher under spruce than under beech. This difference is due to the much higher leaf area index in the spruce stand and the greater efficiency of the spruce canopy in filtering particulates from the air. This may clean the air to some extent, but it also means higher pollution loads for the forest.

For phosphorus, the Oregon stand shows an enrichment in phosphorus in throughfall plus stemflow, while this flux is lower than precipitation in the Solling stands. Net removal of a nutrient from precipitation as it passes over foliage has been measured in several systems. In general, this signifies less-than-optimal availability of this nutrient to plants and the existence of a strong sink for the nutrient in foliage. Excess nitrogen input to some high-elevation forests has been shown to create low phosphorus-to-nitrogen ratios in foliage, indicating phosphorus limitations. The removal of P from precipitation in the Solling stands may be the result of this kind of N–P imbalance.

Potassium content is much higher in throughfall plus stemflow than in precipitation in all stands, in line with its high mobility in plants. The high content of calcium in throughfall in the Solling stands may also be indicative of high particulate inputs to the canopies. As with sulfur, calcium in throughfall is also much higher in the spruce stand than in the beech stand, suggesting the higher air-filtering capacity of the spruce canopy as an important factor.

All nutrient contents are highest in all stands in the leachate just below the organic horizon. Even the very high levels of sulfur in throughfall plus stemflow in the Solling spruce stand are further increased by passing through this horizon. This increase results from the rapid decomposition and mineralization that occur in the organic horizon. Even though roots are present in this horizon, plant uptake can be less than mineralization, resulting in net exports of nutrients to lower horizons through leachate.

Large changes in nutrient content occur within the mineral soil. Nitrogen content decreases, since plant uptake by roots in this horizon is greater than mineralization. Sulfur losses below the rooting zone in the Solling stands are reduced by a combination of biological uptake and sulfate sorption but remain very high as the high levels of S input have begun to saturate sulfur retention mechanisms in this system.

Phosphorus losses are actually lower for the Solling stands than for the Oregon stand, perhaps because of higher plant demand, as mentioned before, and also because the Solling stands are more strongly podzolized and so have higher concentrations of iron and aluminum in the mineral soil to sorb phosphate. Potassium losses are also lower in the Solling stands. One concern regarding acid rain is that it may seriously reduce cation availability by acidifying soils and reducing base saturation. The low K losses may reflect reduced availability as well as generally slow potassium release from weathering sandstones. In contrast, the very high calcium losses from the Oregon stand reflect the young, rapidly weathering parent material, which is rich in calcium (but not in potassium).

SPECIES EFFECTS ON NUTRIENT DISTRIBUTION AND CYCLING

As a final approach to summarizing nutrient interactions in different ecosystems, we will discuss the effects of having the dominant plants in adjacent ecosystems be very different from each other in terms of a basic physiological process. Two examples can be presented: nitrogen fixation and "cation pumping." The first alters the input rate of nitrogen to the system; the second affects the distribution and cycling of metal cations.

Nitrogen Fixation Red alder, a nonleguminous nitrogen-fixing species, often grows in mixture with Douglas fir in young, successional forests in the Pacific Northwest. It has often been considered a weedy species that interferes with the natural regeneration and early growth of new Douglas fir stands. More recently, there has been increased interest in the role of this species in increasing nitrogen content of soils and nitrogen availability to plants.

Table 14.2 reports data on soil properties for two pairs of stands, both planted with Douglas fir 23 years before the measurements were made. One site is of low fertility, as judged by growth of Douglas fir and also foliar nutrient concentrations and soil nutrient availability indices, and one is of high fertility. Red alder seeded naturally into portions of both of these plantations, allowing a comparison of effects of the nitrogen-fixer on soils and productivity in rich and poor sites.

Table 14.2 Differences in soil properties in fertile and infertile Douglas fir stands growing with and without alder*†

PROPERTY	DEPTH (cm)	INFERTILE MT. BENSON		FERTILE SKYKOMISH	
		No-Alder Site	Red Alder Site	No-Alder Site	Red Alder Site
Coarse-fragment-free bulk density (kg l^{-1})	0-15	0.48	0.49	0.57	0.55
pH	0-10	4.5	4.4	4.2	4.1
	10-20	4.8	4.9	4.7	4.6
	20-35	4.9	5.0	4.9	4.5*
	35-50	4.9	5.1	4.6	4.6
N (%)	0-10	0.09	0.19****	0.31	0.29
	10-20	0.07	0.11****	0.21	0.25***
	20-35	0.06	0.08*	0.13	0.16*
	35-50	0.05	0.08**	0.09	0.14*
Available-N index ($\mu g g^{-1}$)	0-10	23	77****	82	61*
	10-20	15	52****	29	29
	20-35	39	48*	9	8
	35-50	35	56***	4	5
C (%)	0-10	2.05	3.92****	4.98	4.72
	10-20	1.61	2.13*	3.74	5.03****
	20-35	1.15	1.52*	2.11	2.85
	35-50	1.09	1.44	1.26	2.33**
Extractable Ca (meq kg^{-1})	0-10	19.9	38.9**	12.50	5.25***
	10-20	7.2	21.0*	3.41	6.35**
	20-35	7.2	16.5*	1.74	1.44
	35-50	6.6	16.8**	2.00	1.29**
Extractable Mg (meq kg^{-1})	0-10	4.4	6.8	2.10	1.03**
	10-20	1.8	3.9**	0.98	0.99
	20-35	1.5	3.0*	0.78	0.20***
	35-50	1.6	3.9**	1.10	0.43****
Extractable K (meq kg^{-1})	0-10	2.4	3.7	1.32	1.21
	10-20	2.2	2.5	0.86	1.46*
	20-35	0.9	1.4	0.85	0.92
	35-50	1.0	0.9	0.83	0.77

*From Binkley 1983.
†Asterisks in table body indicate significant differences with and without alder.

In the infertile site, the presence of the alder has increased nitrogen and carbon content of soils (Table 14.2). Levels of extractable calcium and magnesium are also higher when alder is present, along with a measured index of nitrogen availability. This has in turn increased nitrogen concentrations in Douglas fir foliage while lowering phosphorus concentrations significantly (Table 14.3). Net primary productivity above ground for the Douglas fir alone is similar in the stands with and without alder, but the

Table 14.3 Differences in foliar nutrient concentrations and total net primary productivity by species in fertile and infertile Douglas fir stands growing with and without alder*†

	INFERTILE		FERTILE	
	Without Alder	With Alder	Without Alder	With Alder
Foliar N (%)				
Douglas fir	0.93	1.41*	1.54	1.55
Alder	—	3.05	—	2.35
Foliar P (%)				
Douglas fir	0.22	0.09*	0.14	0.16
Net primary production (T/ha/yr)				
Douglas fir	6.9	6.4	23.2	15.5*
Alder	—	9.3	—	7.0
Total	6.9	15.7*	23.2	22.5

*From Binkley 1983.
†Asterisks in table body indicate significant differences with and without alder.

added productivity of the alder makes total net primary production more than twice that of the stand without alder (Table 14.3).

In contrast, on the fertile site the presence of alder has had much less effect on soil carbon and nitrogen, and extractable cations are actually lower. Foliar nutrient content of N and P is unaffected, as is total net primary production, but that productivity is split between alder and Douglas fir when the alder is present. This means an actual reduction in Douglas fir growth in the presence of alder on the rich site.

These results show that nitrogen fixation can significantly alter soil characteristics, but the degree of alteration is related to initial site conditions. On the infertile site, N fixation increases N availability and tree growth. This has the secondary effects of increasing total stand and soil organic matter content but apparently reducing relative phosphorus availability. As the nitrogen limitation on growth is relieved, phosphorus limitation becomes enhanced. It is not clear how N fixation increases cation availability, but some of the interactions between nutrient sorption, weathering, and cation exchange discussed in Chapter 9 may be at work. These changes in soil properties would be expected to increase growth of Douglas fir even long after the short-lived alder has disappeared from the stand.

On the fertile site, soil nitrogen storage and nitrogen availability are already adequate for Douglas fir, and the addition of alder does little to alter productivity or soil characteristics. In fact, competition between alder and Douglas fir reduces Douglas fir growth. In this stand, the alder does act as a weed, competing with the crop species and reducing its growth.

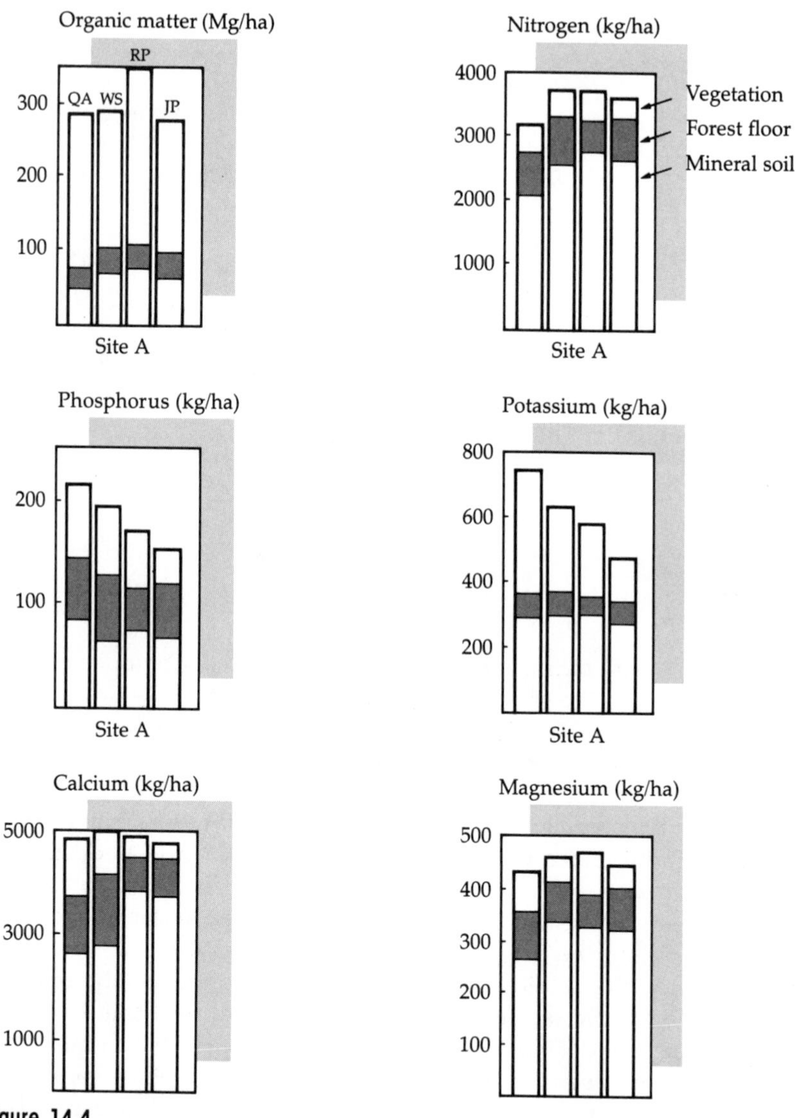

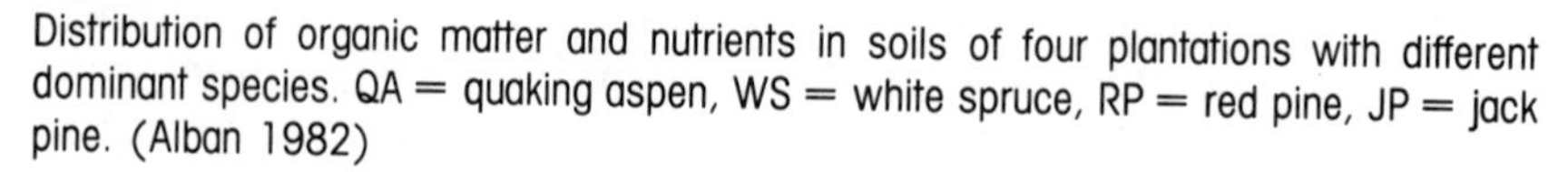

Figure 14.4
Distribution of organic matter and nutrients in soils of four plantations with different dominant species. QA = quaking aspen, WS = white spruce, RP = red pine, JP = jack pine. (Alban 1982)

Cation Pumping Nitrogen fixation is a well-known example of a process with important implications for nutrient cycling in ecosystems, but there are other, less obvious differences between species, which can have similarly impressive effects. One of these is **"cation pumping,"** a term used to describe the tendency in some species to take up and cycle large quantities of the macronutrient cations calcium, magnesium, and potassium.

The effects of cation-pump species can be seen in Figure 14.4. This figure summarizes changes in the distribution of several nutrients and organic matter following 40 years of growth by four different species planted on old agricultural fields in Minnesota. The four different species planted were quaking aspen, white spruce, red pine, and jack pine. The first two are known as high-metal-cation accumulators (cation pumps; white spruce differs significantly from red and black spruce in this respect), while the pines are not.

In Figure 14.4, the total nutrient content in the ecosystem, excluding that present in unavailable mineral forms, is divided into vegetation, forest floor, and mineral soil compartments. All four systems had the same initial content of all nutrients. After 40 years, the total content of calcium remains the same for all four systems. However, under aspen and spruce, considerably more of the calcium has been redistributed to the vegetation and forest floor, while less remains in the mineral soil.

The effect of this redistribution on soil pH can be seen in Figure 14.5. The aspen and spruce stands have more basic forest floors (L+F+H horizons) because of the higher calcium content, but the A horizons are more acidic. This is because of the removal of calcium from this horizon, which reduces base saturation.

The results for potassium and phosphorus are also intriguing. The total measurable amounts of these nutrients have apparently increased in the aspen and spruce stands relative to the pine stands. This could be a secondary effect of lower soil pH in the A horizon, which could lead to

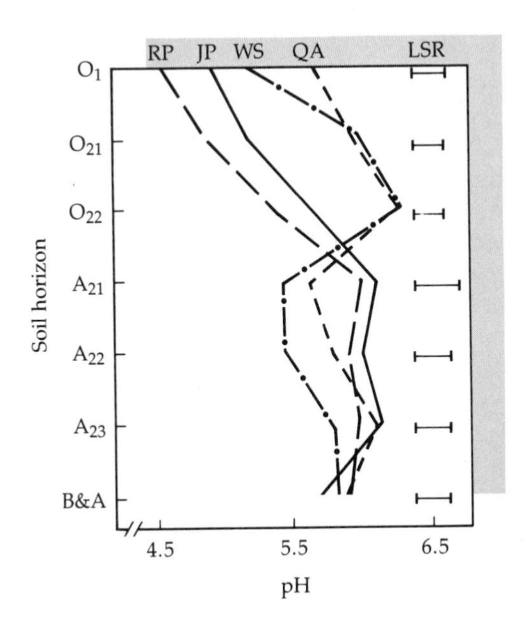

Figure 14.5
Differences in soil pH in four plantations with different dominant species. RP = red pine, JP = jack pine, WS = white spruce, QA = quaking aspen. (Alban 1982)

slightly higher weathering rates and release of these elements from primary and secondary minerals. Other direct physiological processes for biological extraction of these elements have also been identified and may be at work here.

REFERENCES CITED

Alban, D. H. 1982. Effects of nutrient accumulation by aspen, spruce and pine on soil properties. *Soil Science Society of America Journal* 46:853–861.

Binkley, D. 1983. Ecosystem production in Douglas fir plantations: Interactions of red alder and site fertility. *Forest Ecology and Management* 5:215–227.

Heinrich, H., and R. Mayer. 1977. Distribution and cycling of major and trace elements in two central European forest ecosystems. *Journal of Environmental Quality* 6:402–407.

Marquis, D.A., and R. Brenneman. 1981. The impact of deer on forest vegetation of Pennsylvania. USDA U.S. Forest Service General Technical Report NE-65. 7 pp.

Nadelhoffer, K. J., J. D. Aber, and J. M. Melillo. 1983. Leaf litter production and soil organic matter dynamics along a nitrogen mineralization gradient in southern Wisconsin (USA). *Canadian Journal of Forest Research* 13:12–21.

Pastor, J., et al. 1984. Above-ground production and N and P cycling along a nitrogen mineralization gradient on Blackhawk Island, Wisconsin. *Ecology* 65:256–268.

Sollins, P., et al. 1980. The internal element cycles of an old-growth Douglas-fir ecosystem in western Oregon. *Ecological Monographs* 50:261–285.

ADDITIONAL REFERENCES

Bolin, B., and R. B. Cook (eds). 1983. *SCOPE 21: The Major Biogeochemical Cycles and Their Interactions.* John Wiley and Sons, New York.

Clark, F. E., and T. Rosswall. 1979. Terrestrial nitrogen cycles. *Ecological Bulletin* (Stockholm) 33.

Johnson, D. W., and R. I. Van Hook. 1989. *Analysis of Biogeochemical Cycling Processes in Walker Branch Watershed.* Springer-Verlag, New York.

Likens, G. E., et al. 1977. *Biogeochemistry of a Forested Ecosystem.* Springer-Verlag, New York.

Robertson, G. P. 1982. Nitrification in forested ecosystems. *Transactions of the Royal Society of London B* 296:445–457.

Chapter 15 Factors Limiting Consumption: Plant–Herbivore Interactions

H. Armstrong Roberts

INTRODUCTION

The transfer from plant to litter can be short-circuited by the consumption of live tissues by herbivores. In ecosystems dominated by woody plants, consumption generally involves a much smaller transfer than senescence and litterfall. However, in systems dominated by low-growing, herbaceous species, such as the heavily grazed grasslands of Africa, large populations of herbivores regularly consume a significant fraction of the net annual production above ground. In other systems, such as the tundra, there may be large cyclic changes in herbivory. Even forests and shrublands may be subjected to irregular irruptions of herbivore populations due to alterations in food webs through the introduction or removal of species or to stagnation (reduced vigor) in the dominant plant species.

The purpose of this chapter and the next is to discuss the processes that control the levels of herbivory in terrestrial ecosystems. In this chapter we will discuss the chemical interactions occurring between plants and herbivores. We will see that many chemical compounds that affect decomposition also affect herbivores and that symbioses between higher animals and microbes have evolved that increase the potential energy gain from ingested plant material. We will also expand on the discussion in Chapter 11 dealing with the relative availability of resources for plant growth and the allocation of those resources for herbivore defense compounds.

This discussion could be extended to include plant defenses against parasites and disease-causing organisms. Many of the same principles apply. However, a full discussion of the chemical interactions at the cellular level that affect the success of disease organisms is beyond the scope of this book.

CONSUMPTION AS A FRACTION OF NET PRIMARY PRODUCTIVITY

Defoliation of large areas of forest by gypsy moths (Figure 15.1), large-scale tree death due to mountain pine beetle, and overbrowsing of forests by deer are conspicuous examples of consumption of a significant fraction of ecosystem net primary production. Yet such examples are newsworthy partly because they do not represent the norm in these forested ecosystems. Herbivory in forests and shrublands generally consumes less than 10% of annual net primary production above ground.

In contrast with these low average levels, irregular irruptions of herbivore populations can completely strip a canopy of all of its foliage. These large pulses of consumption, even if they occur only rarely, have the potential to alter the species composition and structure of the affected ecosystem. They also demonstrate that the factors generally maintaining herbivory at low levels in forests can, under certain circumstances, be overcome.

When populations of consumers are maintained at artificially high levels for long periods of time, the resulting changes in structure and

Figure 15.1
An example of extreme herbivory in a forested ecosystem: defoliation of a temperate deciduous forest in the northeastern United States by the Gypsy moth. (Van Bucher/Photo Researchers)

function can appear permanent. An extreme example of this is the conversion of deciduous forests in central Pennsylvania to brush or grassland by deer. This region contains one of the highest deer densities in the United States because of the elimination of predators, restrictions on hunting, and the presence of an ideal mixture of forest and farmland habitat. When clearcuttings occur in this region, the resulting lush growth of highly palatable species can attract large numbers of deer, which browse so completely that all trees are killed or severely cut back. In the most extreme cases, no tree growth occurs and the area eventually becomes dominated by bracken fern and other species less palatable to deer. The effect of deer can be seen in a comparison of a small fenced plot that excludes deer (exclosure plot) with an adjacent open area (Figure 15.2). Species composition, production, and all other ecosystem processes are significantly different between two such plots.

Unlike these forest systems where high levels of consumption are unusual, many grassland systems sustain very high levels of herbivory year after year. In such cases, natural selection has favored plant species with morphologic and physiologic traits that allow rapid recovery from cropping by herbivores. Under these conditions, the *removal* of herbivores,

Figure 15.2
An example of the effects of severe deer browsing on the structure of a forest ecosystem. The area outside the exclosure has been subjected to deer activity for several years following a clearcutting, while the area inside has been protected. This site is in the Allegheny Plateau of Pennsylvania. (Marquis and Brenneman 1981)

for example by experimental exclosures, causes significant changes in species composition and ecosystem structure and function.

PLANT–HERBIVORE INTERACTIONS

One major reason for the generally low level of consumption in nongrassland ecosystems is the production of chemical feeding inhibitors by plants. There are few areas in which the intricacy and diversity of the workings of evolution can be seen with greater clarity than in these chemical interactions between plants and the herbivores that would consume them. Plants have evolved the metabolic pathways to produce a tremendous number of secondary products or compounds whose apparent main purpose is to discourage consumption. Over 4000 compounds of the single family of substances called alkaloids have been described. Estimates of the total number awaiting discovery are as high as 10,000.

Herbivore defense compounds can be divided into two categories. The first, and least diverse, are the **"quantitative"** or **general feeding inhibitors**, which reduce the potential energy gain from digestion. The second and much larger group is the **"qualitative" inhibitors**, toxins that are very effective in small concentrations but do not affect all herbivores.

Quantitative Inhibitors

Not surprisingly, the generalized feeding inhibitors that affect the basic energy gain from digestion are the same compounds described before that control decomposition rate. Plant materials that have large amounts of lignin and small amounts of protein and simple carbohydrates make poor-quality food for herbivores, large or small. Even cellulose can be difficult to digest for animals lacking the specialized symbiotic relationships with microbes discussed below. Both cellulose and lignin are considered undigestible "fiber" in the human diet.

Since the concentration of these compounds in plant tissues is what determines their effect on energy gain by consumption, they are called quantitative feeding inhibitors. When plant tissues are high in quantitative inhibitors, herbivores must take in and process larger amounts of material. This means using more energy for movement to reach the larger food mass required as well as for the digestive process. In general, consumers of low-quality plant parts take longer to mature, and this increases the chances for predation, disease, or climatic extremes to reduce survival. Under the most extreme conditions, it may become physically impossible to process the volume of material required to extract enough energy and protein. An example is the "starvation" of certain herbivores of the Serengeti Plain in Africa that were found to have stomachs full of low-quality grass produced because of very dry conditions. Little can be achieved evolutionarily to overcome the basic biochemistry of energy and nutrient yield from plants rich in these general inhibitors of digestion. One evolutionary path available is the development of symbiotic relationships between consumers and microorganisms more closely associated with decomposition.

In general, the digestive systems of insects and mammals cannot produce the enzymes required to degrade polyphenolics or even cellulose. However, diverse groups of animals have evolved a wide variety of symbiotic relationships with microorganisms that can. Perhaps the most familiar example is between ruminant mammals such as cattle, sheep, or deer and the unique symbiotic microflora of their specialized "stomach," the rumen (Figure 15.3). The ruminant provides a constant, optimal environment in which the microbes partially decompose the otherwise indigestible fiber. The simple carbohydrates released by the breakdown of

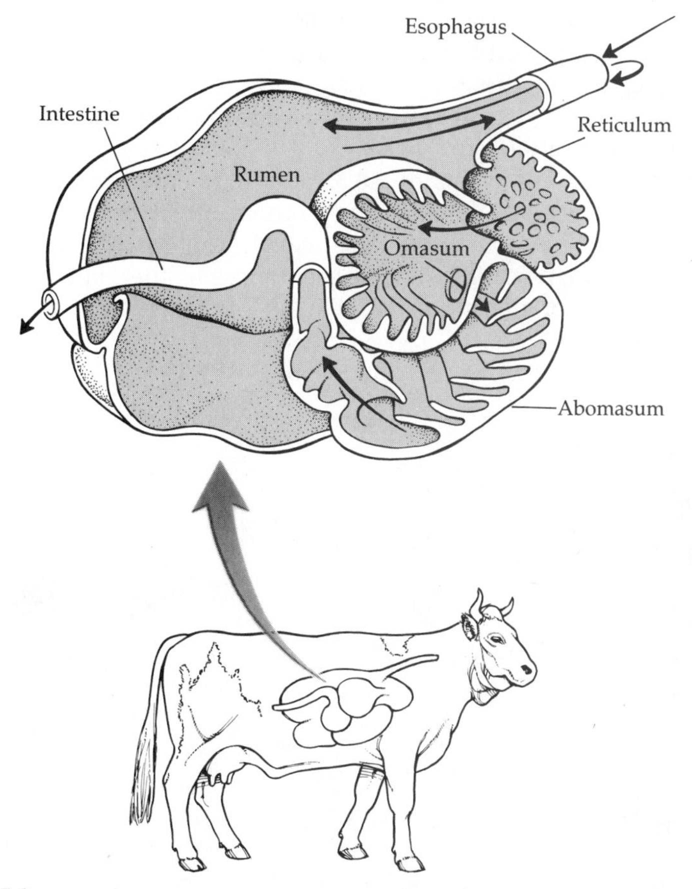

Figure 15.3
Diagram of the structure of the rumen, a specialized digestive tract that supports symbiotic microorganisms, which assist in the digestion of high-fiber plant material. (Botkin and Keller 1982)

cellulose can be absorbed by the ruminant, causing a significant increase in the percentage of energy from the ingested material actually available to the animal.

There are other less pastoral, more exotic examples of this kind of symbiosis. One of the most quantitatively important, especially in tropical ecosystems, is in the termite. Termites support symbiotic microorganisms that both speed the decay of woody or high-lignin substances and fix atmospheric nitrogen. The result can be an astonishingly rapid rate of consumption and degradation of even dry, seasoned wood. Termites can become a major factor in the structure and function of certain ecosystems, as indicated by the high density and unique properties of the large mounds termite colonies produce (Figure 15.4).

In contrast to these physiologic, internal, symbiotic relationships, there are interesting external symbiotic associations that rely on the behavior of the host. One example is the "fungal-farming" tropical ant, which harvests and predigests poor-quality leaf litter, carries the mash to a specially constructed chamber, and regurgitates it into a colony of a specific fungus. The fungus decomposes the material, grows, and is harvested by the ant as its sole food. The chamber is constructed to maintain a constant temperature and humidity (Figure 15.5).

While these three examples are all very different, their effects are the same: to create an optimal micro-environment for the microbiological degradation of poor-quality organic matter. The decay processes themselves are not significantly different from those in litter, but the reactions are controlled for the mutual benefit of host and microbe. These are

Figure 15.4
Termite mounds dominate the landscape in many tropical grassland and woodland sites. (Georg Gerster/Rapho/Photo Researchers)

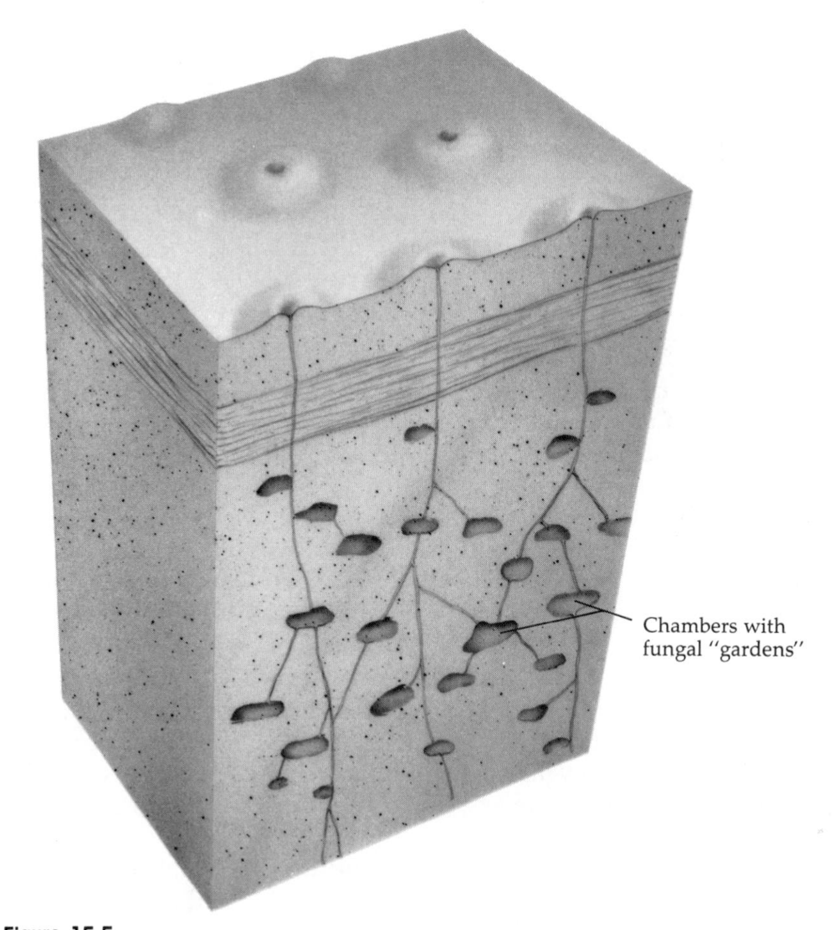

Figure 15.5
Diagram of chambers constructed by the fungal farming ant.

adaptations to overcome the inherent resistance of the material to degradation. Still, with the exception of the heavy grazing of some grasslands by ruminants, even these complex, highly evolved systems do not allow consumption of a sizable fraction of annual net primary production in most systems.

If producing nearly inedible tissues is such an effective barrier to consumption, why hasn't evolution led solely to plants with very high fiber and polyphenol content? One important answer is that such products are expensive to synthesize. More than 3 grams of glucose are required to produce a single gram of lignin. Coniferous trees may have as much as 25–35% lignin in their leaves and even higher concentrations of similarly complex material (suberin) in their roots. This represents a sizable investment of photosynthate that could have gone into height growth, new leaves or stems, or reproduction, all of which would, in the absence of serious herbivore pressure, increase the competitive advantage of one plant over its neighbors. Once again a tradeoff is involved. A heavy investment in these general consumption inhibitors would confer a selective, evolutionary advantage only if the protected tissues—for example, leaves—were thereby able to last longer and so produce enough photosynthate to more than "repay" the "cost" of protection.

Qualitative Inhibitors

It is with the qualitative rather than the quantitative feeding inhibitors that the diversity of the products of evolution is so apparent. The tremendous diversity of compounds arises because each toxin does not affect the basic extractable energy content of the material but rather protects it with a dilute concentration of a toxic substance. For every toxic substance evolved, a herbivore eventually evolves a detoxifying mechanism that makes that plant edible again. This in turn increases the selective pressure for the synthesis of a new protective compound. Plants face numerous herbivorous species simultaneously, particularly insects. Each may be capable of detoxifying a wide array of previously evolved substances. Thus, a picture emerges of plants and herbivores in constant evolutionary movement, with consumers continually pressuring plants employing qualitative inhibitors to develop novel combinations of toxins. Always the advantages of protection are weighed against the cost of synthesizing the protective substance.

Detoxification is not the only method for handling qualitative inhibitors. Herbivores can, instead, develop a tolerance for the chemical and then actually incorporate it into their own tissues. The monarch butterfly feeds primarily on different species of milkweed plants. Milkweeds produce a qualitative inhibitor that effectively reduces herbivory by other insects. The feeding monarch pupa (caterpillar stage), however, incorporates the toxin into its tissues, where it is stored and maintained even into the butterfly form, which is then unpalatable to most predators.

There may be 10,000 qualitative inhibitors. Many have similar basic structures with relatively minor modifications. The importance of such minor changes indicates the extreme selectivity and sensitivity of both the biological processes they disrupt and the detoxifying reactions that confer protection.

The vast array of herbivore defense chemicals can be separated into two groups, depending on whether or not they contain nitrogen. Carbon-based defense compounds, those without nitrogen, include isoprenoids such as terpenes (Figure 15.6) and sterols, in addition to the quantitative inhibitors lignin and cellulose. Nitrogen-based defense compounds include alkaloids, among which are several potent medicinal drugs (Figure 15.7), and glucosinates and cyanogenic glycosides.

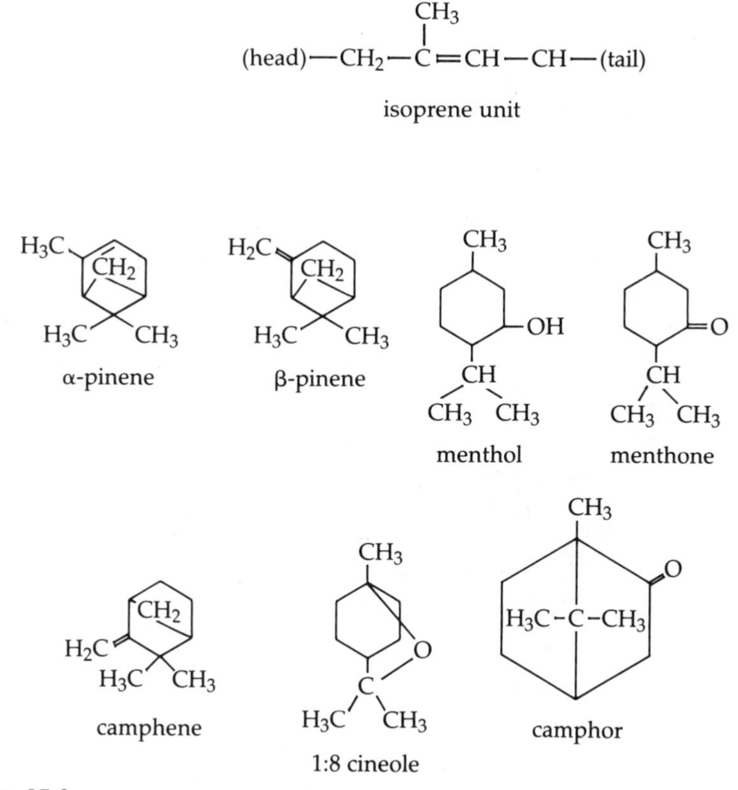

Figure 15.6
Generalized structure of isoprenoid compounds, and several examples of terpenes, one class of carbon-based, qualitative herbivore defense compounds.

PATTERNS OF HERBIVORE INHIBITOR PRODUCTION IN PLANTS

In Chapter 11 we presented some interactions between site quality, leaf nutrient content, leaf longevity, and the amount of resources invested in protection against herbivory (Figure 11.12). These interactions suggest that plants growing in nutrient-poor environments will have lower leaf nitrogen concentrations and hence slower photosynthetic rates. This means that it takes longer to "repay" the carbon "cost" of building the leaf. In order for the leaf to survive herbivory for this longer period, it must be protected by chemical inhibitors.

We have now seen that chemical defenses against herbivory can be either quantitative or qualitative and that the qualitative defenses can be either carbon- or nitrogen-based. There are two theories that predict the

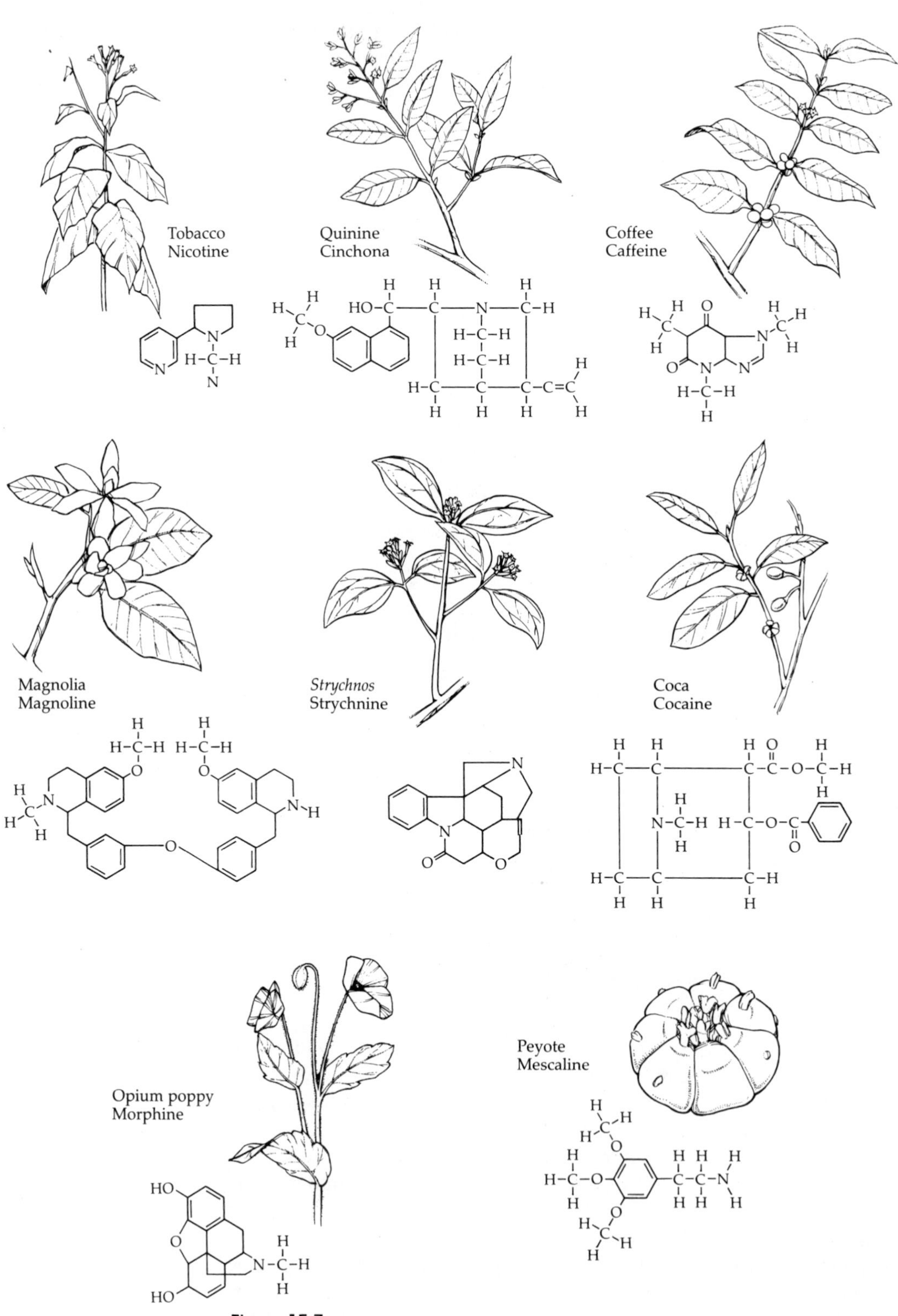

Figure 15.7

Structure of selected alkaloids—nitrogen-based, qualitative, herbivore defense compounds—and the plants that produce them.

distribution of herbivore defense strategies in plants: "apparency," or the relative dominance and longevity of a species in a given location, and a resource availability theory.

Apparency theory suggests that the quantitative inhibitors (lignin, cellulose) will predominate in species that are long-lived or occur in high densities over large areas. This is because of the large numbers of herbivore species that a plant of such a species would encounter during its lifetime. Many of these herbivores would be preselected to detoxify any set of qualitative inhibitors that might be produced, and it would be impossible to simultaneously and continuously evolve new sets of qualitative inhibitors effective against all of the potential herbivores.

Alternatively, rare species, or those that are important for only short periods during the development of the plant community, gain a measure of defense by being hard to find (less "apparent"). There may be relatively few herbivores adapted to seeking out rare species or the disturbed sites in which many of the less apparent plant species grow. Producing a restricted set of qualitative inhibitors to reduce the efficiency of these few herbivores may be less of an energy drain than producing large amounts of the general compounds and thus may confer a selective advantage in the evolution of these species.

The **resource availability theory** suggests that site quality and leaf longevity, rather than apparency, determine whether quantitative or qualitative defenses are used. Qualitative defenses are less expensive to produce because they are required in such low concentrations. However, they are not chemically stable and need to be replaced continuously. In contrast, the quantitative defenses are stable and need to be produced only once. As leaf longevity increases, it becomes less expensive to produce quantitative inhibitors once, during leaf development, than to continuously produce qualitative inhibitors (Figure 15.8). As leaf longevity increases with declining site quality (Chapter 11), low-quality, low-productivity ecosystems tend to have species that rely on the quantitative inhibitors. Table 15.1 summarizes the growth and herbivore defense characteristics of plants from rich (fast-growing) and poor (slow-growing) sites. Interestingly, the high contents of foliar lignin implied by the reliance on quantitative inhibitors in poor sites should also reduce decomposition rates and perhaps cause further reductions in site quality.

The resource availability theory also predicts that the use of carbon- or nitrogen-based inhibitors will depend on the relative availability of carbon and nitrogen to plants. Tropical forests in which phosphorus may be more limiting than nitrogen and in which there may be a significant component of nitrogen-fixing species have a higher proportion of species that use the nitrogen-based defenses. In systems where light and water are more available than nitrogen, the carbon-based defenses predominate. For example, large amounts of terpenes are produced in many coniferous forest ecosystems.

In Part III we will discuss ways in which the availability of resources,

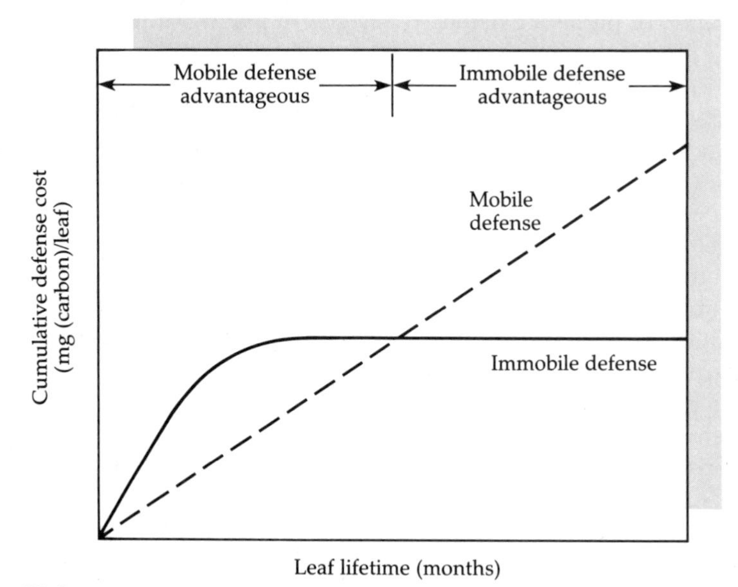

Figure 15.8

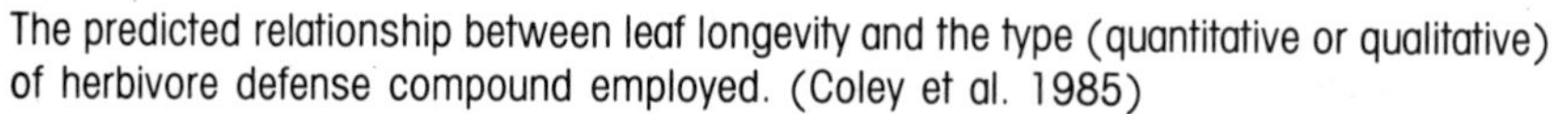

The predicted relationship between leaf longevity and the type (quantitative or qualitative) of herbivore defense compound employed. (Coley et al. 1985)

particularly nutrients, can decrease significantly with time in many ecosystems. This can reduce site quality and plant vigor to the extent that herbivore defense compounds cannot be produced in sufficient quantity. An example of this is in the ponderosa pine ecosystems of the Rocky Mountains. For these systems it has been shown that reductions in tree vigor (measured as the ratio of diameter increment to foliage surface area) significantly increase the probability of mortality due to mountain pine beetle attacks (Figure 15.9). Experimental work has also shown that thinning and fertilizing the stand, reducing the number of trees, and increasing nutrient availability can increase tree vigor and reduce mortality.

Both apparency and resource availability effects may be at work in natural ecosystems. Since recently disturbed sites tend also to have high resource availability and support rapid plant growth, those early successional species that are less apparent also tend to be those with fast growth rates and fast leaf turnover.

From this discussion it should not be surprising that by far the greatest number of specific defense compounds are found in plants from the tropical rainforests. These systems typically have very large numbers of rare species. As many as 100 species of canopy trees may occur per hectare, with perhaps three or four individuals per species. These same

Table 15.1 Generalized relationship between plant growth rate (site quality), growth characteristics, and type of herbivore defense employed*

VARIABLE	FAST-GROWING SPECIES	SLOW-GROWING SPECIES
Growth characteristics		
Resource availability in preferred habitat	High	Low
Maximum plant growth rates	High	Low
Maximum photosynthetic rates	High	Low
Dark respiration rates	High	Low
Leaf protein content	High	Low
Responses to pulses in resources	Flexible	Inflexible
Leaf lifetimes	Short	Long
Successional status	Often early	Often late
Antiherbivore characteristics		
Rates of herbivory	High	Low
Amount of defense metabolites	Low	High
Type of defense	Qualitative	Quantitative
Turnover rate of defense	High	Low
Flexibility of defense expression	More flexible	Less flexible

*From Coley et al. 1985.

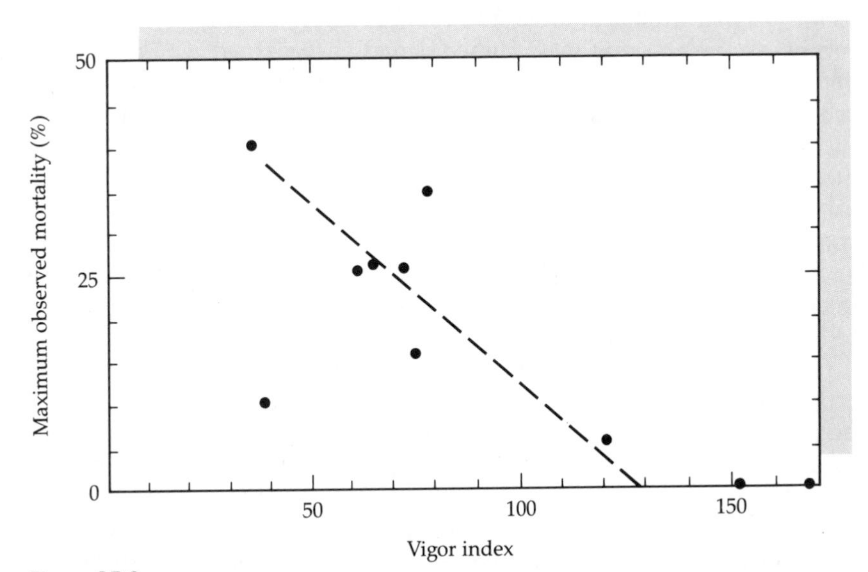

Figure 15.9
Measured probability of mortality in ponderosa pine due to mountain pine beetle attack, as a function of tree vigor in the stand. (After Mitchell et al. 1983)

systems also tend to have high densities of nitrogen-fixing species, which increases the reliance on nitrogen-based defenses.

There are two important implications of the trends discussed superficially in this chapter. The first is the potential importance of the tropical forest regions of the earth as a source of biochemicals of value to society. Evolution has created a vast repertoire of compounds that can affect the human body in unpredictable ways. Some of these compounds are well known, such as the alkaloids nicotine, morphine, and cocaine. Numerous synthetic drugs are simple modifications of natural products.

Thousands of additional secondary products of plant metabolism remain undiscovered in the tropical forests. The potential for these to reduce (or increase) human suffering is a genuine concern of many whose goal is to preserve the species diversity of the tropical rainforest by preserving large areas intact. Because of the low numbers of individuals per species in an area and the large total number of species, reducing the area in rainforest by conversion to forest plantations or permanent agriculture could result in the loss of hundreds, perhaps thousands, of distinct plant species plus their associated insect species. Each species could carry with it the genetic capability of producing a specific compound of great benefit.

A second implication arises from the fact that many of our important crop plants, such as corn, are derived from tropical or semitropical species of disturbed habitats. They are also fast-growing and low in quantitative feeding inhibitors. Under natural conditions, rarity probably conferred some protection from attack, along with qualitative inhibitors for specific pests. We have taken these species and bred much of their natural genetic diversity and energy-consuming defense mechanisms out to increase yield. We plant them in large fields dominating whole regions. It is not surprising that the crop yield gained comes at the expense of substantial investments in pesticides to protect the growing plants. Fossil fuels are used as an energy subsidy so that the metabolic energy from photosynthate in plants may be genetically directed to yield. As fossil fuels become more expensive, society may find it cheaper and more beneficial to make use of natural diversity and the natural defense mechanisms of plants to control herbivory in crops.

All this is much removed from those natural ecosystems in which the interplay between chemical defenses and herbivores causes consumption to be generally a small proportion of net primary productivity. Of course these same ecosystems also produce little food for human consumption! Still, systems with generally low rates of herbivory can experience cyclic or irregular irruptions of herbivore populations that decimate the canopy. In grasslands, herbivory can claim a sizable fraction of above-ground net primary production each year. In the next chapter we will examine the characteristics of those systems that support high rates of herbivory, the evolutionary adaptations that have been made by plants in heavily grazed systems, and factors causing occasional outbreaks of herbivores in systems where their effect is usually minimal.

REFERENCES CITED

Botkin, D. B., and E. A. Keller. 1982. *Environmental Sciences: The Earth as a Living Planet.* Merrill, Columbus, Ohio.

Coley, P. D., J. P. Bryant, and F. S. Chapin III. 1985. Resource availability and plant herbivore defense. *Science* 230:895–899.

Marquis, D. A., and R. Brenneman. 1981. The impact of deer on forest vegetation of Pennsylvania. U.S. Forest Service General Technical Report NE-65. U.S.D.A., Washington, D.C.

Mitchell, R. G., R. H. Waring, and G. B. Pitman. 1983. Thinning lodgepole pine increases tree vigor and resistance to mountain pine beetle. *Forest Science* 29:204–211.

ADDITIONAL REFERENCES

Chabot, B. F., and D. J. Hicks. 1982. The ecology of leaf life spans. *Annual Reviews of Ecology and Systematics* 13:229–259.

Coley, P. D. 1983. Herbivory and defensive characteristics of tree species in a lowland tropical forest. *Ecological Monographs* 53:209–233.

Feeney, P. 1976. Plant apparency and chemical defense. *Recent Advances in Phytochemistry* 10:168–213.

Levin, D. A. 1976. The chemical defenses of plants to pathogens and herbivores. *Annual Reviews of Ecology and Systematics* 7:121–159.

Matson, W. J. 1980. Herbivory in relation to plant nitrogen content. *Annual Reviews of Ecology and Systematics* 11:119–161.

Mooney, H. A., and S. L. Gulman. 1982. Constraints on leaf structure and function in reference to herbivory. *BioScience* 32:198–206.

Penning de Vries, F. W. T., A. H. M. Brunsting, and H. M. Van Laar. 1974. Products, requirements and efficiency of biosynthesis: A quantitative approach. *Journal of Theoretical Biology* 45:358–377.

Price, P. W., et al. 1980. Interactions among three thropic levels: Influence of plants on interactions between insect herbivores and natural enemies. *Annual Reviews of Ecology and Systematics* 11:41–65.

Salisbury, F. B., and C. W. Ross. 1979. *Plant Physiology.* Wadsworth, Belmont, California.

Chapter 16

Characteristics of Ecosystems with High Herbivore Consumption Rates

H. Armstrong Roberts

INTRODUCTION

The chemical defenses discussed in Chapter 15 reduce plant losses to herbivory, especially in systems dominated by woody plants in which insects are the dominant herbivores. In systems where herbaceous plants are dominant, and in which mammals are the major herbivores, consumption can be very high, and the fraction of net primary production consumed can be either fairly constant from year to year or wildly variable, depending on the system.

Irregular irruptions of insects also occur, and when they do they can cause herbivory to become an important transfer in woodland systems as well. These irruptions are often triggered by the introduction or removal of hosts, herbivores, and predators from the system, by declining vigor in dominant plants, reducing the production of herbivore defense compounds, and by climatic fluctuations.

The purpose of this chapter is to examine the causes and consequences of these herbivore "success stories" and to discuss the mechanisms that introduce and/or sustain high levels of herbivory in certain ecosystems.

UNGULATES AND GRASSES: COEVOLUTION?

The Serengeti Plain of Africa supports the highest density and greatest biomass of large herbivores in the world. Herbivory is such a dominant process that even the extent of the Serengeti system is defined by the movement of herds of the major herbivores in response to wet and dry seasons and the resulting changes in plant growth. Herbivores can locally consume close to 100% of annual above-ground primary production in parts of this region. This represents a very strong selective force for the development of effective herbivore defense compounds in the plants, and yet the dominant grasses in grassland systems are often *more* palatable than the foliage of most woody plants.

An alternative to the evolution of defenses against herbivory is the development of effective physiologic responses to cropping. If a species could evolve a mechanism for rapid growth or reproduction following this form of disturbance, one which gave it a competitive advantage over other species, then the value of discouraging herbivory would be lost. It has been noted that the evolutionary appearance and spread of large mammalian herbivores and species of grass are very closely associated. This suggests that the two have evolved in concert.

Some of these adaptive characteristics can be seen even in comparisons between closely related species that have evolved under different degrees of grazing pressure. For example, a species of bunchgrass, *Agropyron desertorum*, has been introduced to the short-grass prairies of the United States from Eurasia, where it evolved under conditions of heavy grazing. A native species, *A. spicatum*, has evolved under lower grazing intensities. In grazed mixtures, the introduced species has been effective in replacing the native species.

Comparisons of responses to simulated grazing (clipping) show that the introduced species can mobilize more carbon and nitrogen for the production of new leaves following this disturbance. Figure 16.1*a* shows a large difference between the two species in the production of new leaves, stems, and sheaths following a severe defoliation. Figure 16.1*b* indicates that total nitrogen content in these tissues is also higher in *A. desertorum*, especially in May and June, resulting in higher total net photosynthesis. Water use efficiency is also higher in the Eurasian species. The physiologic capacity to replace the removed canopy more rapidly and restore photosynthetic rates should be a major factor in the success of *A. desertorum* in grazed areas.

A second factor is the reproductive response to clipping. *A. desertorum* responds to clipping by initiating a number of new tillers, a means of vegetative reproduction (Figure 16.2). So even though both species experienced high mortality following clipping, the introduced species was more effective in replacing the killed plants with new ones.

In summary, the species that evolved under heavy grazing pressure is better adapted to mobilizing the very large reserves of carbon and nitrogen stored in the root systems for both reproduction and the reestablishment of a functional canopy. If grazing gives a species a competitive advantage, then there should actually be a selective advantage in encouraging this form of disturbance to the plant community, perhaps by becoming more, rather than less, palatable.

It has been argued that the relationship between herbivores and the

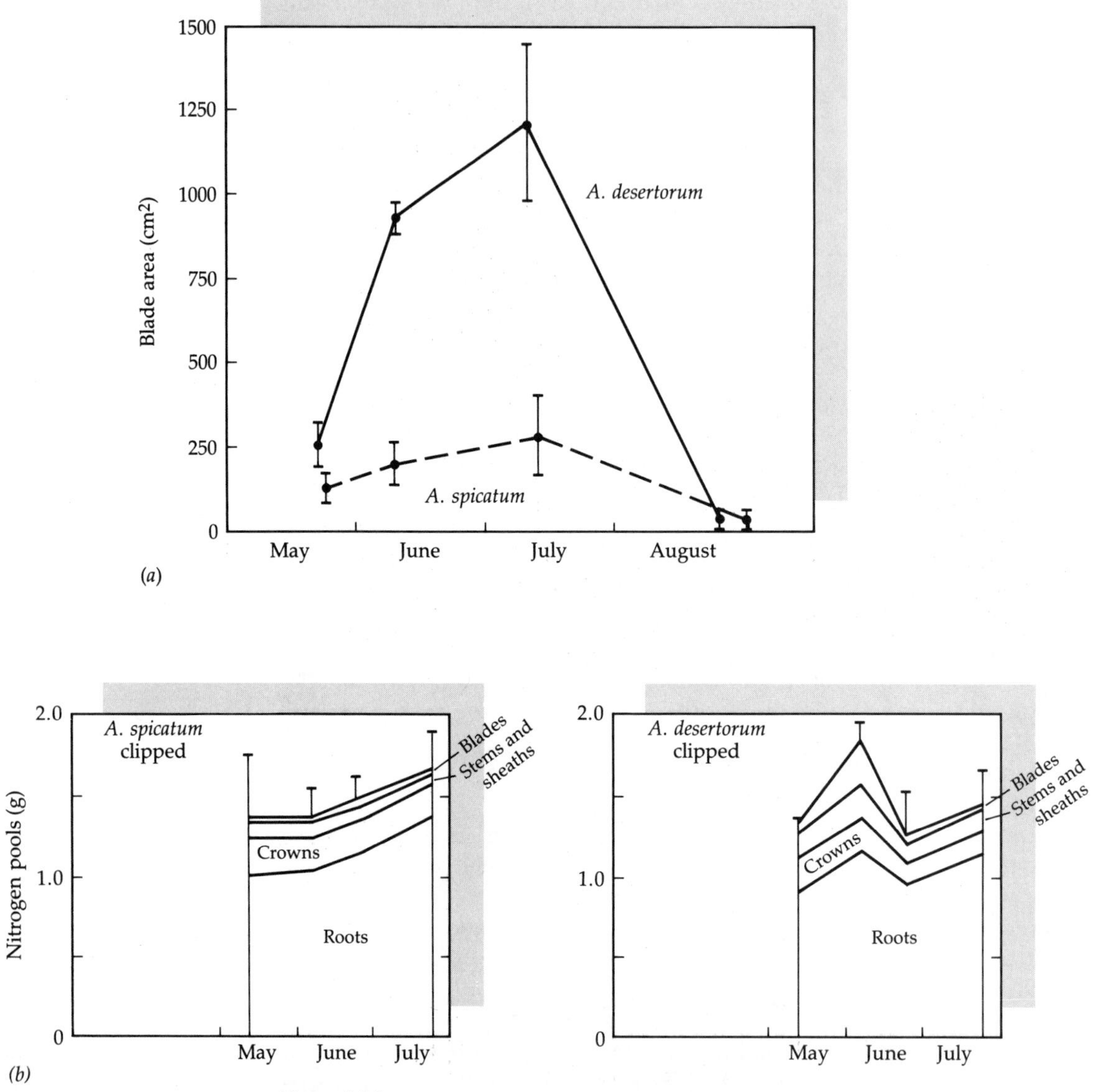

Figure 16.1

Comparison of responses to clipping in two similar species of bunchgrass, on a per-plant basis. *Agropyron desertorum* has evolved under heavier grazing pressures, *A. Spicatum* under lower grazing pressures. *(a)* Rate of new leaf production. *(b)* Total nitrogen content of recreated canopy. (Caldwell et al. 1981)

dominant plant species in grassland systems has actually evolved to become one of mutual advantage to both (a mutualistic, if not symbiotic, relationship). Arguments for this include higher rates of photosynthesis in leaves produced following herbivory, and in younger foliage in general, lower respiration rates due to reduced standing biomass, less self-shading

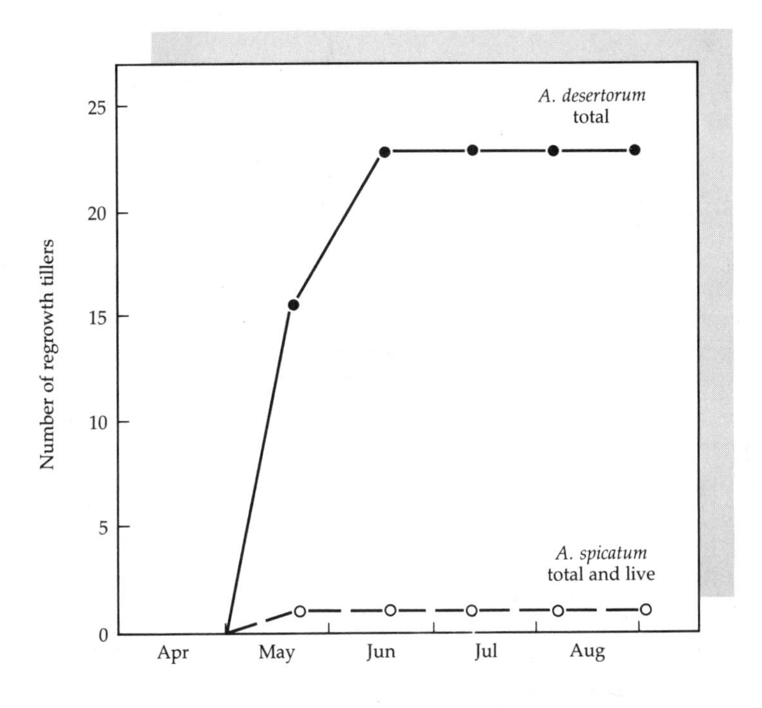

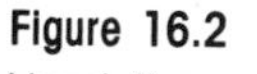

Figure 16.2
Vegetative reproduction following clipping in two species of bunchgrass. (Caldwell et al. 1981)

of foliage, and even the possibility that herbivore saliva contains chemicals that stimulate plant growth. These effects are usually summarized in a "herbivore optimization curve" (Figure 16.3) describing total net primary productivity above-ground as a function of grazing intensity. The apparent increase in total plant growth with moderate grazing intensity is called compensatory growth. While it is clear that herbivory alters the competitive balance between species, and may increase net primary production

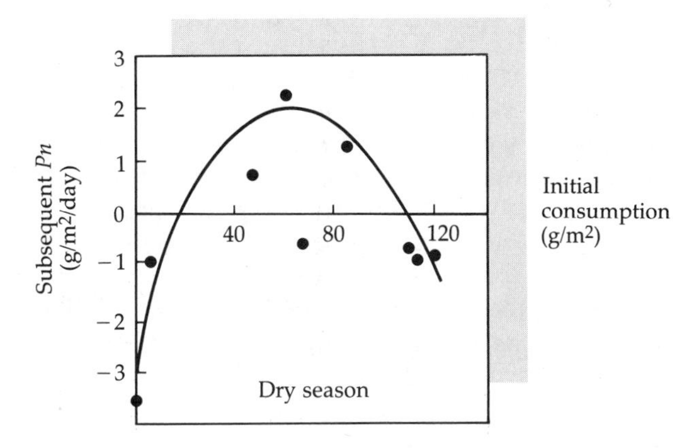

Figure 16.3
An "herbivore optimization curve" showing the effect of different intensities of grazing (as biomass of material removed) on total net primary productivity. (McNaughton 1979)

above-ground, it is still unclear whether actual stimulation of total net primary productivity, both above and below ground, is a general occurrence.

TEMPORAL CHANGES IN HERBIVORY

In ecosystems where herbivory is a major pathway of energy transfer, it is rarely constant from year to year. This is in contrast to decomposition rates, which, as we saw in Chapter 12, are predictable from climatic conditions and initial litter quality. Rates of herbivory are closely tied to population levels of herbivores, and these can change in response to climate, alterations in the community of competitors and predators, and changes in the vigor and chemical quality of the dominant plants. Each of these factors can either increase or reduce the fluctuations in the herbivore populations, depending on their timing and effect on herbivore population levels.

Lemmings in the Tundra

We can begin with a discussion of a system in which many of the feedbacks (Chapter 5) between population change and population control mechanisms are positive rather than negative. That is, the mechanisms tend to destabilize rather than stabilize population levels.

A famous example of drastic, cyclic changes in densities of a herbivore is provided by the lemming, a small rodent of the arctic tundra. The image of hordes of lemmings rushing into the sea to drown has become part of both ancient folklore and modern social commentary. While the actual mass drownings may be rare, very large changes in population density of this species are part of a regular cycle of population explosions and collapse.

Figure 16.4 is a 20-year record of lemming population densities at Point Barrow in northern Alaska. Levels change nearly 100-fold within the two- to four-year recurring cycles. As a classic example of a cyclic population, several system interactions tend to exaggerate, rather than diminish, population fluctuations.

First, the lemmings are the only major herbivore in a very simple food web (Figure 16.5). Therefore, when lemming populations are high, herbivory severely reduces plant biomass and growth. Compensatory growth, if it occurs in this system, is not effective in maintaining plant biomass.

Second, all of the major predators in this system reproduce more slowly than the lemmings, so increases in predator populations lag behind lemming increases. These time lags may actually increase the magnitude of lemming cycles as predator populations will be highest when the rodents are decreasing in number. Heavy predation would then push lemming numbers lower than they would otherwise go, allowing plant growth to

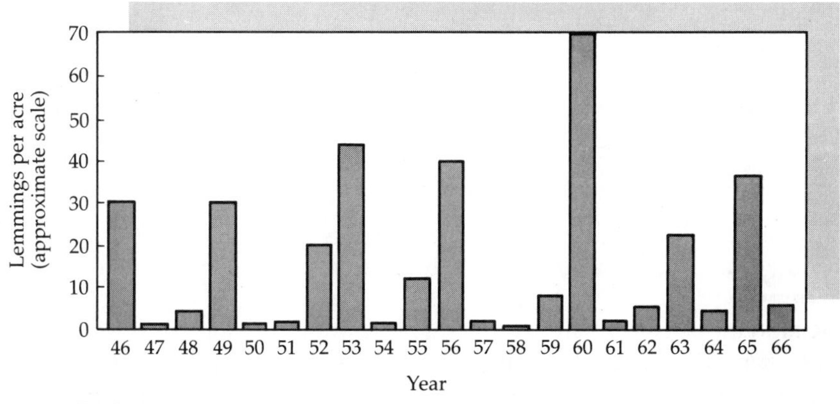

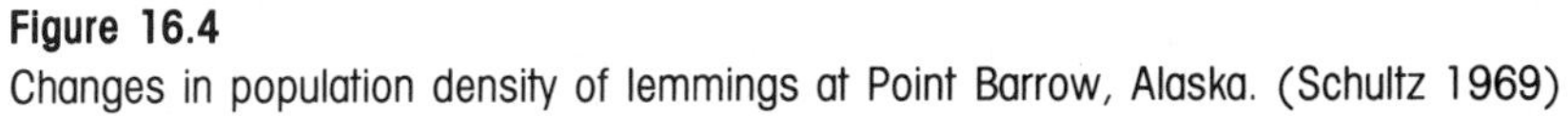

Figure 16.4
Changes in population density of lemmings at Point Barrow, Alaska. (Schultz 1969)

rebound more vigorously, and stimulating another large increase in lemming numbers.

Third, lemmings lack territoriality. Territoriality is a mechanism for reducing reproduction from total biological potential through social interactions that limit reproduction to a few individuals in the population (see discussion of moose and wolf interactions below). However, lemmings do show evidence of social interaction. At high densities, several physiologic and behavioral indicators of stress are evident. There is indeed evidence that reproduction is fast enough, and selective pressure great enough, that

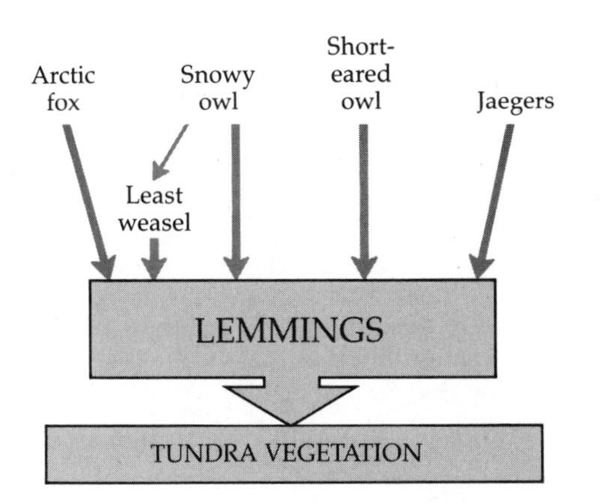

Figure 16.5
The food web of the tundra systems in which lemmings are the primary herbivores. (Pitelka et al. 1955)

the genetics of the population are altered between peak and crash population levels, with individuals tolerant of stress and showing more aggressive behavior becoming more prevalent at high densities. Genetic change is another time-lag factor that tends to increase the amplitude of cycles. Similar changes in genetics, physiology, and behavior have been shown for other cyclic species.

Fluctuations in lemming numbers induce significant changes in plants and soils, which may also tend to exaggerate population cycles. Tundra plants grow in a thin soil layer, which is thawed for only a short time during the summer. The depth to the permanently frozen layer varies with the amount of vegetation and organic litter on the soil surface. The more plant and litter cover, the shallower the thawed soil layer will be, due to the insulating effect of the cover. Heavy grazing and burrowing by lemmings reduces the plant and litter layer, increasing the depth of thawed soil. Associated with this is a reduction in the nutrient content of forage available to lemmings in the following year (Figure 16.6). Reduced consumption in low-population years allows a recovery in both the quantity and quality of vegetation available for consumption. This in turn can help trigger the next population increase.

Figure 16.7 summarizes the set of interactions affecting lemming numbers. There are certainly several mechanisms for limiting population growth. However, as realized for this species, the major checks respond slowly to lemming increases and tend more to increase cycle amplitude than to decrease it.

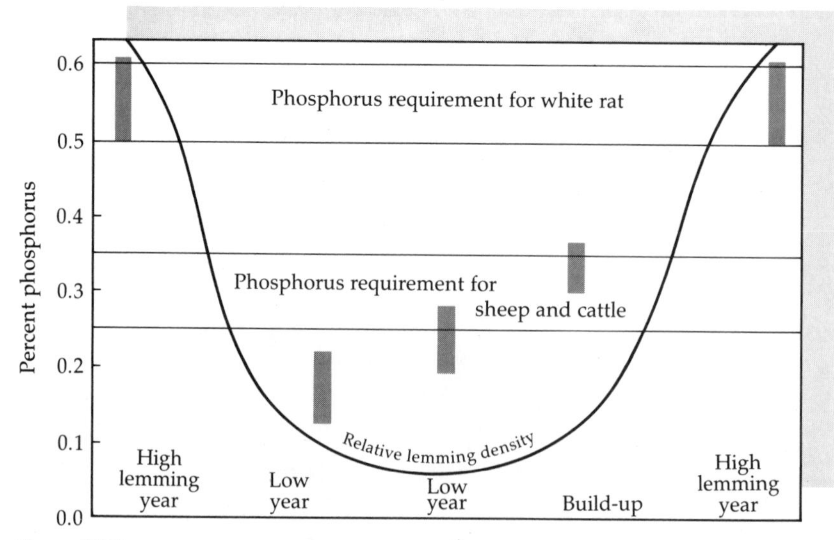

Figure 16.6
Changes in the phosphorus content of herbage over the course of a four-year lemming population cycle, in relation to the requirements of several species. (Schultz 1969)

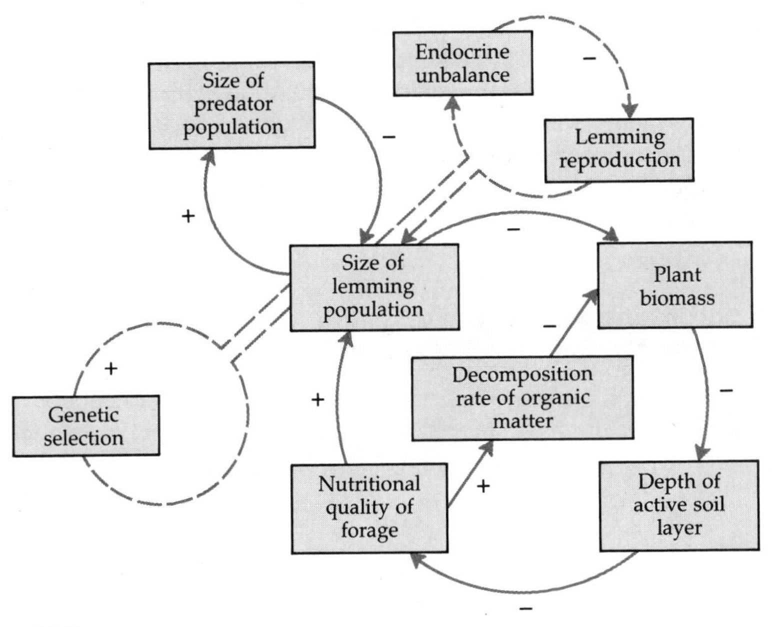

Figure 16.7
Interactions of factors controlling lemming populations in the tundra. (After Schultz 1969)

Other cycles in northern latitudes are well documented. The historical pattern of lynx and hare in the boreal forest zone, as inferred from the number of pelts of each sold to the Hudson Bay Company in Canada, is an often cited example (Figure 16.8). In this case the cycles are repeated at ten-year intervals and, as before, the predator generally lags behind the prey by one or two seasons, suggesting that predation may have a destabilizing effect, as with lemmings. There is also evidence that heavy grazing of dwarf birch by hare may induce increased production of resins,

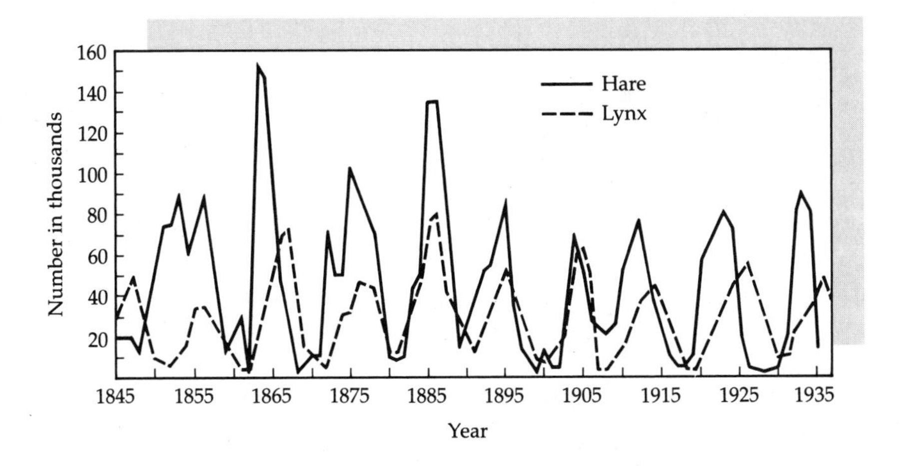

Figure 16.8
Changes with time in the sales of pelts of hare and lynx to the Hudsons Bay Company in Canada. Cycles of sales are assumed related to cycles in population levels. (Odum 1971)

which reduce the palatability of twigs (Figure 16.9), causing reduced quality of plant material at the same time that populations are falling. This will also tend to increase population fluctuations. Heavy browsing can also induce many woody plant species to revert to a juvenile growth form in which thorns or spines may be present to inhibit vertebrate herbivores. This adaptation is lost in the mature plant form, when the plant is tall enough to be beyond reach.

The Stabilizing Effects of Territoriality and Predation: Moose and Wolves on Isle Royale

Predation need not always have a destabilizing effect. Particularly when the major predators have important social interactions that control population density, predation can dampen herbivore population fluctuations. A

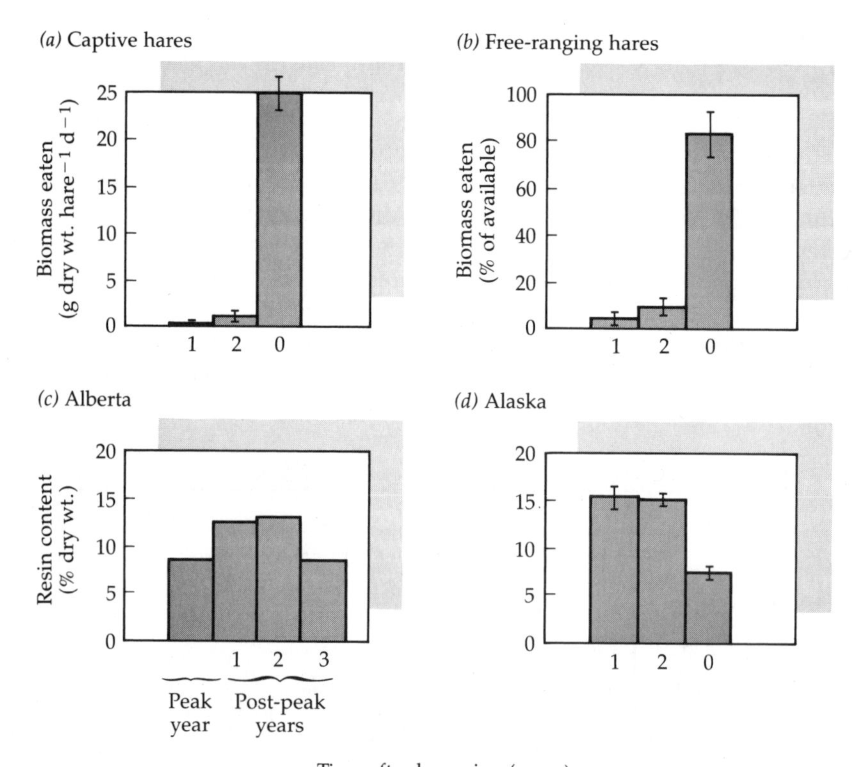

Figure 16.9
Changes in the resin content of twigs of dwarf birch in relation to the occurrence of peak populations of hare. (Bryant et al. 1983)

rare opportunity to study this interaction in operation has occurred over the last 80 years on Isle Royale in Lake Superior.

Isle Royale is a large (210-square-mile) island supporting a mixture of forest and wetland habitat required by moose (Figure 16.10). In 1905, a small number of moose apparently swam from the Canadian side of Lake Superior to Isle Royale, about 20 miles offshore. As new arrivals, the moose found abundant food and cover and an absence of their main predator, the timber wolf. From a small beginning, the population grew exponentially (increasing at an increasing rate) for 23 years (Figure 16.11). By 1926–1927 the moose had modified the vegetative structure of the island to such an extent that a shortage of browse, highly palatable leaves and twigs of certain woody species, was developing.

Increasing moose populations and declining browse production resulted in a massive die-off between 1933 and 1937. The moose population declined to perhaps less than 100 animals. A physiologist examining recovered moose corpses determined starvation to be the sole source of mortality.

Low moose densities allowed the vegetation to recover somewhat and another population irruption began. However, the vegetation on the island had been altered during the first irruption and what remained could

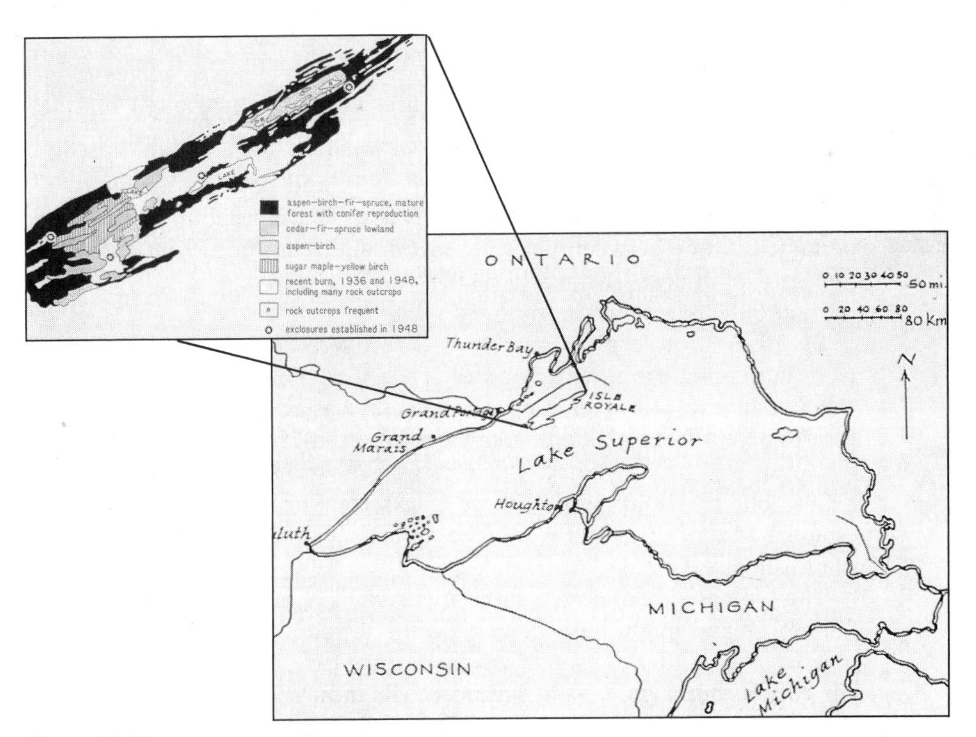

Figure 16.10
Map of Isle Royale and its location in Lake Superior. (From Krefting 1974, Peterson 1977)

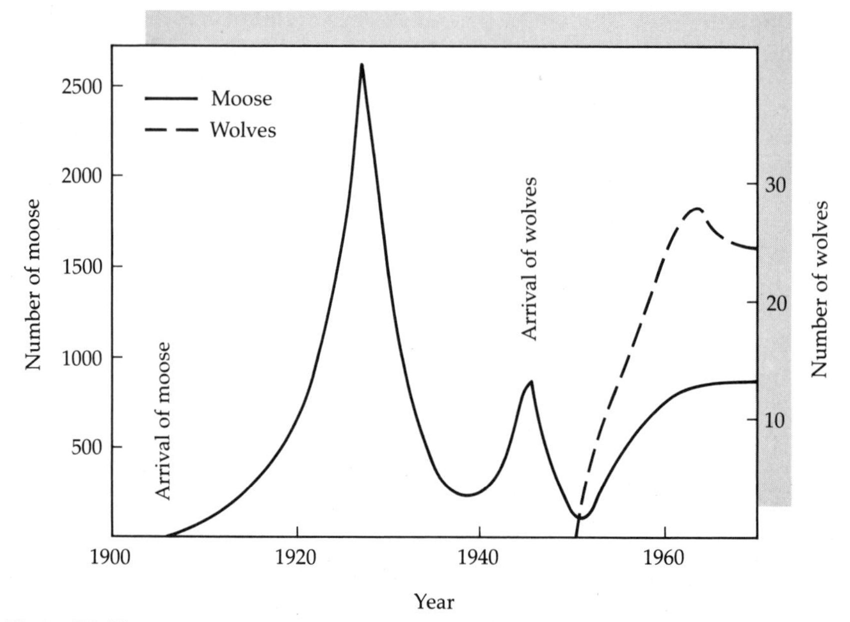

Figure 16.11

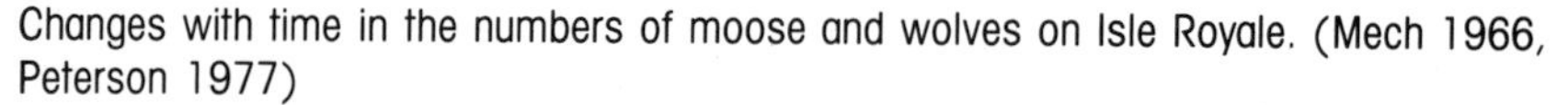
Changes with time in the numbers of moose and wolves on Isle Royale. (Mech 1966, Peterson 1977)

only support smaller moose populations. By 1948, a population of only 800 moose was severely reducing the available browse. A second population crash occurred from 1948 through 1950 and possibly longer.

This pattern is not unlike that described for lemmings, but with a longer periodicity (time between peak populations) due to the slower reproductive rate of moose. Yet such wholesale overbrowsing is not common throughout the rest of the moose's range.

By 1956, browse plant species had recovered somewhat, the moose were increasing again, and another cycle of population boom and crash was under way. However, a new dimension was added with the arrival of wolves across frozen lake ice sometime between 1945 and 1950. The low wolf numbers probably had little effect on the moose decline in the early 1950s, but by 1960 two to three packs totaling over 20 wolves had established territories and were feeding mainly on moose throughout the island (Figure 16.11).

The wolves experienced a rapid increase in numbers similar to that of the moose and might have been expected to reduce the moose to extinction by the kind of "overgrazing" perpetuated by the moose on the vegetation. But there was a significant difference. Wolves have a highly organized social structure. Packs establish distinct territories. Within packs, social interactions limit reproduction mainly to the dominant pair in each pack. Thus the addition of new pups to the pack each year is limited more by social and territorial conditions than by biological potential.

A stabilized predator population (wolves) has helped stabilize the herbivore population (moose). From 1965 to 1974, populations fluctuated moderately and browse was never fully depleted as in the late 1920s. The moose population was lower than during the first population rise, suggesting that the removal of predation was a key contributor to the "herbivore irruption," which followed the arrival of the moose on Isle Royale.

This type of population stability may be more typical of vegetation–grazer–predator interactions than is the lemming example. In the Serengeti region of Africa (Chapter 19), short-term cycles of herbivore populations have not been noticed, and long-term changes appear to result from changes in climate and human intervention. The Serengeti system also contains several herbivore and predator species, some of which exhibit significant social interactions. We will apparently never know if the large buffalo herds of the American Plains experienced the relative stability of the Serengeti populations or showed significant cyclic variation in abundance.

Effects of Vegetative Change and Climate on Irruptions of Insect Populations

Spruce Budworm

The eastern spruce budworm is an economically important pest of commercial forests in North America. It feeds on mature fir and spruce trees in the boreal forest of the eastern United States and Canada and has been the object of intense efforts at population control, through pesticides, by forest managers. As with lemmings, irruptions of spruce budworm are also tied to changes in habitat structure. As the boreal forests change more slowly than tundra, the indications of the cyclic nature of these irruptions has emerged more slowly. An analysis of tree rings in the budworm's range suggests a 30- to 45-year periodicity to irruptions. When irruptions occur, they can spread over thousands of square kilometers in a period of 7 to 16 years.

The trigger for an irruption of the budworm is the existence of large areas of mature fir and spruce trees, the budworm's preferred food. In this vegetation type, fir increases in dominance with time, replacing the less tolerant birches and other deciduous species, which dominate following a disturbance. If succession is allowed to continue, fir tends to be replaced by spruce. However, this succession can be cut short at the point where the concentration of fir is sufficient to stimulate an outbreak of the budworm. This devastates both fir and spruce and reinitiates succession with a new stand dominated by deciduous species.

The human response to budworm irruption has been to spray with insecticides to preserve the economic value of the forest. While this postpones the defoliation caused by the budworm, it also allows further maturation of the forest and increasingly ideal conditions for a major irruption. The situation has been likened to the exclusion of small ground-fires from coniferous forests in the western United States where they would naturally occur (see Chapter 17). The short-term goal of avoiding economic losses is achieved, but the chance for a major loss increases.

Disturbance by fire, defoliation, or other agents is an intrinsic and necessary part of the function of most terrestrial ecosystems—a mechanism for reversing declining rates of nutrient cycling or relieving stand stagnation. The requirement for fire to reverse soil organic matter accumulation and increase nutrient cycling in coniferous forests has been known for some time (see Chapter 17). The ponderosa pine–mountain pine beetle interaction discussed in the last chapter is another example of this. In the cases of the budworm and the mountain pine beetle, stand break up, the reinitiation of succession, and the reversal of stand stagnation are facilitated by herbivory rather than fire.

Locusts

Climate may play as large a role as habitat in the irruptions of many insect species. A famous example is the migratory locusts of semiarid, tropical regions in Africa. These are the "plague" species, which irrupt three to four times per century and travel thousands of miles, increasing their range during the outbreaks from small isolated patches to large portions of the African continent (Figure 16.12). An irruption may last for 20 years. In actuality, small swarms of locusts form almost every year within the permanent resident areas but only occasionally reach densities high enough to induce large-scale movements. As with the lemmings, swarming is accompanied by large changes in morphology and behavior of individuals.

Population levels and swarming seem to be linked to a series of favorable (wet) years leading to high densities in the areas of permanent populations. However, changes in habitat and interactions with food source may also be a factor. The vast scale of the locusts' range and the relatively rare irruptions have hampered detailed research.

On a less dramatic scale, changes in peak populations of thrips on rosebushes in the gardens of an Australian scientist formed the basis for the theory that populations are generally limited by climate alone. While no longer accepted for most species, climatic regulation can be important in triggering outbreaks, as with locusts, or in limiting increases of insects at the edge of their ranges or in extreme climates.

ADDITIONS AND REMOVALS OF HERBIVORES AND PREDATORS

The extreme population fluctuations in moose at Isle Royale are one example of the potential effects of either adding or removing species from an ecosystem. There are many others, particularly in Australia and New

(a)

Sahara
Mauritania
Mali
Niger
Chad
Sudan
Saudi Arabia
Senegal
Burkina Faso
Guinea
Sierra Leone
Ivory Coast
Ghana
Togo
Benin
Liberia
Nigeria
Cameroon
Central African Republic
Ethiopia
Somalia
Gabon
Congo
Uganda
Kenya
Zaire
Malagarasi
Tanzania
Rukwa
Moera
Angola
Zambia
Malawi
Mozambique
Malagasy Republic
Namibia
Zimbabwe
Botswana
Rep. of So. Africa
Outbreak areas
Invasion area
20° 10° 0° 10° 20° 30°
20° 10° 0° 10° 20° 30° 40° 50°

(b)

Figure 16.12

Swarms of the desert locust. *(a)* Photograph showing local density during irruptions. *(b)* Map of the normal areas of locust occupation, and the much larger area over which they swarmed during a population irruption from 1927 to 1944. (*a*, G. Tortoli/H. Armstrong Roberts; *b*, Krebs 1972)

Zealand, where mammals did not exist before their introduction by European settlers. In this region, mammals have displaced many native marsupials and often altered the structure of ecosystems. The cyclic behavior seen in the moose example discussed above is often repeated in the first wave of migration into a new area. A generalized set of changes in animal vigor, vegetative food supply, and reproduction has been developed as a model for these newly arrived populations (Figure 16.13).

Immediately following arrival in the new area, excess food supply leads to high body fat contents, high reproductive rates, higher reproduction by younger age classes, and lower juvenile mortality. All of these characteristics decline as the population increases, until the death rate equals, and

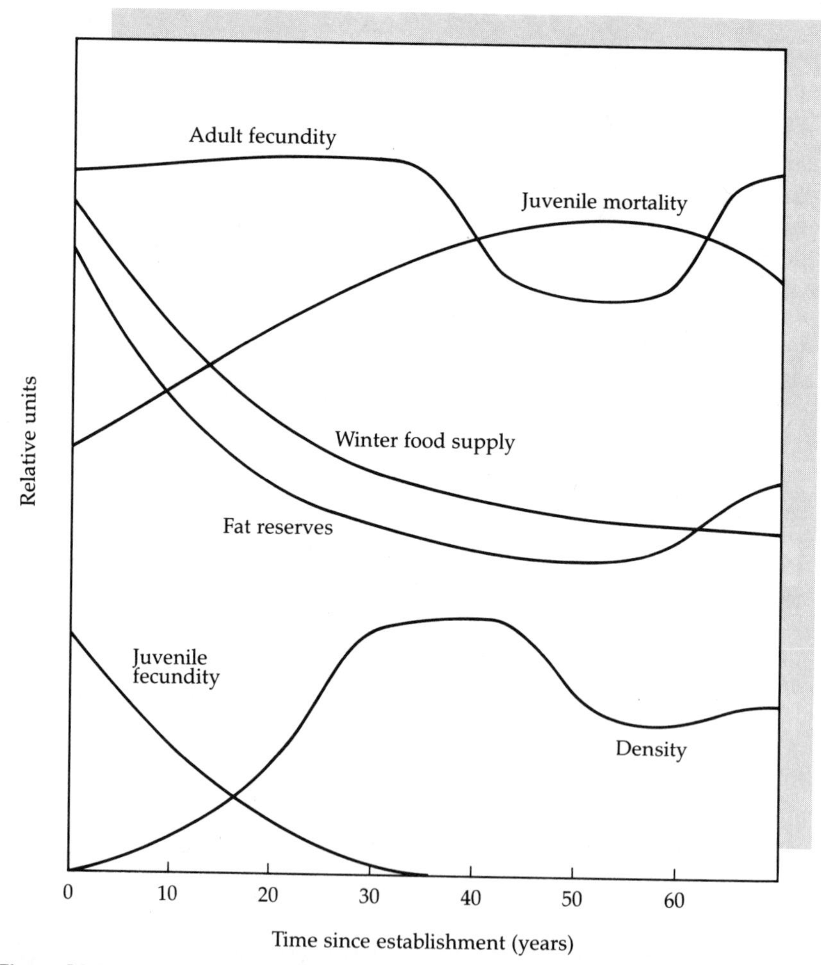

Figure 16.13

Generalized changes in several factors relating to vigor and reproductive success of large, herbivorous mammals introduced into a new area. (After Caughley 1970)

then exceeds, the birthrate. Total number of animals then declines, as in the moose example, until vegetation recovers and the cycle starts again.

The effects of high deer populations on regenerating forests in the Allegheny Plateau of Pennsylvania (Figure 15.2) is an example, not of a newly arrived species, but of predator removal and increased habitat quality caused by the creation of fields mixed with the remaining woods. The effort spent at maintaining and managing this herd, including managing removal by human predators, is a replacement for the population control originally carried out by wolves and other carnivores.

Another example of the effects of escape from "predators" is seen in the population dynamics, not of a herbivore, but of a weedy plant. Klamath weed, an exotic introduced to the United States around 1900 from Eurasia, is toxic to cattle. Following its arrival, it spread to become a major weed species in western grazing lands. In an attempt to control this pest, a natural herbivore (a flea beetle) was introduced in 1945. By 1950, the Klamath weed was reduced to very low population levels and was found only in shaded woodlands where the beetle did not go.

A similar case is the American chestnut tree. Once a dominant and massive species of the forests of eastern North America, it has been reduced to an occasional shrub or sapling by an introduced disease, the chestnut blight. If the disease had arrived before European settlement, we would probably have no idea that these weak and sickly shrubs had the genetic potential to be dominant forest trees.

These last two examples show how populations can be held to rarity by a consumer or parasite that both reproduces more rapidly and is more mobile than the host. There are undoubtedly thousands of such cases throughout the world, but their effect is difficult to see except in such uncontrolled, and often unfortunate, experiments. It is safe to say that little is known, in general, as to why rare species are rare.

We can conclude that most insect herbivores of woody communities are maintained at low population levels by biochemical alterations in leaf quality generated by plants. Irruptions of these herbivores tend to be related to reduced vigor or increased density of host plants. In these cases, the outbreaks may serve an important "restart" function in the ecosystem. In grassland systems, adaptations between the grasses and mammalian herbivores have led to interactions that can support continuously high levels of herbivory. In simple systems such as the tundra, in which the main herbivore reproduces more rapidly than the predators and neither group shows significant territoriality, large cyclic changes in herbivory can be expected. In more complex systems, with more species, and with predators exhibiting territoriality, these large cycles may be controlled. In either forests or grasslands, the addition or removal of important herbivores or predators can significantly alter levels of herbivory and change plant species composition. As we will see in the more complete discussion of the Serengeti system in Chapter 19, these changes can continue long after the added species has disappeared.

REFERENCES CITED

Bryant, J. P., F. S. Chapin III, and D. R. Klein. 1983. Carbon/nutrient balance of boreal plants in relation to vertebrate herbivory. *Oikos* 40:357–368.

Caldwell, M. M., et al. 1981. Coping with herbivory: Photosynthetic capacity and resource allocation in two semi-arid *Agropyron* bunchgrasses. *Oecologia* 50:14–24.

Caughley, G. 1970. Eruption of ungulate populations, with emphasis on Himalayan thar in New Zealand. *Ecology* 51:52–72.

Krebs, C. J. 1972. *Ecology: The Experimental Analysis of Distribution and Abundance.* Harper and Row, New York.

Krefting, L. W. 1974. Ecology of the Isle Royale moose, with special reference to the habitat. University of Minnesota Agricultural Experiment Station Technical Bulletin 297, St. Paul.

McNaughton, S. J. 1979. Grazing as an optimization process: grass–ungulate relationships in the Serengeti. *The American Naturalist* 113:691–703.

Mech, L. D. 1966. *The Wolves of Isle Royale.* Fauna of the National Parks of the United States Series 7. U.S. National Parks Service, Washington, D.C.

Odum, E. P. 1971. *Fundamentals of Ecology.* W. B. Saunders Co., Philadelphia.

Pastor, J., et al. 1988. Moose, microbes and the boreal forest. *BioScience* 38:770–777.

Peterson, R. O. 1977. *Wolf Ecology and Prey Relationships on Isle Royale.* National Parks Service Monograph Series, No. 11, Washington, D.C.

Pitelka, F. A., P. Q. Tomich, and G. W. Treichel. 1955. Ecological relations of jaegers and owls as lemming predators near Barrow, Alaska. *Ecological Monographs* 25:85–117.

Schultz, A. M. 1969. A study of an ecosystem: The arctic tundra. In Van Dyne, G. M. (ed.), *The Ecosystem Concept in Natural Resource Management.* Academic Press, New York.

ADDITIONAL REFERENCES

Andrewartha, H. G., and L. C. Birch. 1954. *The Distribution and Abundance of Animals.* University of Chicago Press, Chicago.

Dempster, J. P. 1963. The population dynamics of grasshoppers and locusts. *Biological Review* 38:490–529.

Good, N. F. 1968. A study of natural replacement of chestnut in six stands in the highlands of New Jersey. *Bulletin of the Torrey Botanical Club* 95:240–253.

Holling, C. S. 1981. Forest insects, forest fires, and resilience. In *Fire Regimes and Ecosystem Properties.* U.S. Department of Agriculture, U.S. Forest Service General Technical Report WO-26.

Huffaker, C. B. 1959. Biological control of weeds with insects. *Annual Review of Entomology* 4:251–276.

Nelson, T. C. 1955. Chestnut replacement in the southern highlands. *Ecology* 36:352–353.

Chapter 17 The Role of Fire in Carbon and Nutrient Balances

H. Armstrong Roberts

INTRODUCTION

Intense wildfires are among the most spectacular occurrences in nature. In a period of minutes to hours, a green and functioning forest, prairie, or shrubland can be reduced to ash. Plant and animal losses can be important both economically and aesthetically.

However, fire can also play an important positive role in the dynamics of ecosystems. Organic matter accumulated over long periods of time is oxidized and the nutrients contained are released. Long-term imbalances in the production/decomposition ratio can be equalized in minutes. While the released nutrients can remain in the system or be lost in smoke and gases, those that remain are generally in forms more available to plants. Cations can be present in the ash in sufficient quantity to increase soil pH significantly. The availability of light and water to the remaining plants, or to the new seedlings and sprouts, is also increased by removal of some or all of the previous vegetation. All of this can lead to increased vigor and growth of plants.

The human response to fire reflects this confusion between positive and negative effects. Fire has been used frequently by aboriginal peoples throughout the world to reduce pests, prepare land for farming, or increase forage available to grazing animals. Fire remains a very major land-clearing practice in the tropics, with consequences at both the local and global levels (see Chapter 24). In contrast, fire suppression has been the accepted forest management practice until recently throughout most of the Western world.

The purpose of this chapter is to discuss fire as it affects ecosystem processes. We will begin by describing the frequency and intensity with which fires occur in different ecosystem types. We will then look at the effects of fire on the availability of resources required for plant growth and how plants respond to the changed conditions, including methods of reproduction to ensure the presence of viable offspring following fire. Finally, we will discuss interactions between fire and herbivory and briefly address some of the important management questions regarding the use of fire in ecosystems.

FIRE FREQUENCY AND INTENSITY IN DIFFERENT TYPES OF ECOSYSTEMS

Fires in forest ecosystems can be separated into three major types (Figure 17.1*a*). **Surface fires** burn only along the soil surface, consuming accumulated litter on the forest floor and part of the organic matter in the forest floor. Shrubs and herbaceous plants in the understory are usually killed along with seedlings and perhaps saplings of dominant tree species. Surface fires are relatively cool and do not cause major changes in the visible structure of the forest.

In contrast, **crown fires** reach up into the crowns of the dominant trees, burning live foliage, branches, and stems as well as the forest floor. Crown fires are much hotter than surface fires and can kill a large proportion of the dominant trees in the system.

Ground fires occur in forests with particularly large accumulations of organic matter over the mineral soil or in bogs and wetland areas when drought dries out the usually wet surface organic mat. Ground fires generally burn very slowly over long periods of time and can be very effective at consuming accumulated surface organic matter.

In grassland systems, there is no distinction between surface and crown fires. Surface fires also consume the vegetation because there is no discontinuity between the soil surface and plant canopy (Figure 17.1*b*). Shrublands, such as chaparral (Figure 17.1*c*), also tend to burn completely with each fire.

While crown fires are generally more intense than ground fires, there are wide variations in intensity within each type. Fire intensity is described in terms of how hot the fire burns in one spot and for how long. The intensity of a fire depends, in turn, on the amount of organic matter present, how it is structured, and how moist it is.

Ecosystems vary widely in the frequency with which they burn. Tallgrass prairies may burn every 2 to 4 years. Pine forests in the Great Lakes region may experience crown fires every 70 to 100 years. Some of the dry forests of the Rocky Mountains may have frequent surface fires and only rare crown fires. In some systems, there may be important interactions between the frequencies of surface and crown fires. When surface fires occur frequently, they reduce fuel loads and reduce the chance of more severe crown fires.

(a)

(b)

(c)

Figure 17.1
Wildfires in different terrestrial ecosystems: *(a)* The three types of fires that occur in forests—surface fire (top), crown fire (middle), and ground fire (bottom). *(b)* A managed fire in a restored prairie ecosystem. *(c)* A fire in a chaparral system. (Photos by R. Krubner/H. Armstrong Roberts)

Taken together, the frequency, intensity, and types of fires that occur in an ecosystem determine that system's fire regime. A tremendous diversity of fire regimes can result from the interactions between these three factors. Figure 17.2 depicts some of this diversity for ecosystems in North America.

Fire is least common in the deciduous forests of the northeast. Sometimes called the "asbestos forests," these systems are dominated by trees whose broad leaves form compact layers of litter when they fall. This tends to reduce evaporation and retain moisture. Litter also decomposes

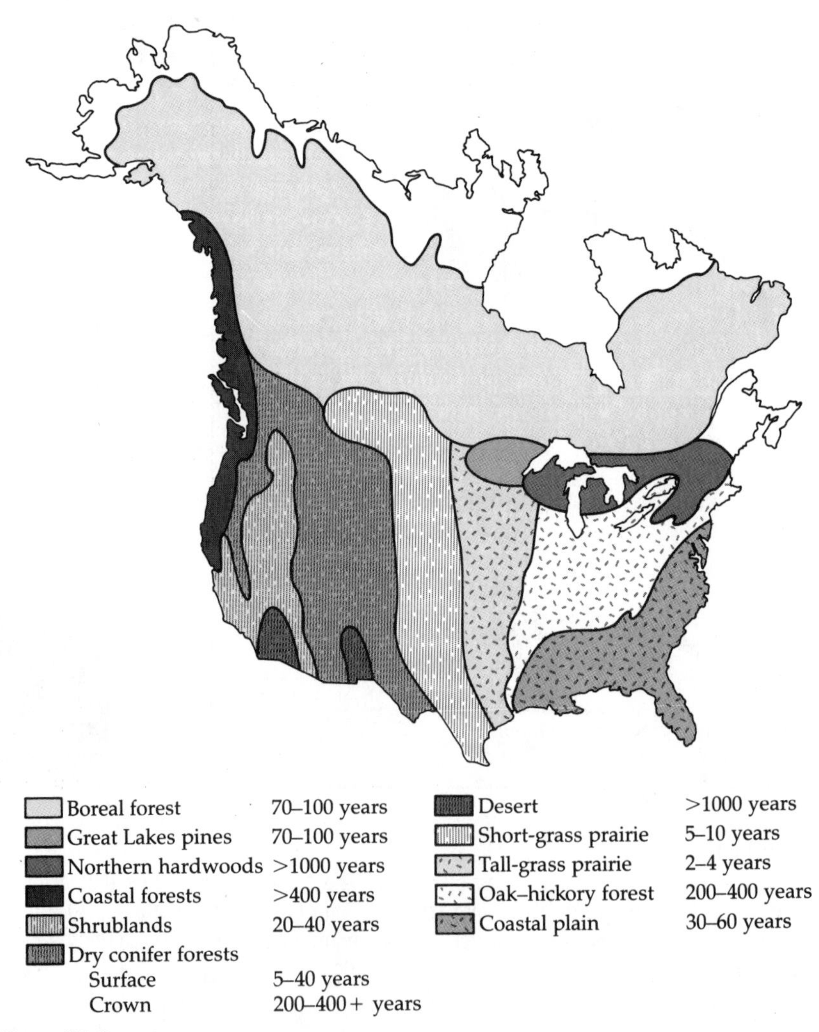

Figure 17.2

Generalized fire frequencies for several major vegetation regions of North America. (Heinselman 1973, 1981; Kilgore 1981; Gray and Schlesinger 1981; Christensen 1981; Bormann and Likens 1979)

relatively rapidly so that large accumulations of litter do not occur (fuel loading is low). These factors are enhanced by the generally humid climate and short warm season. However, fires are not unknown in this region. During particularly dry summers, and during historical periods when large-scale logging has left large quantities of slash (logging residues, branches, rotten stems, etc.) on the forest floor, some very large fires have occurred. Fire is more common in the oak–hickory forest region (Figure 17.2) because of warmer climates and an increased potential for drought. Still, fire frequencies are generally low.

Even within the humid Northeast, locally dry soil conditions have led to the development of ecosystems with high fire frequencies. Soils developed on sandy glacial outwash have very low water-holding capacity. Pitch pine is a dominant species in these pine plain areas. The needle litter of this species decays slowly and forms a loose, aerated litter layer, which dries out quickly following rains. Both litter and live biomass can have relatively high contents of "pitch," highly flammable resins. All these factors combine to cause a relatively high frequency of fires. These fires have severely altered both the appearance (Figure 17.3) and the function of these systems. In the central part of the pine plains area, fires may occur every three to four years. In the border areas surrounding the plains, fire is less frequent and forests of some stature can develop. Forest managers have experimented with the frequent use of deliberately set surface fires as a means of reducing fuel loadings and also reducing the potential for the greater losses caused by crown fires.

Figure 17.3
The pine plains area of New Jersey, where repeated fires have severely modified the structure and function of a forest ecosystem within the humid eastern-deciduous-forest climatic zone. (Fowells 1965)

These pine plain areas in the Northeast are the northernmost extension of a much larger area of pine-dominated forests, which occupy the coastal plain of the southeastern United States. Here, fire is considered instrumental in reversing nutrient stagnation due to organic matter accumulation, and stand-destroying crown fires may occur at intervals of 30 to 60 years. There is some evidence that this fire regime has replaced a presettlement pattern of frequent surface fires.

Fire frequency is somewhat longer in the pine forests of the Great Lakes region. This area was one of the first in which the regular occurrence of natural fires in a forest ecosystem was documented. Sediment cores taken from lakes (Figure 17.4*a*) show distinct layers of increased charcoal input and altered pollen deposition, conditions known to occur following fires. More recent evidence of fire impacts comes from maps of stand age (Figure 17.4*b*), which document the return interval of fire in the landscape. Recognition of this fire frequency led to the development of a theory of ecosystem development in which the "climax," or end state, is never achieved but rather movement toward this end state is continually disrupted and restarted by the occurrence of crown fires (Figure 17.4*c*).

Fire has always been a regular feature of the midwestern prairie and savannah systems. Early explorers' and settlers' accounts are filled with descriptions of the intensity of these fires and the huge areas that could be covered. The extension of tall-grass prairie into the lake states area, where climatic conditions should favor the growth of forests, has been linked to frequent tree-killing fires. Tall-grass prairies require fires at two- to four-year intervals to remain vigorous, while the short-grass prairies of the western plains burn less frequently. In the absence of fire, the build-up of dead grass stems severely restricts light penetration, slowing the warming of the soil in spring and reducing light availability to the sprouting grasses. It also shifts the competitive advantage to taller-growing shrubs and trees. Many forests in the eastern prairie states have developed from savannahs following the removal of fire as a result of the division of the landscape into farms and the plowing of the prairie sod. Bare, plowed fields make excellent firebreaks. These forests still contain the wide-crowned trees that developed in the savannah environment.

In the coniferous forests of the western United States, fires vary from a rare but potentially devastating force in the rainforests of the Olympic Peninsula to a dominant factor in the drier forests of the Cascade, Sierra, and Rocky mountains. It is in the drier forest systems that the interactions between ground fires and crown fires have become a major issue. Total fire exclusion in these forests results in a build-up of fuels, which sets the stage for very large and damaging fires (as discussed below). Several forest types are dominated by long-lived species (e.g., Douglas fir, redwood, giant sequoia, ponderosa pine). While the evolution of such long-lived species may suggest long periods without catastrophic crown fires, many of these systems require frequent ground fires to avoid high fuel loadings, which could lead to stand-destroying events. In the dry forest types, ground fires may occur at five- to ten-year intervals, while catastrophic fires may occur only once in several centuries.

The dry chaparral shrublands of coastal California are similar to prairies in that fires tend to consume the dominant plants as well as dead plant material. However, the accumulation of that material is slower, and fires occur on a 20- to 40-year cycle. Fire has also been thought to play an additional role in this type of system, that of destroying allelopathic chemicals, which accumulate in the soil during succession and which begin to reduce plant vigor over time.

The boreal forest, with its cold climate and low evaporative demand, would seem an unlikely system for fire. Yet the low mean temperature severely restricts decomposition and, together with the low quality of foliage produced by the dominant conifers, leads to large accumulations of organic matter. As we will see in the next chapter, fires tend to occur on a 70- to 100-year rotation and are crucial to maintaining a functional forest ecosystem.

In contrast, the hot and dry deserts, which would seem to have the best combination of climatic conditions for frequent fires, burn only rarely. Climatic conditions are so severe that accumulations of organic matter are too low to support extensive burns.

To summarize, fires are rare, and perhaps unimportant, only at the extremes of the climatic continuum across North America. Neither the most humid temperate forests, nor the driest deserts experience fire as a major factor. However, between these two extremes, in dry forests, cold forests, grasslands, and shrublands, fire exerts a major control on ecosystem structure and function.

EFFECTS OF FIRE ON RESOURCE AVAILABILITY

The simple act of burning organic matter has some fairly complex effects on soil structure and nutrient availability. The most immediate effects are the gaseous loss of carbon and nitrogen from surface organic layers and, in the case of crown fires, from vegetation. Most of the other macronutrients remain in the ash and may be deposited on the soil surface. However, in hot fires, which are often accompanied by strong winds generated by the fire itself, wind-blown ash may remove phosphorus and cations from the site as well. Table 17.1(A) summarizes losses due to fire in two different experimentally burned systems and a major natural wildfire.

Cation nutrients that remain in the ash, such as calcium, magnesium, and potassium, tend to be in both mobile and plant-available forms (Table 17.1(B)). When rainfall follows a burn, these mobile cations move down through the soil profile and tend to displace hydrogen ions from exchange sites in the soil, increasing soil pH. Increases can be dramatic in acid forest soils but tend to be minor and unimportant in grassland and shrubland soils, which are nearly neutral even before fires.

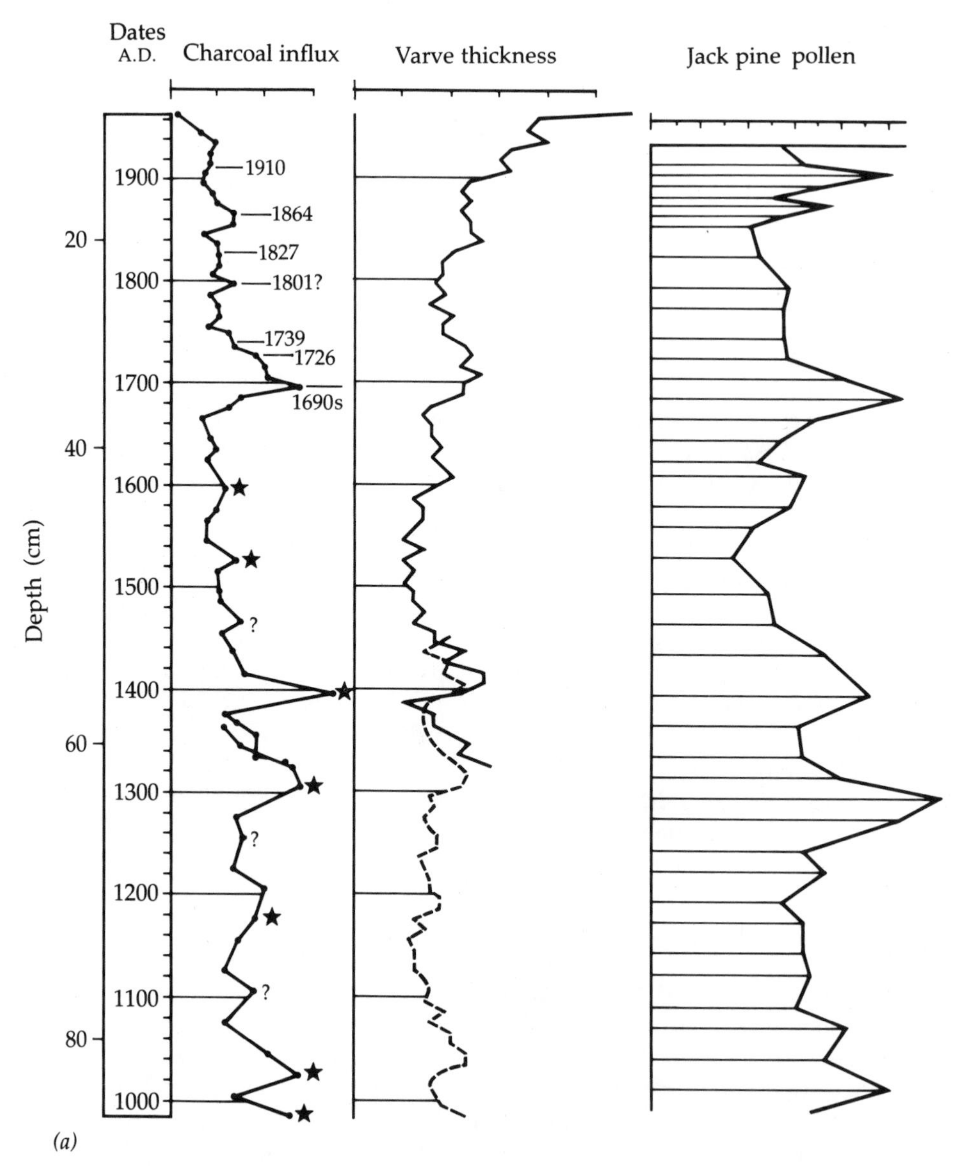

Figure 17.4
Crown fires, which restart ecosystem development, are an intrinsic part of the Great Lakes pine forests. Evidence of repeated fires in the Boundary Waters Canoe Area is present in both *(a)* extracted lake sediment cores, which show repeated layers enriched in charcoal and silt (varve thickness) and altered pollen deposition (stars and dates mark the occurrence of major fires) (Swain 1973), and *(b)* stand maps showing age distribution and time since the last major fire (Heinselman 1973). *(c)* Recurring fires lead to a repeated pattern of truncated succession where ecosystem processes never reach a steady-state, or "climax," condition (after Loucks 1970).

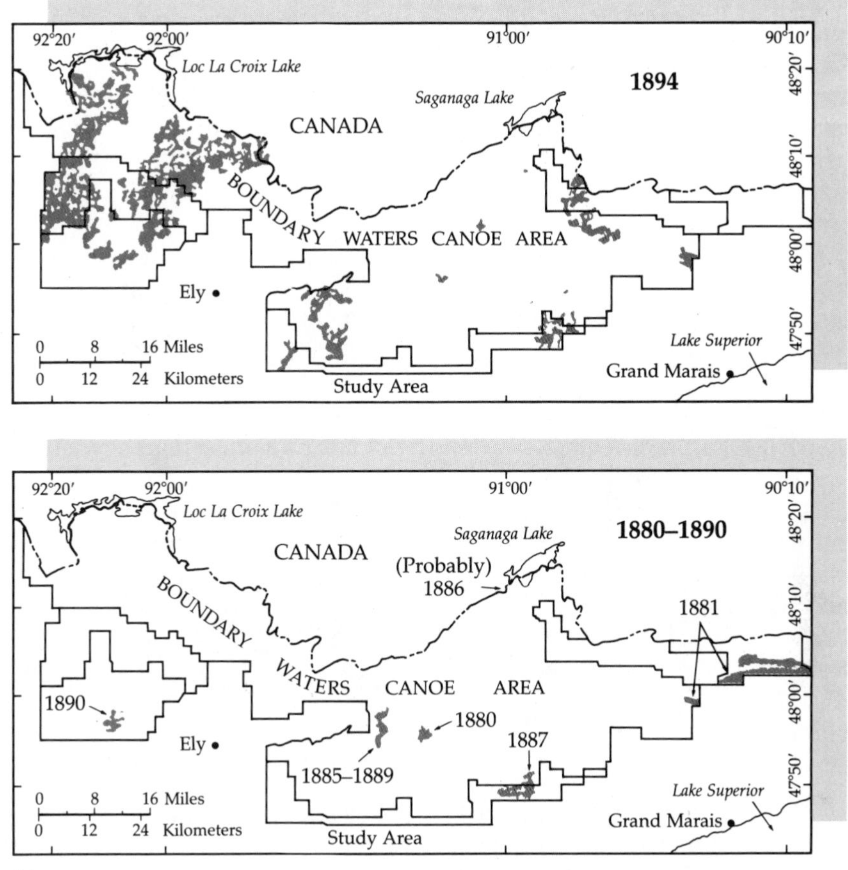
92°20′
92°00′
91°00′
90°10′
Loc La Croix Lake
1894
Saganaga Lake
CANADA
BOUNDARY WATERS CANOE AREA
48°20′
48°10′
48°00′
47°50′
Ely
0 8 16 Miles
0 12 24 Kilometers
Study Area
Lake Superior
Grand Marais
92°20′
92°00′
91°00′
90°10′
Loc La Croix Lake
Saganaga Lake
1880–1890
CANADA
(Probably) 1886
1881
BOUNDARY WATERS CANOE AREA
1890
1880
1887
1885–1889
Ely
0 8 16 Miles
0 12 24 Kilometers
Study Area
Lake Superior
Grand Marais

(b)

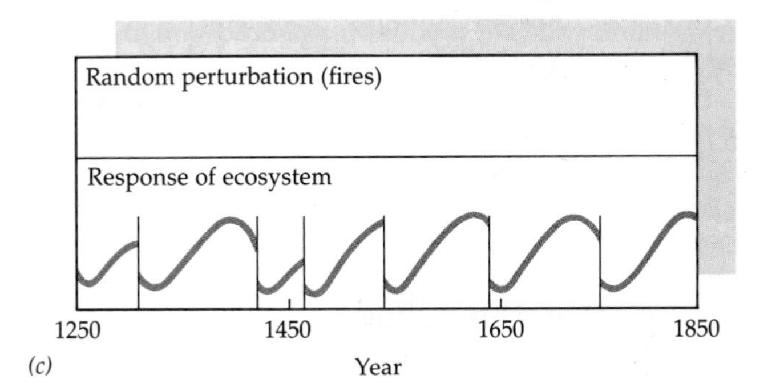
Random perturbation (fires)
Response of ecosystem
1250
1450
1650
1850
Year

(c)

Table 17.1 Changes in nutrient distribution following fire: (A) nutrient losses from experimental surface fire in a longleaf pine stand,* experimental crown fire in chaparral,† and major wildfire in a mixed conifer forest in central Washington;‡ (B) percentage of soluble nutrients in ash from the longleaf pine burn§

	NUTRIENT LOSSES IN kg/ha (% OF TOTAL)			
	N	P	CA	K
A. System				
Longleaf pine	22.3 (66)	1.6 (53)	4.1 (20)	4.2 (40)
Chaparral	146 (10)	11 (2)	35 (.4)	48 (12)
Plants + litter only	110 (39)	0.2 (—)	9 (—)	46 (16)
Mixed conifer	907 (39)		75 (11)	308 (35)

B. Nutrient	
N	0.4%
P	6.7%
K	58.9%
Ca	22.4%
Mg	90.8%

*From Christensen 1977.
†From Debano and Conrad 1978.
‡From Grier 1975.
§Nutrients that were mobile and available to plants; from Christensen 1977.

The dark color of the ash, plus partial or total removal of plant cover, increases absorption of solar energy at the soil surface, increasing soil temperature. Evapotranspiration may also be lower, due to lower foliar biomass, resulting in higher water content in soils. Combined with higher pH, these factors tend to increase microbial activity and may further add to nutrient availability by increasing rates of mineralization of the remaining soil organic matter. This effect may be further augmented by alterations in the chemistry of remaining organic matter by the heat of the fire and also by a narrowing of the C:N ratio. Both higher soil pH and increased availability of ammonium may increase the rate of nitrification and the availability of nitrate. Figure 17.5 shows dramatic increases in the accumulation of both ammonium and nitrate in chaparral soils following fire.

It has also been suggested that volatile organic compounds (terpenoids) produced by plants in ponderosa pine and other dry forest and shrubland ecosystems chemically inhibit both mineralization and nitrification. Rapid and immediate increases in net mineralization and nitrification following fire in these systems may result from destruction of the inhibiting compounds by the heat of the fire.

However, the effects of fire on nutrient availability are not always positive. Short-term increases in nutrient availability may be offset by

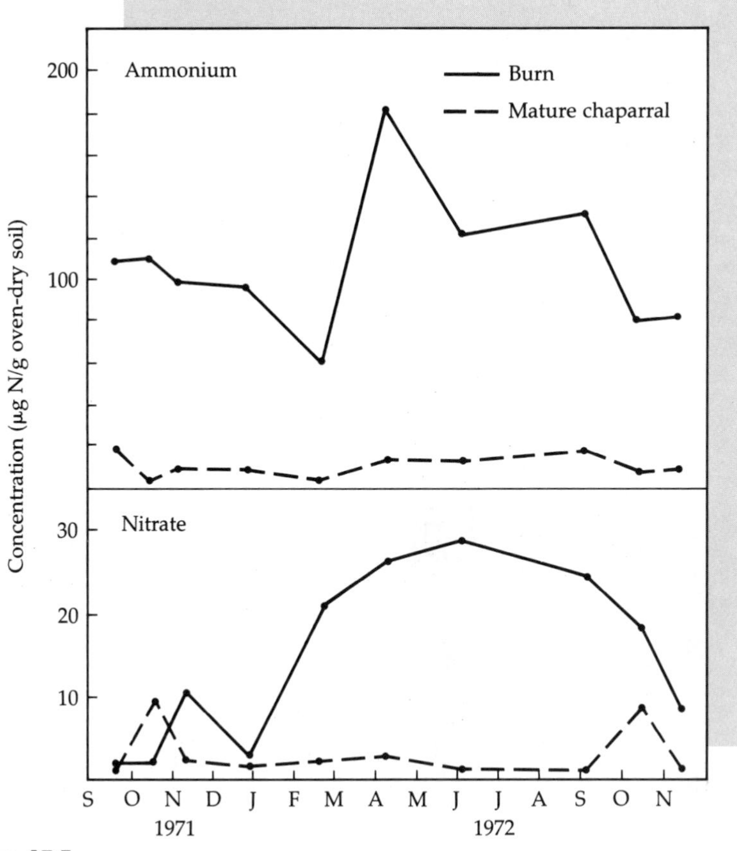

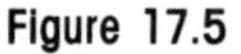

Figure 17.5
Concentrations of ammonium and nitrate in burned and unburned chaparral soils. (Christensen 1973)

long-term decreases in systems where fire frequency is high and inputs to the system between fires are not high enough to replace losses. Both annual and periodic (4-year interval) surface fires repeated for 30 years in an oak–hickory forest resulted in significant decreases in nitrogen mineralization rates (Figure 17.6).

In addition, the increased mobility of nutrients in soils can lead to further losses from the system due to leaching and erosion. Wind erosion may account for some additional losses from burned systems, although it is difficult to judge the net effect of wind erosion, as redeposition in rainfall and by settling out of windblown particles can increase apparent inputs to a burned area.

To summarize, surface fires tend to increase the pH of acid soils and to increase the availability of phosphorus and cations. Significant amounts of carbon are lost and the C : N ratio of soils is lowered. This may result in

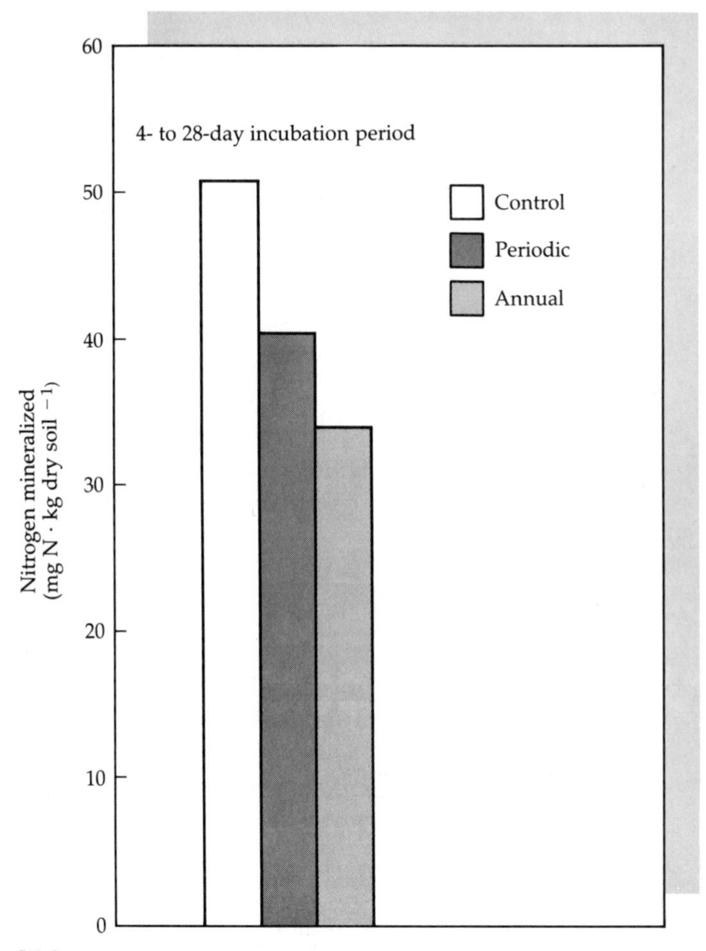

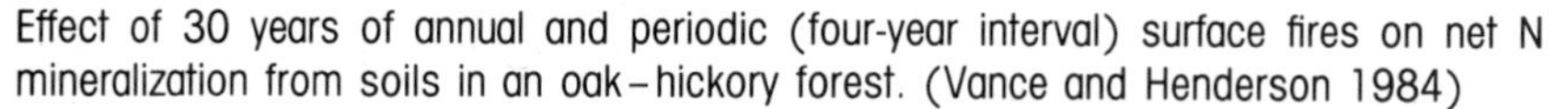

Figure 17.6

Effect of 30 years of annual and periodic (four-year interval) surface fires on net N mineralization from soils in an oak–hickory forest. (Vance and Henderson 1984)

increased N mineralization, but short-term increases in availability may be offset by longer-term decreases in total soil N content.

Fire may also have longer-term effects on the basic structure of soils. Removal of surface organic horizons and the leaching or redistribution of nutrients throughout the soil profile will tend to result in a more even and deeper distribution of fine roots throughout the profile. This in turn affects deposition of root litter, causing a more even distribution of organic matter and future nutrient mineralization.

PLANT RESPONSES TO ALTERED RESOURCE AVAILABILITY

Plant response to the environment following surface fires in forests generally, but not always, includes increased nutrient concentration in foliage. However, these increases may be short-lived if foliar biomass increases. Increased nutrient uptake may be evident more through higher foliage biomass than higher nutrient concentrations. Whether, or under what conditions, surface fires actually increase tree growth is still an open and important economic question.

In grasslands, where rapid and vigorous resprouting of burned vegetation occurs, above-ground productivity is generally increased in the post-fire environment, especially in wet years (Table 17.2). While much of this apparent growth is actually the translocation of carbon and nutrients from storage organs below ground, similar to that which occurs following grazing, the reduction of shading by dead material from previous years removes a major impediment to early vigorous growth.

In very hot ground fires, or the catastrophic crown fires, losses of nitrogen can be great enough to reduce N availability significantly in the post-fire environment. With high availability of other nutrients, as well as of water and light, species with nitrogen-fixing capabilities may have a competitive advantage. Several ecosystem types that experience crown fires have N-fixing species as a major component of the post-fire vegetation. These include species of alder in the Douglas fir region (see Chapter 14), locust (*Robinia* spp.) in the drier forests of the Southwest, and *Ceanothus* in the chaparral. Alder and *Ceanothus* in particular have the capacity to replace nitrogen lost to fire within a few years.

Table 17.2 Increases in above-ground productivity in prairies following fires*

REGION	CONTROL	FIRE
Illinois	3,020	13,210†
	3,610	5,910‡
Missouri	5,090	9,330†
	4,820	5,220‡
Iowa	3,490	7,500
Eastern Kansas	1,860	3,400
Western Kansas	3,800	1,710

*From Kucera 1981.
†Wet years. All values are expressed in $kg/ha^{-1}/yr^{-1}$.
‡Dry years.

PLANT ADAPTATIONS TO DIFFERENT FIRE REGIMES

As with grazing, fire represents an important selective force in the evolution of species in those systems in which it regularly occurs. Thus several widely distributed reproductive characteristics have evolved in response to fire. These can be summarized in three broad categories: seed dispersal in response to fire, seed germination in response to fire, and protection or production of vegetative parts below ground. These three types of adaptations can be found to different degrees in different types of ecosystems.

In forests, the first and third methods are most commonly found. Most of the fire-adapted forest ecosystems are dominated by conifers. Several of the species that dominate these areas have what are called *serotinous cones* (Figure 17.7). Serotinous cones do not open and shed seed in the same year that they are formed, as generally happens with nonserotinous cones. Rather, the scales of the cone are bound together by a resin-like substance that softens only under the conditions of high temperature occurring during crown fires or intense ground fires. Closed, serotinous cones can remain on the tree for up to 15 years in black spruce or 25 years in jack pine. Viable seeds have even been found in cones that have become overgrown by the woody stem or branches of the tree. Serotinous cones cause maximum seed dissemination to occur immediately after a fire, at times when resource availability is high and competition from other plants is low. Seeds within the serotinous cones are generally not damaged by

Figure 17.7
Serotinous cones are held on the tree and kept tightly closed by a resin-like substance. Fire softens this sealant and results in the opening of the cones soon after the burn occurs.

fire unless the cone itself is burned. The cones are resistant to burning and can withstand temperatures as high as 300°C for 60 seconds.

An unusual method of protecting buds (or growing apical tips) from fire occurs in longleaf pine, a species that grows in the fire-prone coastal plain of the southeastern United States. Following germination and seedling establishment, this species remains in a "grass stage" (Figure 17.8) for 5 to 20 years. During this time, a large taproot is produced containing large quantities of stored carbohydrates. The apical bud is embedded in the surrounding erectly displayed needles, which are resistant to burning and protect the bud. After reaching a critical size, the seedling begins very rapid height growth, up to 1 meter per year. This growth pattern is geared to the frequent occurrence of ground fires. The seedling is first protected in the grass stage until sufficient reserves have been accumulated so that rapid height growth will carry the growing tip above the level of ground fires in only a few years.

Many tree species of frequently burned areas also have the ability to root-sprout following fire. Viable below-ground buds are present in most oak species, aspen, black spruce, redwood, and even a few species of pine (e.g., pitch pine). It is thought that the ability of oak to survive and sprout following fire is the reason that it dominates many areas in the western portion of the eastern hardwood forest, where reproduction of oak occurs only sporadically now, in the absence of fire.

Figure 17.8
Stages in the growth of longleaf pine: *(a)* Seedlings in the grass stage are nearly indistinguishable from real grasses. *(b)* Height increase may be more than 1 meter per year after the grass stage is broken. (U.S. Forest Service)

Grassland ecosystems show both bud protection and increased reproduction as responses to fire. Perennial grasses and forbs of prairie systems characteristically maintain large root and rhizome systems from which to reproduce leaves and stems following either fire or grazing. Many species also show very large increases in flowering and seed production following fires. Again, flowering following fire may be a response to increased resource availability and is also adaptive in that fire may create openings in the dense prairie sod, in which seedling establishment is otherwise very difficult.

In chaparral systems, vegetation is reestablished following fire by both sprouting of underground buds and increases in germination of dormant seeds in the soil in response to the elevated temperatures caused by the fire.

In Chapter 16 we presented the idea that successful adaptation to herbivory might then cause further selection for plant traits that would encourage this form of "disturbance." Might this also be true for plants that have adapted successfully to fire? Whether plants actually have evolved characteristics that increase fire frequency or intensity has been argued for some time. There are suggestions that the resin content of conifer plant mass and the open, easily dried structure of conifer forest floors are adaptations that increase the frequency and intensity of fire. Similarly, the retention of dried biomass on certain chaparral plants, as well as the high content of volatile and flammable chemical compounds in these species, may be an adaptation for increasing fire frequency. The large areas of dried, fine-structured plant mass, which occur in prairie systems between midsummer and the following spring, certainly offer an optimal physical structure for promoting burning.

However, all of these characteristics also affect other aspects of plant function. We have discussed the role of leaf and canopy structure in response to energy and water balances (Chapters 6 and 7) and the role of secondary plant chemicals as controllers of herbivory (Chapter 15). It is still unclear to what extent the adaptive advantage of increasing fire frequency has contributed to the evolution of plant structure and biochemistry.

FIRE–HERBIVORY INTERACTIONS

Throughout this part of the book we have discussed alternate pathways for the movement of fixed carbon and energy. It is not surprising that increasing movement along one pathway would decrease movement along another. Thus, inverse relationships have been reported between herbivory and fire occurrence. For example, fire frequency in the grasslands of the Serengeti region of Africa (Chapter 19) has decreased during a period of increasing abundance of major herbivores and higher plant biomass removals through grazing.

FIRE AND THE MANAGEMENT OF ECOSYSTEMS

Fire has proven to be a controversial tool in the management of ecosystems. Early attitudes in the United States toward fire were derived from experiences in European and eastern North American forests, where fire was relatively unimportant in ecosystem development and the effect of fire was a reduction in economic and aesthetic value. Transplanting these ideas into the drier ecosystems of the western United States has caused significant problems.

One important example is with the management of the giant sequoia forests of California. These groves, which contain the most massive trees in the world, developed under a regime of frequent ground fires. In the period before European settlement, a fire occurred somewhere within the groves in and around Kings Canyon National Park nearly every year. Individual trees experienced minor fire damage, as indicated by the accumulation of fire scars in the stems, on an average of once every 10 to 40 years. There are indications that many fires were deliberately set by Native Americans.

The ground fires served to reduce fuel loads by burning off accumulations of organic matter on the soil surface and by killing saplings and seedlings. Mature sequoias were not seriously harmed by these fires because of their very thick bark, and the chance of a major crown fire was reduced by the tremendous distance between the ground and the lowest branches of the trees.

The incidence of fire was reduced by the displacement of Native Americans by Europeans beginning around 1870. Fire suppression became an official and effective policy in this area around 1900. Fire scars on trees declined markedly after 1875, and have been nearly absent since 1900.

Fire suppression has led to the accumulation of surface fuels, and perhaps more ominously, to a tremendous increase in the abundance of white fir in the understory. White fir is a relatively shade-tolerant, fire-sensitive species that was kept to low levels in the presettlement forest by ground fires. A marked increase in establishment of stems of both white fir and sugar pine accompanied the reduction in fire frequency (Figure 17.9). The development of this understory has altered the appearance of the giant sequoia groves and also provides a potential "fire ladder" for the movement of ground fires into the upper canopy. The fear now is that once started a fire may "crown" and destroy this venerable and majestic stand. The use of both controlled ground fires (prescribed burns) and partial logging of the understory are under consideration as means of reducing this hazard and restoring the park to its presettlement condition. However, this type of human intervention in the national parks remains controversial, even though the policy of fire suppression itself represents a major disturbance to the native ecosystem.

Prescribed burning is also of interest in the management of commercial forests, both in the national forests and on private lands. The same concept, that of fire exclusion leading to large fuel accumulation, and in

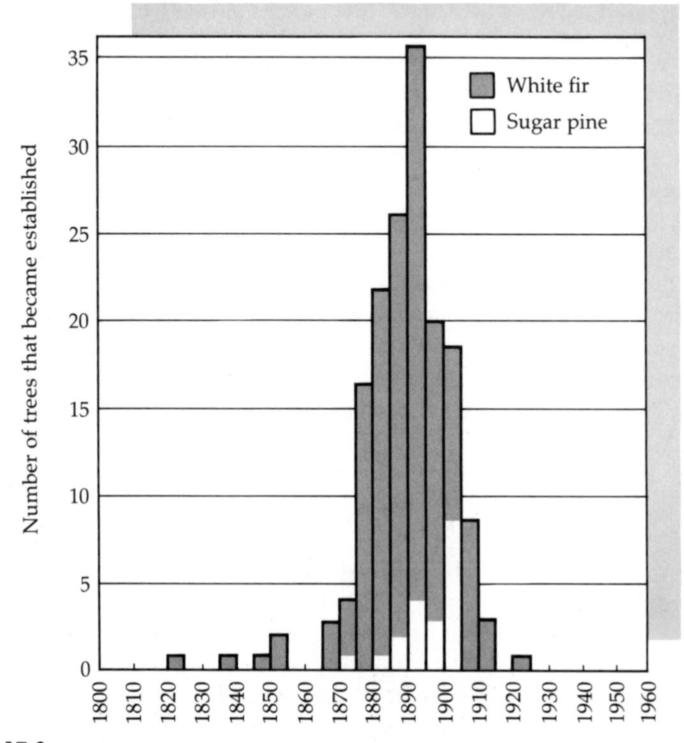

Figure 17.9
Timing of the establishment of pole-sized understory white fir and sugar pine trees in King's Canyon sequoia stands. Note the peak following fire suppression in the late 1800s. (Kilgore and Taylor 1979)

turn to catastrophic fires, is an important concern in the economic management of forests.

The potential for forest protection through prescribed burns was dramatically illustrated in the Coconino National Forest in Arizona. As part of a research program, prescribed burns had been carried out in one part of this forest for several years. In 1980, a major crown fire broke out in the forest and moved toward the experimental site. However, fuel loads had been reduced in the prescribed burn area to such an extent that the fire could not be supported. Figure 17.10 shows how the fire burned up to and around the experimental area, but not through it.

The effects of fire on nutrient availability productivity in forests continues to be debated. Both the immediate effects on productivity and the longer-term effects of forest protection and reduced risk of crown fires are weighed against the risk of damage to the stands from the prescribed burn and the cost of setting and controlling the fires.

Very early work establishing the necessity of fire in the maintenance of

Figure 17.10
Aerial view of part of the Coconino National Forest following a major forest fire. The remaining green (dark) area marks the location of an experimental set of prescribed burns that had significantly reduced fuel loads. (Courtesy of Dr. Wallace Covington)

tall-grass prairies was carried out under a very different type of management plan. Efforts to restore or recreate prairies from abandoned farmland in Wisconsin were only marginally successful until fire was reintroduced on a regular schedule. Now the biennial burns are the major management tool used to perpetuate this once nearly extinct type of ecosystem (Figure 17.1*b*).

If our view of fire has shifted away from one of suppression at all costs, are there still cases where fire occurrence has negative effects and ought to be reduced? Surely in areas where fire is rare but climatic conditions may occasionally be conducive to fire, and where human activity has greatly increased the ignition of fires due to camping, cigarettes, etc., the argument can be made that fire suppression is a way of reducing human impacts on the ecosystem. Prevention of crown fires in the giant sequoia groves may be an example of a case where suppression is required to reduce the potentially catastrophic loss of those stands until the understory and fuel loads can be reduced.

If there is an emerging principle, it is that the economically and aesthetically important species in native ecosystems are well adapted to the naturally occurring fire regimes under which they have evolved. Maintaining the integrity and sustainable productivity of these systems may depend on mimicking or only modifying those regimes.

REFERENCES CITED

Bormann, F. H., and G. E. Likens. 1979. *Pattern and Process in a Forested Ecosystem.* Springer-Verlag, New York.

Christensen, N. L. 1973. Fire and the nitrogen cycle in California chaparral. *Science* 181:66–68.

Christensen, N. L. 1977. Fire and soil–plant nutrient relations in a pine-wiregrass savanna on the coastal plain of North Carolina. *Oecologia* 31:27–44.

Christensen, N. L. 1981. Fire regimes in southeastern ecosystems. In *Fire Regimes and Ecosystem Properties.* U.S. Department of Agriculture, U.S. Forest Service General Technical Report WO-26.

Debano, L. F., and C. E. Conrad. 1978. The effect of fire on nutrients in a chaparral ecosystem. *Ecology* 59:489–497.

Fowells, H. A. (ed.). 1965. *Silvics of Forest Trees of the United States.* Agricultural Handbook No. 271. U.S.D.A., Washington, D.C.

Gray, J. T., and W. H. Schlesinger. 1981. Nutrient cycling in Mediterranean type ecosystems. In Miller, P. C. (ed.), *Resource Use by Chaparral and Matorral.* Springer-Verlag, New York.

Grier, C. C. 1975. Wildfire effects on nutrient distribution and leaching in a coniferous ecosystem. *Canadian Journal of Forest Research* 5:599–607.

Heinselman, M. L. 1973. Fire in the virgin forest of the Boundary Waters Canoe Area, Minnesota. *Quarternary Research* 3:329–382.

Heinselman, M. L. 1981. Fire intensity and frequency as factors in the distribution and structure of northern ecosystems. In *Fire Regimes and Ecosystem Properties.* U.S. Department of Agriculture, U.S. Forest Service General Technical Report WO-26.

Kilgore, B. M. 1981. Fire in ecosystem distribution and structure: Western forests and shrublands. In *Fire Regimes and Ecosystem Properties.* U.S. Department of Agriculture, U.S. Forest Service General Technical Report WO-26.

Kilgore, B. M., and D. Taylor. 1979. Fire history of a sequoia–mixed conifer forest. *Ecology* 60:129–142.

Kucera, C. L. 1981. Grasslands and fire. In *Fire Regimes and Ecosystem Properties.* U.S. Department of Agriculture, U.S. Forest Service General Technical Report WO-26.

Loucks, O. L. 1970. Evolution of diversity, efficiency and community stability. *American Zoologist* 10:17–25.

Swain, A. M. 1973. A history of fire and vegetation in northeastern Minnesota as recorded in lake sediments. *Quarternary Research* 3:383–396.

Vance, E. D., and G. S. Henderson. 1984. Soil nitrogen availability following long-term burning in an oak–hickory forest. *Soil Science Society of America Journal* 48:184–190.

ADDITIONAL REFERENCES

Knapp, A. K., and T. R. Seastadt. 1986. Detritus accumulation limits productivity of tallgrass prairie. *BioScience* 36:662–668.

Schoch, P., and D. Binkley. 1986. Prescribed burning increased nitrogen availability in a mature loblolly pine stand. *Forest Ecology and Management* 14:13–22.

U.S.D.A. 1981. *Fire Regimes and Ecosystems Properties.* U.S. Forest Service General Technical Report WO-26.

White, C. S. 1986. Volatile and water-soluble inhibitors of nitrogen mineralization and nitrification in a ponderosa pine ecosystem. *Biology and Fertility of Soils* 2:97–104.

Part III

SYNTHESIS Disturbance, Succession, and Ecosystem Function in Selected Systems

Sven-Olof Lindblad / Photo Researchers

In Parts I and II of this book, we introduced concepts and patterns related to the study of ecosystems and then took those systems apart to look at the specific mechanisms that drive ecosystem function. In this third section we will try to put the pieces back together again through the analysis of specific systems.

It is impossible here to attempt even a cursory review of all of the major terrestrial ecosystem types. Rather, we will select three systems that differ widely in terms of disturbance history and the relative importance of herbivory, fire, and decomposition in the oxidation of organic matter. We will present a synthetic view of function in a fire-dominated system (the taiga forests of interior Alaska), an herbivory-dominated system (the Serengeti plain of Africa), and a system where these two forces are relatively minor factors (the northern hardwood forests of eastern North America). In each case we will focus on a single, large-scale, ecosystem-level study carried out in the system of interest.

Over most parts of the Earth, there is a fourth and distinctly different factor that both alters productivity and accounts for important removals of organic matter and nutrients from ecosystems: human activity. The impacts of this factor are the subject of Part IV. In this part we will concentrate on the function of relatively undisturbed systems while discussing the implications of ecosystem function for human management.

Chapter 18 A Fire-Dominated Ecosystem: The Taiga Forests of Interior Alaska

H. Armstrong Roberts

INTRODUCTION

The taiga, or boreal forest, covers a large, circumpolar area of the northern hemisphere (Figure 18.1). Extreme cold and a short growing season are the dominant environmental factors. The dominant tree genera include spruce and pine, with deciduous genera such as birch and aspen (poplar) being important in disturbed, warm, or geologically young sites. In newly exposed areas created by the retreat of glaciers or the shifting of riverbanks, nitrogen-fixing species of alder can play an important role. Soils are generally Inceptisols, tending toward Spodosol because of the cold, humid climate. Much of the region contains flat to gently sloping terrain. Small differences in slope and elevation can be associated with large changes in soil water content and the presence of water-saturated conditions. Permafrost (permanently frozen soil) occurs under much of the boreal forest (Figure 18.1), and the depth to which soils thaw in summer can be an important determinant of the rate at which processes occur and can in turn be determined by the degree of organic accumulation over soils. In parts of the boreal forest, fire plays an important role in correcting the imbalance between production and decomposition due to the slow decay rates caused by low temperatures and low-quality litter.

The purpose of this chapter is to present results from a long-term, detailed study of ecosystem dynamics in the taiga of interior Alaska. We will examine how both site conditions and the occurrence of fire affect the distribution of tree species and the operation of important ecosystem processes.

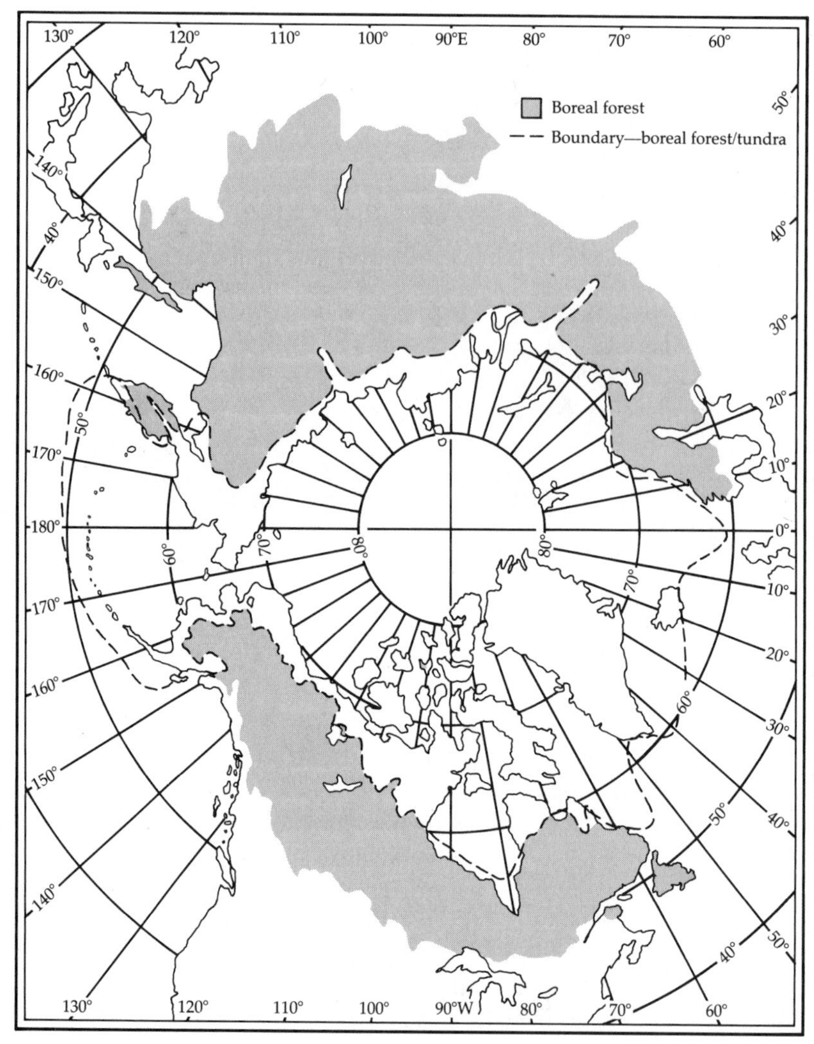

Figure 18.1
Circumpolar view of the distribution of taiga forests and of permafrost. (Van Cleve and Dyrness 1983)

THE TAIGA FORESTS OF INTERIOR ALASKA

The taiga system of interior Alaska is representative of the larger boreal forest region. It lies between the Alaska and Brooks mountain ranges (Figure 18.2) and is quite distinct from the more productive forests of the south coast and the tundra systems to the north. The climate in the taiga zone is continental, meaning that the moderating effect of marine air masses is minimal and that temperature is −25°C. In July it is +16°C.

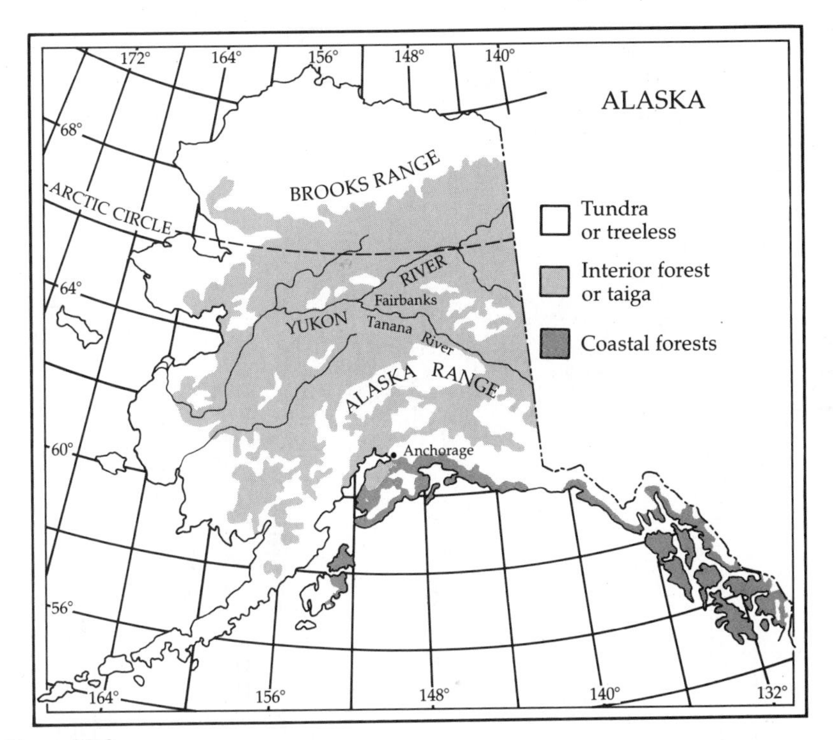

Figure 18.2
Distribution of taiga forests in Alaska. (Van Cleve et al. 1983, © 1983 by the American Institute of Biological Sciences)

Average rainfall at Fairbanks is a relatively low 28.6 cm, but the low average annual temperature reduces total annual evaporative demand such that water availability is still sufficient to support the growth of forests. A full 90% of the young, slowly weathering soils are Entisols or Inceptisols and 7% are Histosols, leaving only 3% that show significant profile development.

Within the Alaskan taiga, topographic location plays a large role in determining the relative importance of the four major tree species through interactions with soil temperature and forest floor depth (Figure 18.3). Ridge tops and south-facing slopes with well-drained soils and warmer microclimates are dominated by the deciduous species quaking aspen and paper birch. These stands may be successional to white spruce on deeper, midslope soils. Poorly drained sites are dominated by black spruce with an understory of feathermoss. On very poorly drained sites, black spruce stands become open and a continuous cover of sphagnum moss develops over the forest floor. In the oldest and most open of these stands, the productivity and nutrient cycling through the moss can be greater than through the spruce trees. Floodplains adjacent to rivers are dominated by the deciduous balsam poplar.

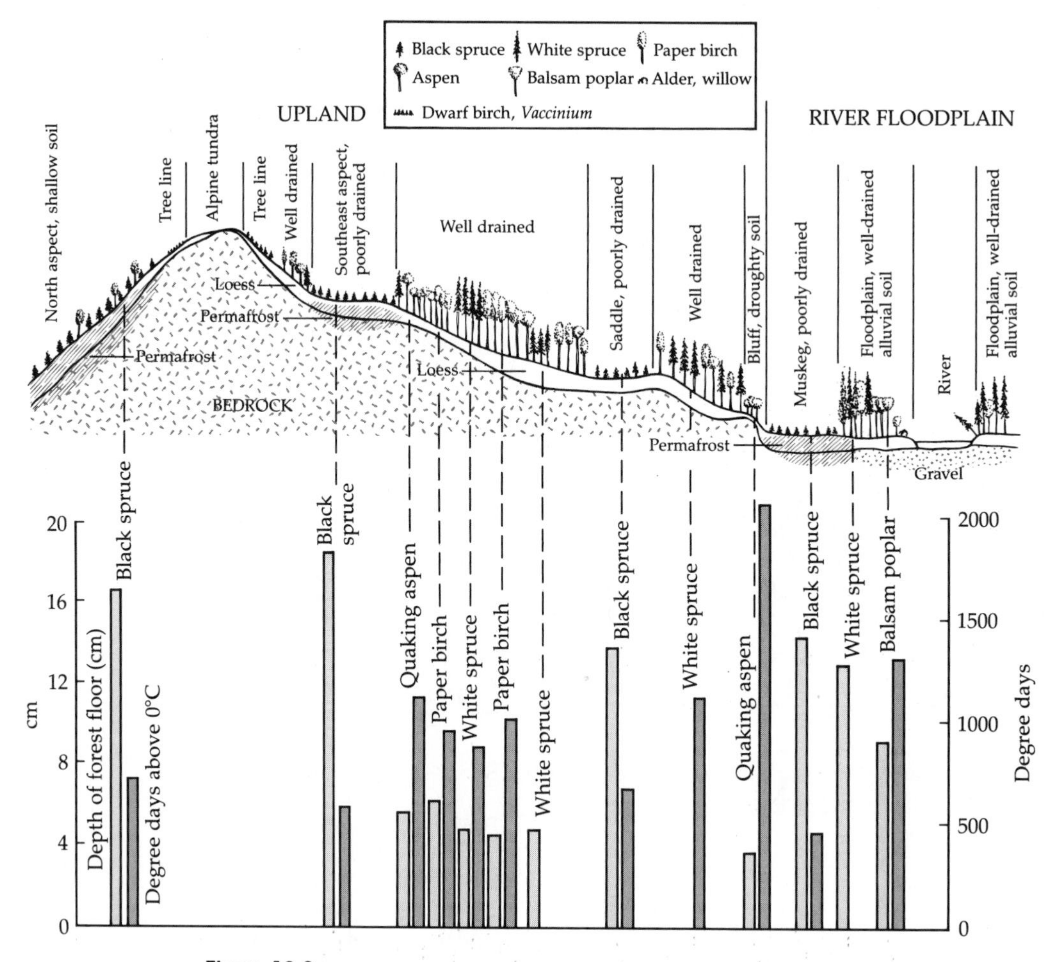

Figure 18.3
Distribution of major vegetation types in the Alaskan taiga zone in relation to topography, forest floor depth, and soil degree days above 0°C. (Van Cleve et al. 1983, © 1983 by the American Institute of Biological Sciences)

Above-ground net primary productivity varies widely between these forest types (Figure 18.4). The deciduous forests are roughly equal in productivity and tend to be somewhat higher than white spruce, which in turn is much higher than black spruce. Currently, more than 40% of the interior taiga forest in Alaska is dominated by black spruce.

There are some very strong correlations between the species that dominate a site and important environmental and ecosystem parameters. Black spruce stands have both higher mean water content in soils and lower mean soil temperature, measured as total degree days above 0°C at 10 cm depth (Figure 18.5). (A degree day is the mean temperature for that day minus the index temperature, in this case 0°C. A day with mean soil

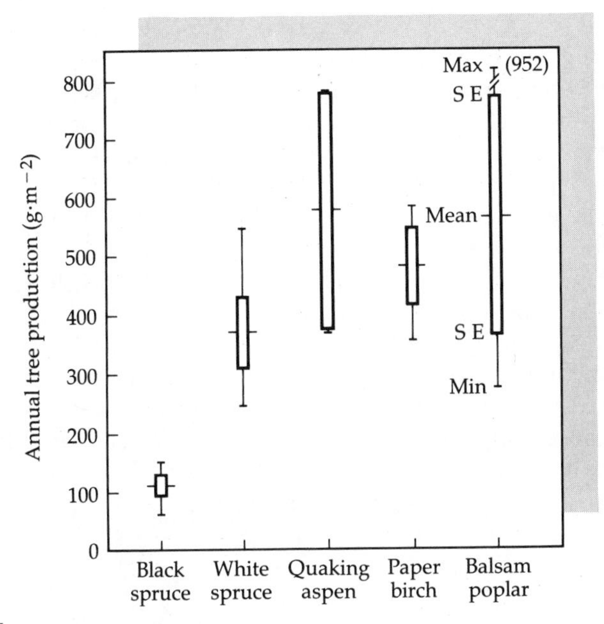

Figure 18.4
Range in above-ground net primary production values for the five major forest types in the Alaskan taiga. (Viereck et al. 1983)

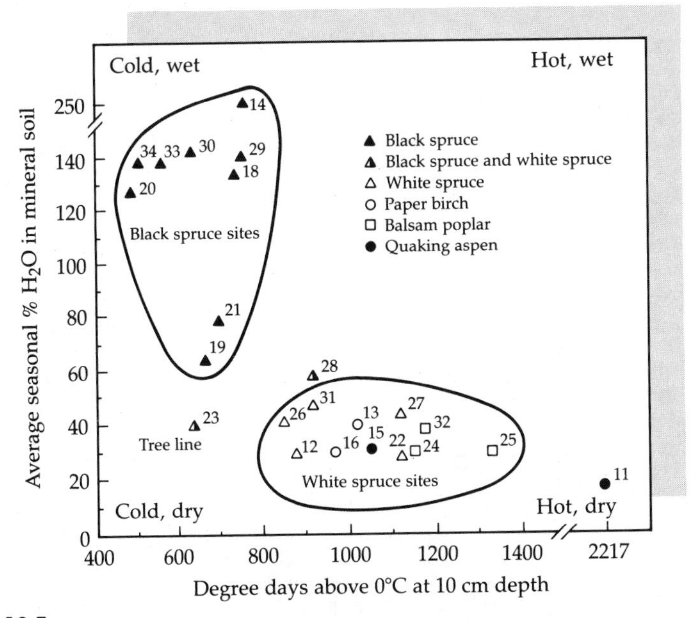

Figure 18.5
Distribution of white and black spruce sites with respect to average annual soil water content and total soil degree days. (Viereck et al. 1983)

temperature of 5°C yields 5 degree days. Three consecutive days at that temperature give a total of 15 degree days, etc.) There is, in turn, a strong correlation between soil degree days and both mean soil water content and the depth of the forest floor (Figure 18.6). A final important factor is that low average soil temperature generally relates to a shallower depth to permafrost during the growing season.

It could be argued from this either that black spruce is adapted to cold and wet sites and so grows in places with low soil temperature and high water content or that because black spruce litter decays slowly, it causes accumulation of a thick forest floor and thus creates low-temperature, high-water-content soils. This type of cross-correlation is the rule rather than the exception in ecosystem studies and makes the separation of cause and effect very difficult on the basis only of observations of existing stands. A more complete understanding of the interactions between species and site can usually be obtained by some sort of experimental modification, manipulating the system in some way and then measuring its response.

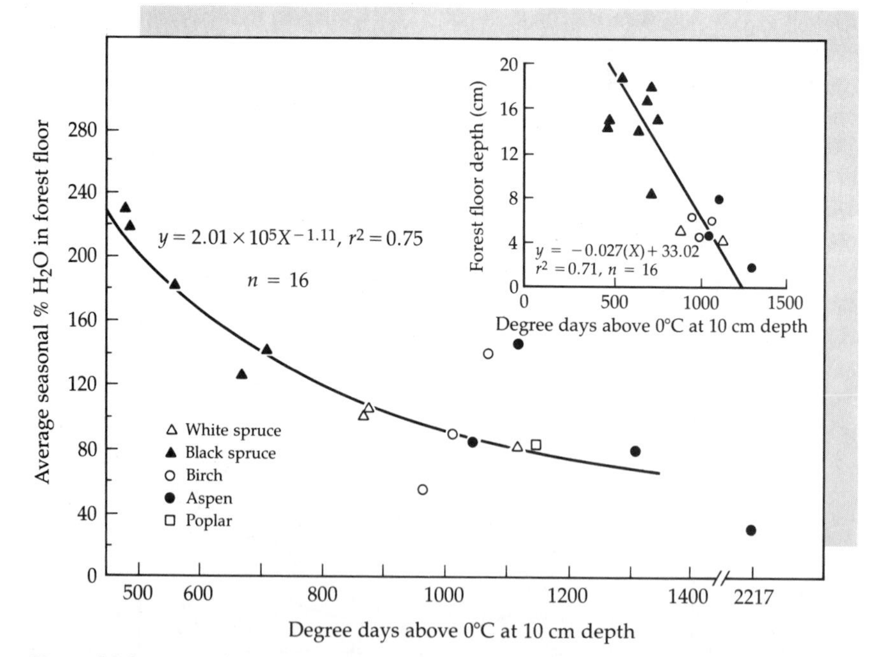

Figure 18.6
Correlation between total soil degree days above 0°C and both average seasonal soil water content and forest floor depth. (Van Cleve et al. 1983, © 1983 by the American Institute of Biological Sciences)

FIRE AND SUCCESSION IN TAIGA FORESTS

Fires provide a natural disturbance, which temporarily alters the close correlations between species and site in taiga forests. Studying the changes in burned stands through time allows a clearer look at the dynamics that lead to the final distribution of mature stands and their related environmental conditions.

Fire is a common component of the interior taiga forests of Alaska.

> *It would be hard to overestimate the importance of fires in shaping the vegetation pattern on the upland of interior Alaska. The distribution of aspen and paper birch can often be traced to patterns of past fires. In the mosaic of burned conditions, the hardwoods generally invade areas where the entire forest floor has been consumed, whereas black spruce replaces itself in areas where a less intense fire left the forest floor intact. This mosaic of highly diverse conditions in the wake of a fire makes it difficult to generalize about the effects of fire on the ecosystem. However, due to increased soil temperature and rates of nutrient cycling, the net effect of periodic fire in black spruce ecosystems is a warmer, more productive site, at least for a period of 10 to 20 years after the fire. (Van Cleve et al. 1983)*

Some forest types experience fire on a 30- to 50-year cycle, while the average return interval over the whole region is approximately 100 years. Forests over 170 years old are rare.

Figure 18.7 shows the integrated changes in species composition, forest structure, nutrient availability, forest floor and moss biomass, primary productivity, soil temperature, and depth to permafrost for a hypothetical 150-year succession following a crown fire in a mature black spruce forest on poorly drained soils.

Immediately following the fire, the forest floor and moss layers have been greatly reduced by burning, and the shading effect of the evergreen spruce canopy is removed. More sunlight reaches the forest floor, increasing soil temperatures and causing a retreat of permafrost to lower levels during the growing season. This increase in available soil volume interacts with higher temperatures to increase mineralization rate of soil organic matter, further increasing nutrient availability above that already caused by the deposition of ash.

Fire also affects species composition. While black spruce is present immediately after fire because of seed input from serotinous cones, there is an increase in deciduous species that both sprout vigorously and seed in from outside the burned area. The enriched soil conditions interact with the higher growth potential of aspen and birch to favor dominance by these species.

This shift in resource availability and species composition sets in motion a series of changes in ecosystem structure and function. The decidu-

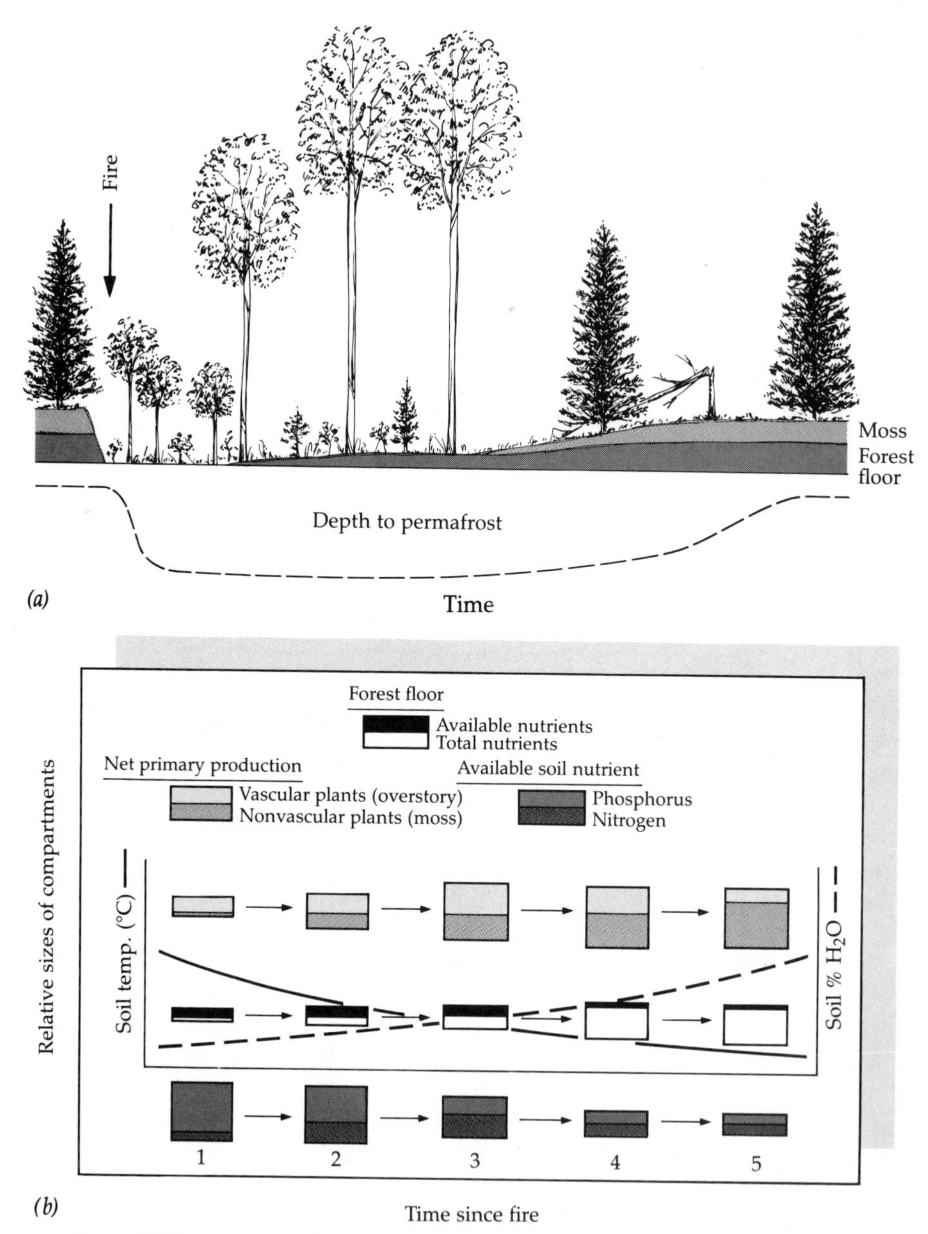

Figure 18.7

Integrated response of taiga ecosystems to fire and succession following fire. *(a)* Changes in species composition and ecosystem structure. *(b)* Changes in environmental conditions, resource availability, and net primary production. (Van Cleve et al. 1983, © 1983 by the American Institute of Biological Sciences)

ous species do not provide leaf cover in spring, allowing greater light penetration and earlier soil warming. In addition, the leaf litter produced is higher in nitrogen and lower in lignin content. The forest floor thus increases in N content and decreases in lignin content. Together, these three factors lead to further increases in mineralization rates (Figure 18.7). Faster mineralization and higher total N mineralization rates are, in turn, linked to greater leaf (Figure 18.8) and total tree net primary production.

The system would now seem to be in a positive feedback condition (Chapter 5), with changes in the system, such as high decay rates, fostering changes in species composition and function that will further increase production, etc. However, further increases in productivity, in conjunction with low temperatures, may reverse the decline in forest floor biomass, causing reductions in forest floor temperature, decomposition, and nutrient availability. More important, black spruce is still present in the system, growing slowly and perhaps reproducing vegetatively under the deciduous overstory.

Black spruce is more tolerant of shade than aspen and birch and increases in importance through successional time (Figure 18.7). As it does, its evergreen foliage decreases average annual soil temperatures by shading the soil surface in spring and fall. The foliage of black spruce is also high in lignin content. So as its litter becomes an increasingly important component of the total litterfall, decay rates decline. Declining decay rates lead to increasing forest floor depth and organic matter content, lower soil temperatures, and slower mineralization rates. As nutrient

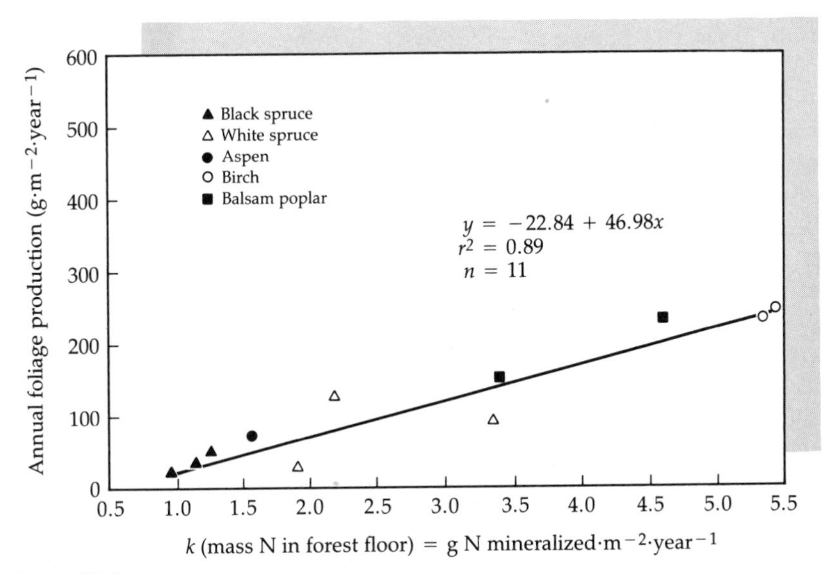

Figure 18.8
Net annual production of foliage as a function of estimated annual nitrogen mineralization in taiga forests. (Flanagan and Van Cleve 1983)

availability declines, the species in this system increase nutrient retranslocation from senescing foliage (Figure 18.9), leading to lower nutrient content in litterfall, still lower decomposition rates, and further reductions in nutrient availability.

As soil temperature declines, the maximum depth to permafrost also declines. This both restricts access to nutrients frozen in permafrost and restricts drainage of water from the soil. Higher water content increases the energy required to raise soil temperatures, so mean soil temperature continues to decline. Finally, high soil water content favors the establishment and spread of first feathermosses and then sphagnum moss, further insulating the forest floor and reducing soil temperatures. The mosses can actually compete effectively with black spruce for the scarce nutrient supply, reducing spruce growth and leading to a "stand" dominated functionally, if not structurally, by moss. The very large accumulations of organic matter over the mineral soil surface that occur at this stage make the system very susceptible to fire during periods of drought, which leave both the moss and tree components tinder-dry.

From this description it is still not clear where the "cause" is and where the "effect." This is because ecosystems, like all systems, are not controlled by linear, one-way interactions. Rather, they are a series of interactions between biological, physical, and chemical processes. In the taiga example, interactions between climate, soils, and species tend to lead the system toward a decadent end stage of low nutrient cycling and low productivity, which then is conducive to fire. Fire is the restart mechanism

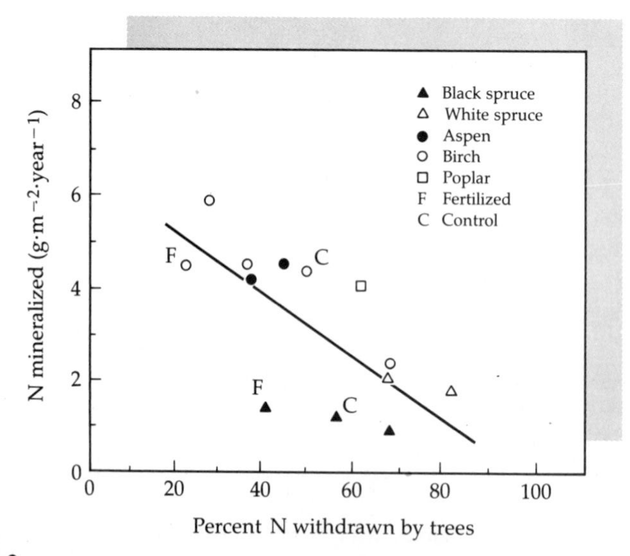

Figure 18.9

Percent of foliar nitrogen withdrawn from foliage as a function of annual nitrogen mineralization rate in taiga forests. (Flanagan and Van Cleve 1983)

in the taiga, which increases rates of several processes simultaneously but which also reinitiates succession to the low-productivity end stage. Forest floor organic matter depth, soil water content, and soil temperature are critical environmental parameters in these interactions.

Underlying site conditions can alter the scenario described above. On south-facing, sloping sites with good internal soil drainage, the deciduous forest stage may be prolonged, appearing permanent on extreme sites. On moderate sites, the more productive and nutrient-demanding white spruce may predominate over black spruce. In this case succession to the decadent moss-dominated end stage may be retarded, and the moss stage may never be reached before the next fire event. There is a clear distinction between sites dominated by white and black spruce in terms of soil temperature and mean soil water content (Figure 18.5). On wetland, muskeg, or bog soils, the importance of deciduous species in this scenario is reduced, movement away from the moss end stage is less, and succession toward it more direct.

EXPERIMENTAL MODIFICATION OF TAIGA ECOSYSTEMS

While natural fires offer an uncontrolled look at system response to disturbance, direct experimental manipulations may also help break up the correlations that occur, for example, between soil water content, forest floor accumulation, and soil temperature. An experimental heating of the forest floor in a black spruce stand was carried out in conjunction with the other studies summarized here. Heating cables were embedded in the forest floor and used to raise the temperature of the lowest layer in the forest floor by 9 °C above control levels. The heating resulted in a 20% reduction in the forest floor biomass and significant increases in ammonium and phosphorus availability. Black spruce foliage showed a roughly 40% increase in N and P concentrations. This demonstrates that soil temperature is indeed limiting decomposition rates, independent of other factors, in this stand.

SUMMARY OF INTERACTIONS AND RELATION TO GENERAL THEORY

Figure 18.10 is a summary view of the interactions between dominant plant species, environmental conditions, and rates of ecosystem processes in the Alaskan taiga. It reemphasizes the role of species changes in altering the nature of biomass produced and affecting rates of decomposition. These changes then feed back to reduce nutrient availability, to increase retranslocation, and to favor species with low nutrient demands, such as black spruce, and eventual dominance by moss. Is there a generalization about ecosystem dynamics that can be drawn from this?

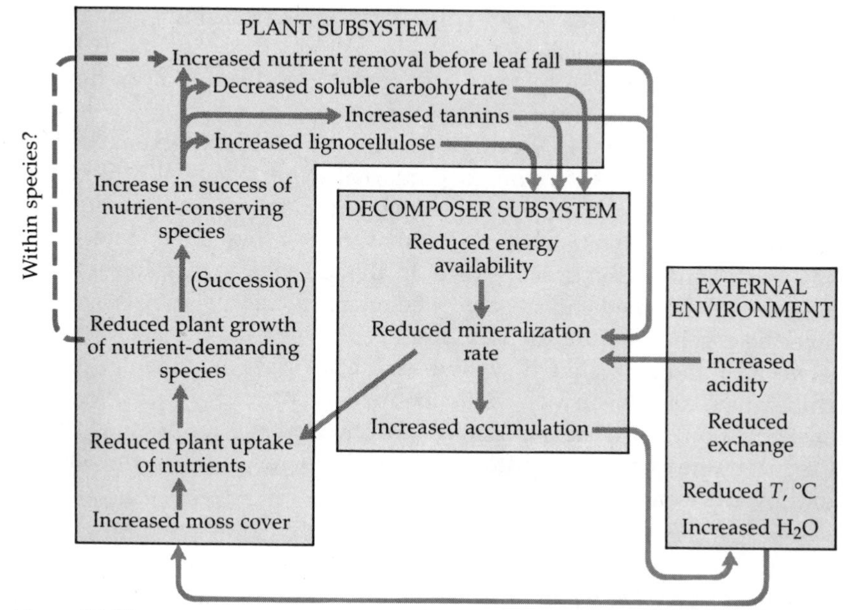

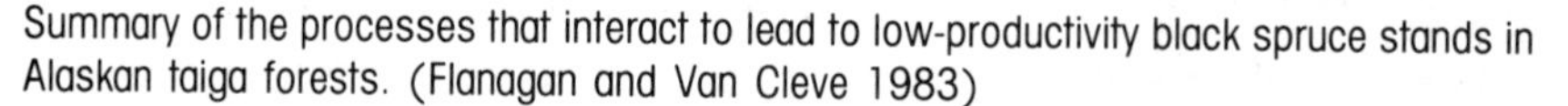

Figure 18.10
Summary of the processes that interact to lead to low-productivity black spruce stands in Alaskan taiga forests. (Flanagan and Van Cleve 1983)

In Chapter 11 we presented an emerging theory on the relationship between resource availability, nutrient content of foliage, rates of leaf function and turnover, and the amount of plant resource allocated to herbivore defense (Figure 11.3). In this theory, enriched sites favor the establishment and growth of species with high foliar nutrient content, rapid foliar turnover, and low allocation to defense. The successional sequence described here for the taiga fits this scheme very well. Aspen and birch represent the rich-site species, and black spruce the classic poor-site species. Foliar nutrient levels are higher in the deciduous species and lignin levels are lower. Foliage is replaced annually in birch and aspen, while the evergreen spruce can hold foliage for up to 20 years, with retention time generally increasing on lower-quality sites. This sequence also fits the generalization that nutrient availability and production are higher in recently disturbed areas and decrease with succession. As is typical of fire-dominated ecosystems, succession leads to the generation of stagnant systems, which then are ripe for fire. Fire restarts the cycle of enrichment, succession, and stagnation.

IMPLICATIONS FOR HUMAN USE OF THE TAIGA

Present human use of the Alaskan taiga zone is minimal but increasing.

> *Large scale development is just beginning in the taiga of interior Alaska. The trans-Alaska oil pipeline extends through a broad band of taiga. . . . The pipeline has brought roads and settlements. There are plans for a gas pipeline and increased mineral, coal and oil exploration are underway. Forestry and agriculture in the taiga are in the initial phases of development. We must learn more about the ecosystem in which much of the future development in Alaska will take place. (Van Cleve et al. 1983)*

Only 7% of the total forested area in this zone is considered commercial, with the rest consisting mainly of open, low-productivity black spruce forests. The results of the intensive study summarized here suggest that the judicious use of fire could serve to increase the productive potential of some of the remaining area. However, if timber production is to become important in the taiga, it must be remembered that logging and fire are very different in terms of their effects on the system. While both forms of disturbance remove vegetation and increase soil temperature, fire returns most of the nutrients in biomass to the soil, while logging removes them.

The actual manipulation of such large natural areas requires consideration of many other factors as well. Wildlife values and recreational and aesthetic values are important in the management of the landscape, as well as the value of wilderness lands, where natural processes are allowed to continue to operate. If, for example, the taiga region had been heavily settled and modified before the study summarized in this chapter was undertaken, we would have no idea how the dominant ecosystem type in this region functioned in its natural state.

REFERENCES CITED

Flanagan, P. W., and K. Van Cleve. 1983. Nutrient cycling in relation to decomposition and organic matter quality in taiga ecosystems. *Canadian Journal of Forest Research* 13:795–817.

Van Cleve, K., and C. T. Dyrness. 1983. Introduction and overview of a multidisciplinary research project: The structure and function of a black spruce. (*Picea mariana*) forest in relation to other fire-affected taiga ecosystems. *Canadian Journal of Forest Research* 13:695–702.

Van Cleve, K., et al. 1983. Ecosystems in interior Alaska. *BioScience* 33:39–44.

Viereck, L. A., C. T. Dyrness, and K. Van Cleve. 1983. Vegetation, soils, and forest productivity in selected forest types in interior Alaska. *Canadian Journal of Forest Research* 13:703–720.

ADDITIONAL REFERENCES

Van Cleve, K., et al. (eds.). 1986. *Forest Ecosystems in the Alaskan Taiga.* Springer-Verlag, New York.

See also: A special issue of *Canadian Journal of Forest Research* 13:695–915.

Chapter 19

The Serengeti: An Herbivore-Dominated Ecosystem

H. Armstrong Roberts

INTRODUCTION

It would be difficult to find two terrestrial ecosystems more different than the boreal forests of the taiga and the grasslands and plains of the Serengeti region of Africa. Processes in the boreal forest are limited by low soil temperatures and high soil water content. In the Serengeti, the severity and timing of seasonal drought determine the amount and timing of plant growth, which in turn determine the carrying capacity and seasonal distribution of the large populations of herbivores. The tremendous diversity and abundance of both large mammal herbivores and their predators make the Serengeti a focal point for ecological research and biological conservation efforts.

The purpose of this chapter is to introduce some of the interactions among climate, vegetation, herbivores, and predators that affect the dynamics of the Serengeti ecosystem. Because of the biological significance of, and human interest in, the herbivore–predator interactions, much more information is available on this aspect of the system than on the plant–soil–nutrient cycle interactions emphasized in other chapters. Still, important interactions between animal and plant populations occur and will be presented.

ENVIRONMENT OF THE SERENGETI REGION

The Serengeti area includes a variety of vegetation and habitat types, but the dynamics of all of these are intertwined through the grazing activities of the dominant mammalian herbivores. The extent of the Serengeti ecosystem is operationally defined by the area covered by the large migratory herds, especially of wildebeest (Figure 19.1). While the image of the Serengeti may be one of endless plains teeming with wildlife, the area covered by both the migratory herds and the Tanzanian National Park created to protect them is not large, extending less than 200 km at both its longest and widest points. The area contains diverse types of plant communities, ranging from short-grass systems in the east, to mid- and tall-grass systems in the central plains area, to open and occasionally closed woodlands in the far north and west.

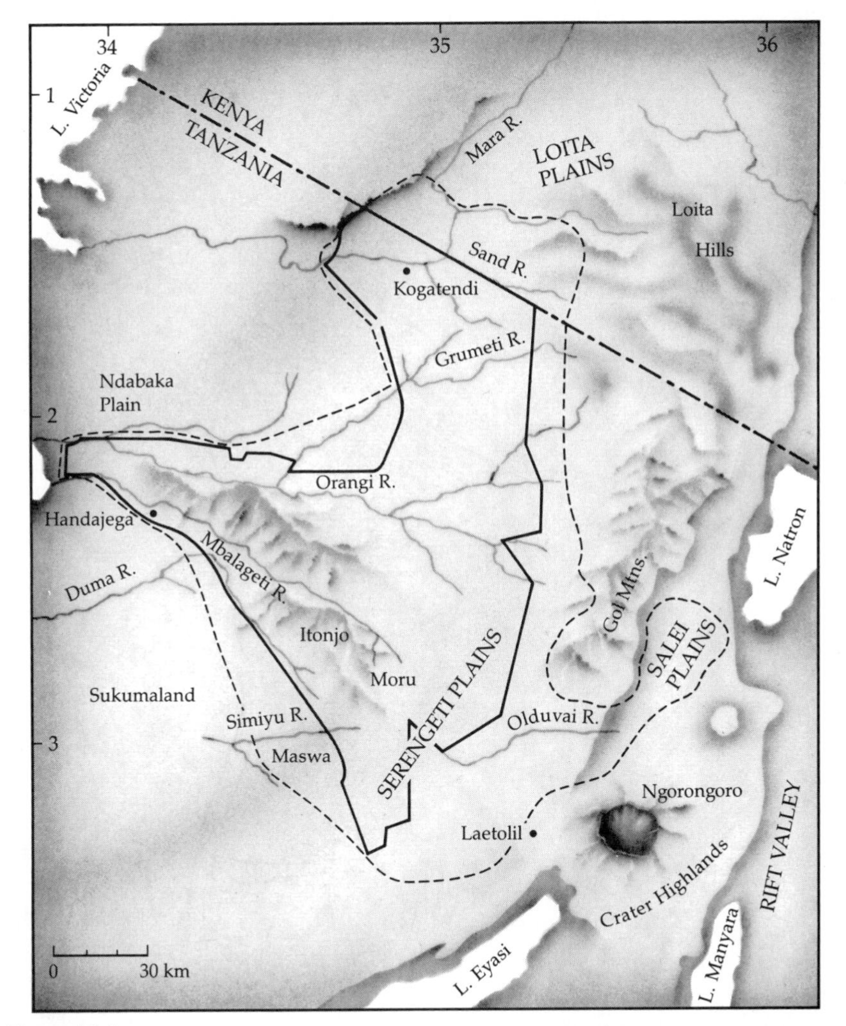

Figure 19.1
Map of the Serengeti region showing the boundaries of the park (solid line) and the range of migration of the wildebeest herd (dashed line). (Sinclair 1979a)

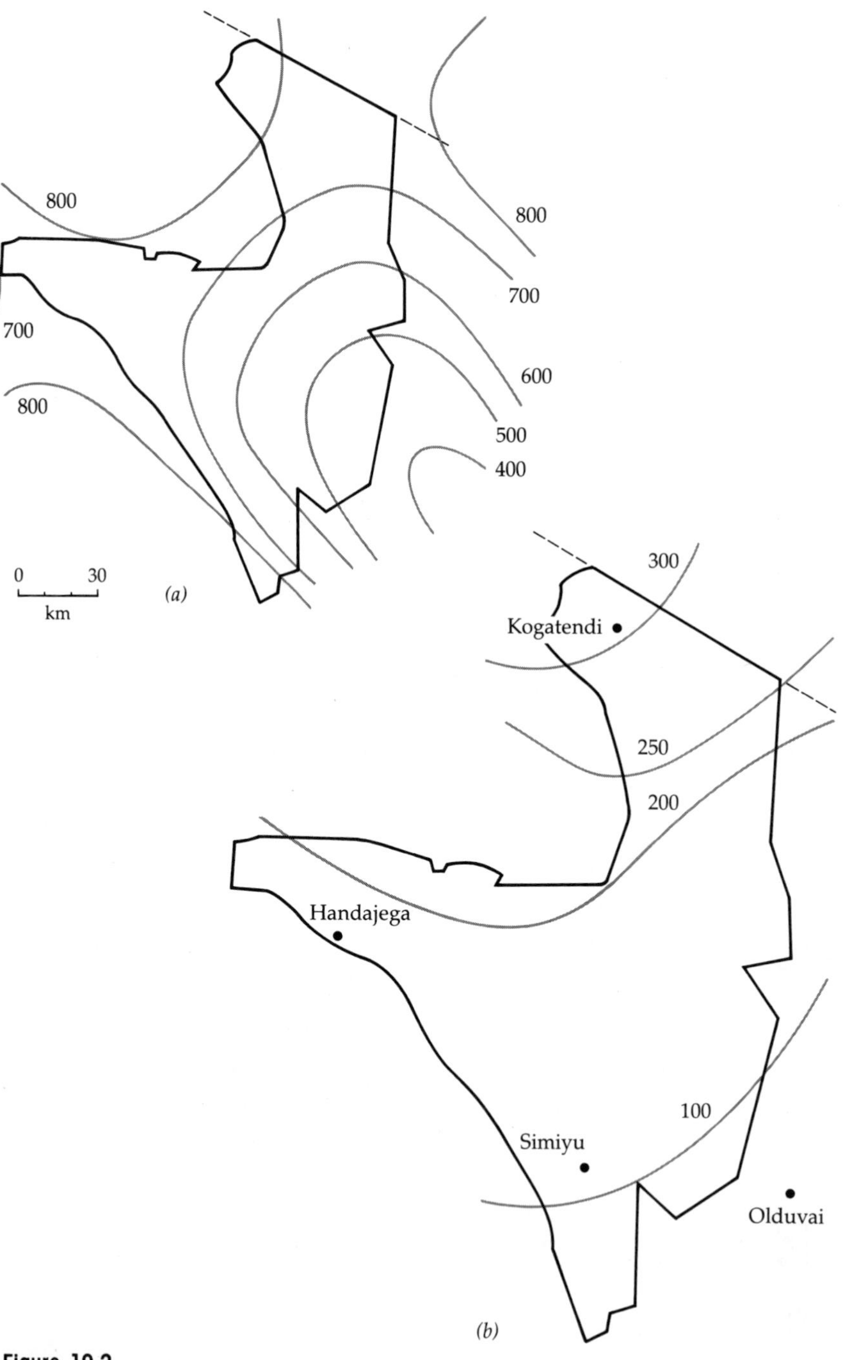

Figure 19.2
Distribution of rainfall over the Serengeti region in *(a)* wet season and *(b)* dry season. (Sinclair 1977)

Seasons are determined by the timing of rainfall in this tropical region, rather than by changes in temperature. The average annual climatic regime includes a marked dry season from June to October, followed by light rains in November through February, and a major wet season from March to May. Rain arrives on winds from the southeast. The Ngorongoro highlands at the southeast corner of the Serengeti area, rising to 3000 m in elevation, create a strong rain-shadow effect, so that rainfall in both the dry and wet seasons increases from southeast to northwest across the area (Figure 19.2).

The variability in rainfall is an important part of the Serengeti environment. The average values in Figure 19.2 mask important year-to-year variation. The importance of this variation is evident in the effects of a recent sustained increase in dry-season precipitation on herbivore populations discussed below. Small-scale spatial variation is also important in the timing of movement of many species in the region.

The geologic substrate of the region includes mainly ancient granitic rock formations, overlain in the southeast by ashfall from past eruptions of Ngorongoro, which ceased to be active about 2 million years ago, and both older and younger volcanoes to the east of the area. One nearby active volcano, Lengai, erupted four times between 1917 and 1966, with some ashfall reaching the Serengeti during one of these occasions.

Soil development shows a regional pattern expected from the range of precipitation conditions (Figure 19.3). Soils at the eastern end of the region are derived from alkaline volcanic ash. Lower rainfall has resulted in

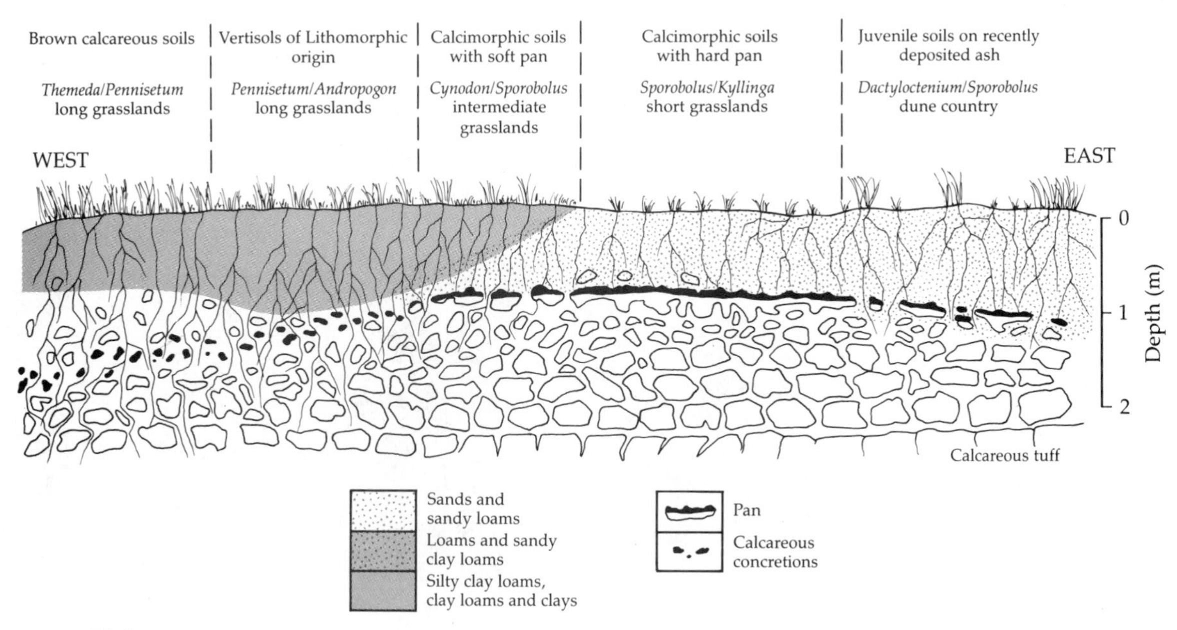

Figure 19.3
Schematic view of changes in soil structure and vegetation along the east–west gradient of total annual rainfall in the Serengeti. (Anderson and Talbot 1965)

minimal soil weathering and development. Calcium carbonates have been leached only from the top horizon and reprecipitated at a depth of about 1 meter. This forms a hardpan (caliche layer; Chapter 9), which further reduces downward water movement and maintains high soil alkalinity and pH (above 8). All of these characteristics are typical of an Aridisol. The vegetation in this area is mostly sparse grasses and sedges, and productivity and biomass are very low (Figure 19.4).

Moving to the west, rainfall increases, increasing both soil weathering and leaching. The calcium carbonate hardpan diminishes as alkalinity and pH decline. Vegetation grades from short grass to tall grass. Soils are described as Vertisols, meaning that they are rich in expandable 2 : 1 clays, which cause soil cracking under dry conditions (see Chapter 9).

Moving farther to the west or north, further increases in rainfall support a transition to woodlands and savannahs. Species of *Acacia*, a genus often associated with symbiotic nitrogen fixation, are a dominant part of the woodland vegetation (Figure 19.5). In adjacent areas of East Africa, there is a very close relationship between the alkalinity of the soil and the density of trees (Figure 19.6), suggesting that soil development and climate may play a large role in determining the areas in which trees can grow. We shall see that the extent of woodland development in the Serengeti area is greatly affected by interactions among herbivores, fire, and vegetation.

Figure 19.4
The short-grass vegetation of the drier plains area of the Serengeti. (Leonard Lee Rue III/Photo Researchers)

Figure 19.5
Acacia woodland of the wetter north and west portions of the Serengeti. (Mary Thacher/ Photo Researchers)

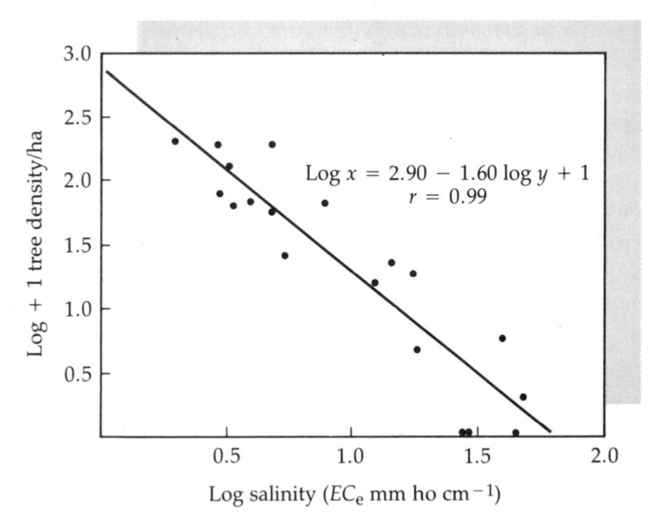

Figure 19.6
Relationship between soil alkalinity and percent of healthy trees in East Africa. (Western and Van Praet 1973)

In general, the transition from dry to subhumid environments described here repeats the trends presented in Chapter 2 for tropical regions. This same transition also occurs on a more local scale within the region along topographic sequences called **catenas.** Even the gently rolling topography of the Serengeti can produce significant upper- to lower-slope erosion of fine soil particles. Lower-slope soils tend to be enriched in silt and clay and have higher water retention capacity. Grass growth will be greater in lower-slope positions.

RESOURCE PARTITIONING AND USE BY HERBIVORES

The grasslands and savannahs that dominate the Serengeti region appear remarkably uniform. Niche theory (Chapter 5) suggests that no two species can occupy the same physical or ecological range unless there is some distinction between the two in how the resources in the area are used. Recent intensive study of the Serengeti as an ecosystem has revealed some of the ways in which this apparently uniform environment is divided or partitioned by herbivores.

> *The Serengeti's ungulates impress the visitor by their huge numbers, the vast extent of the country they occupy, and the number of species that occur together. These combined impressions pose the classic ecological question: how do so many individuals of so many species coexist in such an apparently homogeneous plant community? Studies have now provided some evidence that interspecific competition is taking place and is likely to be the process leading to the pattern of resource partitioning shown by the ungulates. . . . Thomson's gazelle, wildebeest and zebra . . . all face the same problems of obtaining sufficient quantity of good quality food, and they overcome these problems by methods dictated by body size, metabolic rate, and shape of mouth parts. (Jarman and Sinclair 1979)*

There are three major migratory species in the Serengeti: zebra, wildebeest, and Thomson's gazelle (Figure 19.7). They differ in the species and stages of growth of plants they can process most efficiently, and they tend to use the grasslands of the region in different ways and at different times. The timing is related to the nutritional quality of the grasses in different stages of growth.

Young, short, rapidly growing grass has the highest content of protein and nutrients and the lowest content of cell wall material or fiber. In the wet season, slower grass growth on the upper soil catena positions prolongs the period of short-grass availability, and all three of the herbivores concentrate on this position. Food is abundant during the wet season and competition and niche separation less stringent.

As the dry season approaches, zebra move down the soil catena first,

(a)

(b)

(c)

Figure 19.7
The three major species of herbivores in the Serengeti: *(a)* wildebeest, *(b)* zebra, and *(c)* Thomson's gazelle (foreground). (*a*, Stephen J. Krasemann/Peter Arnold, Inc.; *b*, Donald Paterson/Photo Researchers; *c*, Kruuk 1972)

into the taller grass at the foot of the slopes. Zebra are adapted to processing large amounts of relatively poor-quality herbage. They selectively remove stems and sheaths, opening up the vegetation. The remaining material is higher in leaf biomass, which is preferred by wildebeest. The very large number of wildebeest in the single herd that ranges over the Serengeti remove up to 85% of the vegetation, leaving a short sward, which is relatively enriched in low-lying, broad-leaved herbaceous plants.

Thomson's gazelle avoids tall-grass areas, selecting the high-quality herbs made more apparent by the removal of grass by wildebeest. In addition, the removal of the tall, mature grass vegetation by the wildebeest actually stimulates further grass production (Figure 19.8), increasing the total above-ground productivity of the grassland (see discussion of compensatory growth, Chapter 16). Two additional species with smaller populations, Grant's gazelle and eland, appear to utilize the longer-lived

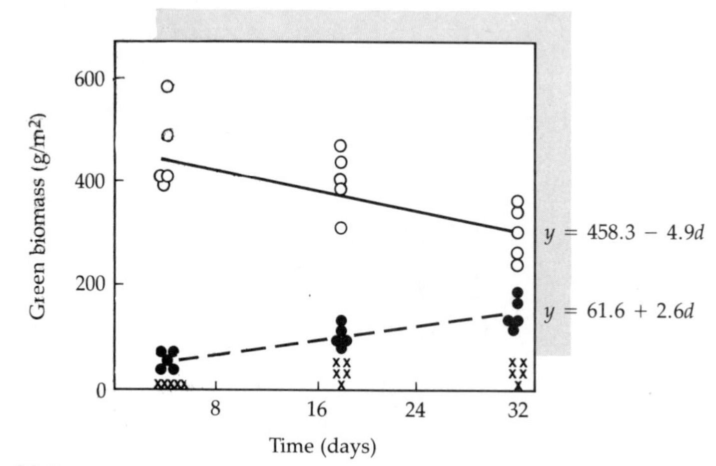

Figure 19.8
Comparison of changes in plant biomass in grazed and ungrazed areas. The open circles describe a loss in total plant biomass over time in ungrazed plots, indicating respiration and mortality in excess of growth. At the same time, grazed plots (closed circles) show net increase in green biomass, indicating further plant growth. (McNaughton 1976)

leaves of woody species (browsers) as well as grasses, thus prolonging their stay in the upper catena areas.

This topographic movement is repeated on a far grander scale with the movement of the migratory herds across the whole of the Serengeti region in response to cyclic occurrence of grass growth driven by the wet season–dry season cycle. The more than 1,500,000 wildebeest constitute a single herd, moving much like a prairie fire across the open grasslands of the plains in the wet season and the savannahs and woodlands in the dry season. While the wildebeest are not dependent on the zebra to modify the grass cover to allow grazing, it appears that grazing by wildebeest does have a positive effect on further grazing by Thomson's gazelle.

The timing of mortality indicates that total grass growth and availability, particularly during the dry season, is a major control on herbivore population levels (Figure 19.9). The quality and abundance of dry-season grass are directly related to the rate at which wildebeest lose, during the dry season, the weight put on during the wet season. This weight loss, in turn, relates to malnutrition and mortality. Most mortality occurs during the dry season.

In addition to these migratory species, there are several nonmigratory herbivores, including water buffalo, topi, impala and other gazelles, giraffes, and elephants, which reside in the more humid western and northern reaches of the Serengeti area. During the wet season when the migratory species are in the plains area to the southeast, the resident species suffer little competition. In the dry season, the migrants return and divide the existing plant resources more finely.

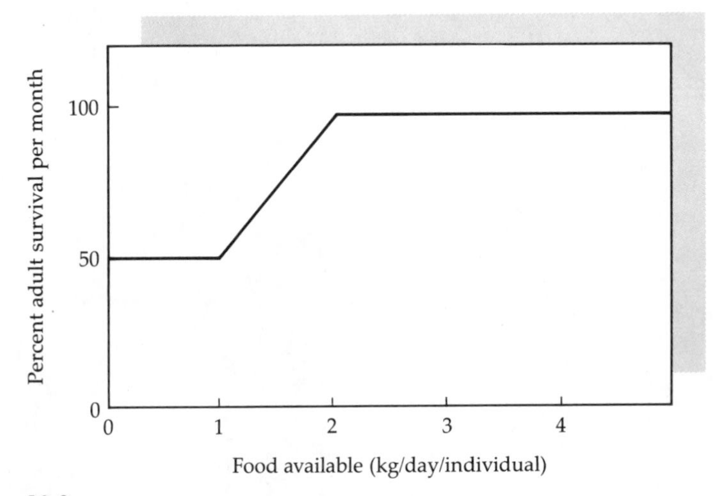

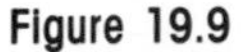

Figure 19.9
Assumed relationship between the monthly percentage mortality of wildebeest and available food per individual. (Hilborn and Sinclair 1979)

At this time the buffalo favors the woods and grasslands of riverine habitats, affected only by the trampling of wildebeest through these areas in search of water. The impala switches to browsing in addition to grazing of herbs at this time and lies somewhere between wildebeest and Thomgazelle in its use of plant foods. Topi are adapted to the partial use of long-grass resources, putting them between zebra and wildebeest. The giraffe represents an extreme adaptation for browsing, which confers a unique niche space not directly affected by other herbivores. However, indirect effects of herbivory on fire and tree reproduction can seriously alter tree density and availability of food for giraffes. Elephants, with their penchant for knocking down small trees to obtain browse (Figure 19.10), can be particularly important.

This finer division of niches and greater pressure on food resources occurs in many other ecosystems in the season of lowest food availability. This is the season when competition is at its peak, and so is mortality. It is the time of year when animal populations are most sensitive to unfavorable shifts in climate. For example, when total annual rainfall in the Serengeti is low, the wildebeest herd spends a greater proportion of the year in the tall-grass and woodland areas of the north and less in the drier, short-grass plains. This increases the pressure on the northern areas and reduces food availability to resident herbivores as well as other migratory species. As discussed below, populations of herbivores in the Serengeti have responded markedly to recent changes in dry-season rainfall.

Figure 19.10
Elephants destroying acacia trees in a Serengeti woodland. (Sven-Olof Lindblad/Photo Researchers)

RESOURCE PARTITIONING AMONG PREDATORS

Limitations on niche overlap apply to predators as well as herbivores. The five major predators in the Serengeti, lions, hyenas, leopards, cheetahs, and wild dogs, should have to divide the herbivore resource by size, condition, location, or season of use. Table 19.1 suggests that this is the case. The major species have different preferences for habitat type, time of day, species taken, and method of hunting. The closest competition between any two pairs of these five species appears to be between hyenas and wild dogs. Recent increases in the hyena population have been linked to reductions in wild dogs, which are nearly extinct in the Serengeti.

Lions account for over half of the predation in the Serengeti. They are territorial and nonmigratory, living in groups, or prides, consisting of 4 to 15 adults plus a number of subadults and cubs, all interrelated. As the dominant predator, they feed on the dominant migratory herbivores during their passage through the territory and on the less numerous resident herbivores at other times of the year. Much like the wolf populations described in Chapter 16, lion populations are stabilized by social interactions within the pride.

In contrast, leopards and cheetahs do not have large social groupings. They are solitary hunters, feeding on the smaller and swifter herbivores, particularly gazelles, which they catch by speed. This is in contrast to the group hunting of the larger wildebeest practiced by lions, hyenas, and

Table 19.1 Differences in hunting and social behavior between major predator species in the Serengeti*

SPECIES	CHEETAH	LEOPARD	LION	HYENA	WILD DOG
Approximate weight of adult (kg)	40–60 kg	35–60 kg	100–200 kg	45–60 kg	17–20 kg
Number in Serengeti ecosystem	220–500	800–1200	2000–2400	3000^{R}–4500	150^{L}–300
Habitat in Serengeti	Especially plains, but woodlands too	Woodlands only	Mainly woodlands, but plains too	Mainly plains	Plains and woodlands
Hunting—time of day	Entirely by day	Mainly at night	Mainly at night	Night and dawn	Mainly by day
Number of animals hunting together	1	1	1–5	1–3 (for wildebeest & gazelles), 4–20 (for zebras)	Whole pack, i.e., 2–19
Method of hunting	Stalk, then long fast sprint	Stalk to close range, then short sprint	Spread out; stalk then short sprint	Long-distance pursuit; others join in	Long-distance pursuit
Distance from prey when chase starts	10–50 m 50–70 m	5–20 m	10–50 m	20–100 m	50–200 m
Speed of pursuit	Up to 95 km/hr	Up to 60 km/hr	50–60 km/hr	Up to 65 km/hr	Up to 70 km/hr
Distance of pursuit	Up to 350 m	Up to 50 m	Up to 200 m	0.2–3.0 km	0.5–2.5 km
Measured success rate	37–70%	5%	15–30%	35%	50–70%
Commonest species taken (in order of frequency in diet)	Mainly Thomson's gazelle, also hare, Grant's gazelle, impala	Impala, Thomson's gazelle, dik dik, reedbuck, many others	Zebra, wildebeest, buffalo, Thomson's gazelle, warthog, others	Wildebeest, Thomson's gazelle, zebra	Thomson's gazelle, wildebeest, others
Health of prey	Healthy	Healthy	Healthy	Sick and healthy	Sick and healthy
Age and sex of prey	Especially small fawns	All ages; only the young of topi, wildebeest, zebra	All ages, but dispro- portionately more young than old	Especially males and young of wildebeest and gazelles; female zebra	Esp. gazelle adult males; wildebeest calves; zebra adult females
% of kills that are partially or wholly lost to other carnivores	10–12%	5–10%	Almost none	5% (20% in Ngoron- goro)	50%
% of diet obtained by scavenging	None	5–10%	10–15%	33%	3%
Important interference competitors	Hyenas; possibly lions	Possibly lions	None	Possibly lions	Hyenas

*From Bertram 1979.

wild dogs. These latter two also show pack or group organization, which is greater in the wild dogs (Table 19.1).

INTERACTIONS AMONG VEGETATION, HERBIVORES, AND PREDATORS

Important constraints in resource acquisition affect the size, distribution, and abundance of both predators and their prey, as demonstrated in the Serengeti. The efficiency of digestion of grasses by herbivores increases with body size, as food ingested is held for a longer period of time. This means that the larger herbivores will have an advantage over smaller ones in areas where grasses predominate. In contrast, smaller species with less efficient digestion and also a higher metabolic rate (higher respiration per unit body weight) are constrained to consume higher-quality plant material. Thus, it is the gazelles that specialize by eating the broad-leaved herbs and new, short-grass growth, which has the highest nutritional value.

High-quality plant material is a more dispersed resource, requiring greater areal coverage per animal. This, in turn, dictates smaller herd size. Herding is an effective mechanism against predation for any individual animal. The absence of large herds places increasing value on alertness, speed, and protective coloration as predator defenses.

Large herbivore size and large herd size both increase the value of both large size and group hunting behaviors in predators. Thus the predators on zebra and wildebeest are large and/or hunt in groups. In contrast, the dispersed nature of the herbivores feeding on the dispersed, high-quality food source requires stealth and speed in would-be predators. The cheetahs and leopards, which hunt the smaller and less abundant species, also show camouflaging coloration and extreme speed and hunt alone. The risk of losing a captured prey to a larger group of lions or hyenas is answered in the leopard by carrying the capture up into a tree. Cheetahs do not have this advantage.

One could see in this division between the generalist and specialist herbivore patterns similar to those described for apparent and less apparent plant species in relation to protection from herbivory. The generalists, or dominant species, rely on generalized defenses (lignin and cellulose for plants, size and herding for herbivores). The specialists, or less apparent species, show a variety of defenses (alkaloids in plants, the wide variation in coloration and behavior in animals).

Predation versus Food as Limiting Factors in Herbivore Populations

Predation is a highly visible process in the Serengeti, but is it the dominant factor determining herbivore population levels? Apparently not. About one third of the mortality of herbivores in the Serengeti is due to predation, with the rest due largely to malnutrition during the dry season. Two mechanisms may interact to cause this: territoriality of predators and the patterns of migration of the major herbivores.

The dominant herbivores in the Serengeti are migratory, while the major predators are resident and territorial. Roughly two-thirds of the total herbivore biomass is in the migratory zebra, wildebeest, and Thomson's gazelle. The effect of this transitory movement of large numbers of animals through stationary territories is a very uneven distribution of food for predators throughout the year.

As the migratory herds pass through a territory, food abundance is very high for a short period of time. However, just as the herbivore population levels are determined by food availability during the dry season, resident populations in territories are determined by resident game populations when the migrating herds are gone. Thus, predator levels throughout the Serengeti are lower than they would be if the herbivores were nonmigratory. Conversely, predator pressure on the migratory herbivores is less than it would be if they were nonmigratory, so starvation becomes a more important cause of mortality. There is some evidence that herbivore populations are lower, and predation is higher, in other, smaller areas of East Africa where such seasonal migrations of herbivores do not occur.

Herding and migration are two aspects of a larger strategy known as "swamping" of predators — the synchronized appearance and disappearance of prey. Another example of this in the Serengeti is the highly synchronized birth of new wildebeests during the wet season on the plains. Not only is this the period of lowest competition for food, but the plains also afford the clearest long-range vision for detection of predators. The synchronized birth produces large quantities of young at one time, numbers far too large to be captured by predators. There is a significant advantage to giving birth during this time, as young born at other times would be part of a smaller, more vulnerable prey pool.

PERTURBATIONS, SUCCESSION, AND THE DYNAMICS OF THE SERENGETI SYSTEM

As discussed in the preceding chapter, the description of static patterns of distribution, abundance, and function do not allow an understanding of the factors that control those patterns. The value and uniqueness of the Serengeti, as well as its size, rule out many potential experimental perturbations. However, the region has experienced two major uncontrolled perturbations in this century, in addition to changes in the degree of human use and modification. These "natural" experiments afford an opportunity to observe the interaction of potentially important processes.

The first perturbation was the outbreak of rinderpest, a virus that strikes both wild and domesticated cattle. It was introduced to northern

Africa in domestic cattle in the 1880s. It appeared in East Africa in 1890, and by 1892 95% of buffalo and wildebeest had died, along with most of the domesticated cattle of the pastoral and nomadic tribes in the region. The resulting loss of human life and depopulation of the area is a tragic tale of human misery. Such apparently unassociated phenomena as the disappearance of the tsetse fly and the appearance of people-eating lions in the region can be traced to the decimation of the ruminant species.

Reduced human population led to abandonment of agricultural fields, and many areas previously cultivated in the more humid areas returned to brush and woods. As some wildebeest became immune to rinderpest and the population became stabilized, the tsetse fly, which uses wildebeest as a host, returned to the area. Increased brush coverage also served to increase the tsetse fly population, and malaria, carried by the fly, returned with a vengeance to the remaining human population.

In the 1930s mechanical brush clearing and a vaccination against rinderpest reversed the trends, and both human and cattle populations have increased since. Wholesale inoculation of all domestic cattle surrounding the Serengeti area eliminated rinderpest from the population, at least temporarily, by 1963. The establishment of the Serengeti National Park has limited, if not eliminated, direct human impacts on the Serengeti since that time. Still, increasing human populations in the surrounding area have caused some shifts in animal abundance, as we will see below.

The effects of rinderpest and its eradication have been profound. Populations of wildebeest and buffalo have increased tremendously, while

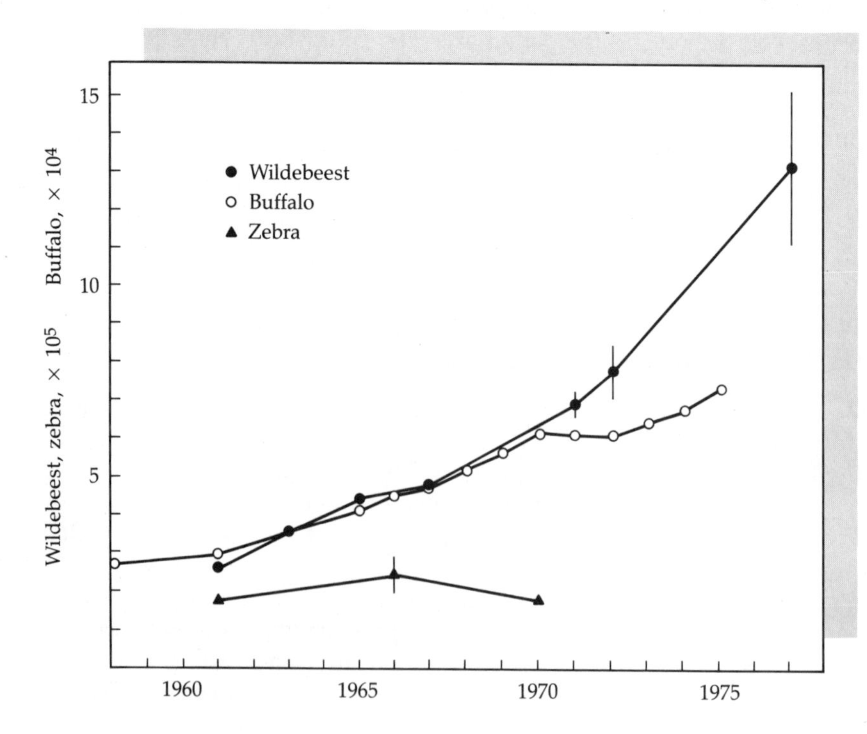

Figure 19.11
Population changes of zebra, wildebeest, and buffalo since the removal of rinderpest from the Serengeti. (Sinclair 1979b)

zebra, a nonruminant unaffected by rinderpest, has remained constant (Figure 19.11). Computer simulations suggest that both rinderpest and predation played a role in keeping the wildebeest population at around 300,000. With rinderpest removed, the population exploded, and it now far exceeds the demand by predators.

The second "natural" perturbation was an increase in dry-season rainfall between the 1967–1971 and 1972–1976 periods (Figure 19.12). Dry-season rainfall delays the movement of herds into the woodland systems of the north and west and also increases total grass production throughout the region (e.g., Figure 19.13).

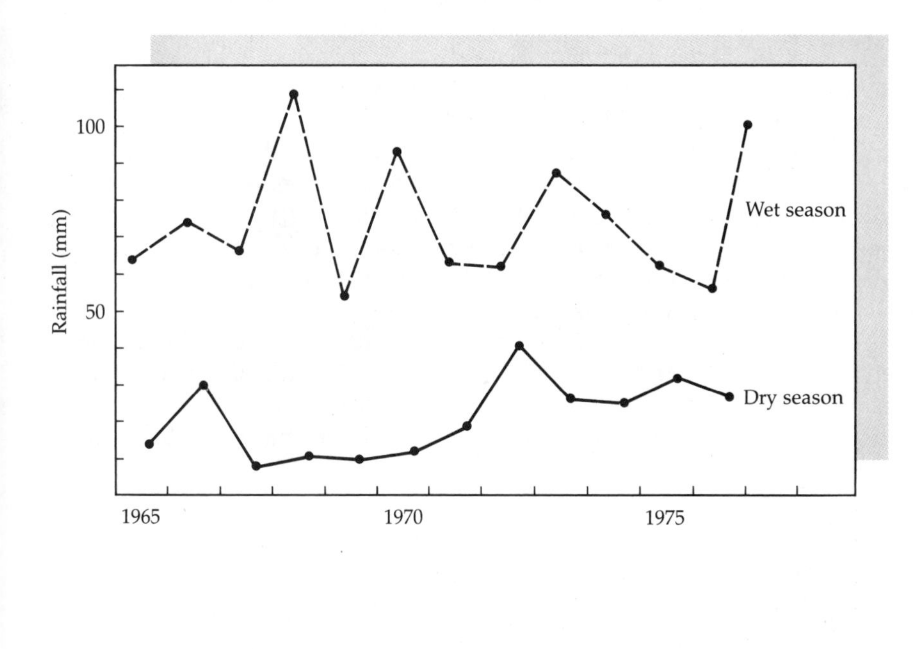

Figure 19.12
Changes in the amount of rainfall in the Serengeti between 1965 and 1977 (note increase in dry-season rainfall over the period). (Hanby and Bygott 1979)

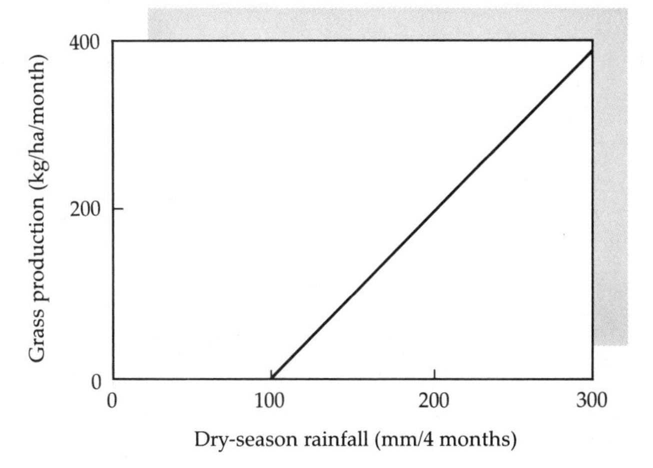

Figure 19.13
Summary relationship between dry-season rainfall in the Serengeti and dry-season grass production. (Hilborn and Sinclair 1979)

Rinderpest control and increasing rainfall have fostered an increase in numbers of wildebeest, which has had important secondary consequences throughout the park. The ways in which the populations of different large mammals have responded to changes in rainfall and wildebeest numbers are summarized in Figure 19.14. This summary diagram is the distillation of years of research in the Serengeti. It represents a first attempt at a systems analysis of this complex ecosystem. We will step through parts of this diagram to explain briefly the effects of perturbations on the Serengeti system.

Heavier rainfall means more grass production in the plains and longer residence there by the migratory herds. Heavier grazing here alters the relative abundance of grass and herb species. This is in comparison with the northern and western areas grazed during the dry season when plants are senescent and consumption is not a selective force. Greater concentrations of herbs lead to increases in populations of Grant's gazelle, which feed on herbs in the plains for most of the year. This may, in turn, have caused an increase in numbers of cheetah, which feed on the gazelle. In contrast, feeding and trampling by wildebeest of grasses in the buffalo's riverine habitat have reduced the increase in that species due to the removal of rinderpest. The increased dry-season rainfall has also increased the populations of relatively rare species such as topi, kongoni, and warthog, which live on the plains year-round.

Both the increase in wildebeest, through greater consumption of grass in the woodlands, and increased precipitation, by maintaining green grass for a longer period, have reduced the incidence of fire throughout the park (Figure 19.15). Fire and consumption are similar in that they both reduce

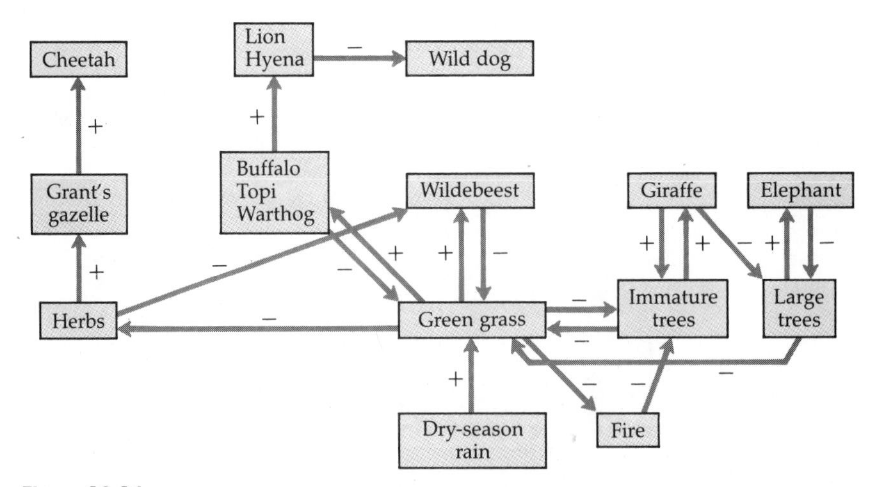

Figure 19.14
Schematic model of interactions among rainfall, fire, vegetation, and major species of mammals in the Serengeti. (Sinclair 1979a)

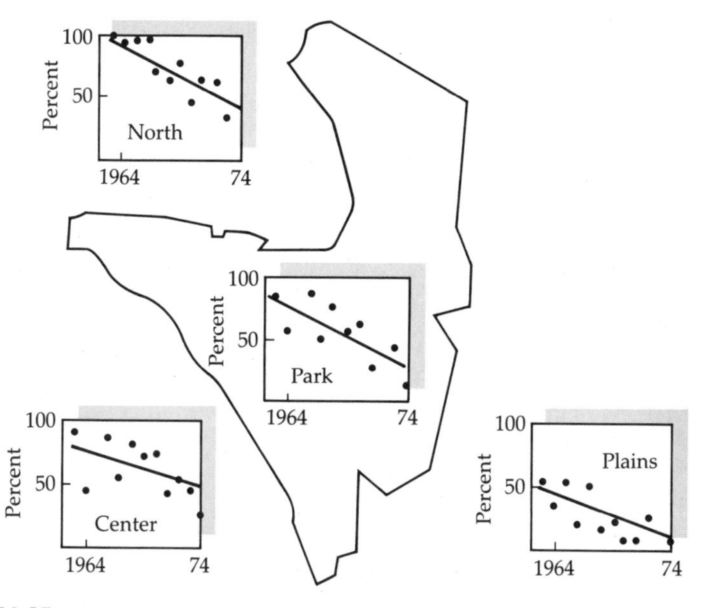

Figure 19.15
Changes in the percentage of area burned in different areas of the Serengeti from 1964 to 1974. (Norton-Griffiths 1979).

storage of carbon in plants and soils and recycle nutrients, but they differ in their effects on competitive balance between grasses and trees. By selecting against herbaceous growth, herbivory promotes the occurrence of trees. Conversion of grasslands to arid shrublands, or deserts, through overgrazing has been a general occurrence in many parts of the world. In contrast, fire kills many woody species, or at least suppresses them relative to the rapidly sprouting grasses, which are adapted to dying back to ground level each year.

The presence or absence of trees affects browsers, particularly giraffe. The presence of immature trees increases giraffe populations, which in turn tend to kill mature trees. In the absence of fire, this further increases the importance of immature trees.

The giraffe–tree interaction has been affected by the increase in human use around the park, which has herded large numbers of elephants into the woodlands. Elephants are very effective at removing mature trees (Figure 19.10). They can be effective enough to hinder regeneration of trees, reducing giraffe habitat.

HUMAN USE AND CONSERVATION CONCERNS IN THE SERENGETI

Several serious problems confront the managers of the Serengeti National Park and its surroundings. The human vacuum created by the rinderpest disaster made the establishment of the park possible. Now, increasing population density of humans in the area surrounding the park is increasing the pressures of poaching and other intrusions.

One of the major problems lies in managing an ecosystem that is still recovering from several major disruptions. Large shifts in species abundance might be expected in such a system, and it is difficult to know exactly what the management goals should be. Management for population stability is probably not possible and could be harmful.

> *Because ecosystems must be considered dynamic, one must resist the temptation to manage them with a view toward maintaining some arbitrary status quo. The present evidence shows that there are natural negative feedback mechanisms operating between some components in the system. . . . Provided these negative feedbacks are strong enough, the system can absorb disturbances without management. . . . For example . . . there are negative feedbacks operating on the wildebeest and buffalo . . . through a lack of dry season food. . . . Droughts . . . could produce rapid declines in wildebeest and other herbivores, but these species will build up again when the food supply increases after the drought is over. Such declines are desirable and should be expected.* (Sinclair 1979a)

Like all ecosystems, the Serengeti is a complex assemblage of highly interactive parts. It is entirely possible that rinderpest may once again become established in the park, setting off another round of large fluctuations in wild cattle populations. The kind of systems analysis outlined in Figure 19.14 is a first step toward understanding the dynamics of such a system, assessing the implications of changes that may occur in fire frequency, population levels, and other state variables, and developing management plans that work with the processes active in the system.

REFERENCES CITED

Anderson, G. D., and L. M. Talbot. 1965. Soil factors affecting the distribution of the grassland types and their utilization by wild animals on the Serengeti plains, Tanganyika. *Journal of Ecology* 53:33–56.

Bertram, B. C. R. 1979. Serengeti predators and their social systems. In Sinclair, A. R. E., and M. Norton-Griffiths (eds.), *Serengeti: Dynamics of an Ecosystem.* University of Chicago Press.

Hanby, J. P., and J. D. Bygott. 1979. Population changes in lions and other predators. In Sinclair, A. R. E., and M. Norton-Griffiths (eds.), *Serengeti: Dynamics of an Ecosystem.* University of Chicago Press.

Hilborn, R., and A. R. E. Sinclair. 1979. A simulation of the wildebeest population, other ungulates and their predators. In Sinclair, A. R. E., and M. Norton-Griffiths (eds.), *Serengeti: Dynamics of an Ecosystem.* University of Chicago Press.

Jarman, P. J., and A. R. E. Sinclair. 1979. Feeding strategy and the pattern of resource partitioning in ungulates. In Sinclair, A. R. E., and M. Norton-Griffiths (eds.), *Serengeti: Dynamics of an Ecosystem.* University of Chicago Press.

Kruuk, H. 1972. *The Spotted Hyena: A Study of Predation and Social Behavior.* University of Chicago Press.

McNaughton, S. J. 1976. Serengeti migratory wildebeest: Facilitation of energy flow by grazing. *Science* 191:92–94.

Norton-Griffiths, M. 1979. The influence of grazing, browsing, and fire on the vegetation dynamics of the Serengeti. In Sinclair, A. R. E., and M. Norton-Griffiths (eds.), *Serengeti: Dynamics of an Ecosystem.* University of Chicago Press.

Sinclair, A. R. E. 1977. *The African Buffalo.* University of Chicago Press.

Sinclair A. R. E. 1979a. Dynamics of the Serengeti ecosystem: Process and pattern. In Sinclair, A. R. E., and M. Norton-Griffiths (eds.), *Serengeti: Dynamics of an Ecosystem.* University of Chicago Press.

Sinclair, A. R. E. 1979b. The eruption of the ruminants. In Sinclair, A. R. E., and M. Norton-Griffiths (eds.), *Serengeti: Dynamics of an Ecosystem.* University of Chicago Press.

Sinclair, A. R. E., and M. Norton-Griffiths (eds.). 1979. *Serengeti: Dynamics of an Ecosystem.* University of Chicago Press.

Western, D., and C. Van Praet. 1973. Cyclical changes in the habitat and climate of an east African ecosystem. *Nature* 241:104–106.

ADDITIONAL REFERENCES

McNaughton, S. J., and N. J. Georgiadis. 1986. Ecology of African grazing and browsing mammals. *Annual Review of Ecology and Systematics* 17:39–65.

Chapter 20 Dynamics in the Absence of Fire and Herbivory: The Northern Hardwood Forest

H. Armstrong Roberts

INTRODUCTION

The northern hardwood forest region of North America is a less extreme environment than either of those presented in the previous two chapters. Located primarily in a band along the border between the northeastern United States and southern Canada (Figure 20.1), the area is neither as cold as the taiga nor as dry as the Serengeti. The climate is a combination of cold winters with significant snow accumulation and warm summers. Climate and characteristics of the major species combine to produce a balance between production and decomposition, which does not lead to the continuous build-up of organic matter seen in the taiga. Herbivory is dominated by insects rather than mammals and is generally much less of a factor in ecosystem function than in the Serengeti. In the relative absence of fire and herbivory, the population dynamics of the dominant tree species, along with occasional pulse disturbance due to windstorms and hurricanes, largely determine the structure and dynamics of the ecosystem.

The purpose of this chapter is to describe the dynamics of vegetation and the changes in ecosystem processes that occur in one area representative of the eastern portion of the northern hardwood forest. We will discuss disturbance patterns, how disturbance alters resource availability, and how the major plant species vary in their response to disturbance. Combining these processes will yield an integrated view of the function of this ecosystem type.

THE NORTHERN HARDWOOD ECOSYSTEMS OF NEW ENGLAND

The northern hardwood forests in northern New England are representative of the eastern portion of this ecosystem type. They occur mainly at mid-elevation in the major mountain ranges of the area, the northernmost extension of the Appalachian Mountains. The climate of the region is determined by a mixture of continental and maritime influences, providing fairly wide temperature extremes and abundant precipitation. A typical mean July air temperature is 19°C; for January it is −9°C. The frost-free season lasts about 160 days, and precipitation occurs fairly evenly throughout the year.

The region supports closed forests dominated by sugar maple and yellow birch, along with beech in the eastern portion and basswood in the western portion of the range. Evergreen conifers such as hemlock and white and red pine also occur. At the northern edge and at high elevations, boreal species such as spruce and fir mix with the hardwoods. At the southern edge, oaks, red maple, white ash, and other species of the oak–hickory forest region occur. Early successional species include the fast-

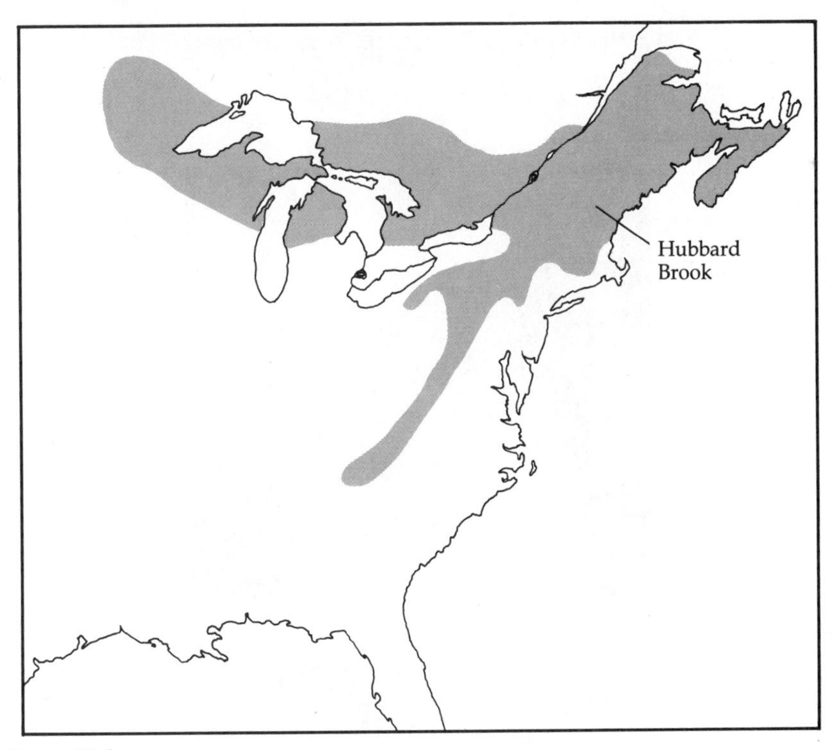

Figure 20.1
Distribution of northern hardwood forests in North America and the location of the Hubbard Brook Ecosystem Study.

growing deciduous species pin cherry and aspen, and the oldest stands can contain a thick understory of slow-growing, low-stature shrub species.

The entire northern hardwood region was stripped by glaciers in the last ice age, and soils have developed on materials left behind or deposited since the glacial retreat. In the east, soils have developed mainly on thin glacial till on mountain slopes. This is in contrast to the midwestern portion of the northern hardwood region, where deposition of wind-blown loess has produced deeper and richer soils. In both cases, soil depth and the texture of the material (sand, silt, clay, and larger rocks) play a major role in determining site quality and productive potential. In the east, the cool, humid climate and shallow, coarse-textured soils favor podzolization as the dominant soil-forming process. Soils are somewhat older than in the boreal zone and subject to greater weathering intensity (Figure 20.2). In the west, deeper, less acidic, and finer-textured soils favor melanization and lessivage as soil-forming processes, and Alfisols are also an important soil type.

Although the Spodosols include an organic surface horizon (Figure 20.2), this need not indicate that litter decomposition is too slow or incomplete to maintain adequate rates of nutrient cycling. Rather, this layer

Figure 20.2
Spodosol that developed in a northern hardwood forest. (Likens et al. 1977)

of humus indicates only the lack of mixing of this end product of active decomposition into the lower soil horizons. Unlike the taiga system, moderate temperature and moisture conditions, along with the high quality of litter produced by the dominant hardwood species (see discussion below), drive relatively rapid decomposition. This means that fire is not required to maintain nutrient cycles or to balance the production : decomposition ratio. Herbivory is also only rarely important in altering forest structure and ecosystem processes.

When fire and herbivory are not major forces shaping ecosystems, changes in structure and function are driven more by processes of natural tree death and by wind and storm damage—and the interactions between species in the competition to fill the gaps these create (Figure 20.3). Much of the research into vegetation dynamics in northern hardwoods has looked into processes that cause these gaps, those that favor the entrance of the different species into different-size gaps, and those that determine the relative competitive abilities of the different species in gaps of different sizes.

Important characteristics related to the movement of species into gaps include distance of seed dispersal and requirements for seed germination

Figure 20.3
A gap created by the loss of a large canopy dominant in a northern hardwood forest. (Bormann and Likens 1979)

(both related to seed size), and the extent of vegetative reproduction—the sprouting of new stems from root systems of mature trees. Related life-history phenomena include maximum tree size and age and the age at which seed production begins. Maximum tree size also affects the size of gaps created by tree death.

Success in the gap environment depends mainly on the rate of height growth that can be achieved and the ability to tolerate partial or deep shade. Secondary effects of nutrient requirements and alterations in nutrient cycling rates by the quality of litter produced may also be important.

Patterns of Disturbance in Northern Hardwoods and Effects on Resource Availability

How frequent is disturbance in a forest in which fire- and herbivory-related tree mortality are rare? The occurrence of disturbance in such systems is tied to the life expectancy of the dominant species and to the occurrence of major windstorms and blowdowns. Small-scale disturbances caused by individual tree death occur regularly and randomly over the landscape, while the large-scale blowdowns take place in major storms and occur less frequently. Taking a plot of a given size—for example, one-half hectare—there is an inverse relationship between the frequency of disturbances and the percentage of trees in the plot that are removed (Figure 20.4). Disturbances removing less than 20% of the trees occur, on average, every 200 years. This size and frequency of disturbance would be tied to the loss of individual trees. A disturbance removing two-thirds of the trees, such a major windstorm blowdown, occurs only once every 1000 years. Within this 1000-year return time for catastrophic disturbance, the stand could go through several cycles of gap-phase replacement.

Blowdowns in the northern hardwood forest are similar to disturbances in other ecosystems in that plant demand for resources is reduced and availability is increased. This is particularly true for elements such as nitrogen that are made available mainly through decomposition and mineralization. Decay is limited more by temperature than by moisture in this area. Removal of the forest canopy increases the penetration of sunlight to the forest floor, increasing soil temperatures and decomposition rates. Because of the rapid decay of litter produced by most northern hardwood species and the high nutrient demand of these deciduous species, nitrogen cycling rates are high even before disturbance.

Combining all these factors, we can understand how the large-scale removal of vegetation from the watershed at Hubbard Brook (Chapter 4), located in the northern hardwood region, resulted in very large increases in the loss of nitrogen and other nutrients to streams. Similar measurements of nitrogen loss from northern hardwoods, using trenched plots over which the canopy remained intact (also Chapter 4), suggest very high

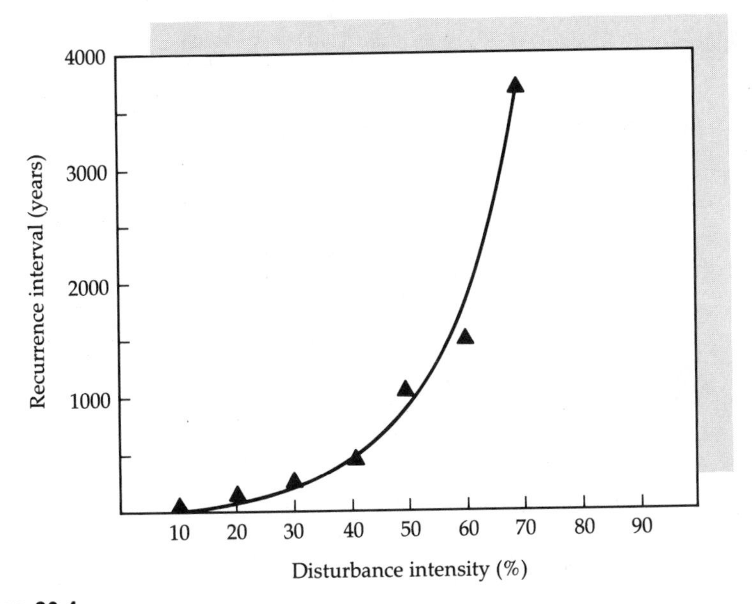

Figure 20.4
Relationship between frequency of disturbance and the percentage of trees removed in northern hardwood forests in the Great Lakes region. (Lorimer and Frelich 1989)

availability in small-scale disturbances as well. In the terminology presented in Chapter 5, northern hardwood systems have very low resistance to disturbance, as measured by rates of nitrogen loss. In terms of resource availability to plants, the disturbed areas are like highly fertilized and irrigated (because of reduced transpiration) sites with increased light availability as well.

SPECIES ADAPTATIONS TO THE DISTURBANCE GRADIENT: REPRODUCTIVE AND LIFE HISTORY STRATEGIES

Gaps come in a wide variety of sizes. We can think of gap size in terms of a gradient of resource availability (Chapter 5). The larger the gap, up to a certain size, the greater the enrichment. Larger gaps also take longer to close, and so the duration of the enriched environment will vary with gap size. The major northern hardwood species show a wide variety of reproductive and growth characteristics, combined in different ways (see Table 20.1). According to niche theory (Chapter 5), each combination should have adaptive value for a particular set of environmental conditions, in this case related to disturbance and recovery.

The northern hardwood forest in New England is also called the

Table 20.1 Seed dispersal and growth characteristics of six major northern hardwood species*

SPECIES	BEECH	SUGAR MAPLE	YELLOW BIRCH	PIN CHERRY	ASPEN	STRIPED MAPLE
Seed size (mg/seed)	283	65	1.0	32	0.15	41
Seed dispersed by (A = animal, W = wind)	A	W	W	A	W	W
Soil requirements (U = undisturbed, D = disturbed, M = mineral soil)	U	U	DM	D	DM	?
Vegetative reproduction	+	−	−	−	+	+
Age at first seed production	40	30	40	4	10	15 (tree form)
Maximum longevity	350	350	250	45	80	30
Maximum growth rate (cm height/yr)	30	35	40	100	100	?
Shade tolerance (T = tolerant, I = intermediate, N = intolerant)	T	T	I	N	N	T?
Leaf chemistry and decomposition						
% Nitrogen (green)	2.2	2.1	2.8	2.8		
(litter)	0.9	0.9	1.1	1.2		
% Lignin	24	10	15	19		
Decomposition rate (%/yr)	8	22	30	30		

*After Bormann and Likens 1979, Marks 1974, Melillo et al. 1982, Likens and Bormann 1970, USDA 1974.

Figure 20.5
A typical second-growth northern hardwood stand dominated by sugar maple, beech, and yellow birch. (Bormann and Likens 1979)

beech–maple–birch forest type because of the predominance of beech, sugar maple, and yellow birch in mature stands (Figure 20.5). Of these three, beech is considered the most tolerant of shade. It also has large, heavy seeds and active vegetative reproduction. Sugar maple is somewhat less tolerant of shade and has lighter seeds and no vegetative reproduction. Yellow birch has still lighter seeds, is less tolerant of shade, and also does not reproduce vegetatively. All three species have long life spans.

These three species are thought to form a cyclic pattern involving continuous recovery from small-scale disturbance within the mature forest (Figure 20.6). The creation of a gap by the fall of an individual large tree creates increased light availability on the forest floor and also usually creates patches of bare mineral soil around the upturned root mass (Figure 20.7*a*). The small, light seeds of yellow birch can be dispersed 100 meters or more by wind. In addition, some seeds are shed during the winter onto snow cover over which they may be blown for even greater distances. The small seeds have very limited energy reserves for building the initial root and stem and can suffer drought damage if they germinate on the seasonally dry forest floor. They survive better on mineral soil and will frequently grow on top of the old root mass of a fallen tree, sending roots down to and into the soil. As these roots mature and become woody, and the old root mass on which the seedling became established erodes away,

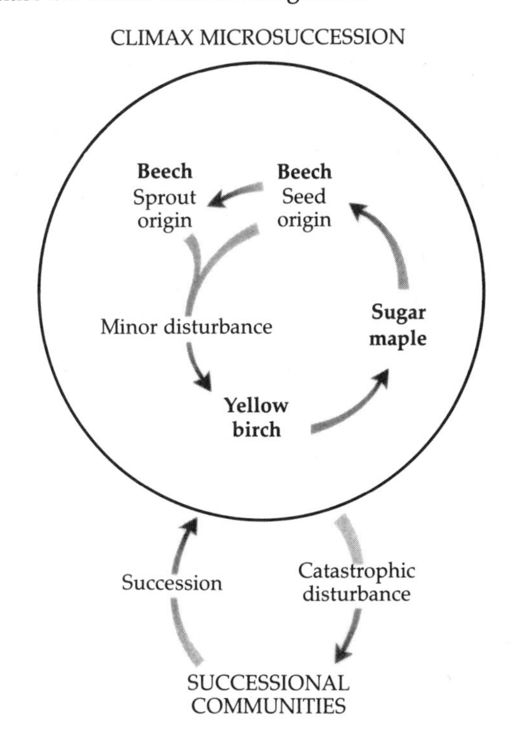

Figure 20.6
A hypothesized cyclic successional sequence of yellow birch, sugar maple, and beech in northern hardwood forests. (Forcier 1975)

Figure 20.7
Early and late stages of colonization of a large tip-up mound. *(a)* A recently created mound that will soon support tree seedlings, including yellow birch. *(b)* A yellow birch that seeded onto a tip-up mound now grows on "stilt roots" after the mound has eroded away.

a stilt-rooted tree characteristic of yellow birch (Figure 20.7*b*) may develop.

The partial shade and developing forest floor produced by yellow birch foliage inhibits the further establishment of yellow birch seedlings but allows germination and establishment of the heavier-seeded sugar maple and beech. However, sugar maple produces many more seeds (as many as 10 million per hectare per year) than beech. The seeds are also lighter and are contained in a samara (Figure 20.8), a winged structure that catches the wind, makes the entire structure spin like a helicopter blade, slows the descent of the seed, and results in greater dispersal distances. All these factors tend to favor more rapid colonization by maple under the birch.

Sugar maples will live and grow slowly under the birch, moving to the canopy after the death of the canopy tree. Under the darker shade of the maple, beech root sprouts, which receive photosynthate from the parent tree, have a competitive advantage and tend to take over the understory, along with several species of very tolerant shrubs. The heavy beechnut reduces the distance that seeds are moved by wind, although limited animal dispersal occurs. The large energy reserves in the seed allow seedling establishment in shade and on a thick forest floor.

At any time in this sequence, the death or blowdown of a large individual tree can reset the pattern by creating an opening large enough to allow the re-entry of yellow birch. The continued occurrence of these small-scale

Figure 20.8
Sugar maple leaf and samara. The wing-like structures cause the seed to whirl in the air, slowing its descent to the ground and allowing it to travel greater distances from the parent tree. (Fowells 1965)

disturbances and the relatively long life span of yellow birch are what allow the continued existence of this relatively intolerant "gap-phase" species even in mature forests.

Large-scale disturbances, such as those caused by major windstorms, hurricanes, and commercial clearcutting, bring a different set of actors to the stage. Pin cherry is a rapidly growing, short-lived species that sprouts from a population of dormant seeds stored in the forest floor. These seeds may have been present in the soil since the last major disturbance and may survive for 60 years or more and still remain viable and capable of germinating after disturbance. It is not clearly known what factor triggers germination, although increased levels of soil nitrate have been implicated in certain fertilizer experiments.

Aspen can also occur in these same areas, but its role is much less important here than in the Alaskan taiga, or even in the western end of the northern hardwood range, perhaps due to the reduced importance of fire. Aspen does not have a buried seed strategy but rather relies on prolific production of very light, wind-borne seeds to reach disturbed areas. Even a minor surface fire could destroy much of a buried seed crop, decreasing the adaptive value of that strategy and increasing the value of aspen's "fugitive" or light seed characteristic.

Several species of shrubs can dominate the understory in mature stands and contribute to the vegetative response to disturbance. One of the most interesting is striped maple, which can persist through vegetative reproduction and slow growth for decades under a dense canopy but grows to tree height, flowers, and fruits only following a major disturbance. After 10 to 20 years, the main stem dies back and the "tree" reverts to vegetative reproduction and a shrub growth form.

There are also species that respond to special conditions of resource availability. At low elevations and on rich sites, white ash can be more important than yellow birch as the "gap-phase" replacement species. Red maple becomes more important on very wet or very dry sites and also seems to occur in place of sugar maple on sites where nitrification does not occur. Red spruce also grows on wet and cool sites within the region and may have been of more general importance as a late successional species prior to 1900 when selective logging of this species was carried out over much of the region.

From this description, we can summarize the different "niches" occupied by the major northern hardwood species in terms of forest conditions under which they reproduce and grow most effectively. Both maple and beech are tolerant, dominant species that maintain both mature individuals capable of producing seed and shade-tolerant seedlings and saplings as advanced regeneration. However, there is a clear distinction between these two species. Beech vigorously reproduces vegetatively by root sprouts, while maple has more effective seed dispersal because of the samara structure.

Yellow birch is the gap-phase species that relies on maintaining long-lived, reproductive individuals in the overstory; these trees can disperse seed to small openings as they occur. Striped maple is also present in the mature forest, but as a low-growing shrub, and also reproduces in gaps, but by converting from long-lived shrub form to short-lived tree form. Pin cherry has the buried seed strategy, while aspen has the "fugitive" strategy.

Successful seed dispersal to disturbed areas is accomplished in all of these species. What differs is the way in which the different species survive or avoid the highly competitive environment between disturbances. Maple and beech are tolerant of these conditions, surviving as seedlings and saplings in the understory. Yellow birch survives in the overstory by being long-lived. Striped maple is also long-lived but grows in the understory, only reaching into the overstory in the first two decades after disturbance. Pin cherry and aspen are not present in the closed forest as live stems. Pin cherry is present as dormant seeds. Aspen relies on the presence of disturbed sites within the larger area to provide a constant source of highly mobile seeds. Both pin cherry and aspen require larger gaps to compete successfully.

Certain life history and growth characteristics are strongly related to mechanisms of seed production and dispersal. Table 20.1 shows that the most intolerant species tend to be short-lived, to grow rapidly, to have low root-to-shoot ratios, and to reproduce at a younger age. All of these traits are consistent with selective pressure to keep foliage above the general level of the developing canopy and to take the fullest advantage of the temporary nature of the enriched environment created by disturbance. Table 20.1 also shows that intolerant species tend to have higher nitrogen concentration in green foliage and in litter. All of these trends are in keeping with the general theory developed in Chapter 11, which suggests that enriched, disturbed environments tend to be filled by rapidly growing, short-lived species with high nutrient concentration in foliage.

In apparent contradiction to this theory is the very high concentration of lignin in pin cherry foliage. One possible explanation for this, involving the effect of pin cherry litter on nitrogen immobilization and retention following disturbance, is presented below.

Sugar maple also appears to be an exception because of the low lignin content in foliage. However, this species is known to have a high concentration of the low-molecular-weight polyphenols known as tannins, which also provide generalized defense against herbivory.

The very large differences in lignin content in the foliage and litter of the two major dominant species, beech and sugar maple, may have important effects on nutrient cycling. Beech leaf litter does decay much more slowly than maple litter and immobilizes more nitrogen. Beech litter acts much like conifer litter in increasing forest floor mass and acidity and reducing litter decomposition. If beech is also less demanding of nutrients than maple, these differences in litter quality could derive from natural selection for litter quality that "creates" site conditions more suitable to the species occupying the site. As with the theories regarding evolution to increase susceptibility to either fire or herbivory presented in previous chapters, this idea is very difficult to test.

PLANT RESPONSES AND SECONDARY SUCCESSION

While all disturbed areas will show increased resource availability, the degree and duration of those increases depend on the size of the gap created. Small gaps surrounded by intact forest close quickly as roots and branches from adjacent trees expand into the opening. The central area in large gaps is unaffected by surrounding plants, experiences this enrichment for a longer period, and tends to be revegetated by intolerant species such as pin cherry. Intermediate-size openings will favor the establishment and growth of the moderately tolerant and moderately fast-growing gap-phase species such as yellow birch.

In the largest openings resulting from windstorms or commercial clearcutting, there is usually, but not always, a remarkable and dramatic response to the enriched environment created. Germination of previously dormant pin cherry seeds can be as dense as 25 to 40 stems per square meter. This rapidly growing intolerant species can create a fully closed canopy within three to four years after disturbance (Figure 20.9). With the additional productivity of other intolerant and some remaining and new stems of tolerant species, annual net primary production can actually be higher in a five-year-old stand than in a mature forest (Figure 20.10). Pin cherry also has very high nutrient content in all tissues, such that at age four a vigorous pin cherry stand can contain a significant fraction of the nutrients that were lost from the devegetation experiment at Hubbard Brook. Thus, pin cherry acts as a major sink for nutrients following disturbance and plays an important role in reducing leaching losses.

Pin cherry leaf litter may also play an important role in minimizing losses. It is high in both nitrogen and lignin content, so rates of decomposition are moderate and considerable nitrogen immobilization occurs. In the fourth year of regrowth, immobilization into pin cherry litter may remove 17 kilograms per hectare per year from the nitrogen cycle, reducing the amount available for leaching losses. While some of this immobilization may be temporary, with mineralization occurring in the following two to three years, the high lignin content of pin cherry leaf litter suggests that a significant fraction of the immobilized nitrogen will be converted into long-term soil organic matter (Chapter 13).

Figure 20.9

A five-year-old pin cherry stand showing the complete recreation of canopy biomass and rapid growth that is typical of this species. (Bormann and Likens 1979)

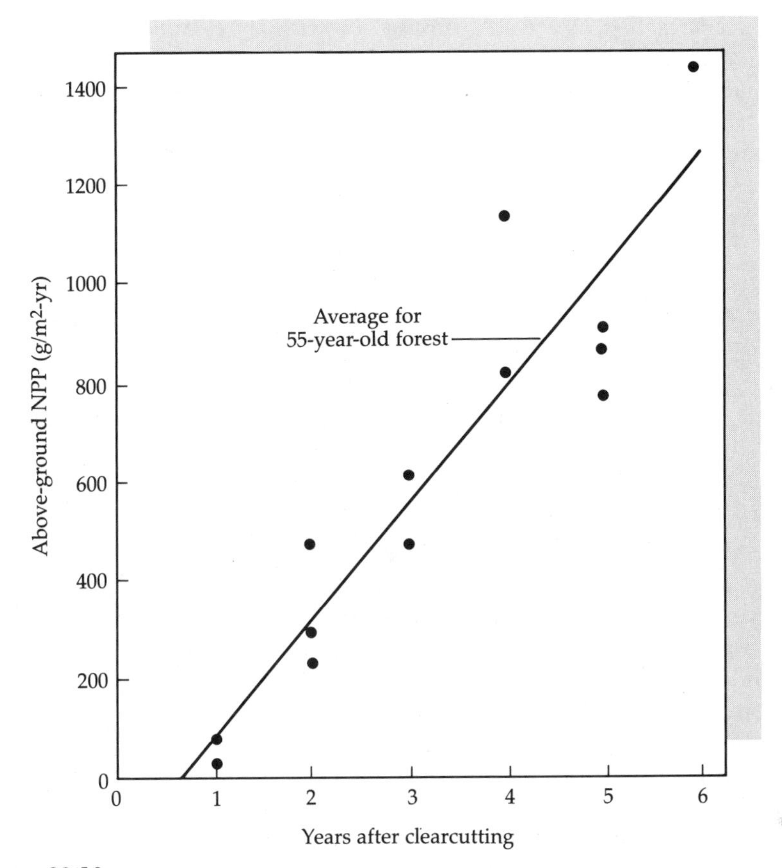

Figure 20.10
Changes in total above-ground net primary productivity for the first six years following clearcutting in northern hardwoods. (Bormann and Likens 1979)

Pin cherry thus contributes substantially to the resilience (Chapter 5) of northern hardwood stands. While initial increases in nutrient losses are high following disturbance, the rapid colonization of disturbed sites by pin cherry soaks up much of the available nutrients. By the fourth year, nutrient losses from disturbed sites in which regeneration occurs is nearly back to predisturbance levels. Other system processes, such as primary productivity and internal nutrient cycling, are also nearly restored.

> A Hypothesis of Homeostasis. *We propose that severe stress initiated by clear cutting not only accelerates the activity of the mechanisms responsible for biotic regulation [in the intact forest], but also calls into action another set of mechanisms largely quiescent during that phase. . . . The cutover ecosystem responds to the conditions of increased resource availability by a burst of primary production, not only by species that characterize the precutting forest but also by a group of species not part of the predistur-*

bance forest (e.g. buried seed species) that may have evolved specifically to fill a niche created by this type of disturbance. . . . The coupling of heterotrophic processes to autotrophic processes is far from perfect. Considerable leakage of nutrient from the ecosystem occurs during the first few years, even in immediately revegetating systems. . . . This suggests that the rapid increase in productivity that follows disturbance is a relatively inefficient activity and is costly in terms of nutrient and biomass storage within the ecosystem. However, the sacrifice of efficiency results in accelerated production, which may be considered an effective strategy of ecosystem stability since it forestalls a still-greater sacrifice in biotic regulation of erosion. (Bormann and Likens 1979)

Pin cherry appears critical to the northern hardwood ecosystem in terms of minimizing the impacts of disturbance and maintaining long-term ecosystem function. Yet its buried-seed reproductive strategy does not ensure vigorous regeneration on all sites. If pin cherry trees have a life expectancy of 30 years, and buried seeds can remain viable for 60, then this strategy will be effective only if disturbance occurs roughly once every 100 years. The disturbance frequency and intensity values in Figure 20.4 suggest that certain areas are likely to remain undisturbed for periods much longer than 100 years. In this case, pin cherry regeneration depends not on dropping of seed in place by previous populations, which occurs at very high densities, but rather on dispersal of seed into the area by birds, which eat the fleshy fruits and drop the seeds. This mechanism results in many fewer viable seeds per unit area. Thus there are both high-density and low-density pin cherry regeneration cycles (Figure 20.11).

Do other species replace pin cherry where it is absent and provide this resilience function? Apparently not. There is a large difference in the rate at which the canopy is re-created in the absence of pin cherry (Figure 20.12). This would suggest that nutrient losses should be greater and more prolonged following disturbance if pin cherry is present only in low densities. Unfortunately, no measurements of nutrient loss from low pin cherry density systems have been made.

Despite the role of pin cherry, there are longer-term changes that occur following disturbance in northern hardwood stands. The total amount of organic matter in the forest floor reflects the balance between litter production and decomposition. Disturbance reduces net primary production (NPP) and increases decomposition in the first few years. In addition, pin cherry allocates a lower proportion of NPP to slowly decomposing fine roots and produces leaf litter that decomposes rapidly.

The net effect of these changes is to cause a longer-term decline in the amount of organic matter in the forest floor (Figure 20.13). It may take 60 years for the forest floor to return to mature forest levels following a severe disturbance such as clearcutting. This has important implications for nutrient availability and future NPP, which are explored further in Chapter 21.

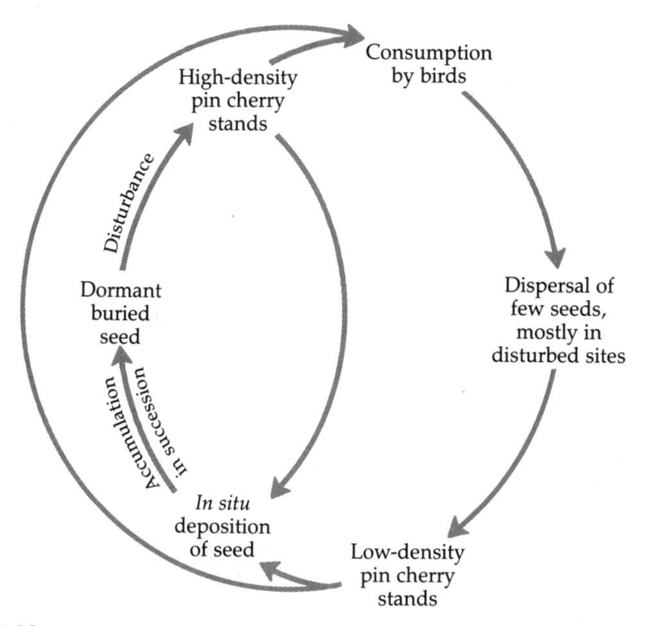

Figure 20.11
Representation of the high- and low-density pin cherry cycles active in northern hardwood forests. The low-density cycle depends on dispersal of seed by birds and results in much lower density of pin cherry stems in disturbed areas. (Marks 1974)

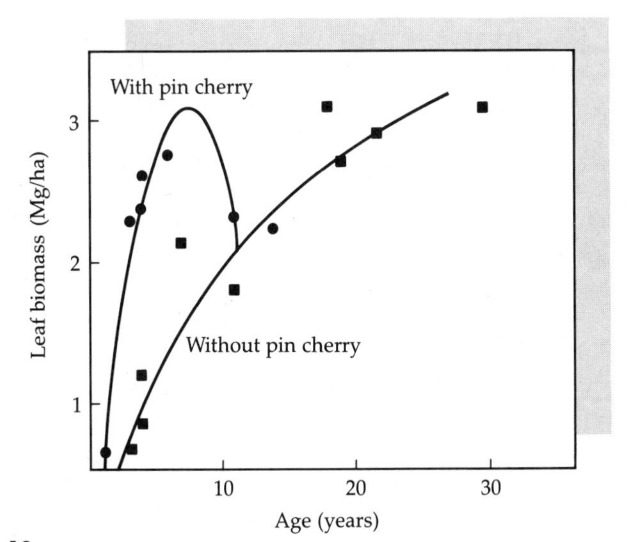

Figure 20.12
Differences in successional changes in total foliar biomass in young northern hardwood stands with and without pin cherry. (Covington and Aber 1980)

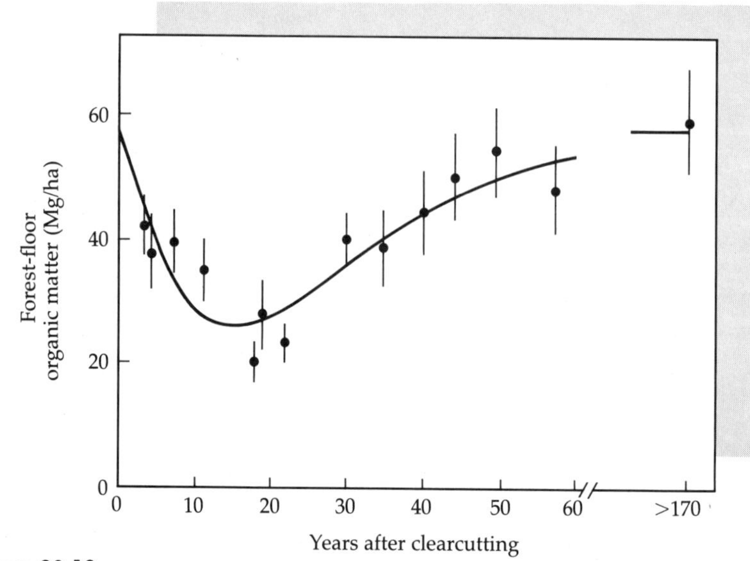

Figure 20.13
Changes in total forest-floor organic matter content in northern hardwood stands following clearcutting. (Covington 1981)

HUMAN USE OF THE NORTHERN HARDWOODS REGION

The eastern portion of the northern hardwoods region has passed through a period of heavy cutting for lumber and is now almost entirely second- and third-growth forest. In the Great Lakes area, much of the forest had been both harvested and cleared for agriculture before it was found that these soils and climates were less suitable for crops than the prairie areas just to the south.

The northern hardwood forest type is now adjacent to some of the most densely populated areas in North America. While wood harvesting continues in the national forests and in the relatively small areas owned by wood products companies, most of the land still in forest is owned by individuals who value the forest more for recreation and aesthetics than for wood production or economic return. Direct human use tends to involve the conflicting demands of wood production, recreation, and wildland preservation. Hotly contested practices, such as clearcutting in the national forests, tend to bring groups representing very divergent populations into court to settle differences. The large number of small landowners may be very important to the environmental stability of the region and inhibits rapid, large-scale shifts in land use. In addition, the system appears to contain mechanisms that confer resiliency, the ability to recover normal function, in the face of normal forest harvesting practices (but see further discussion in Chapter 21).

Perhaps more important at present are the indirect uses of the northern hardwood forest resulting from proximity to an industrial region (Figure 20.14).

> *The natural ecosystems of the United States, using solar energy, help to provide the stable and predictable environment on which we all depend. Yet, they are being subjected to increasing stress and destruction as our population and economy grow, and as our profligate use of fossil energy continues. As these natural ecosystems are degraded or destroyed, they lose their capacity to carry on basic biogeochemical functions [which control] climate, hydrology, circulation of nutrients, and the cleansing of air and water. (Bormann 1976)*

Air pollutants, such as acid rain, can reach into these largely rural areas, presenting a novel and important form of stress to the ecosystem. The integrity of these systems, and their ability to perform the services of air and water purification, which we generally take for granted and on which we rely, may be jeopardized (see Chapter 23).

Figure 20.14
Satellite view of eastern North America at night. Bright areas represent urban centers. Note that the northern hardwood forest region (Figure 20.1) lies amid the region with the highest density of urban settlement. (U.S. Air Force)

REFERENCES CITED

Bormann, F. H., and G. E. Likens. 1979. *Pattern and Process in a Forested Ecosystem.* Springer-Verlag, New York.

Bormann, F. H. 1976. An inseparable linkage: Conservation of natural ecosystems and the conservation of fossil energy. *BioScience* 26:754–760.

Covington, W. W. 1981. Changes in forest floor organic matter and nutrient content following clear cutting in northern hardwoods. *Ecology* 62:41–48.

Covington, W. W., and J. D. Aber. 1980. Leaf production during secondary succession in northern hardwoods. *Ecology* 61:200–204.

Forcier, L. K. 1975. Reproductive strategies and the co-occurrence of climax tree species. *Science* 189:808–810.

Fowells, H. A. (ed.). 1965. *Silvics of Forest Trees of the United States.* Agricultural Handbook No. 271. U.S.D.A., Washington, D.C.

Likens, G. E., et al. 1977. *Biogeochemistry of a Forested Ecosystem.* Springer-Verlag, New York.

Likens, G. E., and F. H. Bormann. 1970. Chemical analyses of plant tissues from the Hubbard Brook ecosystem in New Hampshire. Yale University: School of Forestry Bulletin No. 79.

Lorimer, C. G., and L. E. Frelich. 1989. A methodology for estimating canopy disturbance frequency and intensity in dense temperate forests. *Canadian Journal of Forest Research* 19:651–663.

Marks, P. L. 1974. The role of pin cherry (*Prunus pensylvanica L.*) in the maintenance of stability in northern hardwood ecosystems. *Ecological Monographs* 44:73–88.

Melillo, J. M., J. D. Aber, and J. F. Muratore, 1982. Nitrogen and lignin control of hardwood leaf litter decomposition dynamics. *Ecology* 63:621–626.

U.S. Department of Agriculture. 1974. Seeds of woody plants of the United States. USDA Agricultural Handbook No. 450.

Part IV APPLICATION Human Impacts on Local, Regional, and Global Ecosystems

Paolo Koch/Rapho/Photo Researchers

Parts II and III dealt mainly with the function of ecosystems in which human impacts have been minimal. Such systems are increasingly rare. Over much of the globe, direct human use, such as for agriculture or forestry, constitutes a fourth major fate for organic matter produced in terrestrial ecosystems. In addition, indirect or secondary effects of human activity, such as the deposition of pollutants in ecosystems far from the sources of those pollutants, can also have profound effects on terrestrial ecosystem function.

The purpose of this section is to present an overview of human impacts on ecosystems at three levels. The first two chapters deal with traditional uses for the production of food and fiber by looking at the management of forest and agricultural ecosystems. The first chapter also briefly introduces and discusses the application of computer modeling to ecosystem management. The third chapter deals with the regional effects of air pollution on ecosystems in a discussion of "acid rain" and the more general problems of the deposition of air pollutants in systems distant from the source of those pollutants. In the final chapter we move up to the global level, using concerns over global climate change due to increases in carbon dioxide and other trace gases in the Earth's atmosphere as a vehicle for discussing the interactions between ecosystem dynamics and human use of the Earth.

Chapter 21 Computer Models and Forest Ecosystem Management

H. Armstrong Roberts

INTRODUCTION

Agencies responsible for management or protection of ecosystems require quantitative predictions of the potential effects of different management schemes or pollution control programs. Of particular importance are predictions of long-term effects of a given practice. It is often not possible or practical to test for these effects by carrying out research in the field. As an example, "acid rain" (Chapter 23) is suspected of damaging forest ecosystems in several parts of the world. Performing field experiments that accurately mimic the effects of long-term, low-level additions of pollutants to ecosystems is difficult at best. Yet we would like to be able to predict the effects of air pollution, especially if they are really negative, before they occur—before the "experiment" has been concluded.

One approach to predicting these long-term effects is to construct mathematical representations, or "models," of the affected ecosystems and to run the experiments of interest on these models. A computer model of an ecosystem is just such a mathematical representation, built from a series of equations written in a computer language. Such models have been widely used to predict management or pollution effects because the predicted effects of the treatment are available in minutes rather than decades. The limitation of this approach is that the model is not the real system, and erroneous predictions may result. This generally limits the application of models to questions that would be too difficult, too complex, or too expensive, or would require too much time to answer through direct field research.

The purpose of this chapter is to discuss the process of computer modeling of terrestrial ecosystems and its application to a relatively straightforward example drawn from forest management: the "whole-tree harvesting" question. Whole-tree harvesting is the practice of removing whole stems with branches, twigs, and leaves still attached, rather than taking just the trunk. While this practice increases the amount of biomass harvested, it results in much greater nutrient removals as well. There is general concern that greater nutrient removals in harvested biomass could compromise the long-term productive capacity of the ecosystem.

THE WHOLE-TREE HARVESTING QUESTION

There is a wide range in the intensity with which forest ecosystems are managed (Figure 21.1). At one end are the unmanaged wilderness or national park areas in which human impacts are kept to an absolute minimum. Near the other end are the commercial forests of the southeastern United States where nursery seedlings are planted in plowed ground, treated with fertilizers and herbicides to ensure early growth, and harvested after 20 to 40 years. Two systems that are still experimental represent the ultimate in intensively managed forests: the "sycamore silage" system, where resprouting stumps of sycamore on floodplain soils in the southern United States are "mowed" every two to four years, and experimental energy plantations of fast-growing hybrid poplars, which are both fertilized and irrigated for maximum yield. This last example approaches standard agricultural practices, except that the "crop" grows for more than one year and the harvested product is fiber rather than food.

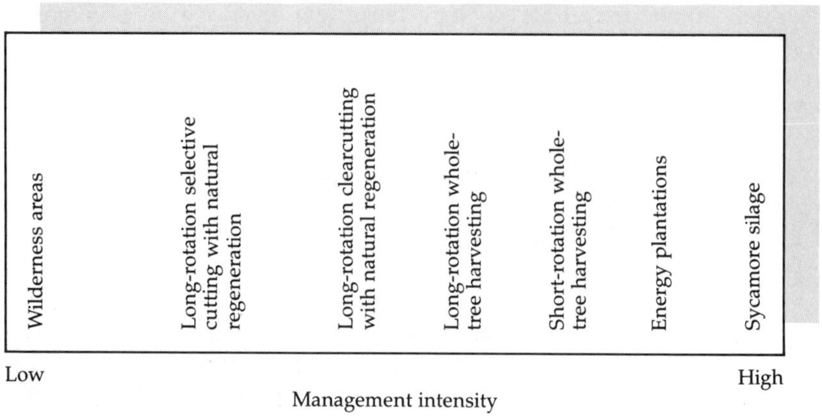

Wilderness areas
Long-rotation selective cutting with natural regeneration
Long-rotation clearcutting with natural regeneration
Long-rotation whole-tree harvesting
Short-rotation whole-tree harvesting
Energy plantations
Sycamore silage

Low High
Management intensity

Figure 21.1
A spectrum of management intensity in forest ecosystems. Intensity is defined as the amount of fossil fuel; chemical, mechanical and labor inputs required; and the frequency of harvest.

elapsing between cuttings. This time between harvests is called the **rotation length**. Following the harvest, stands are either replanted or left to regenerate and regrow naturally until the next harvest. In this situation, where internal processes are relied upon for maintaining productivity, as opposed to the plowed, fertilized, and protected forests under intensive management, there is a question as to how often biomass can be harvested and how much of the material can be removed without damaging the site's productive capacity.

The 1960s and 1970s saw the development of tractor-based machines, which have largely replaced the chain saw as the major harvesting tool (Figure 21.2). With these devices, whole trees can be clipped off at the base, laid in windrows, dragged to a staging area, and reduced to chips for transport to a mill by truck. Most of the wood products industry is now based on chips rather than large pieces of lumber, so the biomass of small trees, as well as branches, twigs, and even leaves of larger ones, can contribute to the yield from a site. This makes it possible to harvest smaller

(a)

(b)

(c)

Figure 21.2
Machines used in whole-tree harvesting operations: *(a)* a feller–buncher for harvesting trees, *(b)* a skidder for removing trees to the loading area, and *(c)* a chipper, which reduces the entire tree to chips for transport to the processing plant.

trees or to harvest whole stands at a younger age. Whole-tree harvesting was seen at first as a way of increasing the efficiency of the logging operation by removing a larger fraction of the total biomass on a site and also of returning to the site sooner for a second harvest.

However, biomass contains nutrients. Increased rates of biomass removal and shorter rotations could cause substantial increases in rates of nutrient removal as well. Branches, twigs, and leaves have much higher nutrient concentrations than stemwood. By adding these components of stand biomass into the harvest, the overall removal of nutrients is increased by far more than is the yield of usable biomass. These removals must, to some extent, be offset by inputs to the system from precipitation and weathering. Otherwise, serious nutrient limitations on forest growth could develop.

Table 21.1 compares the effects of stem-only and whole-tree harvesting in terms of nutrient and biomass removals from stands covering a wide range of age and species combinations. The first thing to notice is that all nutrients occur in higher concentration in foliage than in branches or stems for all of the species listed. This reflects the higher rates of metabolic activities in foliage and the higher density of enzymes, coenzymes, and other important biochemical compounds. In all cases, adding branches, and particularly foliage, to the harvest increases nutrient losses far more than it increases biomass harvested.

Age is a second important factor. One potential benefit of whole-tree harvesting is that usable amounts of biomass can be harvested more frequently, on a shorter rotation. However, younger stands have more foliage and less wood relative to older stands, and the smaller stems in young stands have higher nutrient concentrations (compare the 6- and 90-year-old northern hardwood stands in Table 21.1). Harvesting younger stands increases the quantity of nutrients removed per unit of biomass harvested.

One might expect that, because of lower nutrient concentrations in foliage, the effects of whole-tree harvesting would be less in conifer stands. However, that lower concentration is generally offset by higher foliar biomass. This means that the foliar nutrient content may be higher as a percentage of total tree content in conifers than in hardwoods (compare the loblolly and jack pine stands in Table 21.1 with the aspen/maple and northern hardwood stands). In addition, deciduous forests lose their foliage in the winter, so nutrient removal impacts can be reduced by winter harvesting.

Regardless of species or site, shorter rotations mean more nutrient removal per unit of biomass harvested. The accumulation of low-nutrient-content, physiologically inactive heartwood in the stems of large trees effectively separates most of the harvestable fiber from the high-nutrient-content, physiologically active foliage, roots, and bark. Stem-only harvesting takes advantage of this by removing mostly wood and leaving

Table 21.1 Comparisons of biomass and nutrient removals in stem-only and whole-tree harvests for selected forests*†

STAND TYPE AND AGE	COMPONENT	TWIGS AND FOLIAGE	BRANCHES	STEMWOOD AND BARK	TOTAL
Northern hardwood (pin cherry/aspen, age 6)	Biomass	3.7 (11)	4.5 (14)	24 (75)	32.2
	N	75 (50)	14 (9)	62 (41)	151
	K	48 (59)	5 (6)	28 (35)	81
	Ca	26 (24)	16 (15)	68 (61)	110
Aspen/maple (age 63)	Biomass	2.5 (2)	19 (13)	125 (85)	146.5
	N	45 (17)	78 (28)	152 (55)	275
	P	6 (16)	12 (32)	20 (52)	38
	K	22 (7)	56 (18)	235 (75)	313
	Ca	35 (5)	192 (28)	465 (67)	692
Northern hardwood (age 90)	Biomass	3.6 (2)	54 (27)	145 (71)	202.6
	N	81 (17)	195 (40)	207 (43)	483
	Ca	25 (5)	229 (42)	293 (53)	547
Loblolly pine (age 16)	Biomass	8 (5)	23 (15)	125 (80	156
	N	82 (32)	60 (23)	115 (45)	257
	P	10 (32)	6 (19)	15 (49)	31
	K	48 (29)	28 (17)	89 (54)	165
Jack pine (age 65)	Biomass	8 (7)	10 (9)	89 (84)	107
	N	65 (38)	31 (19)	71 (43)	167
	P	16 (38)	6 (14)	20 (48)	42
	K	30 (36)	13 (15)	41 (49)	84
	Ca	73 (26)	41 (15)	163 (59)	277

*Biomass values in Mg/ha, nutrient value in kg/ha; numbers in parentheses are percentage of total.
†From Weetman and Weber 1972, Jorgensen et al. 1975, Marks 1974, Pastor and Bockheim 1981, Hornbeck 1977.

most of the nutrients on site. Whole-tree harvesting does not exploit this separation, and greater nutrient removals are the result.

Traditional forest practice has treated the productive capacity of a given site as a constant that is unaffected by harvesting practices. With short rotations and whole-tree harvesting, the potential to alter productive capacity by reducing nutrient availability appears very real. But how can this potential be judged? To test this effect in the field would require most of a century, as reduced productivity may not be significant until the end of the second or third rotation.

One alternative is to bring together existing information on important processes affecting nutrient cycling and productivity in the forest type being considered and to use that information as a basis for judgment. This has been done at several very different levels of complexity, from simple tabular comparisons of nutrient input and output rates to very detailed, process-oriented computer models of forest ecosystem dynamics. We will examine the results of these different approaches later on. First we will discuss the process of constructing a computer model.

COMPUTER MODELS AND THE MODELING PROCESS

What is a computer model? Most readers of this book will have had experience with home or school computers and be familiar with the software packages that come on small disks. When the disk is put into the computer, the program contained on that disk is read by the machine, the instructions are executed, and the game, educational program, or the like appears on the screen (Figure 21.3).

A computer model is identical in concept to the program on the disk. It is a sequence of instructions and computations that play the "game" of predicting the effects of different management practices on the ecosystem. In general, a model consists of the program, which is the series of calculations that the model will make, and one or more sets of data, which specify the type of ecosystem and environmental conditions for which the predictions are desired (Figure 21.3).

Building a computer model serves three general purposes (Figure 21.4). First, the computer statements themselves represent a clear and quantitative description of what is known (or assumed) about how the ecosystem

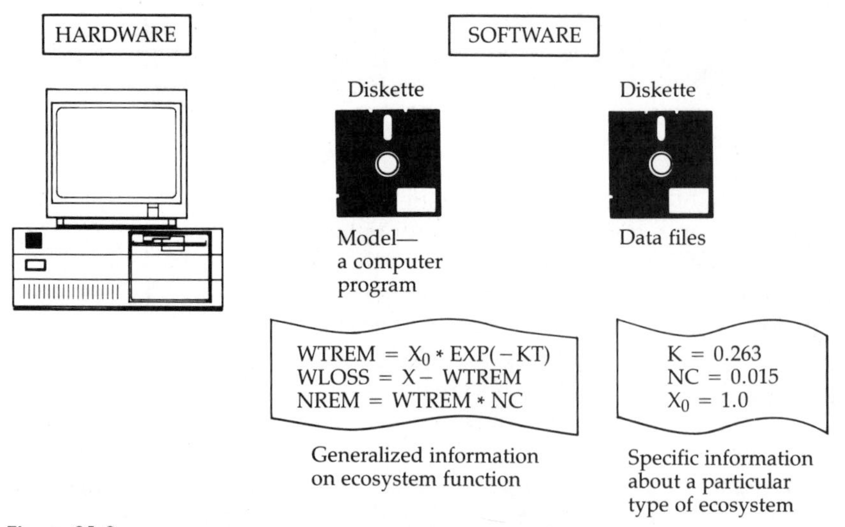

Figure 21.3

Computer models resemble the software packages available for home and school computers. The model consists of a program, which performs the calculations and in which the structure of the model resides, and a data file, which contains specific information about the type of ecosystem to be modeled, providing coefficients for the equations in the program.

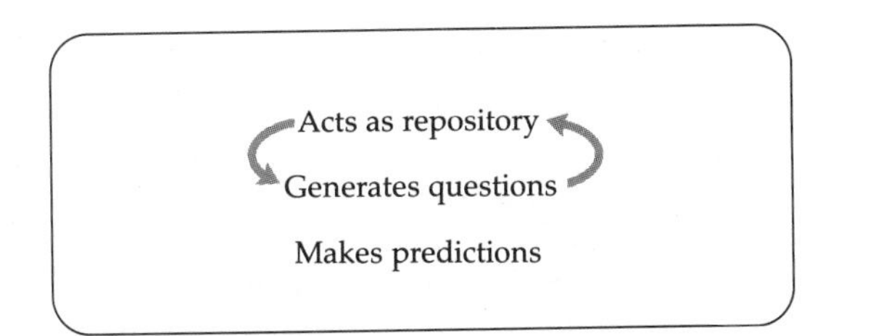

Figure 21.4
The three major purposes of building computer models of ecosystems. Note the cycling interaction between the first two purposes.

works. Any of the relationships discussed in Part II regarding rates of photosynthesis, decomposition, herbivory, and so forth can be included (for example, see Figure 21.5). Equations describing all the relationships included in the model consitute the code that makes up the computer model.

An operational model usually consists of both this code, which expresses the type of equations used to describe each process (e.g., the exponential decay function in Figure 21.5), and a data file, which contains the values of the coefficients to be used with those equations for particular species, sites, etc. Again, in Figure 21.5 the value 0.263 makes the generalized exponential equation specific for Scots pine litter decaying in a Scots pine forest in southern Sweden. Taken together, the code and data files

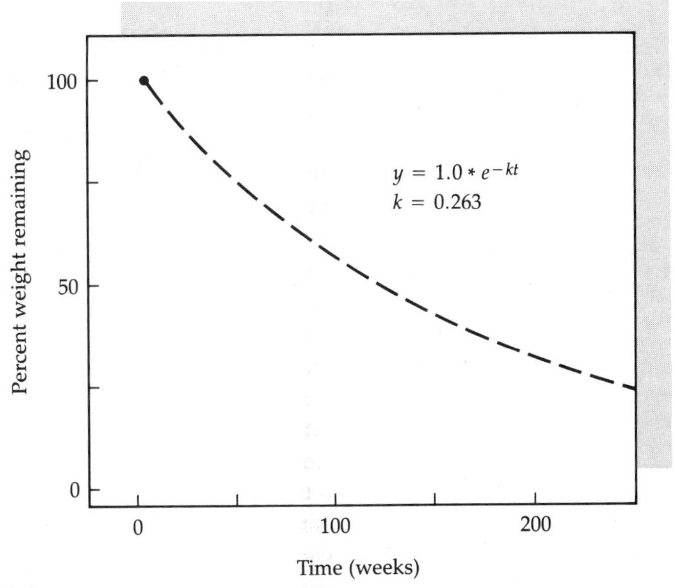

Figure 21.5
The shape of the exponential-decay-type equation used to summarize decomposition dynamics of litter in terrestrial ecosystems and the equation that describes that curve. Note that the coefficient in the equation is *specific* to Scots pine litter decaying in a Scots pine forest in southern Sweden (t = time in years). The general equation would be included in the computer code of the model. The coefficients would reside in the data set describing Scots pine forests. (Staaf and Berg 1982)

constitute a repository of information available about the ecosystem and its function that applies to the question the model is designed to answer.

The second purpose served by computer modeling is to raise questions for future research to answer. Simply attempting to construct a computer model often reveals processes about which too little is known to make quantitative statements. The model will then contain an assumption as to how that particular process works or information borrowed from studies of similar systems elsewhere. These assumptions allow the model to be complete and to function but also introduce important sources of error into the model. There is thus a cyclic interaction between the repository and question-generating functions (Figure 21.4). As questions are answered, the value of the repository grows and new questions are raised. This is not unlike the general application of the scientific method to any field, but the computer acts as the memory and to some extent the logic in the system.

As the repository grows and becomes more complete, the model becomes less valuable for question generation and more valuable for its third purpose: prediction of the outcomes of specific treatments. In even the simplest ecosystem, the interactions between the major processes are too numerous and complex for the human mind to synthesize into quantitative predictions. The model uses the powers of precision and calculation provided by the computer to test the implications of the cumulative information available about the system in the repository.

It is important to realize that no computer model will ever be a *complete* representation of a system, nor should it be. In any system, any number of processes occur that are not important for answering a particular question. For example, including herbivory as a process in systems where it is generally not important only adds to the difficulty of completing the model and the expense of running it. Recurring questions in the modeling process should include: Is this process important enough to be included? Does it need to be represented in this much detail? Are we looking at processes at the time and space scales appropriate to the purposes of the model?

Steps in Building a Computer Model

The steps in building a computer model reflect both the distinction between code and data and the three different purposes discussed above (Figure 21.6).

The first step is to put together the structure of the model. The structure is a generalized expression of how the type of ecosystem under consideration operates. For example, Figure 21.7*a* is a simple box-and-flow diagram of the movement of nitrogen through forest ecosystems. This diagram represents the structure of one of the first computer models created for the prediction of whole-tree harvesting effects on nutrient cycling and pro-

1. **Establish structure (code).**
2. **Parameterize (data files).**
3. **Perform validation/sensitivity analysis.**
4. **Make prediction.**

Figure 21.6
Steps in the construction of a computer model of an ecosystem.

ductivity in forest ecosystems. We have seen similar diagrams several other places in this book. The values in the boxes represent the amount of nitrogen stored in each component, and the arrows and associated numbers mark both the direction and rate of flow between these components.

This is a generalized realization of a forest ecosystem because all forest ecosystems have all of the parts described by the different boxes. However, different numbers of components and flows may be described. Figure 21.7*b* shows a second, more complex structure for describing nitrogen cycling, which could also be applied to any forest ecosystem. The more detailed structure may provide a more accurate vision of ecosystem function, but it also requires a substantial increase in the amount of information needed to run the model. There is a constant tradeoff in the construction of a model between the desire to increase the complexity of the structure, and thereby make the model more complete, and the cost of doing the research to provide the information required to run a more complex model. An additional consideration is the potential for introducing errors into the model's predictions by including a poorly described or understood process.

The second step in building the model is **parameterization**. The computer program constructed to run a model such as the one shown in Figure 21.7 defines the boxes and flows but does not necessarily put in any particular values for storages or rates of movement. Detailing these storages and flows is the parameterization step. The initial values for nitrogen storage in each of the boxes, and the rates at which nitrogen is transferred between boxes, are described in the data files, which are read by the program and used in running the model. The data files thus contain the *particular* knowledge about the system that has been described in a *general* way in the computer program.

This all sounds very tedious and dull! It is an unfortunate fact that there are few things less interesting than describing or discussing computer models. In contrast, actually working with models can be very enlightening and fun. Consider the popularity of video and home computer games, many of which serve educational purposes. The descriptions of the programs that make those games possible would bore to tears all but the most dedicated hacker.

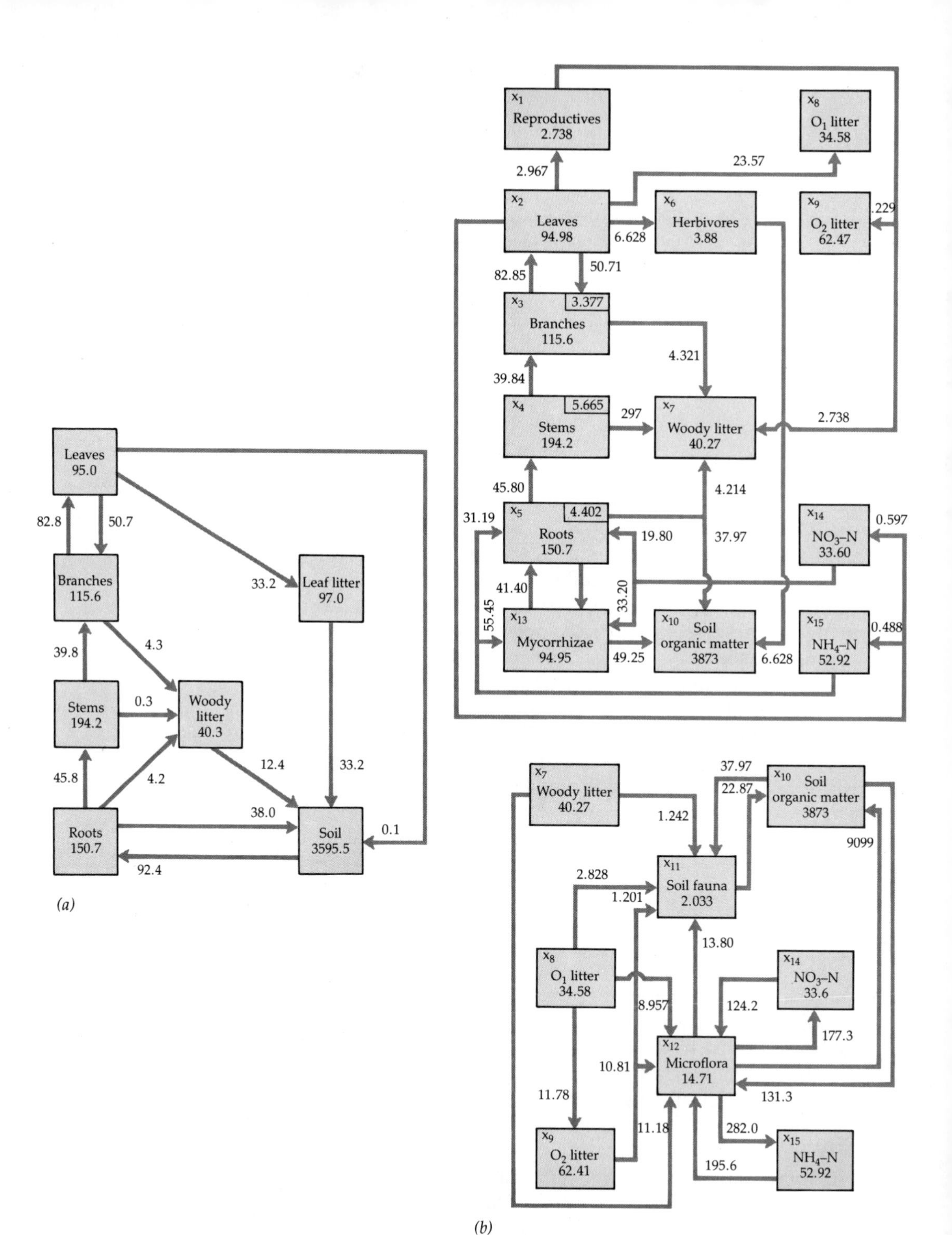

Figure 21.7

Two different models of nitrogen dynamics in an oak–hickory forest: *(a)* a reduced structure model and *(b)* a model with greater detail and more compartments and transfers (presented in two interconnected parts). (Waide and Swanks 1977)

Once the model has been structured and parameterized, the third step is to test, as far as possible, how accurate it is. This step is called **validation**. Validation of the model consists of testing the predictions of the model against some independently measured field data. For example, one of the models used below to discuss the effects of whole-tree harvesting also predicts the amount of dead wood (logs, branches, and so on) lying on the forest floor at a given time. This prediction results from the interactions between tree birth, growth, and death rates and the rate of wood decomposition. It is not defined as part of the model. There is no input parameter that says, "In year 40 there will be 2 megagrams of dead wood on the forest floor." Figure 21.8 compares the model's prediction of changes in dead wood on the forest floor through time following forest cutting with actual values measured in the field. This kind of comparison should be extended to all available data sets. The closer the model comes to accurately predicting these independent variables, the more confidence we can have that it is an accurate representation of the ecosystem. If the model cannot be validated, then it remains valuable as a synthetic tool for understanding the system but is of very limited value in predicting the results of management practices.

Sensitivity analysis is another method for testing the structure of the model. In this step, a specific input parameter is altered, usually by some

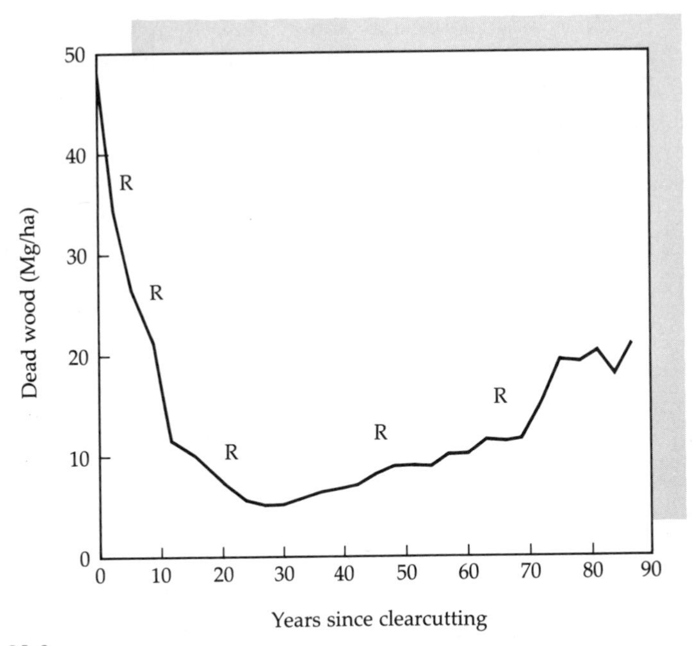

Figure 21.8
An example of validation of a computer model. The model is used to predict changes in a variable as measured in the field. R = field data; line = model prediction. (Aber et al. 1982; field data from Roskoski 1980)

percentage of its original value, and the model is run again. The goal is to see how sensitive the model is to changes in that datum or to the process it represents. For example, if the *k* parameter in Figure 21.5 is doubled or tripled, and this has little effect on the model's predictions, then perhaps the accuracy of that number is not too critical. In the extreme case, sections of code in the model and whole processes might be removed as unimportant, due to the results of sensitivity analysis.

The final step in the modeling process is to predict the outcomes of different treatments. For example, in computer models applied to the whole-tree harvesting question, information is input to the model's data file regarding the workings of the ecosystem in question and the frequency and intensity of harvesting. In a stem-only cut, trees are felled and the branches are cut off before the stem is taken off the site. In the model, the biomass and nutrients in the stem compartment (see Figure 21.7*a*) would be removed, while the branch and leaf components would be transferred to the litter compartment for decomposition. In a whole-tree harvest, all of the biomass and nutrients in the stem, branch, and foliage compartments would be removed from the system.

Structuring Models of Whole-Tree Harvesting

How completely must we understand ecosystem function to answer the questions raised by whole-tree harvesting? How complex do our models have to be to contain all the information that is important in reaching a decision? There are no fixed answers to these questions. Several different types of approaches have been tried, and all enter into the decision-making process. For, while computers can give very precise calculations of the implications of precisely defined equations, they cannot determine which set of equations is best. Human decision making usually occurs with imperfect knowledge. All models are inaccurate to some extent. It is up to people to decide, using intuition and judgment as well as logic, which models or facts are best or most applicable.

It is often best to start with the simplest model, requiring the least amount of information. For whole-tree harvesting, the earliest and simplest approaches simply calculated the rates at which nutrients were removed from the ecosystem under different treatments and compared these with calculations of the time required for precipitation and rock weathering to replace them. This gives an estimate of the shortest rotation length that would result in no net reduction in nutrient content in the entire system.

Figure 21.9 is a hypothetical example describing the amount of nitrogen in biomass that would be harvested by either stem-only (no twigs, branches, or leaves) or whole-tree harvesting as a function of stand age. It also contains the cumulative inputs of nitrogen to the system in precipitation (a straight line, which assumes a constant input of 5 kilograms per hectare per year). The point at which the removal and input lines cross is the age at which inputs will equal removals. For whole-tree harvesting this is 90 years. For stem-only harvesting it is 45 years.

This approach gives some pretty conservative estimates and assumes a management goal of no reduction in total ecosystem nitrogen content. An alternative approach is to calculate the number of harvests that could occur before the nutrients in the ecosystem would be exhausted. For example, if the forest in Figure 21.9 were harvested every 30 years, a net decrease of 150 kilograms per hectare in total ecosystem nitrogen content would occur in each rotation (300 kg/ha removed minus 150 kg/ha in 30 years' worth of precipitation inputs). If this system contains 3000 kilograms of nitrogen stored in the soil, then 20 rotations (600 years) would be required to exhaust the supply!

We can see that these two simplest methods of estimating effects of short-rotation, whole-tree harvesting give very different results. The truth lies somewhere between these two extremes. It is true that not all nutrient losses need to be replaced in order to maintain productivity. However, the second type of calculation described above implicitly assumes that all soil nitrogen can be made available to plants. As we found in Chapter 13, much of this nitrogen is locked away in humus, with a turnover rate of 1000 years or more, and cannot be considered available on the time frame used by the model.

The question of how much of this nitrogen actually *is* available requires a much more complex model. In essence, we need to add to the simple

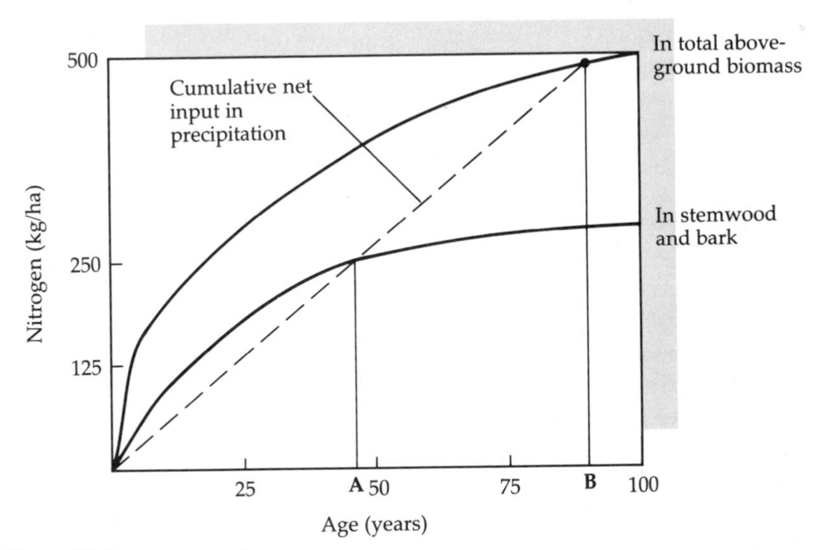

Figure 21.9
Graphic representation of the time required to replace nitrogen removed in stem-only versus whole-tree harvesting. Upper curve is nitrogen accumulation in the biomass of the whole tree. Lower curve is nitrogen accumulation in stem only. Dashed line reflects constant accumulation of nitrogen inputs in precipitation.

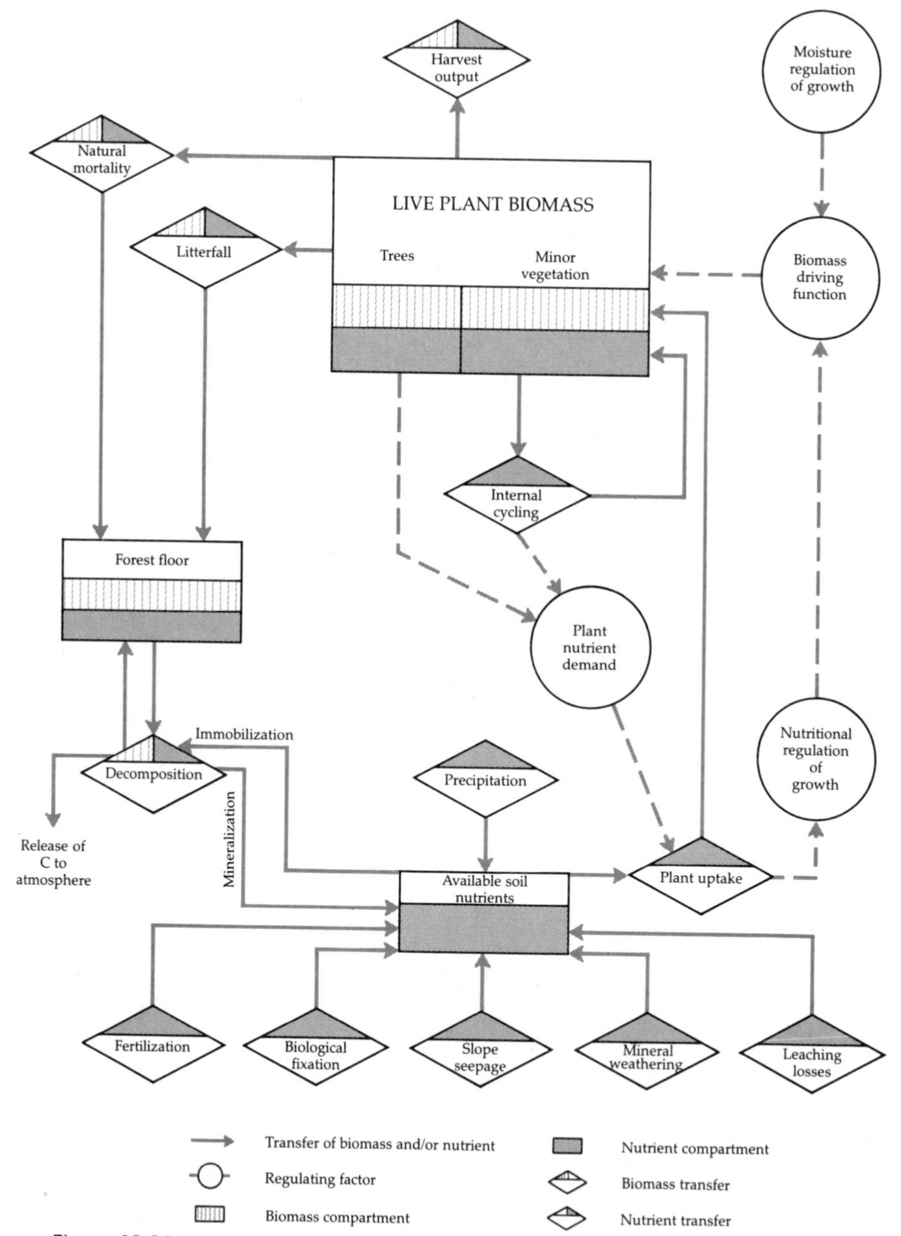

Figure 21.10

Structure of the FORCYTE and FORTNITE models for the prediction of effects of whole-tree harvesting on nitrogen cycling and productivity. FORCYTE was originally developed for use in the Douglas fir forests of British Columbia. FORTNITE was developed for northern hardwood forests. (FORCYTE, Feller et al. 1983; FORTNITE, Aber et al. 1982)

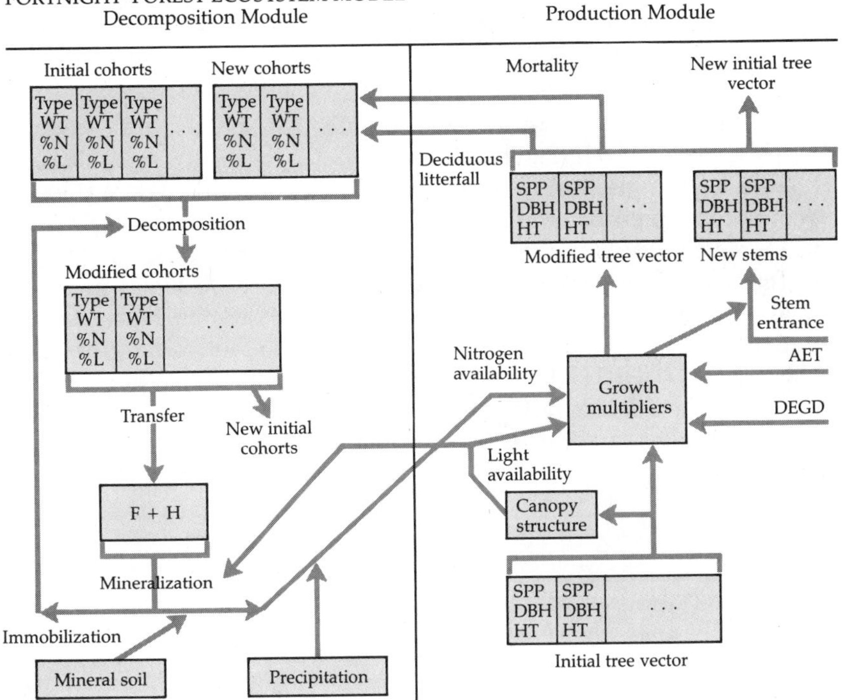

Figure 21.10 ***continued.***

descriptions of quantities stored in different components and nutrient input rates a description of the rate at which nutrients cycle within the system. This opens up the entire Pandora's box of processes described in Part II. We now need to quantify decomposition, production, biomass, and nutrient allocation and retention within plants, along with a whole host of other transfers.

Several models of this type have been applied to the whole-tree harvesting question. These include the relatively simple model in Figure 21.7*a* as well as more complex ones. The structures of two such models are defined in Figure 21.10. These are not included as a test of the reader's powers for deciphering obscure diagrams but to convey, by comparison with Figure 21.9, the increase in complexity required to handle the internal processes affecting nutrient cycling and tree growth in forest ecosystems. With some careful comparisons, however, it is also apparent that all three of these "process" models applied to the whole-tree harvesting question are quite similar in terms of the components of the forest ecosystem treated, reflecting some consensus as to which processes are indeed most important.

Is there also consensus on the effects of whole-tree harvesting? All three of these models have been used to predict the effects of different harvesting intensities and frequencies on different types of forest ecosystems. Their conclusions are startlingly similar in some respects. All three predict significant decreases in the amount of organic matter or nitrogen in the forest floor for whole-tree versus stem-only harvesting. The two models that deal with changes in rotation length show strong effects of this parameter as well (Figure 21.11).

There is less consensus as to changes in total fiber yield. The Waide and Swank model predicts reduced yield with whole-tree harvesting. FORTNITE predicts large reductions in yield with shorter rotation lengths but not for whole-tree versus stem-only harvesting. The Douglas fir model (FORCYTE) predicts a decline in yield with whole-tree harvesting on rotations shorter than 60 years (Figure 21.12). It is interesting that both FORCYTE and FORTNITE predict greater yield with whole-tree harvesting than stem-only harvesting for any given rotation length. This is because removing a larger percentage of the total biomass on the site more

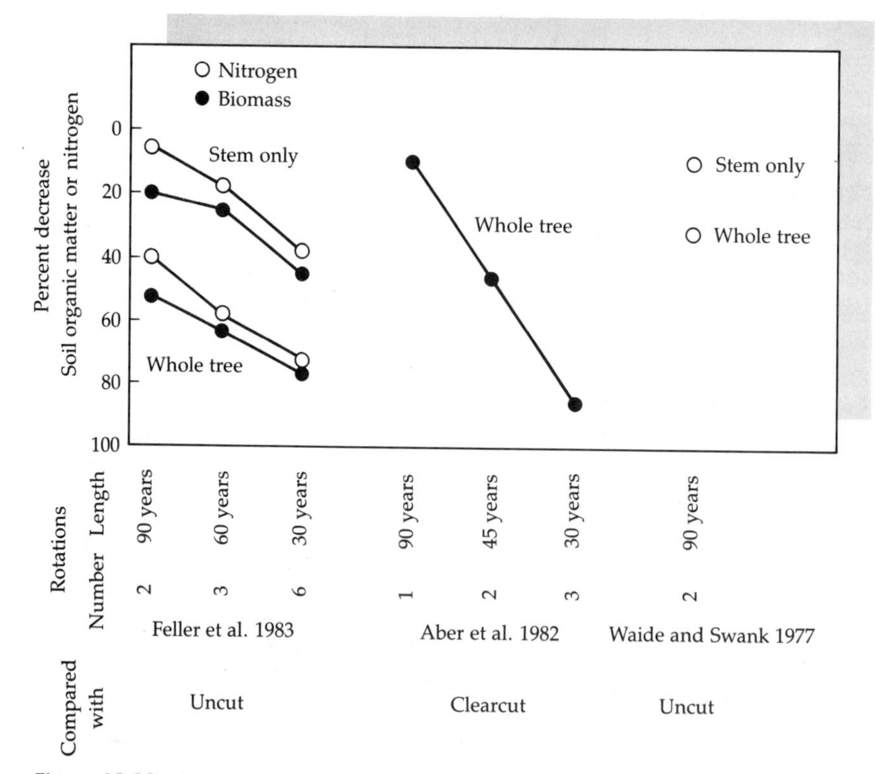

Figure 21.11

Comparison of predictions of whole-tree harvesting effects on forest-floor biomass or nitrogen content in three different forest types by three different models, expressed as a percentage reduction from uncut or clearcut stands. Note the importance of rotation length. (Waide and Swank 1977, Feller et al. 1983, Aber et al. 1982)

than compensates for the reduction in yield due to harvesting intensity alone. In terms of actual yield, the length of the rotation appears to be more important than the intensity of the harvest. The authors of all three of these models concluded that short-rotation, whole-tree harvesting would actually reduce yields in the absence of nutrient replacement by fertilization.

Different models designed to answer the same question do not always agree even as well as these three do. Models dealing with global climate change and atmospheric chemistry (Chapter 24) vary widely in structure and are less complete and often contradictory. This reflects the more complex nature of these problems, and the database is still evolving.

The consensus among forest ecosystem modelers regarding the effects of whole-tree harvesting suggests that a general shift to short rotations will be counterproductive. This view is now widely held (only in part because of the model results), and there is general recognition that intensive harvesting will also require other aspects of intensive management such as fertilization. As forestry moves farther into intensive management (Figure 21.1), it becomes increasingly like agricultural crop management, and many of the principles derived from agricultural sciences will become increasingly relevant.

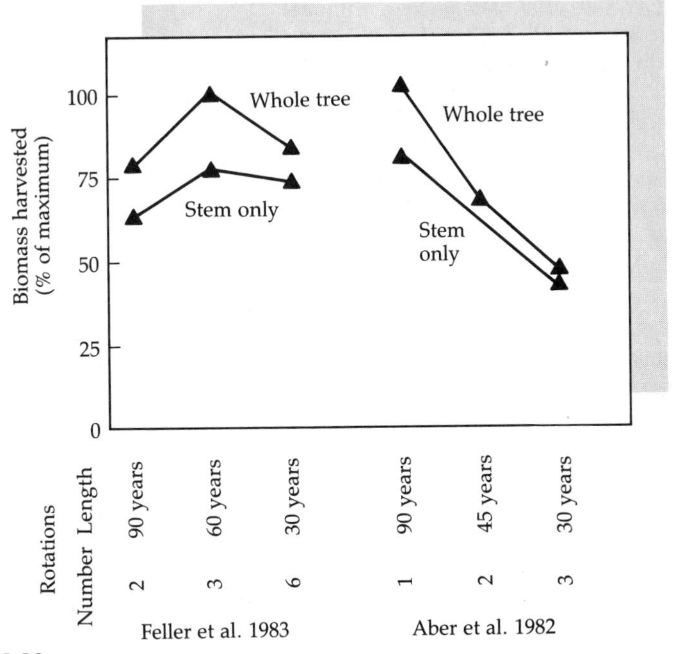

Figure 21.12
Comparison of predictions of the effects of whole-tree harvesting on yield (expressed as a percentage of the maximum value) for two of the three models in Figure 21.11.

REFERENCES CITED

Aber, J. D., J. M. Melillo, and C. A. Federer. 1982. Predicting the effects of rotation length, harvest intensity and fertilization on fiber yield from northern hardwood forests in New England. *Forest Science* 28:31–45.

Feller, M. C., J. P. Kimmins, and K. A. Scoullar. 1983. FORCYTE-10: Calibration data and simulation of potential long-term effects of intensive forest management on site productivity, economic performance, and energy benefit/cost ratio. In *IUFRO Symposium on Forest Site and Continuous Productivity*. U.S. Department of Agriculture, U.S. Forest Service General Technical Report PNW-163.

Hornbeck, J. W. 1977. Nutrients: A major consideration in intensive forest management. In *Proceedings of the Symposium on Intensive Culture of Northern Forest Types*. U.S. Department of Agriculture, U.S. Forest Service General Technical Report NE-29.

Jorgensen, J. R., C. G. Wells, and L. J. Metz. 1975. The nutrient cycle: Key to continuous forest production. *Journal of Forestry* 73:400–403.

Marks, P. L. 1974. The role of pin cherry (*Prunus pensylvanica L.*) in the maintenance of stability in northern hardwood ecosystems. *Ecological Monographs* 44:73–88.

Pastor, J., and J. G. Bockheim. 1981. Biomass and production of an aspen–mixed hardwood Spodosol ecosystem in northern Wisconsin. *Canadian Journal of Forest Research* 11:132–138.

Roskoski, J. P. 1980. Nitrogen fixation in hardwood forests of the northeastern United States. *Plant and Soil* 54:33–44.

Staaf, H., and B. Berg. 1982. Accumulation and release of nutrients in decomposing Scots pine needle litter. Long term decomposition in a Scots pine forest. II. *Canadian Journal of Botany* 60:1561–1568.

Waide, J. B., and W. T. Swank. 1977. Simulation of potential effects of forest utilization on the nitrogen cycle in different southeastern ecosystems. In D. L. Correll (ed.), *Watershed Research in North America*. Smithsonian Institution, Washington, D.C.

Weetman, G. F., and B. Weber. 1972. The influence of wood harvesting on nutrient status of two spruce stands. *Canadian Journal of Forest Research* 2:351–369.

ADDITIONAL REFERENCES: WHOLE-TREE HARVESTING

Boyle, J. R., and A. R. Ek. 1972. An evaluation of some effects of bole and branchwood harvesting on site macronutrients. *Canadian Journal of Forest Research* 2:407–412.

Kimmins, J. P. 1977. Evaluation of the consequences for future tree productivity of the loss of nutrients in whole tree harvesting. *Forest Ecology and Management* 1:169–183.

ADDITIONAL REFERENCES: ECOSYSTEM MODELS

Agren G. I., and E. Bosatta. 1988. Nitrogen saturation of terrestrial ecosystems. *Environmental Pollution* 54:185–197.

Botkin, D. B., J. F. Janak, and J. R. Wallis. 1972. Some ecological consequences of a computer model of forest growth. *Journal of Ecology* 60:849–872.

Cosby, B. J., G. M. Hornberger, and J. N. Galloway. 1985. Modeling the effects of acid deposition: Assessment of a lumped parameter model of soil and stream-water chemistry. *Water Resources Research* 21:51–63.

Parton, W. J., J. W. B. Stewart, and C. V. Cole. 1988. Dynamics of C, N, P and S in grassland soils: A model. *Biogeochemistry* 5:109–132.

Pastor, J., and W. M. Post. 1986. Influence of climate, soil moisture and succession on forest carbon and nitrogen cycles. *Biogeochemistry* 2:3–28.

Shugart, H. H., and D. C. West. 1980. Forest succession models. *BioScience* 30:308–313.

Chapter 22 Agricultural Ecosystems

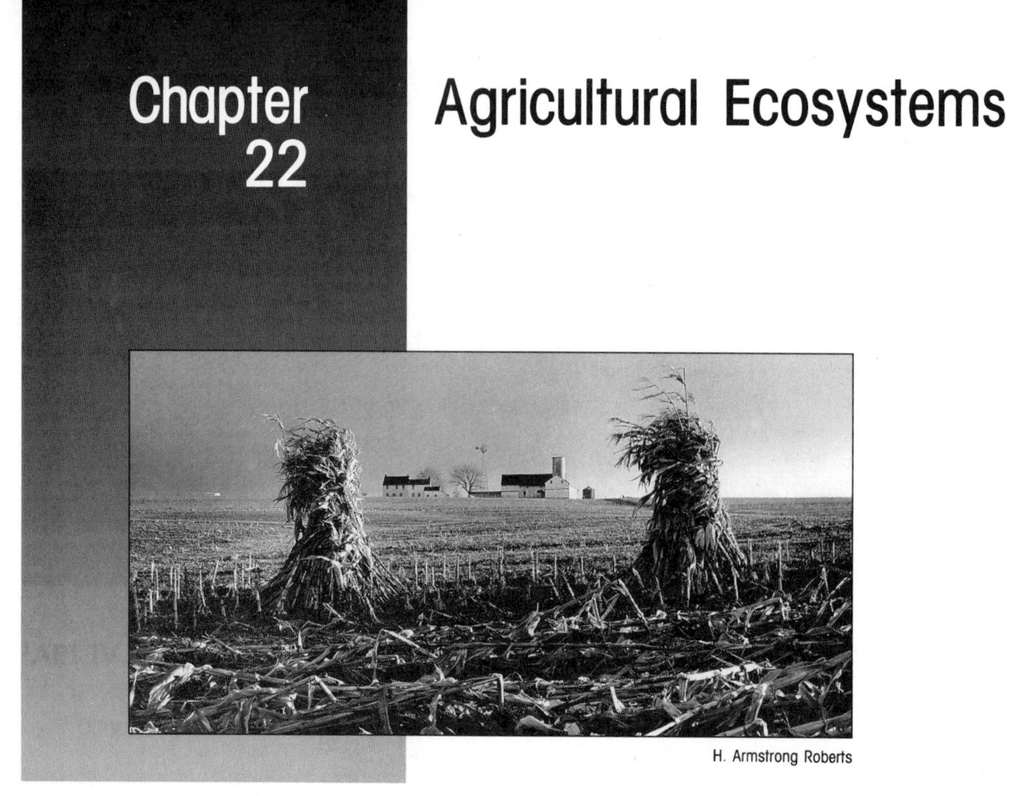

H. Armstrong Roberts

INTRODUCTION

If forestry is becoming more like agriculture, agriculture is, in many parts of the world, becoming more like forestry. Using forests, grasslands, and other natural systems as examples, many agricultural scientists are exploring ways to mimic the efficient use of resources and the low loss rates of nutrients and water that natural systems achieve. These efforts are in response to the large inputs of fertilizers and pesticides required in modern, temperate-zone agriculture and the environmental problems and high costs attendant to these requirements. These efforts also result from the problems that have arisen in trying to export the temperate-zone approach to agriculture to developing nations, largely in tropical zones.

The purpose of this chapter is to discuss different agricultural systems in terms of ecosystem analysis. We will see that modern temperate-zone agricultural ecosystems are very different from the systems we have discussed so far in terms of rates of disturbance, resource requirements, input/output budgets for nutrients, and the allocation of productivity to reproduction versus perennial growth. We will also examine some of the compromises now being explored between modern, high-resource-demanding approaches and less intrusive methods of modifying existing systems for the production of food and fiber. Not all of these approaches are new. Some borrow from ancient systems of agriculture that have proven both productive and stable, even in the tropical environment, since pre-industrial times.

A complete review of the vast literature pertaining to agricultural ecosystems is well beyond the scope of this volume. We will instead provide a cursory overview of the range of methods by which landscapes are managed for the production of food.

MODERN TEMPERATE-ZONE AGRICULTURAL ECOSYSTEMS

These widely different approaches to agriculture can be placed on a continuum relating to the intensity of management similar to the one derived for forestry in the previous chapter (Figure 22.1). A good example of an agricultural system at the most intensive end of the management spectrum is in corn production as practiced in the American Midwest. The "corn belt" occupies a large expanse of flat to rolling land centered on the states of Iowa, Illinois, and Wisconsin. Highest productivity is found on deep, fine-textured soils rich in organic matter formed by millenia of true or tall-grass prairie growth and in the deep and extensive root systems formed by the dominant grasses. Corn cultivation has converted the system from a very diverse mixture of mostly perennial grasses and forbs to a monoculture of a highly productive annual, corn (Figure 22.2).

While the original prairie system and the new corn system may appear similar in terms of late-summer height of vegetation, they are very dissimilar in nearly every other ecosystem parameter.

At the heart of agricultural practice is the maximization of edible yield. Total net annual primary production is increased greatly over native tall-grass prairie systems through fertilization and sometimes irrigation. Production is also allocated very differently among roots, stems, and seeds in the two systems (Table 22.1). In prairies, a large part of the annual

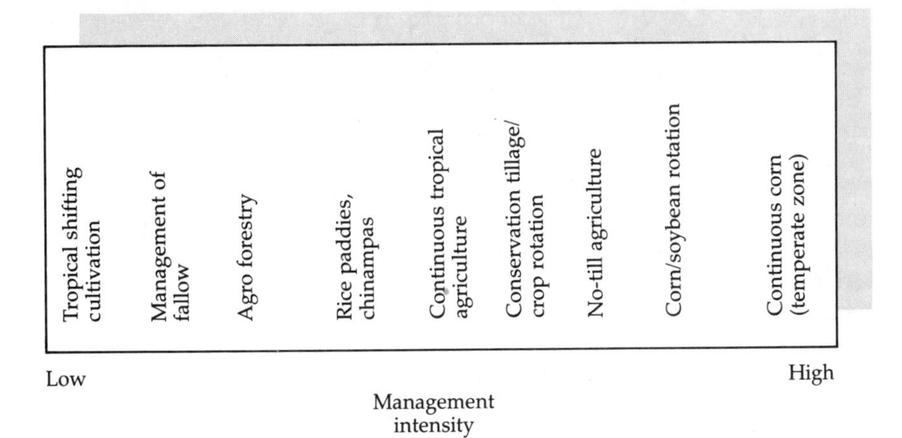

Figure 22.1
Spectrum of intensity of agricultural practices analogous to that of forest management in Figure 21.1. Increasing intensity generally indicates higher inputs of fertilizers and pesticides, greater mechanization, continuous cultivation, and higher consumption of fossil fuel.

(a)

(b)

(c)

Figure 22.2
Structure of the cornfield ecosystem in comparison with the native tall-grass prairies they replaced in the American Midwest. *(a)* A cornfield in summer and *(b)* in winter. *(c)* A prairie in summer. (*a*, *b*, H. Armstrong Roberts; *c*, Tom Riles)

productivity is allocated to root growth and to perpetuating a large root and root crown system. This is partly in response to the occurrence of droughts, but more as an adaptation for survival and regrowth following fires and grazing. In contrast, corn is an annual plant, which even in the wild state allocates much of its annual growth to seed production to ensure successful reproduction in the following year. Genetic selection has accentuated this tendency, producing plants in which over 45% of the total above-ground primary production can be harvested as grain.

Much of the remaining production is also harvested for use as animal fodder or silage (the conversion of low-quality plant residues to high-quality by partial fermentation). A conspicuous part of this type of cultivation is the bare-soil winter aspect produced by the complete removal of all

Table 22.1 Annual net primary productivity in a cornfield and a tall-grass prairie ecosystem and allocation of production to different plant parts*

	PRAIRIE		CORN	
	t/ha/yr	%	t/ha/yr	%
Above ground				
Vegetative	3.46	39	8.0	53
Reproductive	—	—	7.0	47
Below ground				
Perennial	2.86	32	0	
Annual	2.56	29	?	
Total	8.88		16 + roots	

*From Boyer 1982, Sims and Singh 1978.

Table 22.2 Nutrient input and output budgets for a cornfield and a tall-grass prairie*†

	PRECIPITATION (IN)	PRAIRIE		CORN		
		Stream (Out)	Fire (Out)	Fertilizers (In)	Stream (Out)	Harvest (Out)
Nitrogen	11.0	—	10	160	35	60
Calcium	3.2	—	—	190	47	1
Phosphorus	0.03	—	—	30	3	12
Potassium	0.2	—	—	75	15	13

*Inputs are precipitation and fertilizers; outputs are leaching below the rooting zone or to streams (all values in kilograms per hectare per year).
†From Frissel 1978; Gilliam 1987; Woods et al. 1983; Timmons and Dylla 1983; Seastedt, personal communication.

plant residues (Figure 22.2*b*). This also results in a very large reduction of organic matter inputs to soils, which in turn leads to the long-term reductions in soil organic matter levels discussed in Chapter 13. In prairies, much of the above-ground plant material is removed by frequent fires, but with most of the production going below ground, most of the annual production is protected from burning. The accelerated removal of plant materials and the associated loss in soil organic matter under cultivation are very similar in principle to the losses due to intensive forest management discussed in the previous chapter.

The high yield of grain in modern corn culture comes at the cost of very high inputs of nutrients and protective chemicals, as well as intensive manipulation of soils through plowing and cultivation. Table 22.2 compares the inputs and outputs of nutrients in cornfields and native prairie systems. Fertilizer application is well in excess of the total uptake of nitrogen by the growing corn plants. In prairies, only 10 to 20% of the annual nitrogen uptake comes from outside the system through precipitation and dust, with the remaining fraction being supplied by mineralization of soil organic matter.

The application of high doses of fertilizer also increases nutrient losses to groundwater, as availabilities are too high even for the high demands of the growing corn. Such losses are very low to negligible in prairies. In general, nutrient cycles under intensive agriculture are very "open," with inputs and outputs in excess of annual plant uptake. Nutrient cycles in native prairie systems are quite "closed."

Of even greater concern is the loss of soil through erosion under standard agricultural practices. Spring plowing and the absence of plant or plant residue cover during those periods when the crop is not growing leave the soil vulnerable to serious erosion. Images of the dust bowl of the 1930s with worn-out and deeply eroded farms show extreme examples of the types of damage that can occur. Currently, losses of topsoil in the American Midwest are estimated to be 35 tons per hectare per year. While soil organic matter can be rebuilt by changes in tillage and harvesting practices, the chemical and physical weathering of soils to produce topsoil is a much, much slower process.

The overriding concern with the kind of modern agriculture described here is: Is it sustainable? This system is vulnerable in at least two respects. First, the high demands for fertilizers and pesticides and mechanical cultivation assume abundant and relatively cheap fossil fuels. As the cost of these fuels increases, so will the price of food and the proportion of the national economic effort required to produce it. Second, the continuous loss of topsoil is beyond the ability of any economic system to repair. How the decline in topsoil will affect productivity is not fully known. One can imagine that the effects will be greater if applications of fertilizers need to be reduced for economic reasons. If we face a future of higher fossil fuel costs and lowered availability, then conserving topsoil as a means of assuring future agricultural productivity seems at least prudent.

TRADITIONAL PRACTICES IN THE HUMID TROPICS

At the other end of the management intensity spectrum from the modern temperate-zone cornfield are the traditional extensive methods of **shifting cultivation** practiced in humid tropical regions. Also called **"slash-and-burn"** or **swidden-fallow agriculture**, this approach has been alternately despised by temperate-zone observers as a great waster of the tropical forest or held up as the ultimate model of sustainable, low-resource-demand agriculture, depending on political and scientific fashion. To the native peoples of the tropics it has provided subsistence for thousands of years.

Shifting cultivation in the tropical forest zone (Figure 22.3*a*) consists of a relatively long cycle that begins with the cutting and burning of the

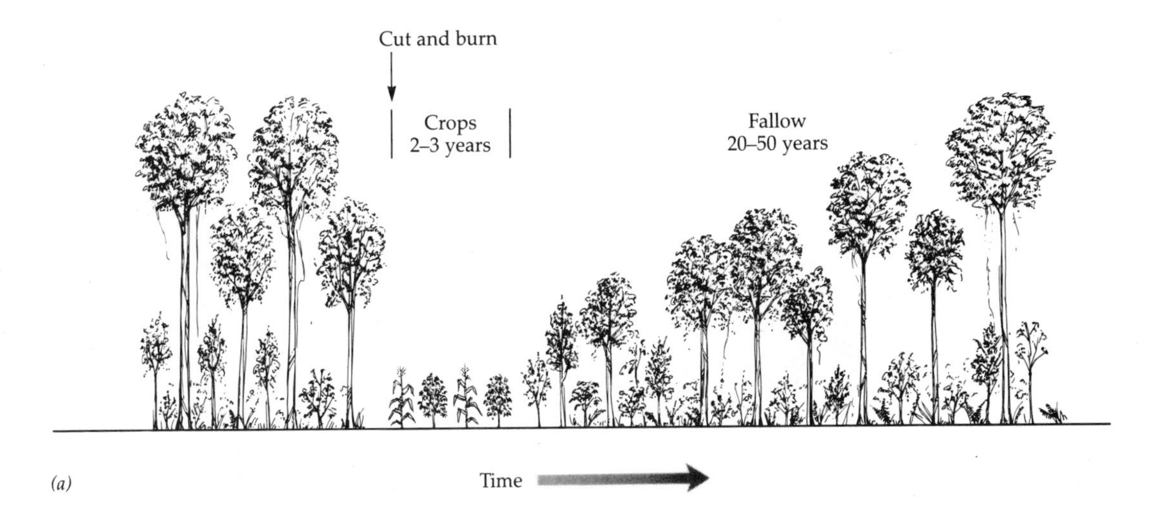

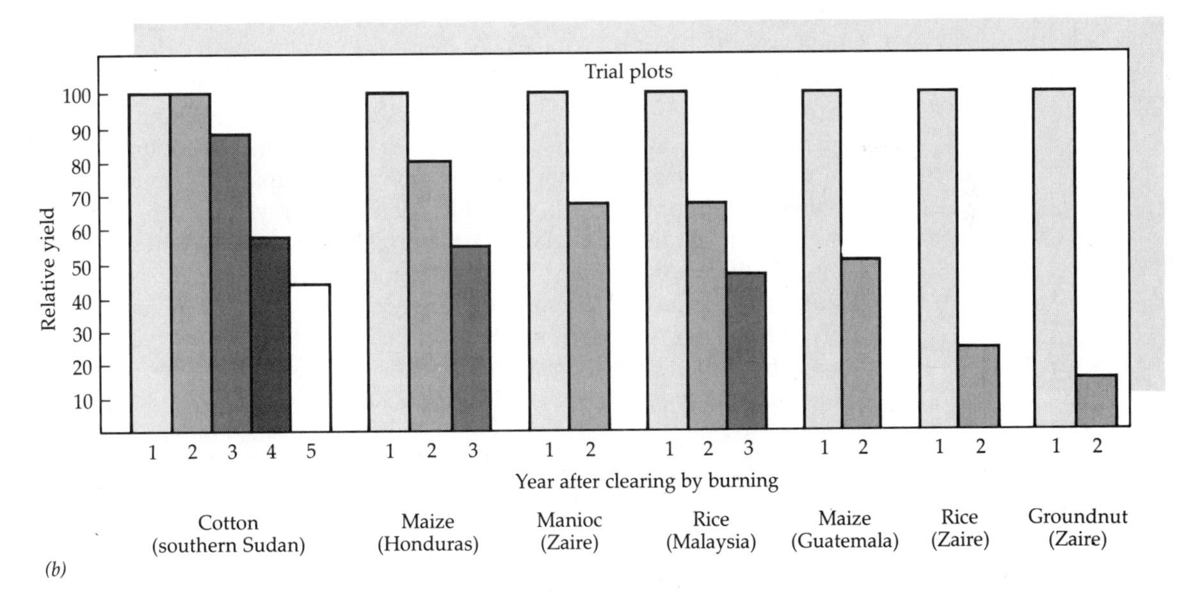

Figure 22.3

Shifting cultivation in the humid tropics. *(a)* Schematic drawing of the structure of different stages of the cycle. *(b)* Relative changes in yield in different years, following cutting and clearing. (Cox and Atkins 1979)

forest. This is followed by short-term cropping and a longer "fallow" period during which the forest is allowed to reclaim the site and rebuild soil fertility.

This process has many features in keeping with the tropical environment. In Chapter 2 we discussed the relatively nutrient-poor nature of many tropical soils and the relatively large proportion of rapidly cycling nutrients that are held in the vegetation. Clearing and burning the forest releases these nutrients in a pulse. This pulse of high nutrient availability supports a brief period of agricultural production. However, rapid leaching by excessive rainfall and reduced plant growth in crops compared with the growth of the original forest cause a steep decline in the productivity of cleared land over a two- to three-year period (Figure 22.3*b*). After the area has lost much of its productive value, it is let go to fallow and a secondary forest invades rapidly. A fallow period of 20 to 30 years is usually required to rebuild the nutrient stores in the vegetation and soil.

Shifting cultivation has other advantages in addition to nutrient self-sufficiency. In the first year of cultivation, weed invasion is relatively low, since the presence of important weed species has been reduced by the highly competitive growing conditions present during the fallow period. This invasion increases in years 2 and 3, reducing yield and increasing the amount of work required. Insects and other pests are also reduced during the fallow period and require some time to locate and proliferate in the new agricultural fields.

Considering that crop plants are generally of high nutrient and protein content and relatively low in general feeding inhibitors, shifting cultivation can be seen as a managed system that mimics the environments in which such species would naturally grow. In Chapters 11 and 15 we discussed the rapid growth, short life span, and prevalence of "qualitative" defense strategies in plants occupying disturbed areas. These plants grow rapidly in the temporary, nutrient-enriched environments created by disturbance and invest less in herbivore defenses. Crop plants tend to be derived from wild species native to disturbed sites. Shifting cultivation is a system by which disturbance frequency is increased and specific high-value crop plants are introduced to the disturbed area.

As a low-input, extensive form of agriculture, shifting cultivation is not highly productive. As an example, total wet-weight yield of manioc over a three-year cultivation period may average only 2.8 t/ha/yr. Dry-weight yield may be half that amount (compare with corn production in Table 22.1 and consider that corn grows for less than half the year). That 2.8 tons becomes only 0.28 t/ha/yr when the production of manioc is spread over the entire 30-year rotation of cropping and fallow. As a result, this system is suitable only for low population densities. As population density increases, the fallow cycle must be shortened, giving the soil and vegetation less time to recover (Figure 22.4). This can result in long-term declines in productivity and an actual reduction in total crop yield (again, the similarity to the case of whole-tree harvesting in the previous chapter should be clear).

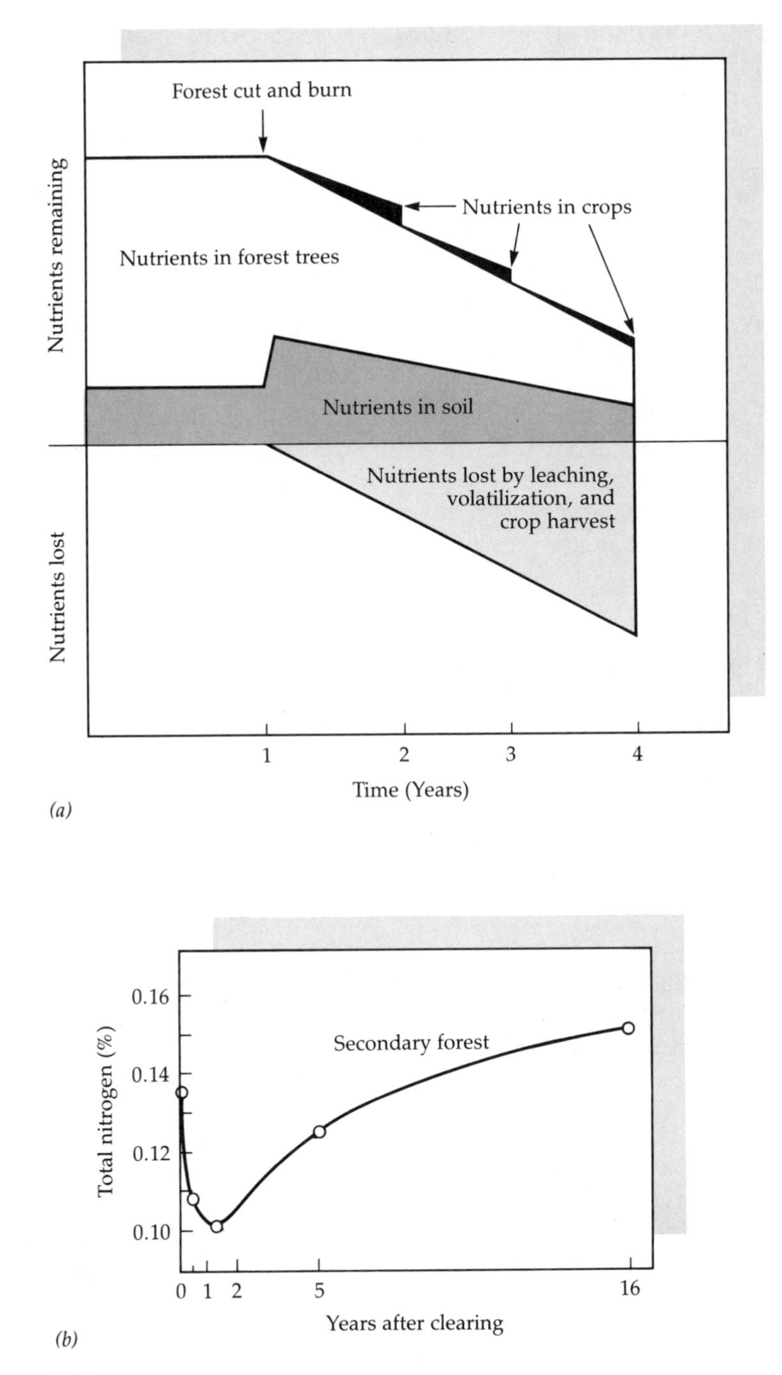

Figure 22.4

Effects of shifting cultivation on the nutrient content of humid tropic ecosystems. *(a)* Generalized pattern of the distribution of nutrients during the first four years of the shifting cultivation cycle (Uhl et al. 1983). *(b)* Changes in nitrogen storage over a full 16-year cycle. (Sanchez 1982)

What are the nutrient balances for a given site over a single cycle under shifting cultivation? The patterns are not dissimilar to those discussed for naturally disturbed ecosystems presented in Part III. Figure 22.4*a* depicts a generalized pattern for nutrient retention and loss over the first four years. Nutrient content in trees is reduced initially by burning and continues to decline as decomposition of the remaining material proceeds. Content in soils increases initially as some of the nutrients stored in vegetation are transferred by decomposition and ash deposition. It then declines due to leaching losses and removals in crops. Once the field is abandoned to fallow, plant regrowth will begin rebuilding the soil until preclearing levels are reached (Figure 22.4*b*; note the similarity between Figures 22.4*b* and 20.13).

A working assumption is that allowing the fallow to continue until soils reach preclearing conditions will ensure the integrity of the system. With increasing population density and increasing demand for food, pressures are generated to reduce the fallow period, jeopardizing the sustainability of shifting cultivation.

ALTERNATIVES TO THE EXTREMES

What we have described so far are the two extremes in terms of productivity and sustainability. Under the pressures of population growth and increasing concerns over environmental quality and the sustainability of agricultural systems, there has recently been quite a bit of research activity into alternative systems. Methods are being explored that will increase the sustainability of intensive temperate-zone agriculture while maintaining yield and increase the yield of tropical agriculture while preserving its sustainability.

In the temperate zone, several practices are already common for reducing environmental impacts and maintaining productivity. Some of these are designed to reduce erosion of topsoil. One example of such a system that has been practiced for quite some time is **crop rotation**. In this system, different crops are raised in different fields in different years. A particularly conservative rotation might be one year of corn followed by one year of wheat, followed by three years of hay. The hay crops could be of perennial grass, which would afford soil protection year-round. In a prairie region, this resembles a short-rotation form of shifting agriculture.

Crop rotation can be combined very easily with **contour tillage** (Figure 22.5), a method of plowing around rather than up and down slopes. This reduces the overland flow of water and hence surface soil erosion. When alternating strips of grass and crops are included, an efficient barrier to erosion is created.

Other methods of soil protection are also occasionally practiced. In **no-till cropping**, the residue from this year's crop is left on the soil surface rather than removed or plowed under. This creates a "litter layer" over the

Figure 22.5
Contour tillage with interspersed strips of cropland and perennial grass. This combination both reduces erosion and provides a fallow period for cropland. (H. Armstrong Roberts)

soil, which reduces water runoff and erosion. Use of mulches and wet manures has also been tried.

Any of these practices can reduce soil erosion significantly (Table 22.3). All of them can be added to conventional farming systems with relatively little cost. However, the complexities and economic pressures of modern farming often argue against even these relatively simple modifications.

Many of these practices can also reduce the demand for nutrients in crop production. A common rotation now in the corn belt is between corn and soybeans. Soybeans are legumes, and as such can fix significant amounts of nitrogen from the atmosphere, reducing the need for nitrogen fertilizer while producing a crop of higher protein (and nitrogen) content than corn. The soybean rotation can increase nitrogen availability for the next crop as well. However, this contribution is diminished if heavy nitrogen fertilization is used during the soybean part of the rotation. High nitrogen availability inhibits nitrogen fixation.

Rotations in which grasses are grown and plowed under (green manuring) increase the organic matter and nutrient content of the soil. So do no-till planting, mulching, and manuring. All of these can decrease reliance on fertilization.

The off-site effects of erosion and nutrient loss can be reduced by landscape-level management practices. For example, forest borders along streams running through an agricultural area can act as important sinks

Table 22.3 Losses of cropland to erosion under different agricultural practices*

TECHNOLOGY	TREATMENT	SOIL LOSS (t/ha/yr)	SLOPE (%)	COUNTRY
Rotation	Corn – wheat – hay – hay – hay – hay	3	12	U.S.
	Continuous corn	44	12	U.S.
Contour planting	Potatoes on contour	0.2		U.S.
	Potatoes, up-and-down hill	32		U.S.
Rotation plus contour planting	Cotton on contour and grass strips	8		U.S.
	Continuous cotton planted up-and-down hill	200		U.S.
Terraces	Peppers on terraces	1.4	35	Malaysia
	Peppers on slope	63	35	Malaysia
Manure	Corn with 36 t/ha of wet manure	11	9	U.S.
	Corn without manure	49	9	U.S.
Mulch	Corn planted on land with 6 t/ha of rice straw	0.1	5	Nigeria
	Continuous corn	148	5	Nigeria
Grass cover	Grass	0.08	10	Tanzania
	Plowed	13.6	10	Tanzania
No-till	Corn	0.14	15	Nigeria
	Conventional corn	24	15	Nigeria
Ridge planting—crop residues left in trenches on land surface	Corn	0.2	2	U.S.
	Conventional corn	10	2	U.S.

*From Pimentel et al. 1987.

for nutrient exported from cropland (Figure 22.6). While it is not yet clear what the capacity of these riparian-zone forests is for absorbing nutrients, it is clear that some are still effective after decades of receiving fertilizer runoff from adjacent croplands. Methods of nutrient loss from these forests, which are not always well measured, such as denitrification (loss of gaseous N_2 and N_2O to the atmosphere), may be important in keeping nutrients out of the stream. Of course, they then represent an input to the atmosphere and may have implications for global climate change (see Chapters 23 and 24).

Methods are also being developed to reduce dependence on chemical pesticides. Integrated pest management is a rapidly growing field that employs a combination of biological and chemical methods for reducing pest damage. One problem with chemical pesticides is that they generally affect both the pest and the predators and parasites that help control the pest. Reintroduction of these predators and parasites can be effective in reducing pest problems. General examples of this approach include the introduction of parasitic wasps and bacterial pathogen sprays, which attack specific groups of organisms and so selectively reduce pest species while leaving natural predator and disease organisms still viable in the system. In practice, chemical pesticides are also used in integrated pest management to ensure success, but the quantities used are generally lower than in chemical-only programs.

A final method of reducing input demands for modern agriculture is to alter the genetic makeup of the plant. This can be either through the standard practices of genetic selection or through genetic engineering. Breeding programs are currently under way to produce strains of the major crop species that are more tolerant of water and nutrient stress and are resistant to disease, while still maintaining high rates of production. This approach essentially involves fine-tuning previous selection efforts to adapt to the changing availability of fossil fuels.

These breeding programs require that the greatest possible degree of genetic variability be available to the breeders. There has been growing concern that rigid selection for high yield, along with the disappearance in the wild of the native species from which today's crop plants have been bred, will severely limit our ability to breed new varieties. In response to this, efforts are now under way to preserve both the genetic information held in the seeds of wild plants and to preserve the ecosystems in which these plants naturally grow.

Other, more experimental approaches to increasing the productivity and sustainability of agriculture in the temperate zone exist. From the total exclusion of all chemicals, as practiced by a limited number of organic farmers, to the introduction of totally new species of crop plants, to the total enclosure of the food production and waste treatment system in a recycling hydroponics system (Figure 22.7), these all aim to reduce inputs and pollution while maintaining yield.

The widespread application of the available and experimental approaches to sustainable agriculture in the temperate zone will be determined largely by the political and economic systems of which they are a significant part.

Modifications of Traditional Tropical Agriculture

The modification of traditional agricultural systems in the humid tropics to increase yield without sacrificing sustainability is currently a very active area of research and application. The various approaches can be described in terms of increasing levels of intensity of management. Many of them fall under the general term **"agroforestry,"** which involves the use of woody plants for the production of food, fiber, and fuel or for soil protection and enrichment.

The least intensive modification involves management of the secondary forest in the fallow period. This can take the form of thinning out and

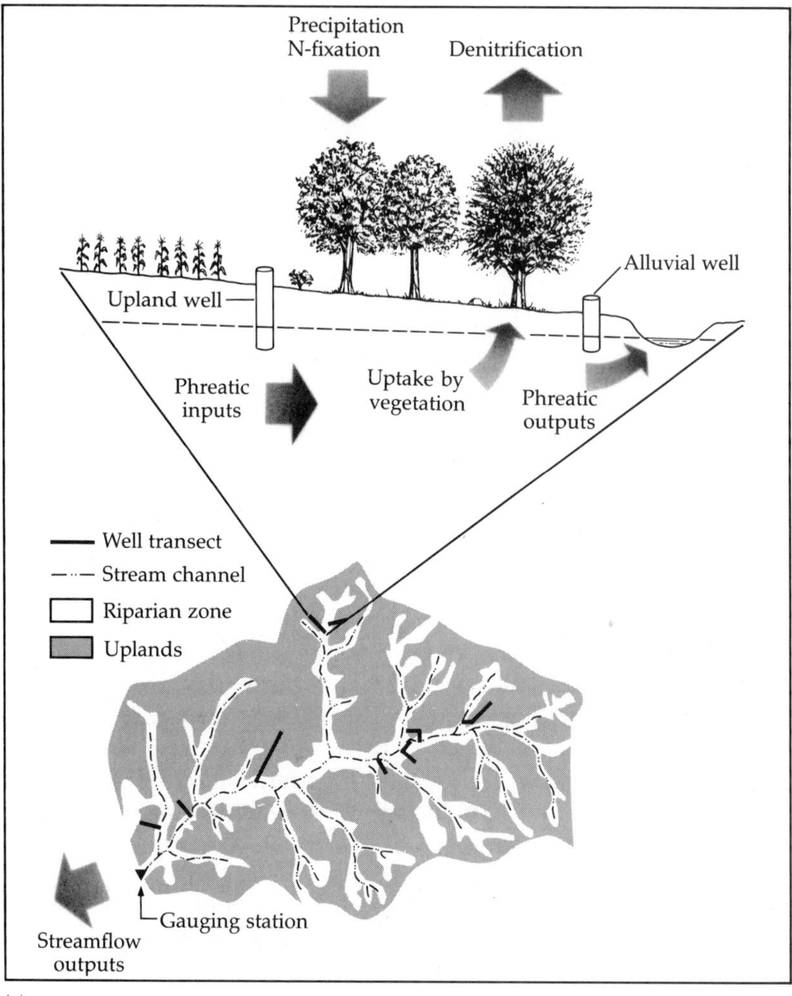

(a)

Figure 22.6
The role of riparian forests in reducing nitrogen losses to streams in agricultural regions. *(a)* Schematic of Watershed N, a subcatchment of the Little River watershed in Georgia (note the presence of forests adjacent to stream channels). *(b)* The nitrogen, phosphorus, and calcium balances over the riparian forest zone. The riparian zone is a net sink for all three nutrients. Including denitrification, the riparian forest removes 37.9 kg of nitrogen per hectare from the water entering the adjacent stream. (Lowrance et al. 1984, © 1984 by the American Institute of Biological Sciences)

selecting useful species from the thick growth that arises naturally or actually planting tree crops in the fallow at the end of the cropping period and tending them as the secondary forest develops. While the species actually involved are as varied as the sites and floras available, combina-

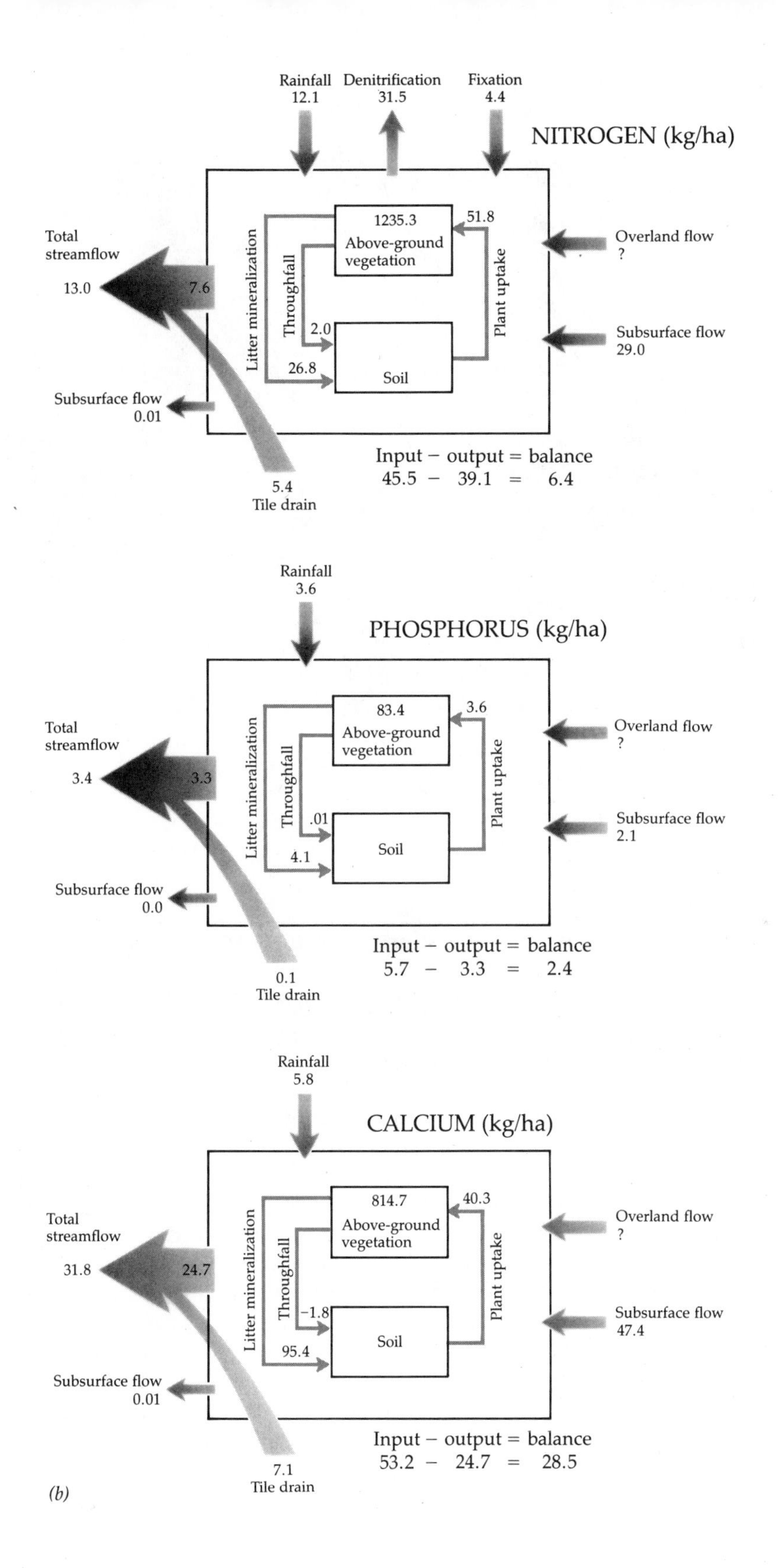

Rainfall
12.1
Denitrification
31.5
Fixation
4.4
NITROGEN (kg/ha)
1235.3
Above-ground vegetation
51.8
Plant uptake
Litter mineralization
Throughfall
2.0
26.8
Soil
Total streamflow
13.0
7.6
Overland flow
?
Subsurface flow
29.0
Subsurface flow
0.01
5.4
Tile drain
Input − output = balance
45.5 − 39.1 = 6.4
Rainfall
3.6
PHOSPHORUS (kg/ha)
83.4
Above-ground vegetation
3.6
Plant uptake
Litter mineralization
Throughfall
.01
4.1
Soil
Total streamflow
3.4
3.3
Overland flow
?
Subsurface flow
2.1
Subsurface flow
0.0
0.1
Tile drain
Input − output = balance
5.7 − 3.3 = 2.4
Rainfall
5.8
CALCIUM (kg/ha)
814.7
Above-ground vegetation
40.3
Plant uptake
Litter mineralization
Throughfall
−1.8
95.4
Soil
Total streamflow
31.8
24.7
Overland flow
?
Subsurface flow
47.4
Subsurface flow
0.01
7.1
Tile drain
Input − output = balance
53.2 − 24.7 = 28.5

(b)

Figure 22.7
The Ark, built by the New Alchemy Institute on Cape Cod in Massachusetts as a demonstration of the potential for sustainable, low-input, self-sufficient plant and fish culture powered by solar energy. (Courtesy of the New Alchemy Institute)

tions of food, fiber, and fuel species are used. The goal of these substitutions is often to replicate the structure of the natural secondary forest while substituting crop species for noncrop species. There is evidence that keeping the structure of the fallow vegetation while substituting economic species for ones that do not produce a useful product does not affect the soil-building processes of the fallow period.

Other approaches involve the establishment of permanent agricultural areas but by using systems that vary significantly from the temperate-zone model. One example is the use of permanent tree crops, such as rubber, tea, oil palm, coffee, and banana. Since tree crops occupy the site constantly, they offer some protection against erosion and leaching of nutrients. Root systems and understory plants provide a constant, if reduced, supply of organic matter to soils. The yield from such systems is low, but the economic value of the product is relatively high. With shade-loving plants such as coffee, selection of nitrogen-fixing species for the overstory of shade trees can improve yields and increase sustainability. Again, methods of agriculture cannot be separated from the social and economic systems of which they are a part. The conversion of land from subsistence shifting agriculture to export-oriented cash crops, such as the tree crops listed here, has profound social implications for the regions involved.

These monocultural or structurally simplified plantations run the risk of increased pest and disease outbreaks. In an experiment designed to test the effects of species diversity on herbivory, a recently cut area in Costa Rica was in part allowed to regenerate naturally and in part modified to contain more or fewer species than the natural regeneration. As species richness in these areas increased, losses to herbivory declined (Figure 22.8). The establishment of large tree crop plantations, as is occurring in several places in the tropics, may lead to increased pesticide requirements.

Permanent agriculture is not impossible in the tropics. The region contains a wide variety of soil and climatic conditions, some of which can be made permanently productive if the methods applied reflect the realities of the tropical environment. An ongoing demonstration of permanent agriculture using moderate chemical inputs in Peru has demonstrated continuous productivity and economic viability (Table 22.4). The systems developed include constant crop rotation and considerable human labor. They represent a compromise between the high-intensity, continuous cultivation practices of the temperate zone and the realities of the tropical environment.

Historical Methods for Sustainable Agriculture in the Tropics

There are ancient precedents for continuous crop production in the tropics. The most notable is paddy rice production. Paddy rice is grown in flooded fields using very labor-intensive methods of planting, cultivation,

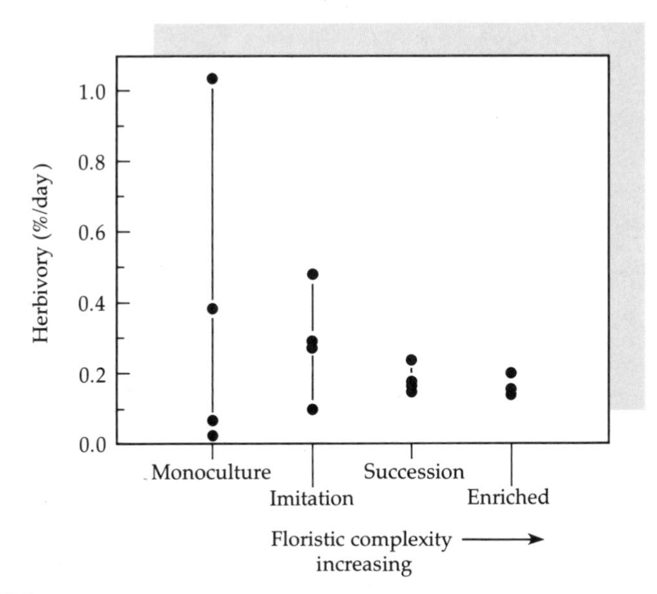

Figure 22.8

Comparison of herbivore consumption of leaves in early successional tropical forests in Costa Rica in which species diversity was experimentally manipulated. Plots with higher species diversity had lower herbivore consumption rates. (Brown and Ewel 1987)

Table 22.4 Crop yields for different types of continuous agriculture near Yurimaguas, Peru*

PRODUCTION SYSTEM	CROP ROTATION											
	Corn–Peanut–Corn			Total	Peanut–Rice–Soybean			Total	Soybean–Rice–Soybean			Total
Traditional	2.44	1.10	1.77	5.31	0.97	1.91	1.34	3.53	1.43	1.91	1.15	4.49
Improved, no lime or fertilizer	3.81	1.36	2.73	7.90	1.22	3.56	1.98	6.76	2.09	2.25	1.89	6.23
Improved, with lime and fertilizer	5.12	1.62	4.66	11.40	1.49	4.53	2.75	8.77	2.73	2.53	2.22	7.48

*From Nicholaides et al. 1985.

and harvest (Figure 22.9*a*). The field can be on level ground in floodplain areas or even on steeply sloping hillsides that have been terraced to produce flat areas (Figure 22.9*b*). The flat and flooded paddies provide excellent protection against erosion and can actually trap sediments from the water used in flooding and result in a build-up of soils. The presence of blue-green algae in the stagnant water is also thought to provide nitrogen inputs to the system through nitrogen fixation from the atmosphere. The combination of sediment trapping, reduced erosion, and ni-

Figure 22.9
Paddy rice culture in Asia. *(a)* Preparing plots with water buffalo and human labor. *(b)* Continuous rice production in terraced hillside slopes in the Philippines. (*a*, May/H. Armstrong Roberts; *b*, H. Armstrong Roberts)

trogen fixation reverses the usual pattern of nutrient loss from cultivated land. Some paddy systems in Asia have been in operation for thousands of years.

A new-world version of the rice paddy is the chinampas system developed in pre-Columbian Mexico and still in use today in low-lying wetland areas (Figure 22.10). Chinampas consist of raised plots surrounded by moats or canals filled with water. The moat traps runoff and erosion and reduces these losses. Plant growth, which is luxuriant in the moat, is harvested regularly and mounded up on to the plot, along with trapped sediments. This system provides a constant supply of organic matter and nutrients to the soil within the plot, including those that would have been lost in runoff. Productivity in these plots is very high.

CHARACTERISTICS OF SUSTAINABLE AGROECOSYSTEMS

It should be apparent from this brief review that sustainable, high-productivity agriculture, no matter where it is practiced, involves a balance between gains and losses in soils and nutrients. Temperate-zone systems replace nutrients lost in harvest and erosion through chemical fertilization and mechanical cultivation of soils. We are living to a certain extent on the historical forces that created rich prairie soils and on the largesse allowed by our current availability of fossil fuels. It has been said that all the most productive soils in the world are on transported material (e.g., the loess soils of the American Midwest and the volcanic soils of both tropical and temperate regions). In the temperate zone, erosion is rapidly retransporting those soils off the farm.

Tropical systems that rely on natural processes for rebuilding soils require long fallow periods and thus produce low yields per unit area. (There are plenty of precedents in the temperate zone for the exhaustion of soils by pre-industrial agriculture, the most dramatic of which might be the cultivation of tobacco in colonial America.) Intensive tropical systems that have proven sustainable for centuries either minimize losses, increase contributions from outside the system, or both.

This discussion has been limited to crop production systems and has not dealt with animal systems. Many of these, such as moderate-intensity pasture systems and extensive grazing systems, are highly sustainable. As long as grazing does not reduce plant cover to the point where erosion becomes important or where palatable species are reduced and replaced by unpalatable ones, they can be permanent systems with low inputs. However, the quantity of food produced is relatively low. The inefficient conversion of plant mass to animal mass (a maximum of 10% and often less) is something of a luxury, except in those areas where grazing animals can convert plant mass that is undigestible by humans into meat. This mainly includes extensive grazing systems in semiarid grasslands and certainly does not include the highly centralized, grain-based feedlots found in the developed countries.

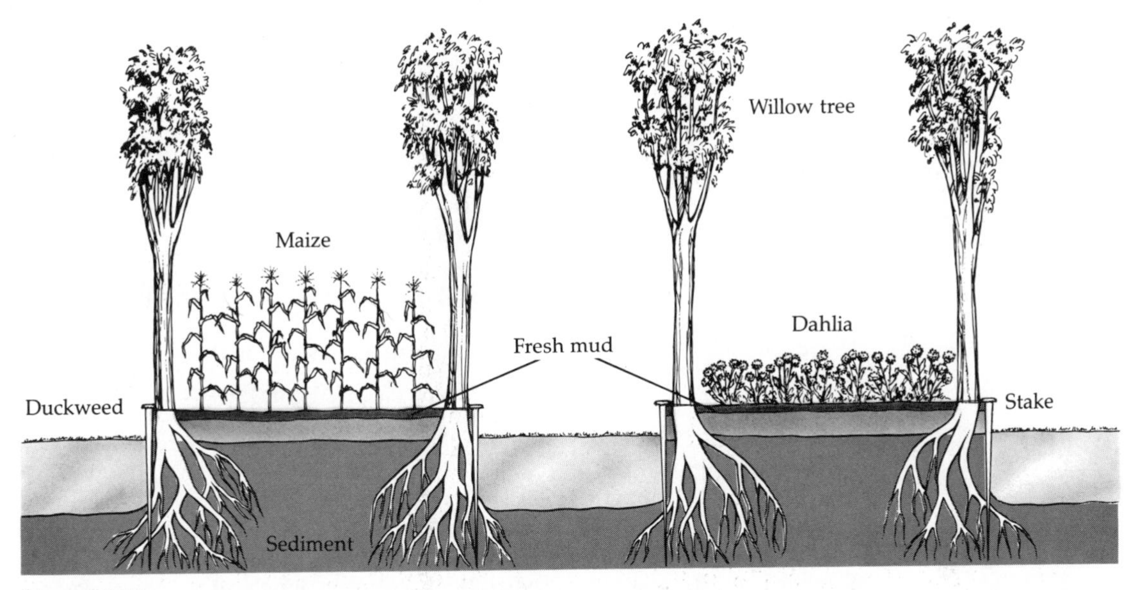

Figure 22.10
The Chinampas system of agriculture in Mexico. (Cox and Atkins 1979)

Relation to Conservation of Native Ecosystems

The management of agricultural ecosystems cannot be separated from discussions of the conservation of wild ecosystems. A growing world population means increasing demands for food. This demand can be met by increasing the intensity of management on existing agricultural lands or by increasing the amount of land in agriculture. Both these approaches can have negative environmental implications. Expanding the base of agricultural land will mean a drain on the small stock of area remaining in native ecosystems. This can even affect areas nominally protected in parks and reserves, as indicated by the increasing poaching pressure in the Serengeti Park (Chapter 19). More intensive agriculture can bring increased pollution due to increased use of agricultural chemicals and fossil fuels, unless new systems less dependent on these inputs are developed and applied.

Three avenues seem open: controlling world population growth, increasing the intensity of agricultural production, particularly in the tropics, and increasing the amount of agricultural land. A fourth option, that of failing to provide enough food to meet the human demand, is already a reality in some areas of the world but is too repugnant to consider. Conservation of natural ecosystems means choosing one of the first two options.

REFERENCES CITED

Boyer, J. S. 1982. Plant productivity and environment. *Science* 218:443–448.

Brown, B. J., and J. J. Ewel. 1987. Herbivory in simple and complex tropical successional ecosystems. *Ecology* 68:108–116.

Cox, G. W., and M. D. Atkins. 1979. *Agricultural Ecology*. W. H. Freeman and Co., San Francisco.

Frissel, M. J. 1978. *Cycling of Mineral Nutrients in Agricultural Ecosystems*. Elsevier Scientific Publishing Co., Amsterdam.

Gilliam, F. S. 1987. The chemistry of wet deposition for a tallgrass prairie ecosystem: Inputs and interactions with plant canopies. *Biogeochemistry* 4:203–218.

Lowrance, R., et al. 1984. Riparian forests as nutrient filters in agricultural watersheds. *BioScience* 34:374–377.

Nicholaides, J. J., et al. 1985. Agricultural alternatives for the Amazon basin. *BioScience* 35:279–285.

Pimentel, D., et al. 1987. World agriculture and erosion. *BioScience* 37:277–283.

Sanchez, P. A. 1982. Nitrogen in shifting cultivation systems of Latin America. *Plant and Soil* 67:91–103.

Sims, P. L., and J. S. Singh. 1978. The structure and function of ten western North American grasslands. III. Net primary production, turnover and efficiency of energy capture and water use. *Journal of Ecology* 66:573–597.

Timmons, D. R., and A. S. Dylla. 1983. Nitrogen inputs and outputs for an irrigated corn ecosystem in the northwest corn belt. In Lowrance, R. R., R. L. Todd, L. E. Asmussen, and R. A. Leonard (eds.), *Nutrient Cycling in Agricultural Ecosystems*. University of Georgia College of Agriculture Special Publication No. 23, Athens, Georgia.

Uhl, C., C. F. Jordan, and F. Montagnini. 1983. Traditional and innovative approaches to agriculture on Amazon Basin tierra firme sites. In Lowrance, R., R. Todd, L. Asmussen, and R. Leonard, (eds.), *Nutrient Cycling in Agricultural Ecosystems*. University of Georgia College of Agriculture Special Publication No. 23, Athens, Georgia.

Woods, L. E., et al. 1983. Nutrient cycling in a southeastern United States agricultural watershed. In Lowrance, R. R., R. L. Todd, L. E. Asmussen, and R. A. Leonard (eds)., *Nutrient Cycling in Agricultural Ecosystems*. University of Georgia College of Agriculture Special Publication No. 23, Athens, Georgia.

ADDITIONAL REFERENCES

Bayliss-Smith, T. P. 1982. *The Ecology of Agricultural Systems*. Cambridge University Press.

Correll, D. L. 1983. N and P in soils and runoff of three coastal plain land uses. In Lowrance, R., R. Todd, L. Asmussen, and R. Leonard (eds.), *Nutrient Cycling in Agricultural Ecosystems*. University of Georgia College of Agriculture Special Publication No. 23, Athens, Georgia.

Ewel, J., C. Berish, and B. Brown. 1981. Slash and burn impacts on a Costa Rican wet forest site. *Ecology* 62:816–829.

Jordan, C. F. 1985. *Nutrient Cycling in Tropical Forest Ecosystems*. John Wiley and Sons, New York.

Knapp, A. K., and T. R. Seastedt. 1986. Detritus accumulation limits productivity of tallgrass prairie. *BioScience* 36:662–669.

Sanchez, P. A. and J. R. Benites. 1987. Low-input cropping for acid soils of the humid tropics. *Science* 238:1521–1527.

Todd R., R. Lowrance, O. Hendrickson, L. Asmussen, R. Leonard, J. Fail, and B. Herrick. 1983. Riparian vegetation as filters of nutrients exported from a coastal plain agricultural watershed. In Lowrance, R., R. Todd, L. Asmussen, and R. Leonard (eds.), *Nutrient Cycling in Agricultural Ecosystems*. University of Georgia College of Agriculture Special Publication No. 23, Athens, Georgia.

Chapter 23 Effects of Air Pollution on Terrestrial Ecosystems

H. Armstrong Roberts

INTRODUCTION

Agricultural practices are one important way in which humans modify terrestrial ecosystems, changing both the productivity of those systems and their nutrient input and output balances. Most of these effects are local in nature and very visible. In contrast, human industrial activity may produce effects that are more subtle and difficult to identify and that may be spread over large regions. One major vector for this effect is through the creation of air pollution, which comes in many forms and may alter terrestrial ecosystems in complex ways.

In combination, human industrial and agricultural activities have altered the movement of major nutrients and pollutants (their biogeochemical cycles) on a global scale. Industrial processes move mineral ores from mines to factories to consumers to landfills (or hopefully to recycling centers and back to consumers). Agricultural practices convert atmospheric nitrogen to fertilizer, which is spread on fields, harvested as crops, and transported to cities and eventually to sewage treatment plants, lakes, rivers, and the ocean. Often, excess fertilizer nitrogen takes a shortcut and moves directly to surface water or groundwater.

All of these processes require energy in the form of heat, electricity, or motion, and much of that energy comes from the burning of fossil fuels (partially decayed organic matter converted, over geologic time, to coal, oil, and gas). Thus the burning of these fuels is a major component of human intervention in natural elemental cycles. One result of the burning is the injection of tremendous amounts of carbon into the atmosphere as CO_2, the potential effects of which will be discussed in Chapter 24.

Sulfur and nitrogen are also present in fossil fuels because of their important role in plant chemistry. By extracting the energy stored in fossil fuels, we release not only carbon but also nitrogen and sulfur into the atmosphere. In addition, substantial amounts of atmospheric nitrogen (N_2, which constitutes 79% of the atmosphere) are converted to reactive oxides of nitrogen by combustion, especially in high-compression automobile engines. Trace amounts of heavy metals (such as zinc, lead, cadmium, and nickel) are also released by combustion, but even larger amounts of these toxic metals are released by the smelting of ores into pure metals and alloys for industrial use.

Once in the atmosphere, these pollutants can react chemically before being deposited onto terrestrial ecosystems. Once in those systems, they can also interact with plants, animals, and soils. In some areas in central Europe and the northeastern United States that have been exposed to high levels of air pollution, serious declines in forest growth, and even the death of whole stands, have prompted vigorous arguments over how and to what extent air pollution should be controlled. Ideally, this decision should be made on the basis of some understanding of the interactions between air pollutants and ecosystems.

The purpose of this chapter is to discuss the origins and distribution patterns of some important forms of air pollution and their potential effects on terrestrial ecosystems. We will concentrate on the components of air pollution that are currently thought to be most likely to cause damage. This discussion will draw on the information presented throughout this book on processes that occur in ecosystems, as almost all of these are affected in some way by air pollution.

SOURCES OF MAJOR FORMS OF AIR POLLUTION

The major forms of air pollution thought to affect terrestrial ecosystems are summarized in Table 23.1. They can be grouped into three categories. Sulfur dioxide, oxides of nitrogen, and ozone are all oxidant gases. Sulfuric and nitric acids, when dissolved in precipitation, are the principal causes of acidity in "acid rain." Lead, cadmium, zinc, nickel, and others are grouped as heavy metals.

The pollutants are also categorized as primary or secondary products of industrial activity. Primary products, such as SO_2 and NO_x, are the immediate result of combustion. Secondary products, such as ozone and nitric and sulfuric acids, are formed in the atmosphere from the primary products or as a result of their effects on atmospheric chemistry. As such, their concentrations in the atmosphere might actually increase at some distance from the source of the primary pollutants because of the time required for the reactions to occur (see Figure 23.6).

For example, ozone is a highly reactive gas that is formed by a complex

Table 23.1 Major components of air pollution thought to affect terrestrial ecosystems

TYPE	1°/2°*	COMPOUNDS	ECOSYSTEM EFFECTS
Oxidant gases	1° 1° 2°	Sulfur dioxide (SO_2) Nitrogen oxides (NO_x) Ozone (O_3)	Reduction of net photosynthesis, damage to leaf cell membranes, formation of toxins (SO_2)
Dissolved acids	2° 2°	Sulfuric acid (H_2SO_4) Nitric acid (HNO_3)	Soil acidification, metal mobilization, cation nutrient deficiencies, nitrogen saturation
Heavy metals	1°	Lead (Pb), nickel (Ni), copper (Cu), zinc (Zn), cadmium (Cd), others	General interference with biochemical reactions

*Specifies whether the compound is primary or secondary production of industrial activity.

set of reactions involving NO_x gases, partially oxidized hydrocarbons from gasoline, and oxygen (Figure 23.1) in the presence of sunlight. It has been said that the basin surrounding Los Angeles is a perfect environment for the production of ozone because of the large number of automobiles (which produce both NO_x and partially oxidized hydrocarbons), the generally sunny weather, and a ring of surrounding mountains, which holds the air over the city, allowing time for the formation of ozone.

Nitric and sulfuric acids are formed by reactions between water in the atmosphere and NO_x and SO_2, which can be generalized as:

$$SO_2 + H_2O \rightarrow H_2SO_4 \leftrightarrow H^+ + HSO_4^- \leftrightarrow 2H^+ + SO_4^{2-}$$
$$NO_x + H_2O \rightarrow HNO_3 \leftrightarrow H^+ + NO_3^-$$

Nitric and sulfuric acids are "strong acids" in that they tend to dissociate completely (the above reactions go strongly to the right), releasing hydrogen ions. Thus, clouds and fog, and eventually rain and snow, are increased in hydrogen ion (H^+) concentration; they are acidified. These two forms of air pollution are the ones that contribute to "acid rain."

Acidity is expressed as the concentration of hydrogen ions in a solution and is summarized by the term pH:

$$pH = -\log(\text{concentration of } H^+)$$

Two things are important about this equation. First, it is a logarithmic scale, so every unit change in pH means a tenfold change in H^+ concentration. Second, the pH value is inversely related to H^+ concentration, so a

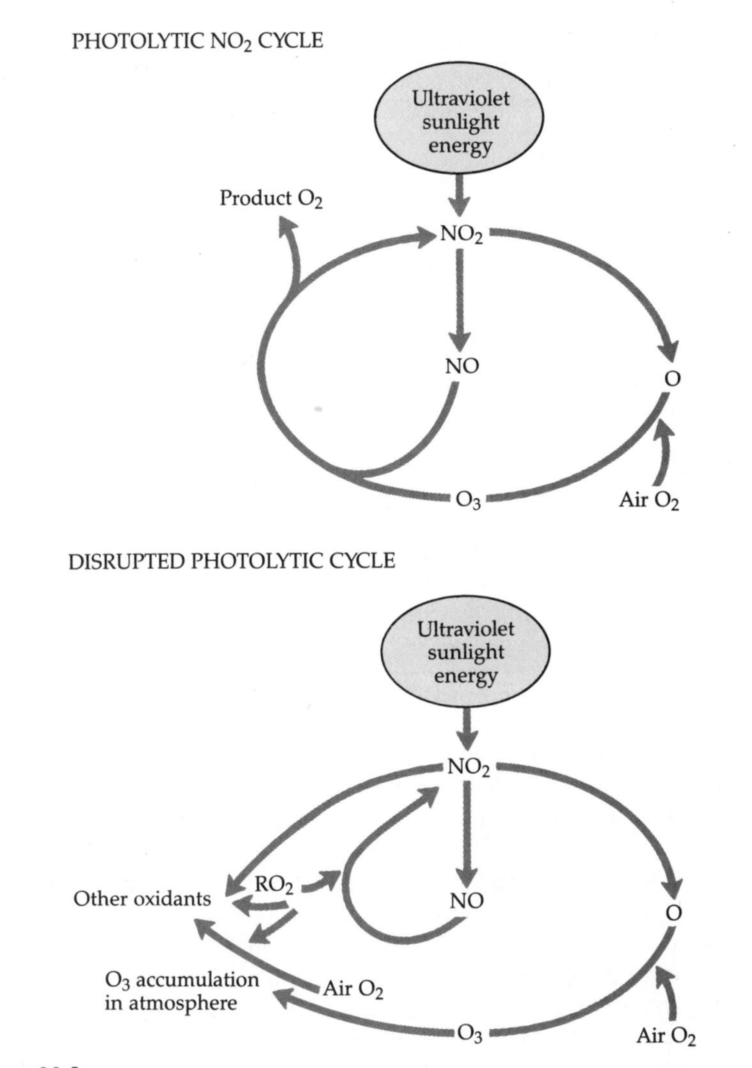

Figure 23.1
The production and breakdown of ozone in the lower atmosphere *(a)* under normal conditions and *(b)* in the presence of partially oxidized hydrocarbons (RO_2).

lower pH value denotes increased acidity. Distilled water has a pH of 7 (it is neutral). Anything below pH 7 is considered acidic; anything above pH 7 is considered alkaline. Figure 23.2 shows representative pH values for different kinds of solutions.

Unpolluted rainfall is not neutral. The presence of CO_2 in the atmosphere leads to the formation of carbonic acid:

$$CO_2 + H_2O \rightarrow H_2CO_3 \leftrightarrow H^+ + HCO_3^- \leftrightarrow 2H^+ + CO_3^2$$

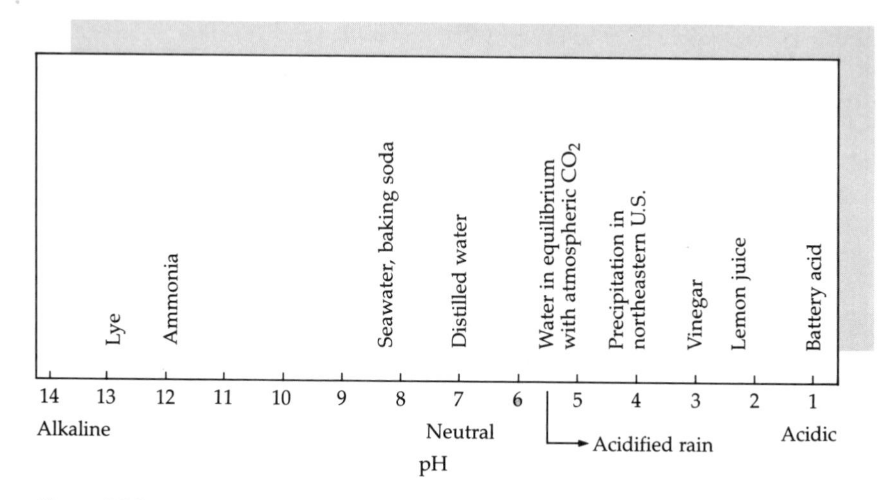

Figure 23.2
The pH scale and pH values of some common substances compared with current values for rainfall in the northeastern United States.

At current levels of atmospheric CO_2, carbonic acid should produce rainfall with a pH of about 5.65. Rainfall pH values below 5.65 are considered to be acidified. However, there are other naturally occurring chemicals that can also lower the pH of rainfall. An extensive study of the acidity of rainfall in places remote from industrial activity (Figure 23.3) suggests that background pH levels may actually be closer to 5.0. However, the same study emphasized the magnitude of change in precipitation chemistry brought about by human activity in the northeastern United States, where annual mean rainfall pH may be as low as 4.0 and individual storm events may reach pH 2.5 (more than 1000 times the hydrogen ion content of "normal" pH 5.65 rain).

Although the term "acid rain" has been used most frequently in popular discussions of the effects of industrial pollutants on forests and other ecosystems, acid rain is actually only one form of air pollution. The scientific terminology has changed over the years from "acid rain" to "acid precipitation" (which includes snow, fog, etc.) to "acid deposition" (with the realization that acidic or acidifying compounds can be deposited directly on surfaces in dry form as dust or aerosols) and finally to "atmospheric deposition" (which includes nonacidifying pollutants such as heavy metals, lead, zinc, etc.). This term still fails to include the gaseous pollutants such as SO_2 and NO_x and ozone (O_3). In dealing with the effects of air pollution on terrestrial ecosystems, we have to address the potential effects of each of these components as well as their potential interactions.

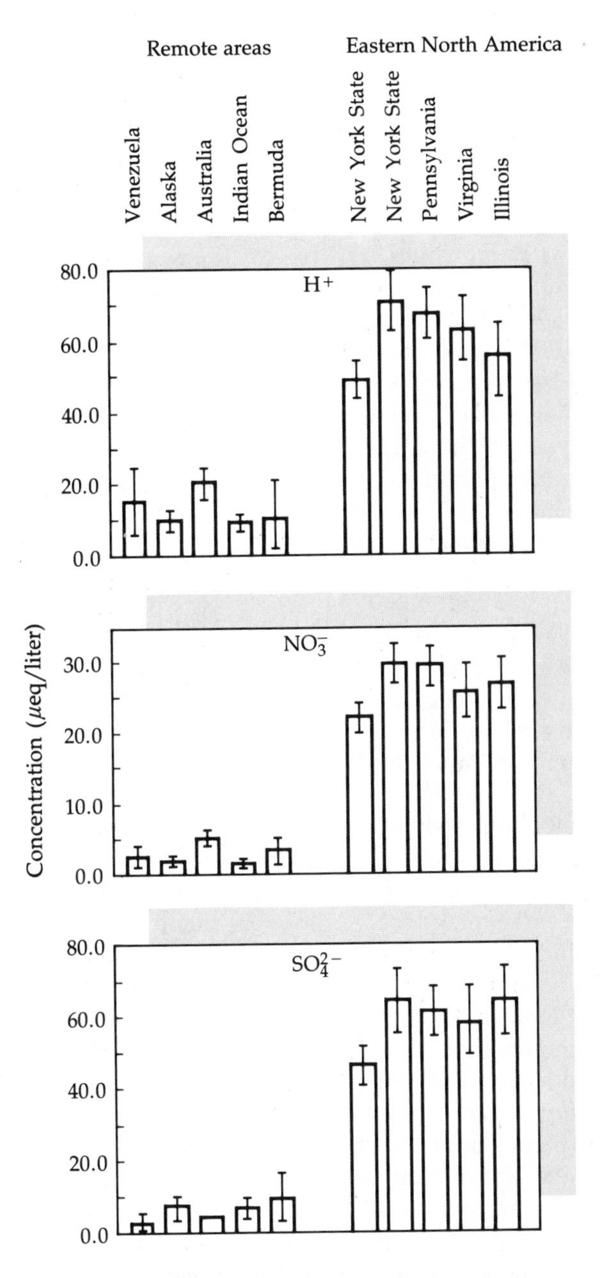

Figure 23.3
Differences in the concentrations of hydrogen ion, nitrate, and excess sulfate (in addition to that provided by sea salt from evaporated sea spray) between five remote locations and five locations in eastern North America. (Galloway et al. 1984)

Sources and Distribution of Elements and Air Pollutants in Precipitation

As early as the mid 1800s, atmospheric chemists were beginning to measure the differences in the chemical content of air samples taken near industrial areas and in remote areas. In 1852 an English scientist named Robert Angus Smith reported that the concentration of sulfuric acid in the air increased with proximity to cities. In 1872 he published a book on the chemistry of air and rain, which outlined many of the reactions involved in the formation of acid rain.

However, widespread and systematic measurement of the chemistry of precipitation has occurred only within the last 20 years. Largely in response to initial, isolated measurements indicating significant changes in the chemistry of precipitation over time, nationwide and continent-wide networks of precipitation collection stations, using standardized methods of analysis, have been established. By summarizing the tremendous amount of information gathered by these networks, large-scale patterns in precipitation chemistry, and their causes, can be determined. Systematic monitoring of the content of important gaseous pollutants like sulfur dioxide (SO_2) and ozone (O_3) has only recently begun outside urban areas.

Figure 23.4 is one attempt at summarizing the patterns in element input or deposition that occur in precipitation (rain and snow) across North America. The areas of highest incidence of different elements are related to the occurrence of important processes that inject those elements into the atmosphere.

For example, both sodium and chloride, the elements present in common salt, are also present in large amounts in the ocean. Aerosols (small airborne particles) of each of these are injected into the atmosphere over the oceans in large quantities by the evaporation of ocean spray and are carried by winds and storms over land, where they are either deposited as dry particles or washed out of the air in precipitation. As a result, coastal areas receive large amounts of these two elements (Figure 23.4*a,b*). The high concentrations right along the coast and the rapid change in deposition away from the coast suggests that these elements do not travel great distances before leaving the atmosphere. Similarly, the deposition of calcium can be higher in areas where limestone is a common geologic substrate or where agricultural practices such as liming and plowing can release calcium-carrying dust for deposition in other areas.

Human activity plays a large role in the distribution of nitrogen deposition as both ammonium and nitrate. The application of urea- or ammonium-based N fertilizers to agricultural fields can result in the volatilization (loss to the atmosphere in gaseous form) of ammonia (NH_3). This is converted to ammonium (NH_4^+) in the atmosphere and deposited in rainfall somewhere downwind (to the east in Figure 23.4*c*) of the major agricultural regions. Feedlots can also be an important regional source of ammonia to the atmosphere.

As discussed above, nitrate and sulfate are formed in the atmosphere from the primary products of combustion, NO_x and SO_2. As a result, their deposition rates are much higher downwind of the heavily industrialized areas of the upper Midwest and Northeast (Figure 23.4*d,e*). There is a general correlation between the deposition rates for nitrate and sulfate, the two strong acids that contribute to the acidification of rainfall, and the hydrogen ion content of precipitation (Figure 23.4*f*).

A critical feature of the deposition distributions for nitrate and sulfate is that they can be transported long distances, hundreds to thousands of miles, before coming back to earth. Thus, industrial activity in the Great Lakes region of the United States affects rainwater quality in eastern Canada. Similarly, pollution produced in central Europe falls on the relatively rural areas of Scandinavia. The political complexity of dealing with a problem of this type should be apparent.

Most heavy metals enter the atmosphere as aerosols and do not travel the large distances indicated for nitrate and sulfate. Areas of high deposition are generally concentrated around large smelters. However, the use of leaded gasolines has proven a very efficient method of distributing large amounts of lead generally throughout urbanized regions. Growing concern over the effects of lead on human health has prompted a long-term shift to lead-free gasolines, resulting in lower concentrations in the urban atmosphere. Reduced concentration of lead in recent growth rings produced by urban trees (Figure 23.5) offers proof that reduced emissions are translated into reduced bioavailability and uptake by plants.

In contrast, partial efforts to reduce local levels of primary pollutants can lead to increased levels of secondary pollutants in other areas. For example, concentrations of oxidant gases in the urban atmosphere have long been identified as serious threats to human health. One early method of "air pollution control" in cities and around big industrial sources was to build taller smokestacks (Figure 23.6). The taller stacks inject the exhausts into higher levels of the atmosphere, where they are more likely to be carried away instead of migrating to ground level in the immediate area. As opposed to the removal of lead from gasoline, which reduced total pollutant emissions, this approach only altered the distribution of pollutants—a classic case of "the solution to pollution is dilution." This tactic has increased the concentrations of both gaseous pollutants and sulfate and nitrate in the atmosphere over remote areas.

EFFECTS OF AIR POLLUTANTS ON TERRESTRIAL ECOSYSTEMS

By far the most dramatic effects of pollutant deposition can be seen in the immediate vicinity of large smelting plants (Figure 23.7*a*). These plants have released vast quantities of metals and acid-forming gases, which combine to virtually eliminate plant growth in the immediate area. The toxic effects decline with distance from the source (Figure 23.7*b*).

For areas not adjacent to large point sources of pollutants, the effects of pollution deposition on the function of terrestrial ecosystems is more complex and subtle. Still, it seems clear that the potential for profound

Figure 23.4
Summary patterns of concentrations of different elements in precipitation across North America, generated from a system of more than 170 collection stations. *(a)* Sodium. *(b)* Chloride. *(c)* Ammonium. *(d)* Nitrate. *(e)* Sulfate. *(f)* Hydrogen. (Data from the U.S. National Atmospheric Deposition Program)

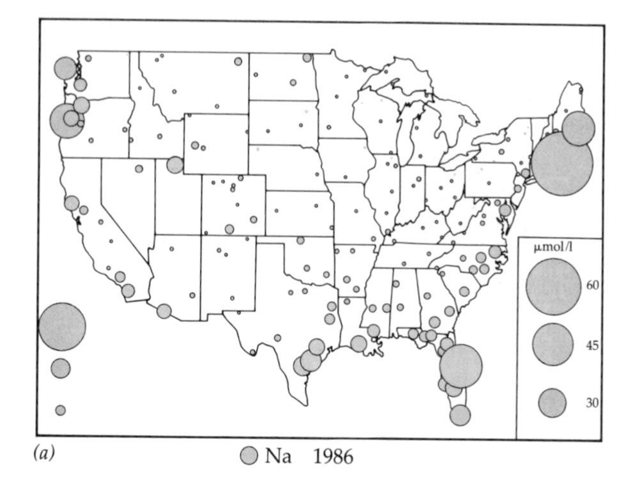

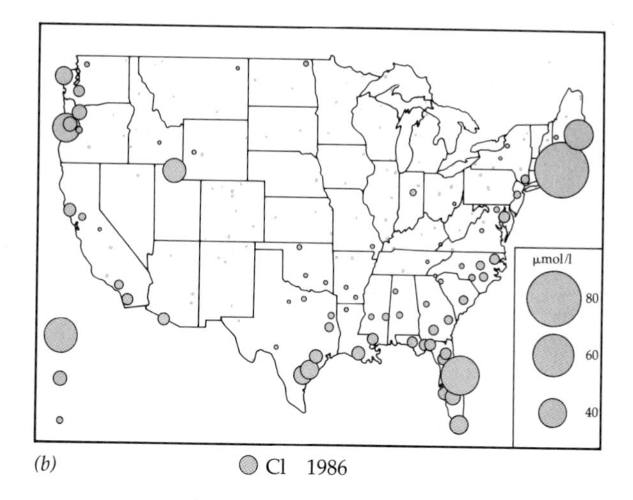

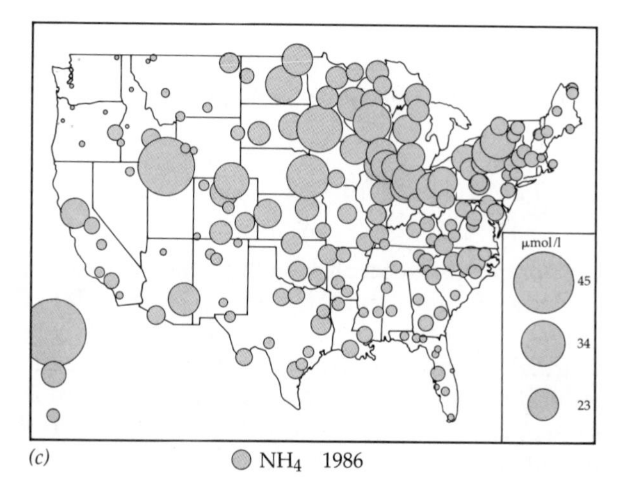

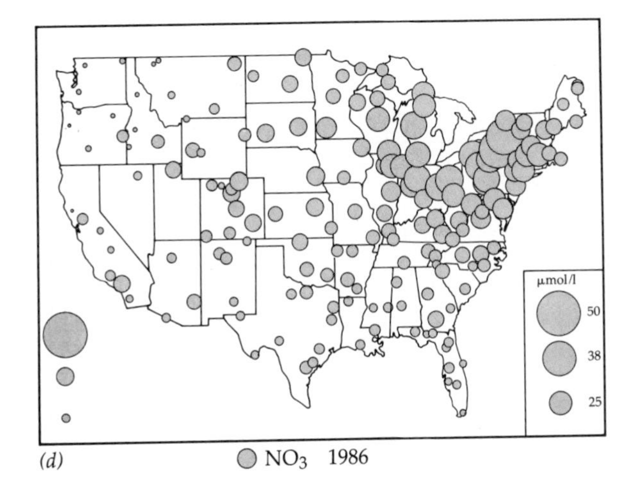

(d) NO_3 1986

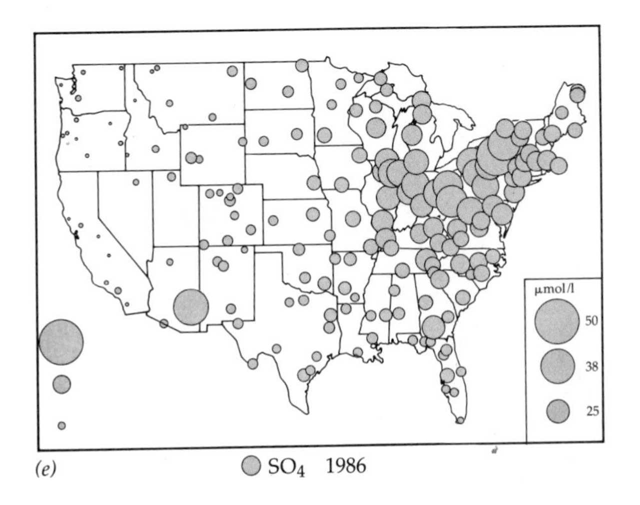

(e) SO_4 1986

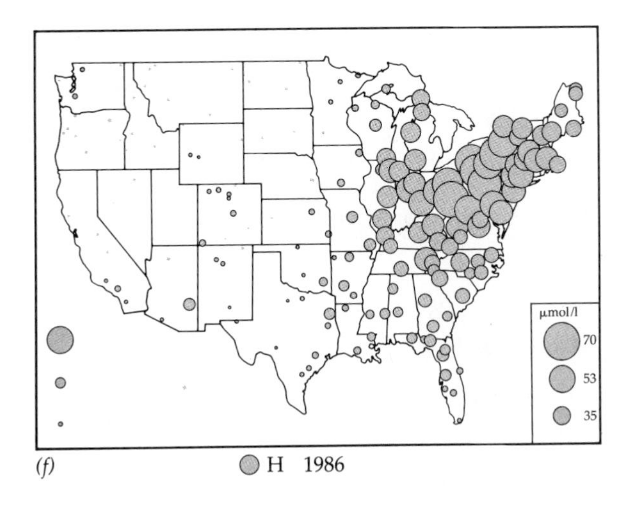

(f) H 1986

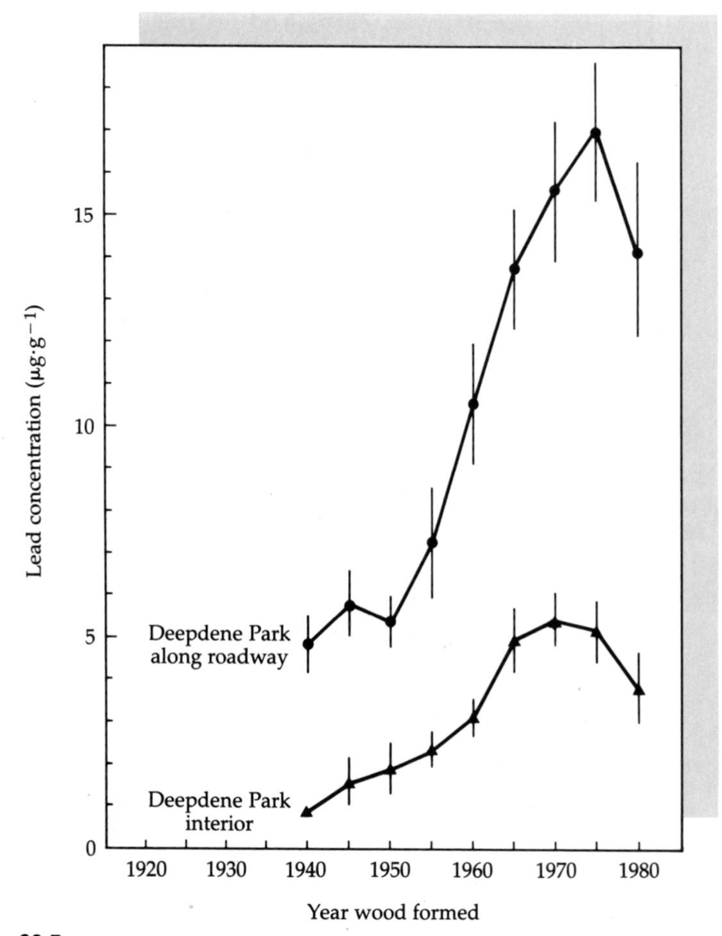

Figure 23.5
Changes in the lead content of tree rings in an urban park in Atlanta, Georgia. The decrease coincides with the introduction of lead-free gasolines. (Ragsdale and Berish 1988)

disruptions is present, as the pollutants listed in Table 23.1 affect nearly every major process that determines resource availability, nutrient cycling, or plant growth. The effects of individual pollutants can be significant. Interactions between pollutants can either offset or enhance individual effects.

Effects of Individual Air Pollutants

Gases Of the gaseous forms of air pollution, ozone appears to have the greatest potential for damaging ecosystems outside urban areas. Its effects are very straightforward. Ozone diffuses into foliage through the stomata.

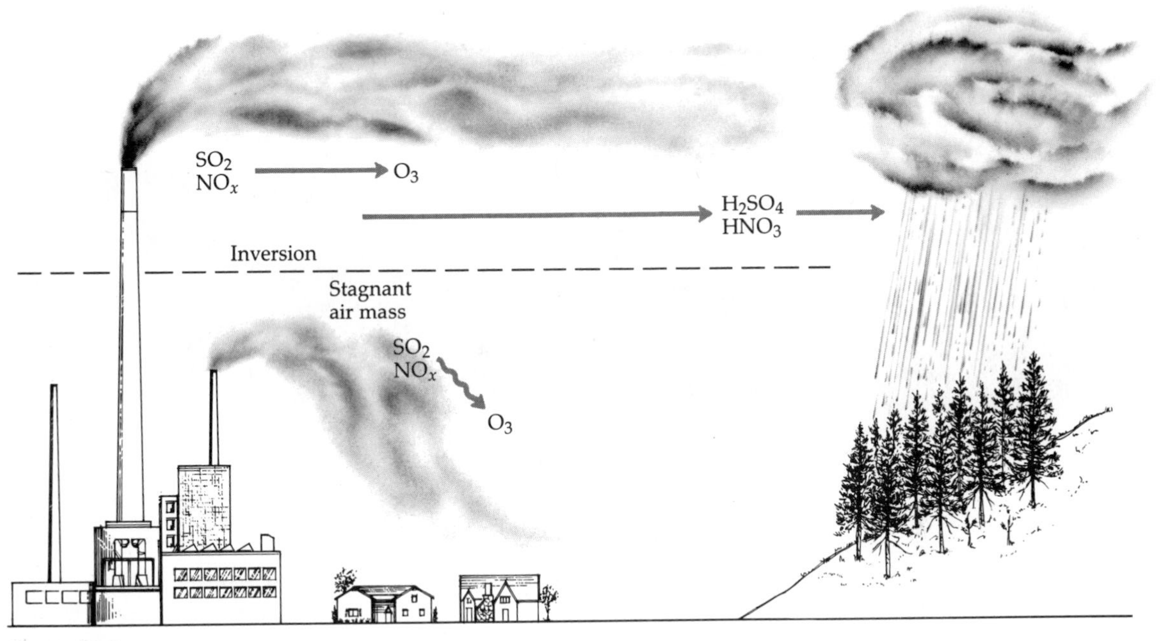

Figure 23.6
The effect of taller smokestacks on the distribution of emitted pollutants. A taller stack increases the chances that emissions will escape lower, possibly stagnant layers and be carried away from the source area.

Inside the leaf, the reactive ozone attacks cell membranes, decreasing the efficiency of membrane function. This increases respiration rate as membrane repair becomes an increased metabolic cost. This, in turn, reduces the rate of net photosynthesis, reducing carbon accumulation in plants. SO_2 may also have oxidizing effects on plants. In addition, both SO_2 and NO_x, once in the leaf, can form toxic compounds, which further inhibit leaf function.

Interestingly, these gaseous pollutants appear to have linear and cumulative effects on plant growth. That is, the response of plants in terms of reduced growth rates, or yield in agricultural crops, can be expressed as a linear function of the cumulative dose of the pollutants (Figure 23.8*a*). This makes determination of effects relatively easy. Rare events of unusually high concentrations do not seem to be crucial.

The susceptibility of different species to gaseous pollutants is in direct relation to the rate at which they exchange gases with the atmosphere. Fast-growing plants have high exchange rates for gases (high conductance, for rapid CO_2 fixation), including pollutants, and a greater reduction in net photosynthesis (Figure 23.8*b*). Ozone and other gases may be the only air pollutants that can seriously affect crop plants, as the damage is directly to the plant rather than through effects on soils. The chemistry

(a)

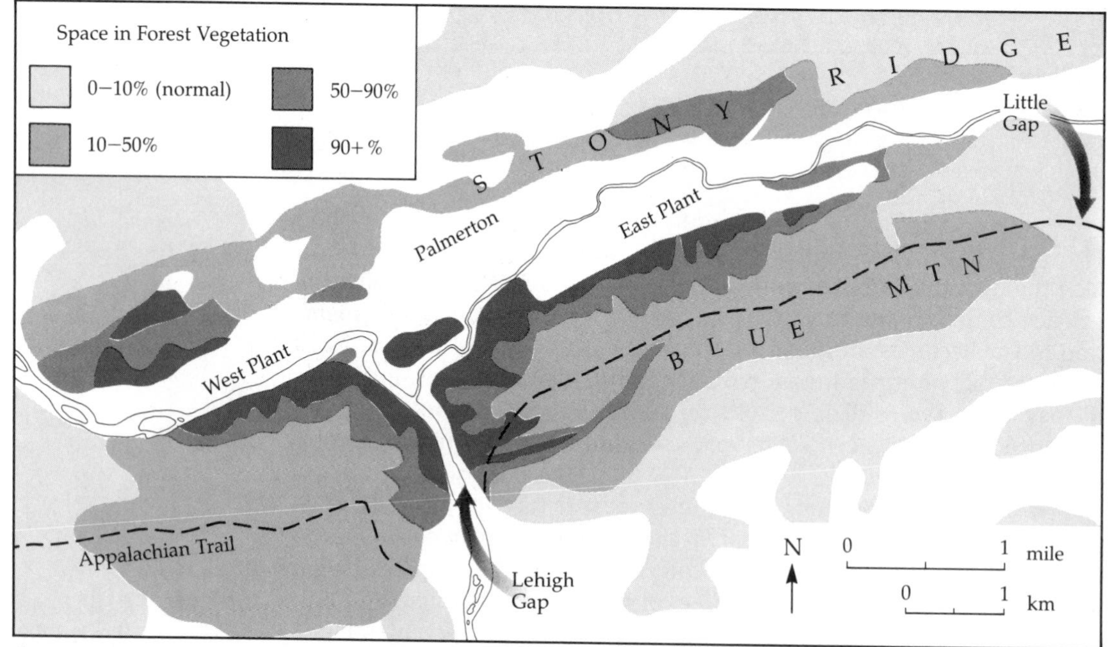

(b)

Figure 23.7

Ecosystem destruction in the vicinity of a zinc smelter. *(a)* Two photos taken from the same spot, looking through the Lehigh Gap near Palmerton, Pennsylvania. The left-hand photo was taken in the 1880s, before the smelters began operating, and the right-hand photo was taken in the 1930s. *(b)* Zones of damage in the Palmerton area showing the effect of proximity to plant on the amount of damage to vegetation. (*a*, John Mankos, courtesy of Marilyn Jorndan; *b*, Jordan 1975)

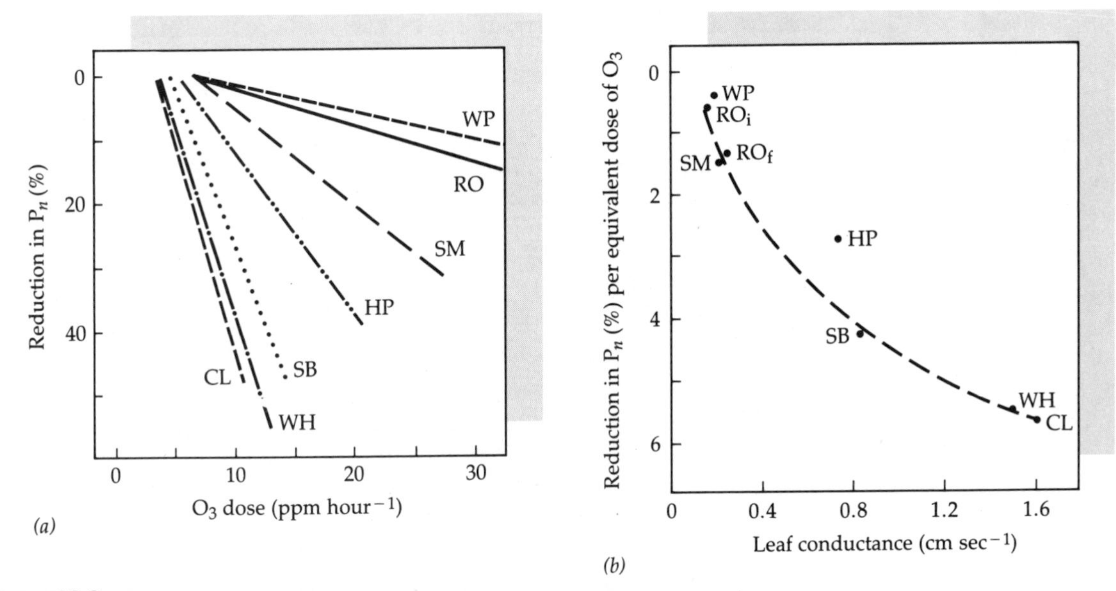

Figure 23.8
The effect of ozone on net photosynthesis of several species of crops and trees. *(a)* Reduction in rate of net photosynthesis as a function of dose. *(b)* Relative reduction in net photosynthesis as a function of the rate of gas exchange between leaf and atmosphere. CL = red clover, WH = wheat, SB = soybean, HP = hybrid poplar, SM = sugar maple, RO = northern red oak, WP = eastern white pine. (Conductance, Reich and Amundson 1985)

of agricultural soils in the temperate zone is much more likely to be altered by management practices than by deposition of pollutants.

Heavy Metals Heavy metals interfere with many critical biochemical reactions and so tend to act as general biocides. Again there are some fairly straightforward relationships between, for example, heavy metal concentration in soil organic matter and the rate at which that organic matter is decomposed (Figure 23.9). Direct effects of metals in artificial soil solutions on seedling root growth and nutrient uptake have been shown, but those same effects are difficult to demonstrate in the field. Heavy metals are "fixed" by organic matter and are relatively inactive. This fixation reduces their concentrations in soil solutions in organic soil horizons but also leads to continuous accumulations in that organic matter, which could result in reduced decomposition and nutrient mineralization.

Nitric and Sulfuric Acids Nitric and sulfuric acids, particularly sulfuric acid, have received the most attention due to their contribution to the acidity of precipitation. In systems where nitrogen availability does not limit biological activity, increased N deposition in the mobile nitrate form can increase

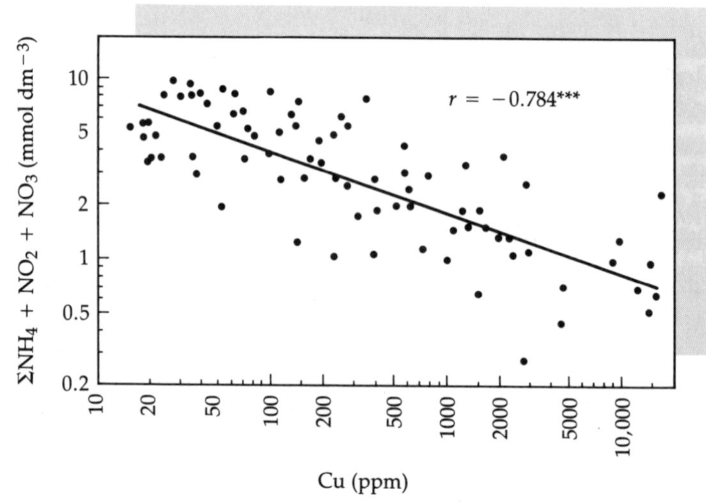

Figure 23.9

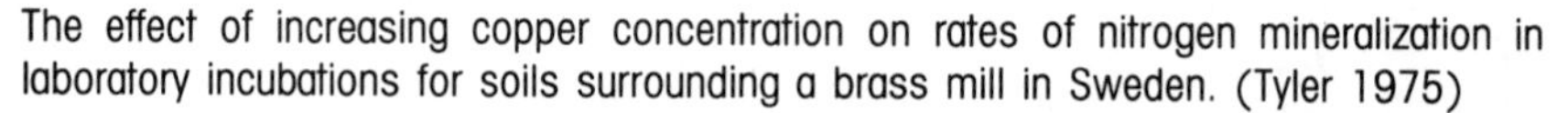
The effect of increasing copper concentration on rates of nitrogen mineralization in laboratory incubations for soils surrounding a brass mill in Sweden. (Tyler 1975)

nitrate leaching from soils. Similarly, sulfate deposition can lead to elevated sulfate leaching from soils without significant sulfate sorption potential (Chapter 9). Anion loss from soils is balanced by cation loss, so mobile nitrate and sulfate tend to increase losses of nutrient cations such as calcium, magnesium, and potassium, reducing base saturation and nutrient cation availability and lowering soil pH. In heavily affected soils where nutrient cation availability has been severely reduced, anion leaching will lead to increased concentrations of aluminum in the soil solution and stream water (Figure 23.10). Elevated aluminum concentrations can cause death in fish by interfering with the operation of gills.

Total Nitrogen Deposition In areas of concentrated agricultural and industrial activity, combined ammonium and nitrate deposition results in very high rates of total N deposition. The cumulative effect of high N deposition over many years in these areas is availability of nitrogen to plants and microbes in excess of their nutritional demands (nitrogen saturation). This may represent a novel form of stress on certain temperate and boreal forest ecosystems in which nitrogen has always been a limiting nutrient. Plants may not be pre-adapted to this situation and may not adjust carbon allocation patterns appropriately. Potential negative effects include pathologically low root and mycorrhizal biomass and loss of frost-hardiness in evergreen foliage. In the most extreme case of forests bordering large feedlot operations in the Netherlands, accumulations of nitrogen in foliage have actually reached toxic levels, as indicated in part by foliage with very

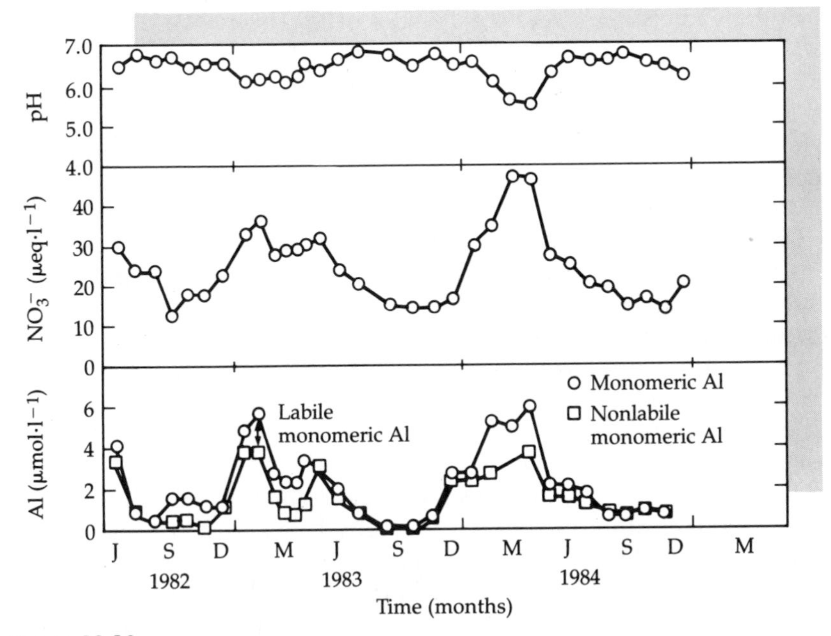

Figure 23.10
The relationship between nitrate leaching, water pH, and aluminum content in a small lake in northern New York State. Increased nitrate losses through a soil with low base saturation, particularly during the spring snowmelt season, result in the acidification of stream and lake water and increased mobility and leaching of aluminum. (Driscoll et al. 1987)

high N content but little chlorophyll and very low rates of net photosynthesis.

Air Pollution and Forest Decline: The Interactive Effects of Individual Forms of Pollution

All of these processes *might* pose serious threats to terrestrial ecosystems in regions where air pollution causes significant changes in the chemistry of the atmosphere and of precipitation. The threat might also be increased by interactions between pollutants, which magnify the effects each might have on its own. Are there systems that show the visible effects of air pollution damage? If so, are these isolated cases or just the early warning signs of more general damage to come?

Throughout Germany and all of central Europe, trees are "declining" at an alarming rate. Noticed first in conifer species, particularly Norway spruce, the reductions in tree vigor, foliage density, and growth have now been detected in several other species as well. In many stands, again particularly in conifers, trees are dying or dead. This region is also one of very concentrated industrial activity, and deposition of several types of air pollutants is quite high. The co-occurrence of tree death and pollution has led to the general conclusion that air pollution has caused or contributed to forest decline, and has given rise to a great deal of research to prove or disprove that conclusion.

Similar symptoms of decline have been seen in high-elevation spruce–fir forests of the Appalachian Mountains in the eastern United States (Figure 23.11). In the northeastern states, both ground surveys and remote sensing from satellites (see Chapter 24) have been used to determine the amount of this forest type that can be classified into different stages of severity of decline (Table 23.2). In general, the degree of forest decline increases from east to west, increasing in areas closer to the major sources of pollution and experiencing higher rates of deposition. Again, this suggests that air pollution plays a role in decline.

However, identifying the causes of tree decline is very difficult. Regional die-backs of tree species have occurred frequently in areas where, or eras when, pollution loading has been low. For example, at least two episodes of spruce decline have occurred in the northeastern United States, each apparently triggered by a period of extreme drought. Reduced growth by spruce in the Northeast began in the 1960s, coincident with another period of drought. Still, this kind of circumstantial evidence cannot prove drought as the cause of the current decline.

Figure 23.11
Mortality of red spruce trees on Camel's Hump in Vermont. (Courtesy of Dr. James Vogelmann)

Table 23.2 Percentage of conifer forests in different damage classes for mountain ranges in New York, Vermont, and New Hampshire*†

SITE	% OF TOTAL HIGH-ELEVATION FORESTS IN DIFFERENT DAMAGE CLASSES		
	Low	Medium	High
New York (Adirondack Mountains)			
Whiteface Mountain	8.6	12.7	78.7
High Peaks area	2.3	6.8	90.9
Vermont (Green Mountains)			
Camel's Hump	26.5	20.7	52.8
Mt. Abraham	25.6	25.6	48.9
Broadloaf	37.9	24.6	37.5
New Hampshire (White Mountains)			
Mt. Moosilauke	72.3	16.8	10.9
Lafayette	63.9	19.0	17.0

*From Rock et al. 1986.
†Damage estimates for Whiteface Mountain result in part from natural formation of "fir-waves" (Spruegel 1984), but the regional trends here have been supported by other studies (e.g., Craig and Friedland 1990).

There may be no single cause. Forest declines have been described, in general, to consist of three stages or levels of response. The first is a predisposing stress that weakens the trees and perhaps reduces growth rates without causing visible damage. The second is the triggering stress, which causes a sudden loss of vigor and perhaps leads to tree death. The third is a secondary or contributing stress, such as insect or pathogen attack on weakened plants, which further reduces vigor or increases mortality. Without being the sole cause of forest decline, air pollution could easily play a role at any of these three levels, with the strength of the air pollution stress being modified by changing climatic or pest infestation conditions.

Combining the complexities of the forest decline phenomenon with the interactions between several sources of air pollution and plant and soil physiology, it is not surprising that the causes of the current declines have not been identified. There are currently several hypotheses as to how air pollution might cause or contribute to forest decline. Most of these are related to interactions between pollutants.

Soil Acidification and Heavy Metals The longest-standing hypothesis is that acidic precipitation will lower soil pH and reduce cation availability, inhibiting biological function and reducing forest vigor. A modification of this theory is that low soil pH will increase the solubility of aluminum, which in turn will have a toxic effect on soil and plant function. While this effect alone is not generally seen as a major threat, reduced soil pH can also bring into solution other heavy metal elements that may have been deposited in the soil. This series of effects is the most difficult to test because of the complex nature of soils and the role of mycorrhizae in both increasing nutrient availability and screening roots from heavy metal uptake. In general, heavy metal toxicity is not seen as the most important source of air pollution damage.

Nitrogen Saturation, Soil Acidification, and Magnesium Deficiencies Lower soil pH means reduced base saturation and lower availability of cations such as potassium and magnesium. One of the most current explanations of forest decline in Europe relies on an induced imbalance between nitrogen and magnesium availability to plants. High nitrogen availability may foster increased production of foliage. However, low availability of magnesium leads to low concentrations in foliage (Figure 23.12) and chlorosis (yellowing) of that foliage, as magnesium is required in the synthesis of chlorophyll. Chlorotic foliage, especially with high nitrogen content, can represent a net carbon drain to the plant, rather than the carbon source provided by healthy foliage. Toxic accumulations of N in foliage in the forests adjacent to feedlots (discussed previously) may also represent this kind of imbalance. It is intriguing that in some forested regions of Germany increases in forest growth have been documented and have been attributed in part to increased N availability. It is likely that N deposition effects on forests are complex and perhaps very nonlinear, stimulating growth at moderate rates of addition and contributing to forest decline at high cumulative doses.

Ozone and Nitrogen Ozone reduces net photosynthesis and so reduces plant growth and vigor. By reducing the amount of mobile carbon available to plants, elevated ozone levels may intensify internal carbon–nitrogen imbalances resulting from excess nitrogen availability. This interaction would also intensify the effects of the nitrogen–magnesium imbalance described previously.

Ozone and Acidification An interaction between ozone and acidified rainfall has also been proposed. By damaging membrane function in leaves, ozone increases the "leakiness" of cells. Cation nutrients can migrate away from cells to the leaf surface and there be removed in cation-exchange-type reactions by the hydrogen ions in rainfall. This could increase nutrient leaching from plants and reduce their ability to retain cations in the system against increased anion leaching rates.

Heavy Metals and Nitrogen The combined effects of ozone and excess nitrogen in reducing root biomass and mycorrhizae may be accentuated by

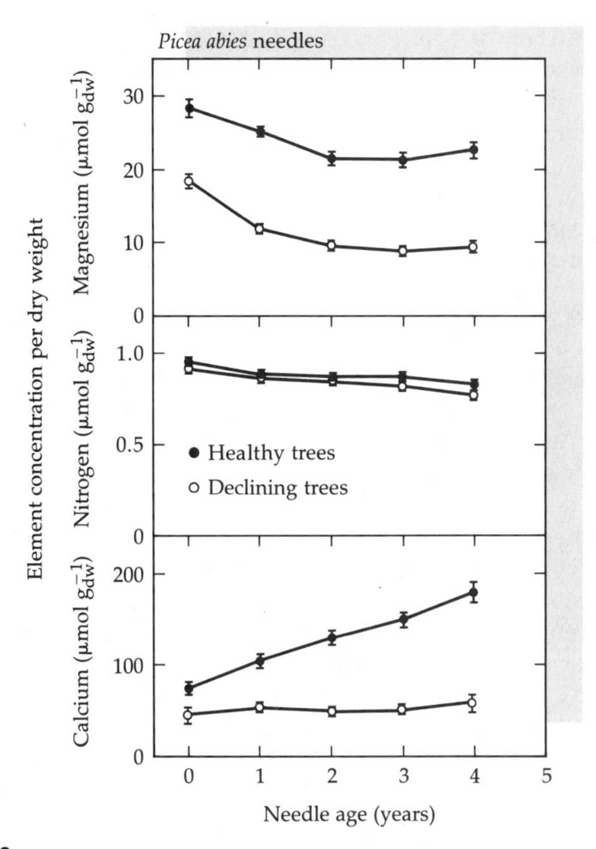

Figure 23.12

Concentrations of magnesium, nitrogen, and calcium in foliage of different ages on healthy and declining Norway spruce trees in Germany. N concentrations are similar in both healthy and declining trees and show no change with age. Magnesium concentrations are very low in foliage of declining trees and decrease with age, indicating active retranslocation to new foliage. Calcium concentrations are also lower in declining trees. (Schulze et al. 1989)

the presence of heavy metals in the soil solution, which could again be increased by soil acidification. In contrast, heavy metals in soil organic matter might reduce nitrogen mineralization rates and actually reduce excess N availability.

At least in the northeastern United States, several factors combine to make the high-elevation spruce–fir forests especially susceptible to air pollution damage. First, these forests tend to occur on very thin, poorly buffered soils, with low sulfate sorption capacity. Second, the cold climate and short growing season limit total annual net photosynthesis and so reduce the potential for nitrogen uptake and allocation. Third, the mountaintop locations in which these forests grow tend to experience both long periods of immersion in cloud layers and high wind speeds due to compression of air movement over ridge tops. These factors interact with the fourth factor, stand structure (high leaf area indices cause canopies to act as very efficient air-filtering systems) (see Table 14.1), to cause large increases in total pollutant deposition. All of these factors combine to increase pollutant inputs and forest sensitivity to those pollutants.

DETERMINING THE MAGNITUDE OF POLLUTION EFFECTS

Controlled experimentation is the proper scientific method for sorting out these effects and for determining just how much of a threat air pollution poses for forest ecosystems. However, it is basically impossible to carry out long-term chronic pollution exposure studies on all the different kinds of ecosystems that occur in heavily polluted regions. The completion of one or two major studies per region can severely tax the financial resources available to do such research. Performing research on smaller portions of the whole forest (seedlings, soil cores, etc.) can isolate effects on individual processes, yet runs the risk of missing important interactions operating within the intact systems over long time intervals.

In the absence of full experimentation, results from single-factor studies in the field or seedling/sapling/soil core studies in the laboratory can be used to derive computer models, which may then be used to predict the effects of pollution (Chapter 21). While this approach is being pursued vigorously, the level of understanding of forest function required to predict responses to this complex set of interacting forces exceeds the scope of any existing model. Existing data and partial models suggest that complex interactions between ozone, excess nitrogen, and soil acidification are involved in reduced tree growth and forest decline. Soil acidification and increased anion leaching are clearly acknowledged as threats to the quality of lake and stream water flowing from affected forests.

The scientific debate over which forms of air pollution contribute to forest decline and reduced water quality will continue. At the same time, an even more intense policy debate continues over whether to regulate pollution emissions now or to wait for more definitive research results. At the heart of this debate is the very visible and disturbing damage that has already occurred to particularly sensitive forest and lake ecosystems. Do these ecosystems constitute our early-warning system, or do they represent the full extent of the damage that will result at current levels of pollution loading? In areas as complex as this, complete scientific understanding may not be available in time, and decision making will involve important uncertainties.

REFERENCES CITED

Craig, B. W., and A. J. Friedland. 1990. Spatial patterns in forest composition and red spruce mortality in montane forests of the Adirondacks and northern Appalachians. *Forest Science* (in press).

Driscoll, C. T., C. P. Yatsko, and F. J. Unangst. 1987. Longitudinal and temporal trends in the water chemistry of the North Branch of Moose River. *Biogeochemistry* 3:37–62.

Galloway, J. N., G. E. Likens, and M. E. Hawley. 1984. Acid precipitation: Natural versus anthropogenic components. *Science* 226:829–831.

Jordan, M. J. 1975. Effects of zinc smelter emissions and fire on a chestnut–oak woodland. *Ecology* 56:78–91.

Ragsdale, H. L., and C. W. Berish. 1988. The decline of lead in tree rings of *Carya spp.* in urban Atlanta, GA, USA. *Biogeochemistry* 6:21–30.

Reich, P. B., and R. G. Amundson. 1985. Ambient levels of ozone reduce net photosynthesis in tree and crop species. *Science* 230:566–570.

Rock, B. N., et al. 1986. Remote detection of forest damage. *BioScience* 36:439–445.

Schulze, E.-D., R. Oren, and O. L. Lange. 1989. Nutrient relations of trees in healthy and declining Norway spruce stands. In Schulze, E.-D., O. L. Lange, and R. Oren. *Forest Decline and Air Pollution: A Study of Spruce* (Picea abies) *on acid soils.* Springer-Verlag, Berlin.

Spruegel, D. G. 1984. Density, biomass, productivity and nutrient cycling changes during stand development in wave-regenerated balsam fir forests. *Ecological Monographs* 54:165–186.

Tyler, G. 1975. Heavy metal pollution and mineralisation of nitrogen in forest soils. *Nature* 255:701–702.

ADDITIONAL REFERENCES

Binkley, D., et al. 1989. *Acid Deposition and Forest Soils: Context and Case Studies in the Southeastern United States.* Springer-Verlag, New York.

Friedland, A. J., et al. 1984. Winter damage as a factor in red spruce decline. *Canadian Journal of Forest Research* 14:963–965.

Gorham, E. 1989. Scientific understanding of ecosystem acidification: A historical review. *Ambio* 18:150–154.

Hornbeck, J. W., and R. B. Smith. 1985. Documentation of red spruce growth decline. *Canadian Journal of Forest Research* 15:1199–1201.

Munger, J. W., and S. J. Eisenreich. 1982. Continental scale variations in precipitation chemistry. *Environmental Science and Technology* 17:32A–42A.

Nihlgard, B. 1985. The ammonium hypothesis—an additional explanation to the forest decline in Europe. *Ambio* 14:2–8.

Reuss, J. O., and D. W. Johnson. 1986. *Acid Deposition and the Acidification of Soils and Waters.* Springer-Verlag, New York.

Schlessinger, W. H., and W. A. Reiners. 1974. Deposition of water and cations on artificial foliar collectors in fir krumholz of New England mountains. *Ecology* 55:382–386.

Schulze, E.-D., O. L. Lange, and R. Oren. 1989. *Forest Decline and Air Pollution: A Study of Spruce* (Pices abies) *on acid soils.* Springer-Verlag, Berlin.

van Breeman, N., and H. F. G. van Dijk. 1988. Ecosystem effects of atmospheric deposition of nitrogen in the Netherlands. *Environmental Pollution* 54:249–274.

Wellburn, A. 1988. *Air Pollution and Acid Rain: The Biological Impact.* Longman Scientific and Technical Pub., New York.

Wolt, J. D., and D. A. Lietzke. 1982. The influence of anthropogenic sulfur inputs upon soil properties in the copper basin region of Tennessee. *Soil Science Society of America Journal* 46:651–656.

Chapter 24 Terrestrial Ecosystems and Global Biogeochemistry

H. Armstrong Roberts

INTRODUCTION

In Chapter 1 we proposed that the largest definition of the boundaries of an ecosystem is the Earth itself. The worldwide spread of DDT, even to penguins in the Antarctic, was offered as an early example of the interconnectedness of local- and regional-scale systems. Other examples have followed, ranging from the global distribution of radioactive fallout from nuclear weapons tests to the patterns of precipitation chemistry discussed in the previous chapter. Influenced by such evidence and by the very graphic images of the finite boundaries of our small blue planet returned by astronauts and satellites, we begin to perceive the potential for human activity to alter the global environment in ways that could be very dangerous to ourselves and to life itself.

Most recently, this concern has focused on our effects on the chemistry of the atmosphere and the potential for serious alterations in climate. Specifically, recent increases in the concentrations of several "radiatively active" trace gases, those that trap long-wave radiation and contribute to global warming of the lower atmosphere (or troposphere), have spurred a major policy debate over proposals to reduce the human production of these gases. In response, the focus of terrestrial ecosystem studies has shifted dramatically from factors affecting internal processes, such as those discussed throughout most of this book, to the interactions of these processes with inputs and outputs of important trace gases. Such studies must deal explicitly with balances at spatial scales of biomes-to-regions.

A major stumbling block to research at these scales is the need to acquire information over very large areas. What are the instruments of choice for this type of work? One rapidly expanding area with great potential is remote sensing with satellite-based instruments. The blending of several traditional disciplines, including geology, meteorology, oceanography, earth science, atmospheric chemistry, and terrestrial ecology with satellite remote sensing, has spawned a new discipline dealing with human intervention into the natural distribution and movement of elements, energy, and water over the earth. This new field has been called **global biogeochemistry**, or **Earth system science**.

Carbon dioxide is an important "greenhouse gas" that has been increasing in concentration in the atmosphere as a result of human activity. The quest for understanding of the global carbon cycle has led to studies of such diverse topics as total combustion of fossil fuels on the earth and changes in soil carbon storage in tropical forests following conversion to agriculture. Concerns over the potential effects of increased atmospheric CO_2 have spurred research on topics as different in scale and process as global climatology and water use efficiency at the plant leaf level. Overall, the scientific response to the "global carbon question" has been a major catalyst for the establishment of global biogeochemistry as a vigorous new field of research.

The purpose of this chapter is to examine the role of terrestrial ecosystems in controlling and changing the concentration of CO_2 in the atmosphere. We will also present some examples of the uses of remote sensing in measuring changes in carbon storage pools and fluxes over the earth. We will conclude with a brief discussion of the role of other important atmospheric gases in controlling global climate and a broader, philosophical discussion of the human occupation of the Earth.

THE GLOBAL CARBON QUESTION

In 1958, scientists at the Mauna Loa Observatory on the Island of Hawaii began regular, detailed sampling of atmospheric chemistry. The site was chosen partly because of its distance from human influence, which reduces the effects of local sources such as factory, car, or home furnace exhausts on air chemistry. The measurements taken are assumed to represent the average or well-mixed condition of the lower atmosphere (the troposphere) in the northern hemisphere.

Those measurements clearly show that the concentration of CO_2 in the atmosphere (Figure 24.1) is not constant. Rather, two patterns can be seen, one of natural origin and one of human origin. The first is an annual cycle in concentration tied mainly to the seasonality of photosynthesis in the northern hemisphere, which causes CO_2 concentrations to decline in spring and summer, as photosynthesis outpaces decomposition, and to rise again in the fall and winter. In this sense the measurements at Mauna Loa are charting the metabolism of a hemisphere.

Imposed on top of this annual cycle is a continuous, long-term increase

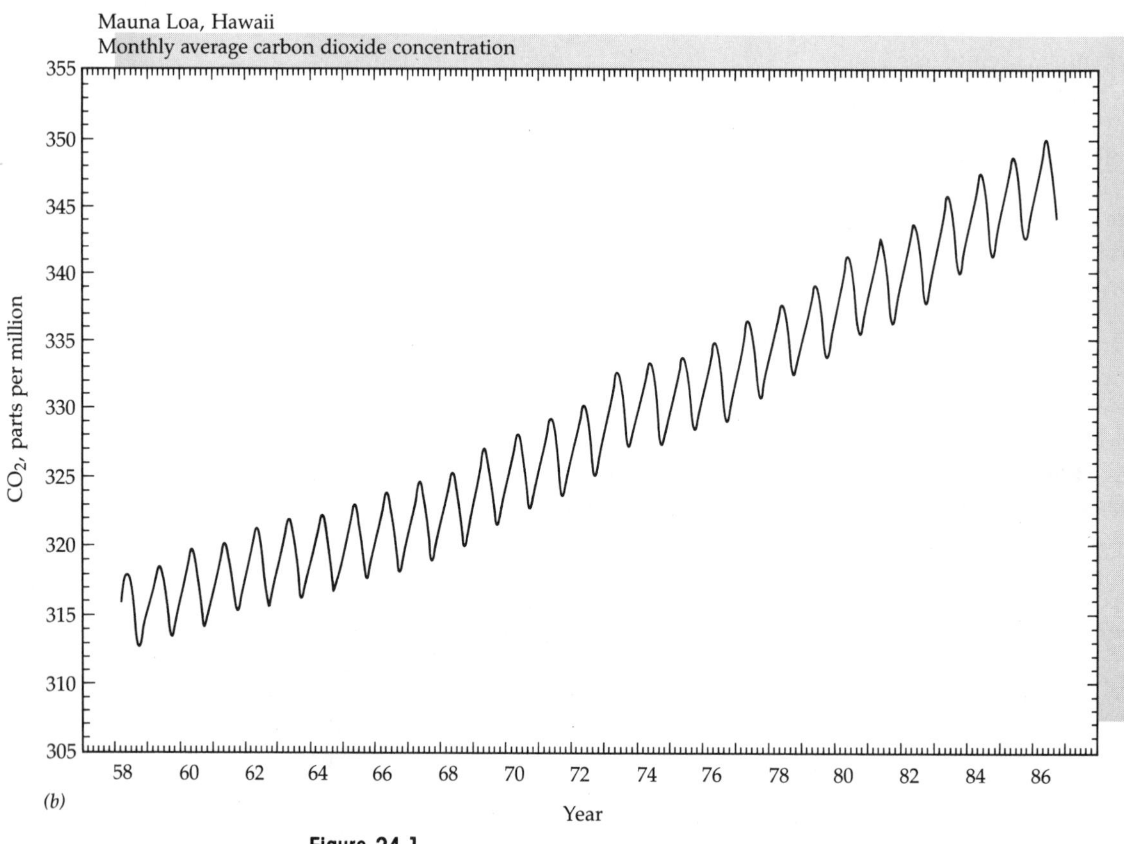

Figure 24.1
Trends in the CO_2 concentration in the Earth's atmosphere as measured at the Mauna Loa observatory in Hawaii. (Moore and Bolin 1987)

in CO_2 concentration, indicating an imbalance between inputs and outputs. It was immediately suspected that this imbalance was due to human intervention in the global cycle of carbon due to industrial and possibly agricultural activity.

This increase has become a major concern to climatologists. Carbon dioxide in the atmosphere is transparent to short-wave (including visible) radiation but is fairly opaque to long-wave radiation. Thus it allows incoming solar radiation, which is mostly short-wave, to penetrate to the Earth's surface but traps long-wave, thermal radiation (Chapter 6) before it can be lost back out into space. This has been called the **"greenhouse effect"** because the resultant warming of the atmosphere is similar to the warming of air in a glass greenhouse on a sunny day (Figure 24.2).

Several different models of global climate have been used to predict the effects of increases in atmospheric CO_2. Because of the complexity of the problem, the models do not all agree on the amount or distribution of effects. There is some consensus that global mean annual air temperature

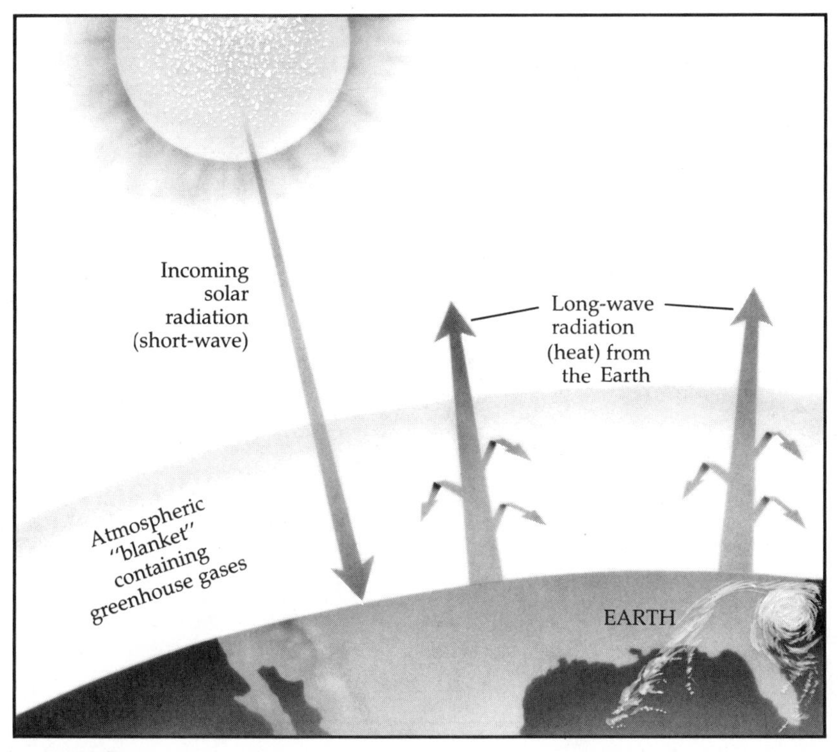

Figure 24.2
Schematic view of the role of "greenhouse gases" in warming the Earth's atmosphere. As concentrations of these gases increase, a larger percentage of out-going long-wave radiation is retained, leading to increased atmospheric temperatures. (Bretherton 1987)

will increase over the next 50 years by 1.5 to 4 degrees centigrade. Moreover, this increase is predicted to be unevenly distributed, with greater increases at the poles and less in the tropics. The implications of these changes for agricultural productivity, energy demand, and sea-level rise are currently under intense study. Predictions range from no effects to worldwide calamity. In the absence of more complete information, we are currently running an uncontrolled global-scale experiment in human effects on the biosphere.

Why has this increase in CO_2 occurred? Figure 24.3 is the current view of the carbon metabolism of the Earth. Before the industrial age, the complementary processes of photosynthesis and respiration were nearly balanced in any one year, as was shown for individual, mature ecosystems in Chapter 3. The high solubility of CO_2 in seawater acted to keep atmospheric CO_2 levels both low and relatively stable (significant changes in atmospheric concentrations of CO_2 have occurred over geologic time scales, but the causes of these changes lie beyond the scope of this discussion).

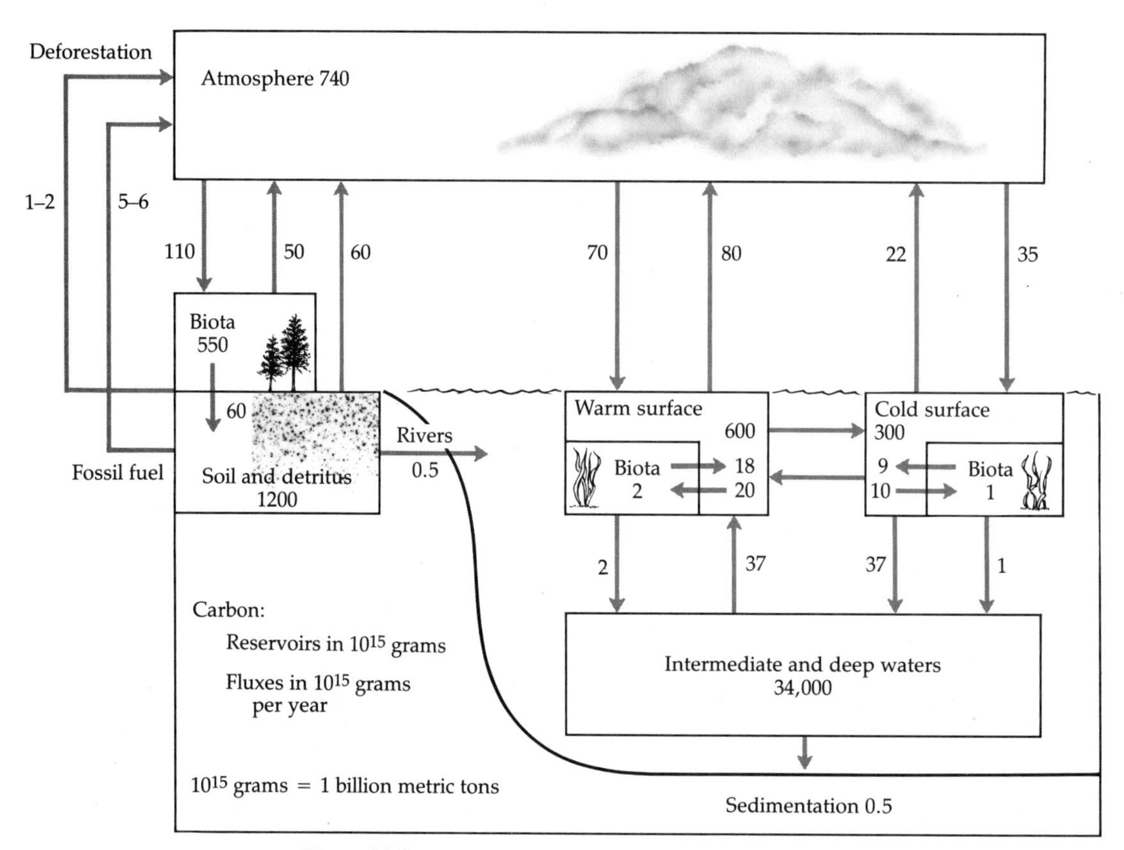

Figure 24.3
Current view of the carbon cycle of the Earth. Human interventions in the cycle through the two processes of fossil fuel combustion and deforestation are evident in the upper left-hand corner. (Moore and Bolin 1987)

Human activity has modified this balance. Perhaps the most obvious cause is the increase in combustion of fossil fuels. The earliest geochemical models that were applied to the problem balanced fossil fuel inputs (roughly 5.2×10^{15} g/yr) against three sinks: increases in the atmosphere (3.1×10^{15} g/yr), increased content in the ocean (2.0×10^{15} g/yr), and increases in total carbon storage in terrestrial ecosystems (0.1×10^{15} g/yr).

To support the idea of greater carbon storage on land, the argument was made that the increased CO_2 concentration in the atmosphere had increased plant photosynthesis and thus caused increased storage of carbon in terrestrial ecosystems. There is some merit to this argument. Consider the concept of water use efficiency discussed in Chapter 8. Water use efficiency was defined as the rate of carbon fixation in photosynthesis divided by the rate of water lost in transpiration. The tradeoff results from the opposite flow of CO_2 and water through the stomates of leaves (Figure

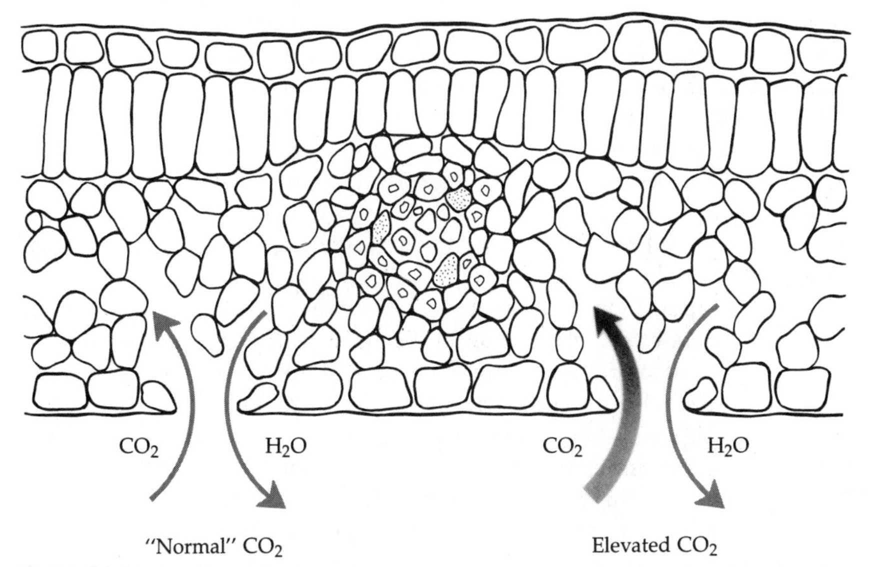

Figure 24.4
The effect of increased CO_2 concentration on water use efficiency at the leaf level. Increased CO_2 in the atmosphere around the leaf results in faster diffusion into the leaf for a given rate of water loss through transpiration. This increases the instantaneous water use efficiency of photosynthesis.

24.4). Since these rates are the result of concentration gradients between the atmosphere and the internal leaf cavities, an increase in atmospheric CO_2 could steepen the gradient and cause a faster CO_2 flux to the leaf. Leaf-level adaptations to increased CO_2 concentrations have been shown to convert this to higher water use efficiency and faster rates of net photosynthesis under laboratory conditions where light and nutrients are not limiting.

Whether these responses will occur in native ecosystems and whether they would lead to increased total carbon storage are major unknowns. Net annual photosynthesis in many ecosystems may be more limited by the availability of nutrients for synthesis of leaf tissues or by the availability of light to drive increased photosynthesis than by water use efficiency. In addition, for ecosystems to be net carbon sinks, any extra carbon fixed in photosynthesis would have to be stored in vegetation or soil rather than respired by plants, animals, or decomposers.

Whole-ecosystem CO_2 fertilization experiments are very difficult to perform in the field, especially since they must be maintained for several years to allow meaningful responses of ecosystem processes to occur. Two studies to date, on an Alaskan tundra system dominated by C3 plants and on a coastal salt marsh with both C3 and C4 plants, have given some different responses (see Chapter 6 for a discussion of C3 and C4 carbon fixation pathways). The cold, wet, C3-dominated tundra showed only very short-term increases in photosynthesis and negligible changes in net ecosystem carbon storage. As expected, the C4-dominated marsh systems also showed little increase in net photosynthesis. However, the C3-dominated marsh systems, growing in a warm environment with continuous access to water and nutrients due to subsurface water movement, showed sustained increases in net photosynthesis. It should be stressed that, complex as these experiments are, they deal only with the CO_2 component of global change and not with changes in temperature or precipitation chemistry, which might also occur.

Acting in opposition to the potential for increased net photosynthesis is the very visible effect of large-scale conversion of native ecosystems to agriculture, particularly in the tropics. This conversion results in very large losses of carbon from vegetation (particularly in forested systems) and from soils. The earliest estimates of the magnitude of this loss, mainly as a result of cutting and conversion of tropical forests, ranged as high as 15×10^{15} g/yr, or three times the atmospheric inputs from the burning of fossil fuels. Many suggested that the CO_2 inputs to the atmosphere resulting from changes in land use (conversion) were about equal to those from the burning of fossil fuels.

This initial discrepancy has been the basis for a decade-long discussion among oceanographers, geochemists, atmospheric scientists, and terrestrial ecologists. Resolving the discrepancy has required significant new interdisciplinary research and a raised level of cooperation among these usually disjunct fields. It has also forced terrestrial ecologists to begin thinking about very large units of the landscape. From this effort has come a clearer understanding of the carbon balances of natural and managed terrestrial ecosystems and of their role in the carbon cycle of the Earth.

The Role of Terrestrial Ecosystems in the Global Carbon Cycle: Effects of Disturbance and Recovery

The impact of fossil fuel burning on atmospheric CO_2 is relatively straightforward. It represents a simple but variable source of carbon. By contrast, terrestrial ecosystems can be either net sources or net sinks for carbon, depending on their stage of development and recent disturbance history.

Remember first that an ecosystem's net CO_2 balance is not related solely to total photosynthesis, respiration by plants, or respiration by decomposers, but rather to the balance among these processes. Table 24.1 demonstrates this principle. The temperate beech forest has a much lower annual rate of gross, or even net, photosynthesis and yet is accumulating more plant biomass than the tropical rainforest. A complete analysis of the carbon balances of these two systems would require the inclusion of soil dynamics as well, but the importance of net versus gross fluxes is still apparent. A highly productive ecosystem, in terms of net primary produc-

Table 24.1 Comparison of gross versus net carbon flux through a tropical forest and a temperate forest*

	TEMPERATE DECIDUOUS FOREST	TROPICAL RAINFOREST
Gross photosynthesis	19.6	127.5
Leaf respiration	4.6	60.1
Net photosynthesis	15.0	67.4
Other respiration	4.2	38.8
Net primary production	10.8	28.6
Litterfall	3.9	25.5
Net biomass accumulation	6.9	3.1

*The temperate forest has a higher rate of net carbon accumulation in above-ground biomass even though the tropical forest has a much higher rate of gross and net photosynthesis (Larcher 1975).

tivity, may be a smaller sink for carbon than a low-productivity system. It may even be a source with a negative overall carbon balance.

A second principle is that all ecosystems tend toward a zero carbon balance at maturity. Some systems, such as the taiga discussed in Chapter 18, tend to show continuous increases in carbon storage until stagnation is reversed by fire, but in the older systems, the net accumulation of carbon is relatively small.

Even agricultural systems that are continuously disturbed in the same way, i.e., through plowing and harvest, eventually reach a new equilibrium or constant carbon content. This new equilibrium may take a century to occur and will certainly be at a much lower carbon content than the native ecosystem (see again Figure 13.10), but once the new equilibrium is reached, the field is no longer a net source of carbon.

On a more complex level, whole regions show net changes in carbon storage only if the relative distribution of land uses is changing. As a simple example (Figure 24.5), envision a region consisting of three large blocks of forest that are harvested at 90-year intervals, but with the harvests staggered by 30 years so that different blocks are cut at different times. Each third of the area continually passes through the full spectrum of carbon content, but as one area is losing carbon through harvest and decay, another is gaining carbon through regrowth. The total carbon storage for the region, the sum of the three curves for the subregions, varies relatively little.

The same concept applies to regions that contain a mix of agricultural and native landscapes. As long as the percentage of the region in each land use category does not change, the total carbon storage for the region does not show large changes. Conversely, large changes in carbon storage

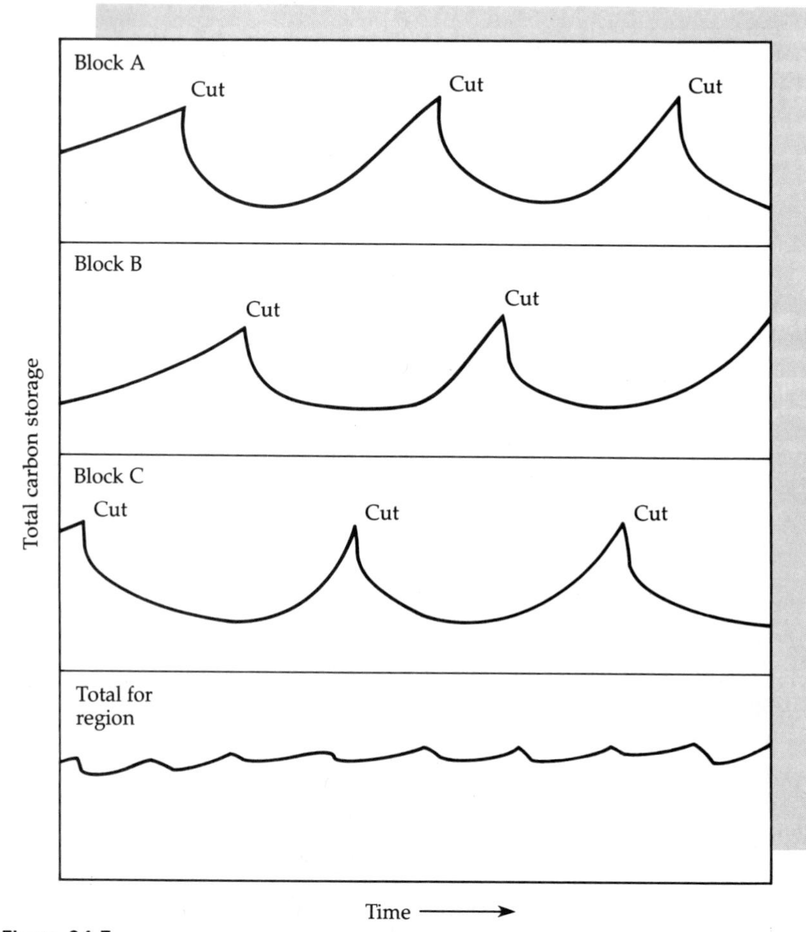

Figure 24.5
The effect of continuous forest cutting, with regrowth, on the net carbon balance of a region. Continuous cycles of cutting and regrowth cause individual units of the landscape to be alternately net sources and sinks for carbon. The combined effect of these is very little net change in total carbon storage for the whole region.

occur in regions where dramatic changes in the distribution of land use is occurring.

Recently several large-scale attempts have been made to derive computer models that estimate the overall carbon balance of all terrestrial ecosystems. One of the most straightforward has applied the type of temporal changes in carbon storage discussed above to "typical" vegetation types throughout the world.

First, the land mass of the world is divided into ten regions (Figure 24.6*a*). Within each zone, several vegetation/land use types are defined (e.g., tropical rainforest, grasslands, cropland). For each, a set of charac-

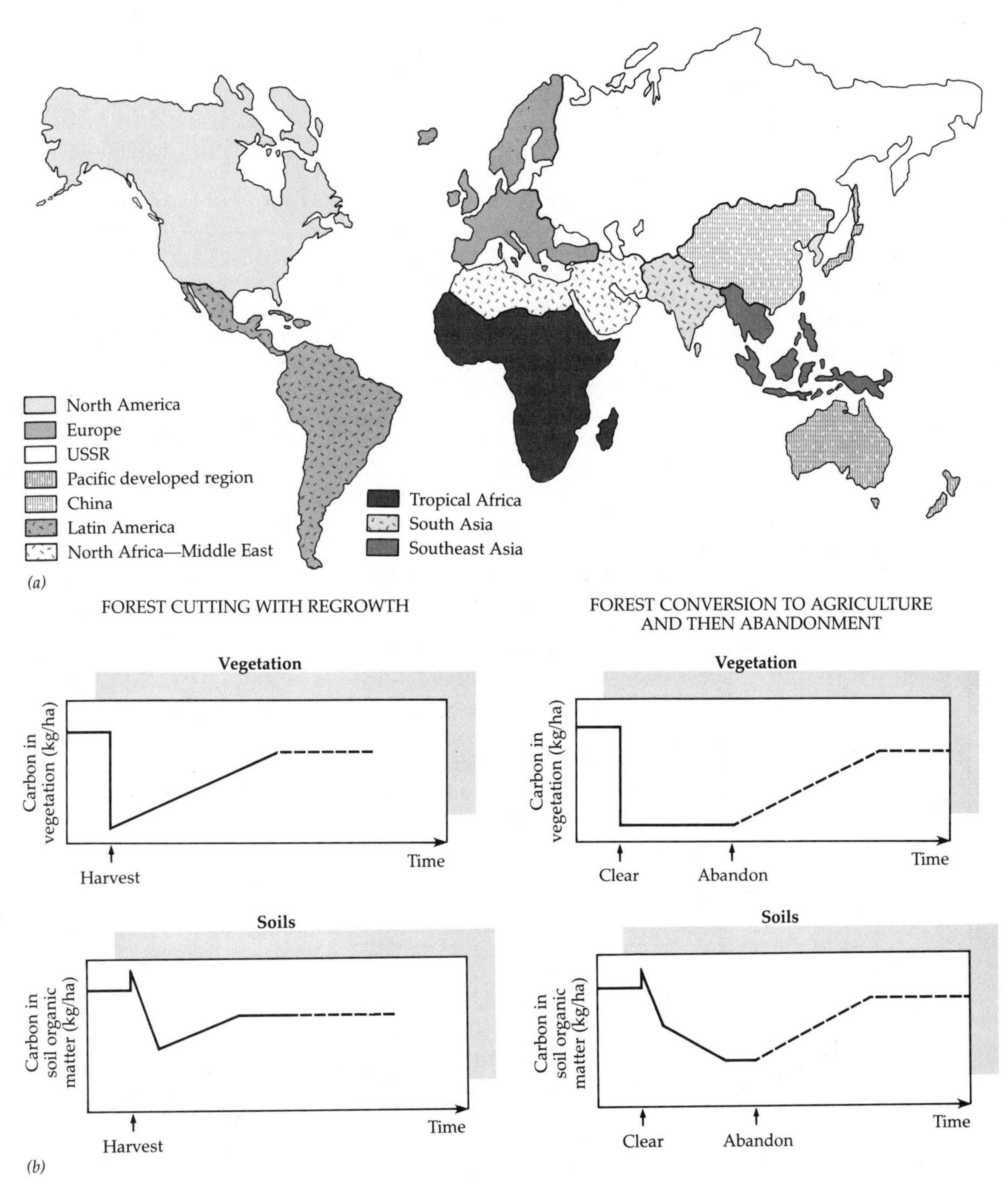

Figure 24.6

One approach to the calculation of the carbon balance over the Earth's terrestrial ecosystems. *(a)* The land areas are divided into ten regions and into several vegetation types within each region. *(b)* For each region and combination of vegetation types, a series of curves describing changes in total carbon storage as a function of land use are employed in conjunction with records of land use history to calculate the total balance for the region. (Houghton et al. 1983)

teristic curves showing changes in carbon storage in soils and in vegetation are defined for each important land use. For forests, this could include commercial cutting with immediate regrowth or with conversion to agriculture either temporarily or permanently (Figure 24.6*b*). Grasslands and shrublands would be defined similarly, with smaller dynamics for the vegetation component of the budget.

With the characteristic curves established, the next step is to determine how much of the land area in each region and vegetation class is being cleared, cut, or converted each year and to follow these acreages through the characteristic curves. The balance of sources and sinks for all of the regions and vegetation types is the carbon balance of the world's terrestrial ecosystems. Using historical records of land use, it is also possible to project backwards, determining carbon fluxes from terrestrial ecosystems over the last century and longer.

Clearly there are difficulties with arriving at either a single characteristic curve for an entire ecosystem type or a cutting and conversion rate for each type. Still, in terms of the modeling discussion in Chapter 21, the complexity of the model reflects the level of information available. It helps frame the question and determine the sensitivity of the estimate to the different parameters.

Historical trends in land settlement and the spread of agriculture are clearly revealed in the model's projections of carbon balances at the continental scale (Figure 24.7). In North America in the late 1800s, mature forests were still being harvested and both forests and grasslands were being converted to agricultural land, mainly in the West and Midwest, as populations increased. As this happened, large amounts of carbon were released. In comparison, the long-settled European continent was only a minor source of carbon at this time.

The tremendous increases in population levels in South America and tropical Africa in the twentieth century have driven the expansion of agricultural activity, including the conversion of forests to cropland and pasture. The resulting increase in carbon release reflects the impacts of both permanent conversion to agriculture and pasture and the increasing amount of land under shifting cultivation.

Another model projection suggests that only within the last 40 years has the input of carbon to the atmosphere from fossil fuels been greater than the input from terrestrial ecosystems (Figure 24.8). This suggests that much of the 30% increase in the CO_2 content of the atmosphere is the result of human use of the landscape, rather than fossil fuel consumption.

The current best estimate derived from this model and others is that terrestrial ecosystems are a source of 1 to 2×10^{15} g of carbon per year to the atmosphere, or 20 to 40% of the fossil fuel input. There is general agreement that harvesting and conversion of tropical forests is by far the largest component of the carbon flux from terrestrial ecosystems.

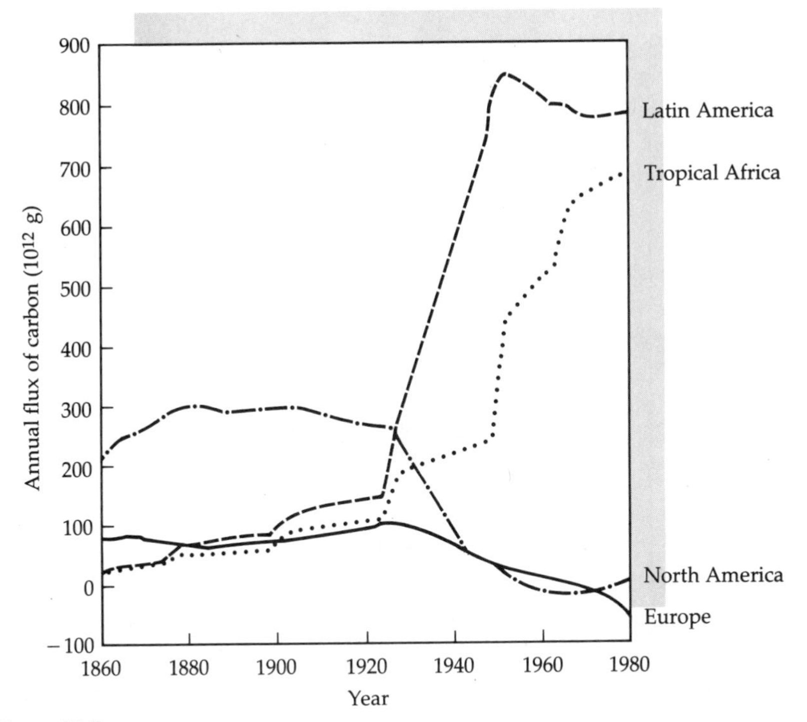

Figure 24.7
Changes in calculated carbon balances for different continents over a 120-year period. (Houghton et al. 1983)

REMOTE SENSING OF REGIONAL AND GLOBAL ENVIRONMENTS

Much of the land use data required to run the global CO_2 model just described came from written sources. Various national and international agencies report on the quantity of wood products harvested, the amounts of agricultural crops produced, and so forth. These compilations are difficult, time-consuming, and not always accurate. In essence, there is a major difference in scale between the data needs of such global assessments and the ability of traditional information-gathering systems to fill those needs.

Satellite remote sensing is, by contrast, an emerging methodology that fits rather closely with the needs of large-scale environmental research. Satellite remote sensing can continuously monitor the condition of the earth's surface through constantly orbiting space platforms containing sensors that measure reflected or emitted energy.

A satellite remote sensing system has several components (Figure 24.9). The platform itself includes the actual sensor for measuring radiation returning to space from the Earth's surface, an optical system for focusing

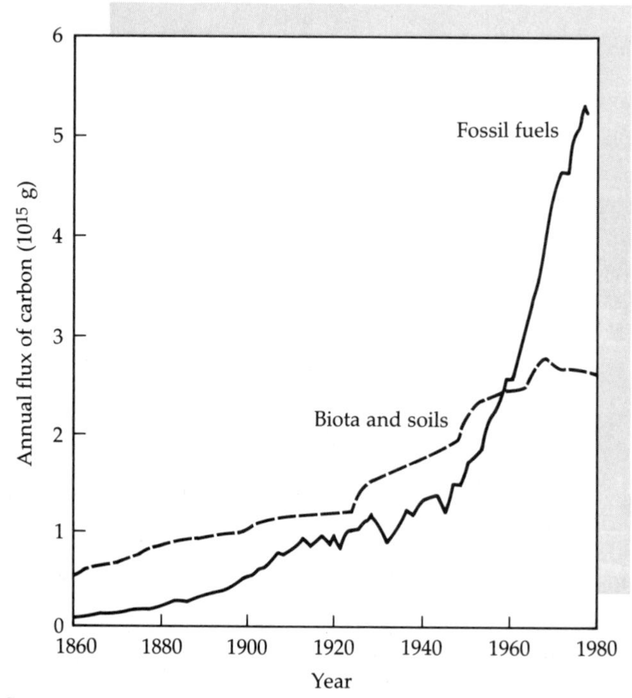

Figure 24.8
The estimated total accumulation of carbon inputs to the atmosphere through fossil fuel combustion and land use changes in terrestrial ecosystems over a 120-year period. (Houghton et al. 1983)

and concentrating this radiation, a computer and data storage system for processing and holding the acquired information until it can be transmitted to receivers on the ground, a solar panel array to provide electricity to run the system, and other physical and electronic systems for temperature control, navigation, and the operation of the sensor. The ground station provides a system for receiving data stored on the platform and for translating it into a computer tape format, which can be stored, copied, and sent to users. A final, important part of the system is computer software, which reads the images stored on tape, translates them back into visual images, and allows mathematical and statistical analyses to be performed to extract the information desired.

A number of different satellite systems sample the energy returning from the Earth's surface in different ways. The differences fall into four categories: (1) the portions of the electromagnetic energy spectrum covered, (2) the size of the smallest picture element (or pixel) that can be seen, (3) the area that can be covered in a single scene (how many pixels), and (4) how often a given portion of the Earth can be resampled. Although it is beyond the scope of this book to offer a full discussion of remote sensing

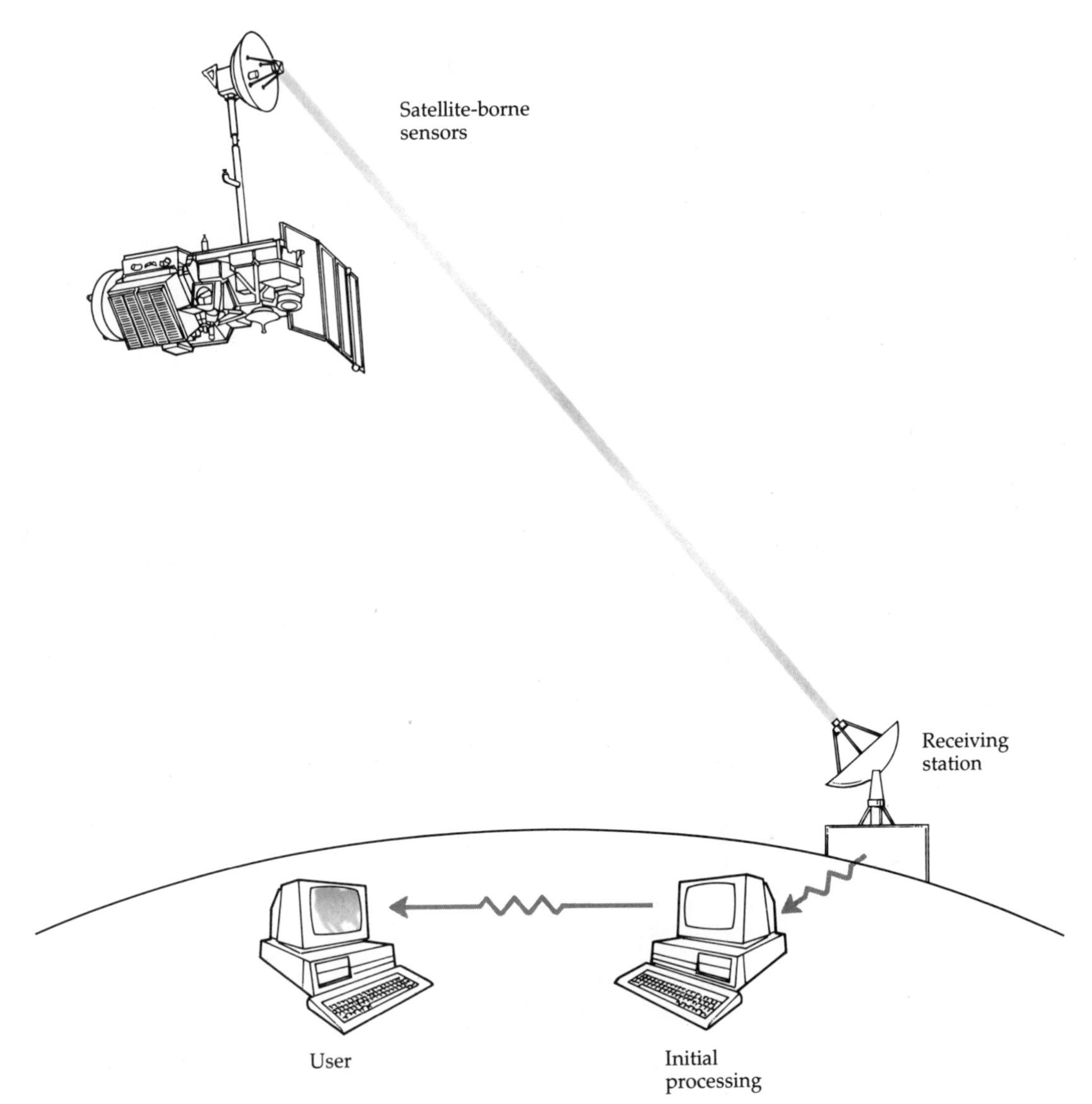

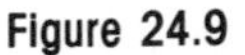
Figure 24.9
The components of a remote sensing system.

systems, we do present some important recent examples of applications of remote sensing systems to pressing environmental problems. These examples also demonstrate the uses of two systems that differ in these four categories.

Two of the most widely used satellite remote sensing systems for terrestrial research are LANDSAT and the AVHRR (Advanced Very High Resolution Radiometer). Figure 24.10 compares the capabilities of these two systems.

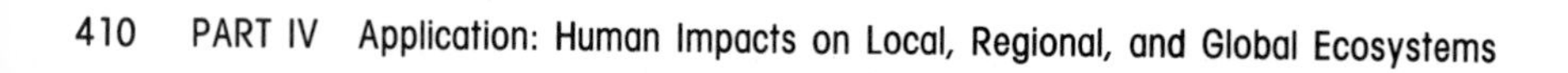

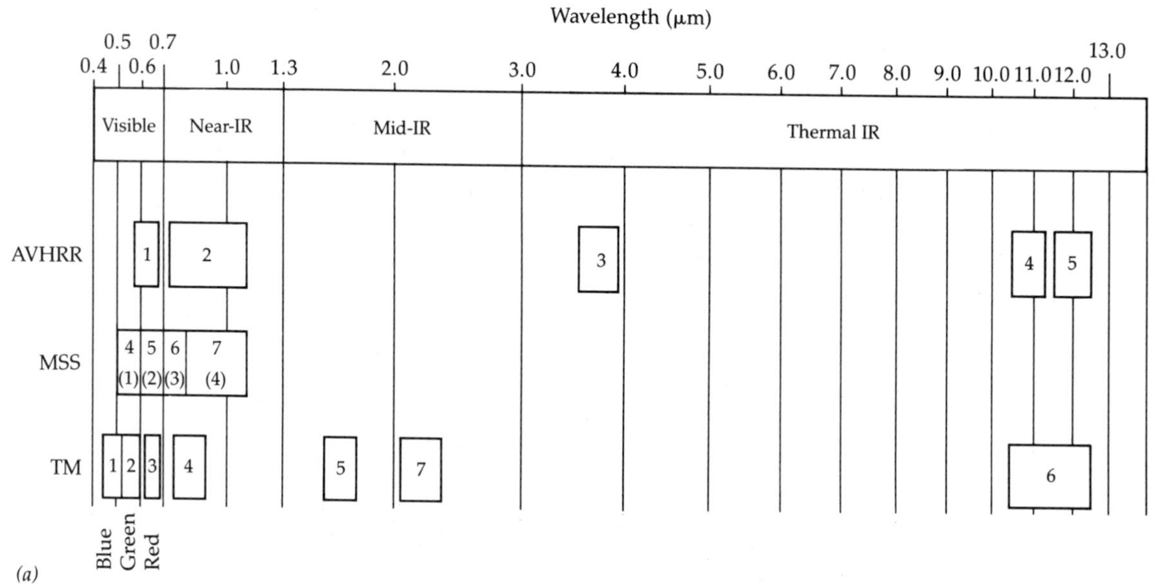

Wavelength (μm)
0.4
0.5
0.6
0.7
1.0
1.3
2.0
3.0
4.0
5.0
6.0
7.0
8.0
9.0
10.0
11.0
12.0
13.0
Visible
Near-IR
Mid-IR
Thermal IR
AVHRR
1
2
3
4
5
MSS
4 (1)
5 (2)
6 (3)
7 (4)
TM
1
2
3
4
5
7
6
Blue
Green
Red
(a)

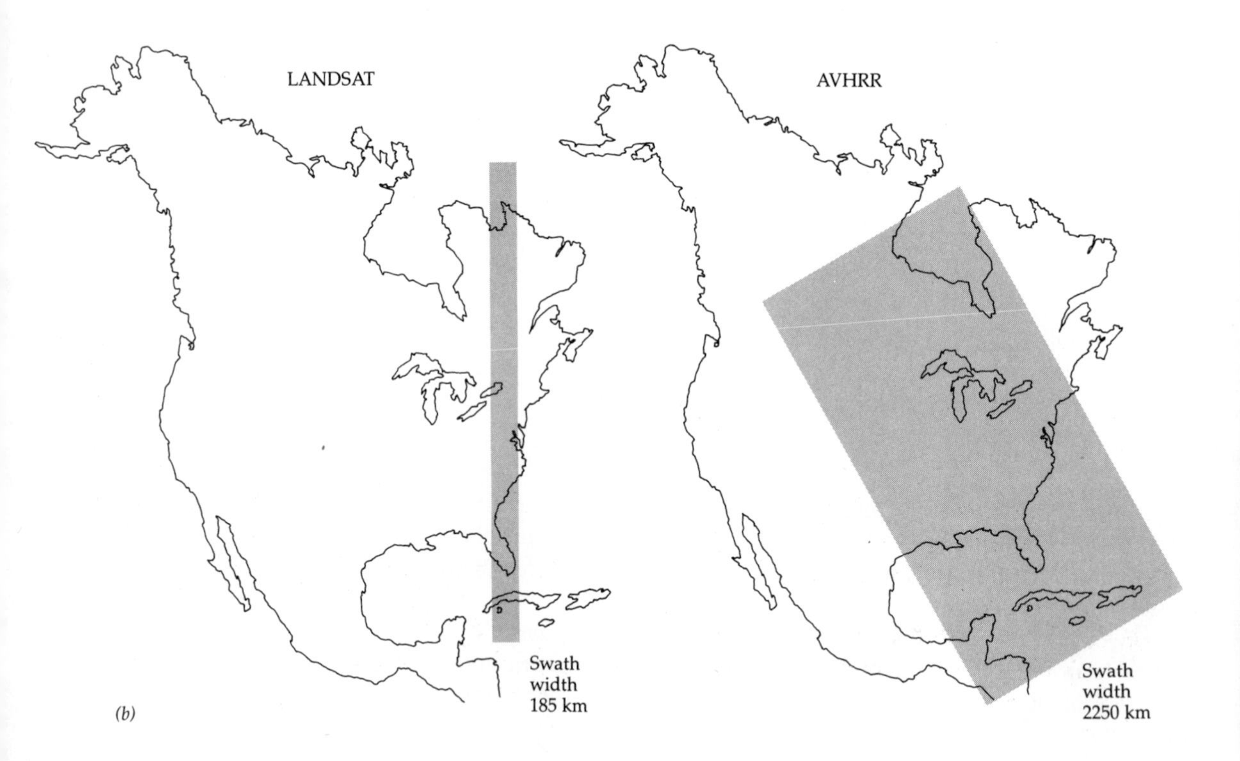

LANDSAT
AVHRR
Swath width 185 km
Swath width 2250 km
(b)

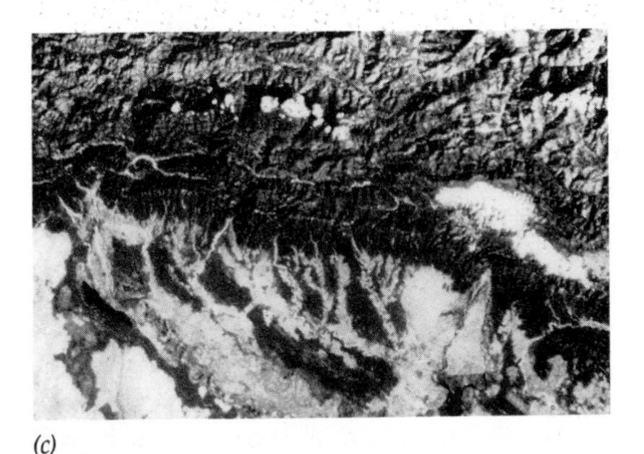

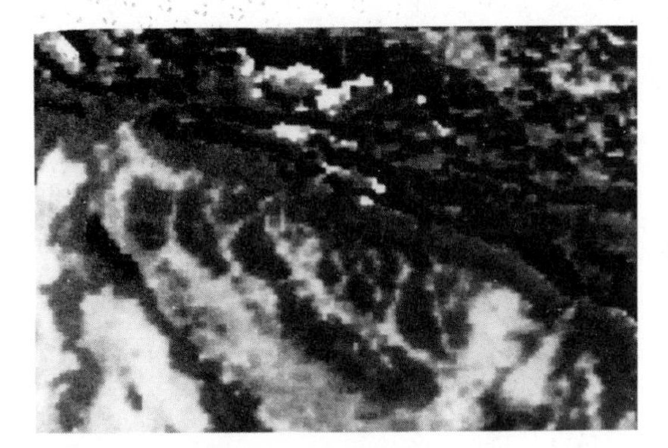

(c)

Figure 24.10

Comparison of two remote sensing systems, LANDSAT and the AVHRR, that are frequently used for analysis of terrestrial ecosystems. *(a)* Comparison of the parts of the visible and infrared spectrum over which data are collected. *(b)* Comparison of swath widths. *(c)* Effect of pixel size on the resolution of the images generated (both images are of the same area, taken 16 days apart). (Lillesand and Kiefer 1988, Baker and Wilson 1987, NASA 1986)

The LANDSAT satellites were designed, as the name implies, for the identification and classification of landscape patterns and land uses. They were built by the National Aeronautic and Space Administration (NASA) and are now operated by the National Oceanic and Atmospheric Administration (NOAA). The earliest versions carried a multispectral scanner (MSS) that sensed radiation in four parts of the electromagnetic spectrum (Figure 24.10): green and red in the visible region and two portions of the near-infrared region. All four channels were for use in vegetation discrimination. The red and green channels are affected by the selective absorption of red light and the reflection of green light by chlorophyll; the infrared channels are responsive to changes in leaf structural characteristics and water content. Later LANDSAT satellites carry the Thematic Mapper (TM) sensor, which has an additional channel in the blue part of the visible range, a different alignment of three near- and mid-infrared channels, and a long-wave or thermal infrared channel designed for sensing surface temperature.

Because land use patterns can change over small distances, the LANDSAT satellites were designed to have a small pixel size, 80 × 80 meters on the MSS and 30 × 30 meters for the TM. One full LANDSAT is just over 6000 pixels wide, allowing data acquisition over an area 185 km wide. Because of this relatively narrow swath width, LANDSAT returns to a given point over the Earth only once every 16 days.

In contrast, the AVHRR sensor was designed and is operated by NOAA for the observation of weather phenomena. As a result of this difference in purpose, the AVHRR uses only 5 of its 46 channels for producing visual images of the surface, the remainder being used for measurements within the atmosphere. Of the six channels, only one is in the visible region, with one in the near-infrared and three more in the long-wave, thermal infrared.

The scale of the imaging is also very different from that in LANDSAT. Important climatic features, such as cloud formation along frontal systems and large changes in surface temperatures, occur over much larger areas than do land use changes. As a result, the AVHRR has a pixel size of 500 to 1000 m, depending on the channel, and covers a swath 2250 km wide, about one-half the width of the United States. This wider swath allows repeat coverage every day.

The scale (pixel size and swatch width) of the imaging makes LANDSAT and AVHRR useful for different applications in the study of the role of terrestrial ecosystems in the biosphere. Figure 24.10*c* emphasizes the loss of resolution of small-scale details caused by moving from the 80-meter resolution of MSS to the 1-kilometer resolution of AVHRR. However, the small swath width and higher data density of MSS and TM make them inappropriate for the study of continental- to global-scale phenomena.

Both of these systems have been used to gather information related to carbon balances and to the changes in carbon storage in terrestrial ecosystems that are associated with the clearing of tropical forests. Figure 24.11 shows three remote-sensing images of the Rondonia region in Brazil. In line with that country's program to open up the Amazon basin for agriculture and permanent settlement, roads and clearings have been made in the tropical forest to speed transport and the establishment of new communities.

Figure 24.11*a* is an AVHRR scene that clearly shows the regional scale pattern of forest clearing. At the LANDSAT scale (Figure 24.11*b*), the distribution of individual fields and roads can be discerned. Figure 24.11*c* shows the even greater spatial resolution that is available from the newer SPOT satellite (a French system; SPOT is an acronym for *satellite pour l'observation de la terre*), which has a 10-meter resolution for black-and-white images.

One of the most dramatic uses of AVHRR data has been to clock the annual pulses of vegetative growth over the globe. Figure 24.12 shows the distribution of "greenness," a value obtained from the ratio of reflectance in the red and in the near-infrared channels, over the African continent, at different times of the year. Particularly striking is the sharp transition in greenness over the Sahel region, the border zone between the Sahara Desert of northern Africa and the tropical forests of central Africa. This is an area of recurring drought and famine. Monitoring of seasonal changes in greenness over this area by remote sensing offers an early-warning system for detecting the vegetative response to drought and offers the hope of speeding international response to relieve famine.

From all of these examples we obtain a view of important interactions between the human population and the landscape we inhabit. Such clear,

(a)

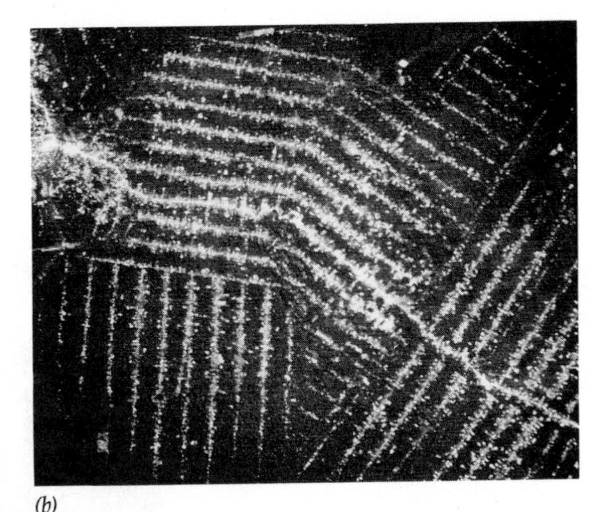

(b)

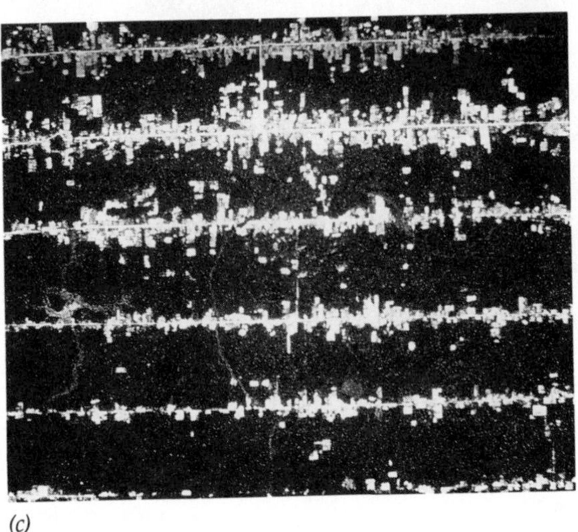

(c)

Figure 24.11
Forest clearing in the Brazilian province of Rondonia. As part of a government program to stimulate settlement in the Amazon basin, roads are constructed in remote areas. These three images show both the pattern of clearing and the spatial resolution of information available from three different satellite systems. *(a)* AVHRR. *(b)* LANDSAT. *(c)* The French SPOT satellite. (Courtesy of Peter and Trish Wolter)

quantitative and persuasive images could not be obtained by land-based methods. As we increasingly recognize the attempt to cope with truly global-scale human occupation of the Earth, the synoptic views of our habitat provided by remote sensing devices will become a more fundamental method for viewing ourselves and our environment.

GLOBAL BIOGEOCHEMISTRY AND GLOBAL CHANGE

Carbon is only one element whose distribution is being altered by human activity, and CO_2 is only one form of carbon with important implications for atmospheric chemistry and global climate. Several other important

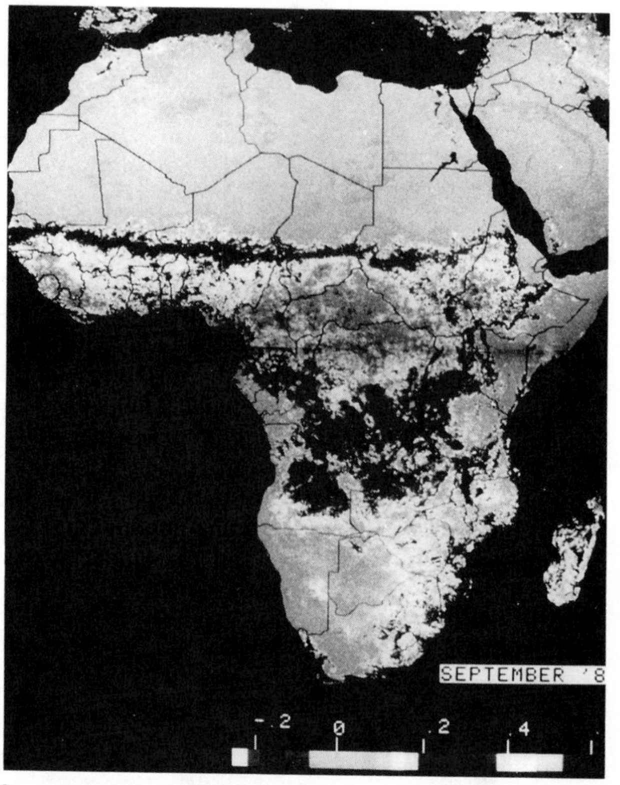

Figure 24.12

Vegetation density in Africa. This image was produced using ratios of the visible and near-infrared channels of the AVHRR sensor, resampled to a spatial resolution of 8 kilometers. The combination of these two channels produces an index related to the amount of plant biomass within each pixel. In this scene, the Sahara Desert of northern Africa is seen to be nearly devoid of vegetation, while the rainforests of central Africa have very high indices. The sharp transition between these two in the Sahel region is also clearly visible. (Tucker et al. 1985)

"greenhouse" gases are present in the atmosphere in trace amounts, the concentrations of which are being increased by human activity. These are summarized in Table 24.2, along with the major sources of such gases from terrestrial ecosystems and their residence times in the atmosphere. Those with the longest atmospheric lifetimes have received the most attention to date because of the potential for increased production to result in lasting and significant increases in atmospheric concentrations. After CO_2, increasing concentrations of methane (CH_4) and nitrous oxide (N_2O) are thought to have had the greatest effect on the ability of the atmosphere to trap long-wave radiation. (An additional set of compounds produced only by human industrial activity, the chlorinated fluorocar-

Table 24.2 The major greenhouse gases, their current concentrations and residence times in the atmosphere, and terrestrial systems and treatments resulting in maximum flux rates*†

SPECIES	NATURAL SYSTEM OR GEOGRAPHIC AREA OF MAXIMUM FLUX	LAND USE FOR MAXIMUM FLUX	APPROXIMATE ANNUAL FLUX (log g)	ATMOSPHERIC LIFETIME (days)
CO_2	Wet tropical forest	Forest, biomass burning	17	~2,500
H_2O	Wet tropical forest	Forest	20.5	10
CH_4	Wetlands	Rice paddies, pasture (animal husbandry), biomass burning	14.5	~3,600
N_2O	(Fertile) tropical forest	Fertilized agriculture, biomass burning	13	~60,000
C_5H_8	Tropical forest	Forest	14.5	<1
$C_{10}H_{16}$	Temperate forest, shrublands	Forest	14.5	<1
CO	Produced from C_xH_x in atmosphere	Biomass burning	15	75
NO_x	Tropical forest	Fertilized agriculture, biomass burning	14	4
NH_3	Temperate grasslands	Pasture (animal husbandry), fertilized agriculture, biomass burning	14	9
COS	Wetlands	Biomass burning	12.5	~900
Sulfur gases*	Wetlands, wet tropical forest, oceans	Fertilized agriculture, biomass burning	13	2

*Includes other reduced sulfur gases such as hydrogen sulfide (H_2S), dimethyl sulfide (DMS), methyl mercaptan (CH_3SH), and carbon disulfide (CS_2).
†From Mooney et al. 1987.

bons, or CFCs, are also thought to be important greenhouse gases. This is in addition to their role in reducing the ozone concentration in the upper atmosphere, or stratosphere.)

The biogeochemistry of methane has received particular attention because, while present in the atmosphere at much lower concentrations than carbon dioxide, it has been increasing at a rate of over 1% per year (compared with 0.2% per year for CO_2; Figure 24.13) and is as much as 20 times as effective as CO_2 in trapping long-wave radiation. Methane is generated by microbial activity associated with the anaerobic environments of wetland areas, including paddy rice agriculture (Chapter 22), and by animal digestive systems. It is consumed by microbes in aerobic soils and by chemical reactions in the atmosphere. Increases in feedlot activity, the extent of paddy rice cultivation, and the mining and movement of

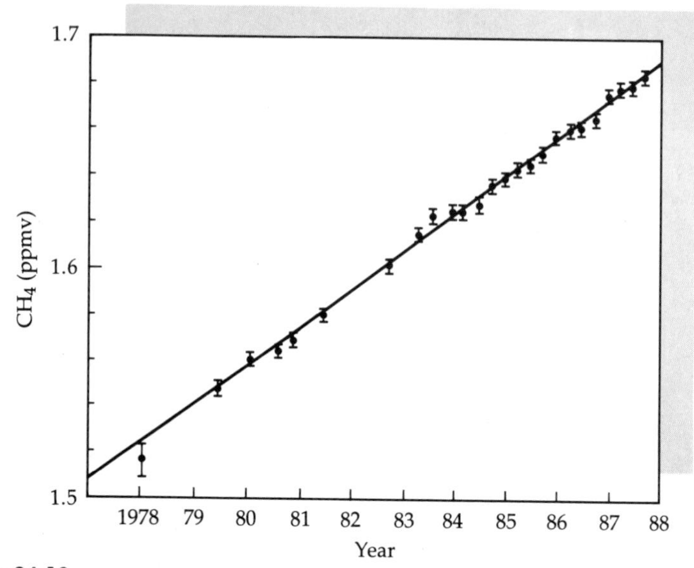

Figure 24.13
Changes in the concentration of methane (CH_4) in the global atmosphere from 1978 to 1987. (Blake and Rowland 1988)

natural gas for energy and industrial use are all processes that may have contributed to the increasing methane content of the atmosphere.

Yet the atmosphere is only one global reservoir being affected by human activity. Regional changes in forest health and water chemistry and quality have been the focus of the "acid rain" debate. Changes in land use affect the CO_2 content of the atmosphere, but they have more immediate and dramatic effects on the ability of the human population to produce food and find shelter. Studies of similar scope and scale are examining human effects on the oceans and their role in climate change. The list of local- to global-scale changes induced by human occupation of the globe is much larger still than these few examples. It is in this larger context that the full scope of global biogeochemistry and global change research, which deals with the Earth as an ecosystem, is carried out (Figure 24.14).

THE STABILITY OF THE GLOBAL ECOSYSTEM

The responses of terrestrial ecosystems to both natural change and human use, which have been the subject of this volume, play a large role in the global ecosystem. We have seen that human activity has altered the extent of different types of vegetation over much of the land mass of the Earth and has also altered the inputs of various elements and compounds into

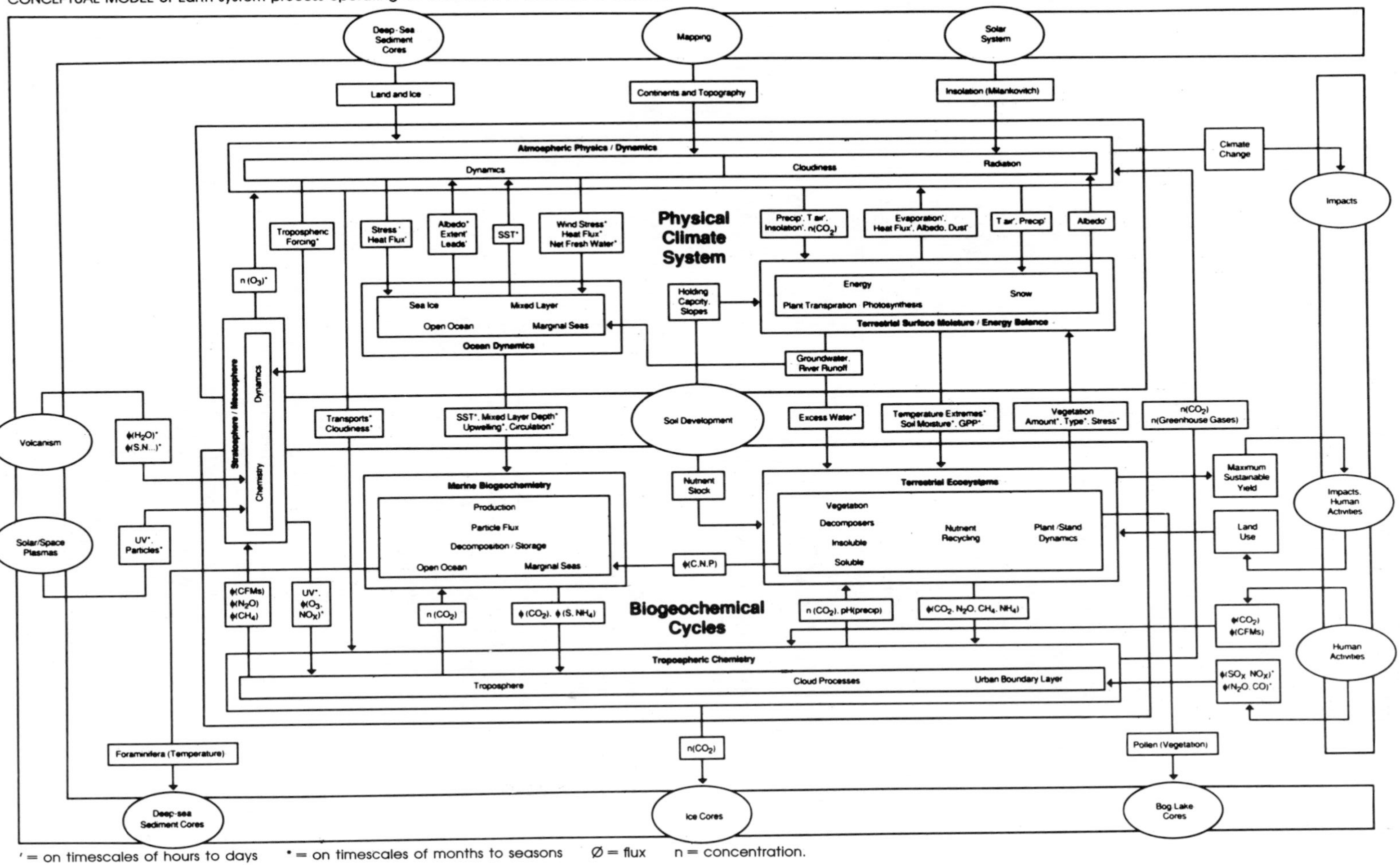

' = on timescales of hours to days * = on timescales of months to seasons Ø = flux n = concentration.

Figure 24.14
A conceptual outline for the study of the Earth as an integrated ecosystem. Human activities act as a major external forcing function on the interactions among land, ocean, and atmospheric processes. The atmosphere acts as a major pathway for exchange of materials among these three spheres. Trace gas concentrations within the atmosphere are seen as a very sensitive part of the feedbacks that are present because of their capacity to alter radiation balances and global temperature through relatively small changes in flux and concentration. (Bretherton 1987)

many of the remaining remote areas. The question that continually arises is: are the Earth's ecosystems, and the Earth itself as an ecosystem, stable or sustainable in the face of human alterations?

One point of view on this question has been very eloquently set forth in a popular book entitled *Gaia*, after the Greek term for "Mother Earth."

> *The chemical composition of the atmosphere bears no relation to the expectations of steady-state chemical equilibrium. The presence of methane, nitrous oxide and even nitrogen in our present oxidizing atmosphere represents violation of the rules of chemistry to be measured in tens of orders of magnitude. Disequilibria on this scale suggest that the atmosphere is not merely a biological product, but more probably a biological construction: not living, but like a cat's fur, a bird's feathers, or the paper of a wasp's nest, an extension of a living system designed to maintain a chosen environment. Thus the atmospheric concentration of gases such as oxygen and ammonia is found to be kept at the optimum value from which even small departures could have disastrous consequences for life. . . .*
>
> *We have since defined Gaia as a complex living entity involving the earth's biosphere, atmosphere, oceans and soil; the totality constituting a feedback or cybernetic system which seeks an optimal physical and chemical environment for life on this planet. . . .*
>
> *Gaia has remained a hypothesis but, like other useful hypotheses, she has already proved her theoretical value, if not her existence, by giving rise to experimental questions and answers which were profitable exercises in themselves. (Lovelock 1982)*

The author of this book hypothesizes the existence of very strong negative feedbacks in the Earth's biogeochemical systems, which have maintained a livable biosphere for over 3 billion years in the face of large changes in total energy input from the sun and dramatic changes in global chemistry. Less comforting, perhaps, is that this seeming permanence is attributed to life in general, life as a process, rather than to the persistence of any one form or species.

The **Gaian hypothesis**, in its scientific rather than philosophical form, is currently an important paradigm for the study of the Earth. If these feedbacks exist, what are the processes and how do they work? Just how fragile are these feedbacks and at what rates can they buffer global change? A prime concern is that rates of change in land cover types and in atmospheric chemistry appear to be far more rapid now, due to human activity, than in any previous age of the Earth.

We have not yet changed the Earth as much as we might. The development of nuclear energy for both weapons and electricity has "freed" us from the limitations of the purely solar-driven world described in these pages. It seems clear that we can make the world uninhabitable for humans, if not for all forms of life, by the reckless use of this power. While the direct effects of nuclear weapons should be painfully clear to all, the secondary effects on the Earth's energy budget have only recently been exposed. The concept of "nuclear winter" holds that the tremendous amount of dust, smoke, and ash produced by a nuclear war and resulting fires would reflect enough sunlight away from the Earth that temperatures would drop by tens of degrees, eliminating food production and intensifying human suffering and mortality. Some postulate that such a severe change could break whatever chain of feedbacks currently operates to moderate the Earth's environment and move it permanently to a new state that would not support life, similar to conditions on our neighboring planets, Mars and Venus.

On a more subtle but no less important level, humans are already altering the Earth's environment. We are living an experiment at the global scale. Whether the Gaian hypothesis holds or not will be a continuing question. As with the discussion of air pollution in the previous chapter, we may be forced to decide whether to let the experiment continue or alter the way we live in order to stop or reverse the changes under way. It is hoped those decisions will be made on the basis of increased understanding of our role in the natural function of the planet, including a better understanding of the function of terrestrial ecosystems.

REFERENCES CITED

Baker, D. J., and W. S. Wilson. 1987. Spaceborne observations in support of Earth Science. *Oceanus* 29:76–85.

Blake, D. R., and F. S. Rowland. 1988. Continuing increase in tropospheric methane 1978 to 1987. *Science* 239:1129–1131.

Bretherton, F. P. 1987. The oceans, climate and technology. *Oceanus* 29:3–8.

Houghton, R. A., J. E. Hobbie, J. M. Melillo, B. Moore, B. J. Peterson, G. R. Shaver, and G. M. Woodwell. 1983. Changes in the carbon content of terrestrial biota and soils between 1860 and 1980: A net release of CO_2 to the atmosphere. *Ecological Monographs* 53:235–262.

Larcher, W. 1975. *Physiological Plant Ecology*. Springer-Verlag, Berlin.

Lillesand, T. M., and R. W. Kiefer. 1988. *Remote Sensing and Image Interpretation*. John Wiley and Sons, New York.

Lovelock, J. 1982. *Gaia: A New Look at Life on Earth*. Oxford University Press.

Mooney, H. A., P. M. Vitousek, and P. A. Matson. 1987. Exchange of materials between terrestrial ecosystems and the atmosphere. *Science* 238:926–932.

Moore, B., and B. Bolin. 1987. The oceans, carbon dioxide and global climate change. *Oceanus* 29:9–15.

National Aeronautics and Space Administration. 1986. Moderate-resolution imaging spectrometer: Instrument panel report. Earth Observing System volume IIb, Washington, D.C.

Tucker, C. J., J. R. G. Townshend, and T. E. Goff. 1985. African landcover classification using satellite data. *Science* 227:369–375.

ADDITIONAL REFERENCES

Bolin, B., and R. B. Cook. 1983. *The Major Biogeochemical Cycles and Their Interactions*. John Wiley and Sons, New York.

Matson, P. A., and P. M. Vitousek. 1987. Cross-system comparisons of soil nitrogen transformations and nitrous oxide flux in tropical forest ecosystems. *Global Biogeochemical Cycles* 1:163–170.

Parton, W. J., A. R. Mosier, and D. S. Schimel. 1988. Rates and pathways of nitrous oxide production in a shortgrass steppe. *Biogeochemistry* 6:45–58.
Rock, B. N., J. E. Vogelmann, D. L. Williams, A. F. Vogelmann, and T. Hoshizaki. 1986. Remote detection of forest damage. *BioScience* 36:439–445.
Scientific American. 1989. Managing Planet Earth. Vol. 261, No. 3 (Special Edition).
Walsh, J. 1988. Famine early warning system wins its spurs. *Science* 239:249–250.
Woodwell, G. M., R. A. Houghton, T. A. Stone, and A. B. Park. 1986. Changes in the area of forests in Rondonia, Amazon basin, measured by satellite imagery. In Trabalka, J. R., and D. E. Reichle (eds.), *The Changing Carbon Cycle, A Global Analysis*. Springer-Verlag, New York.

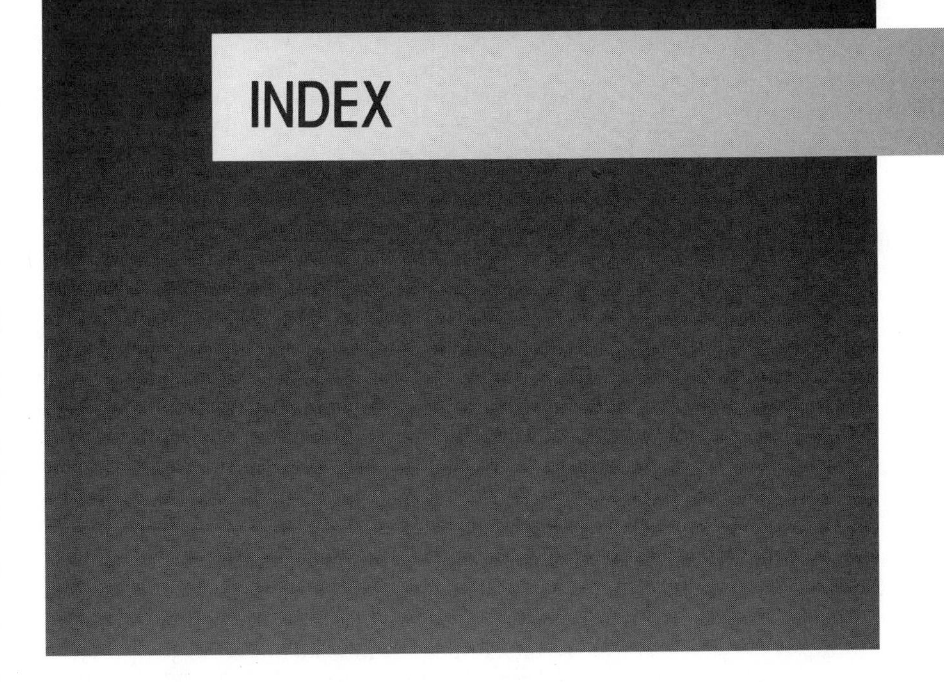

INDEX

Illustration Credits

Academic Press, Inc.
1.1 © 1962; 8.4 © 1985; 12.11 © 1986; 16.4 © 1969; 16.6 © 1969

American Association for the Advancement of Science
8.8 © 1965; 15.8 © 1985; 17.5 © 1973; 19.8 © 1976; 20.5 © 1975; 23.3 © 1984; 23.8 © 1985; 24.13 © 1988

American Geophysical Union
4.4 © 1969

American Institutes of Biological Sciences
11.5 © 1982; 18.2 © 1983; 18.3 © 1983; 18.6 © 1983; 22.6a © 1984; 22.6b © 1983

American Society of Agronomy
Table 10.1 © 1965

American Zoologist
17.4c © 1970

British Ecological Society
19.3 © 1965

Cambridge University Press
6.9 © 1986

CRC Press, Inc.
10.5 © 1985

Ecological Society of America
1.1 © 1942; 2.5 © 1974; 7.7 © 1979; 8.6 © 1976; 11.11 © 1989; 12.9a © 1982; 12.10a © 1978; 16.5 © 1955; 16.13 © 1970; 17.9 © 1979; 20.11 © 1974; 20.12 © 1980; 20.13 © 1981; 22.8 © 1987; 23.7b © 1975; 24.6 © 1983; 24.7 © 1983; 24.8 © 1983

Elsevier Science Publishers
9.13 © 1976; 11.7c © 1985; 13.10 © 1973; 13.11 © 1988; 23.5 © 1988; 23.10 © 1987

Harper and Row
16.12 © 1972

Iowa State University Press
9.6 © 1980

Kluwer Academic Publishers
9.11 © 1984; 22.4b © 1982

MacMillan Magazines, Ldt.
19.6 © 1973; 23.9 © 1975

MacMillan Publishing Company, Inc.
2.1a © 1975; 2.4 © 1975; 8.3 © 1969

Charles E. Merrill Publishing Company
15.3 © 1982

National Research Council of Canada
11.7b © 1983; 11.8 © 1981; 12.8 © 1982; 12.9b © 1981; 18.1 © 1983; 18.4 © 1983; 18.5 © 1983; 18.8 © 1983; 18.9 © 1983; 18.10 © 1983; 20.8 © 1989

NASA
24.10c © 1986; 24.12 © 1985

Oceanus Magazine
24.1 © 1987; 24.2 © 1987; 24.3 © 1987

Oikos
16.9 © 1983

Princeton University Press
6.3 © 1971; 6.5 © 1971; 6.6 © 1971

Quarternary Research
17.4a and b © 1973

The Ronald Press Company
6.2 © 1980

Scandinavian Journal of Forest Research
12.7 © 1986; 12.10b © 1986

Smithsonian Institution
21.7 © 1977

Soil Science Society of America
14.4 © 1982; 14.5 © 1982; 17.6 © 1984

Society of American Foresters
11.7a © 1977; 15.9 © 1983; 21.8 © 1982; 21.10b © 1982

Springer-Verlag, Inc.
3.6 © 1970; 3.7 © 1970; 4.3 © 1977; 4.5 © 1977; 4.6 © 1977; 4.8 © 1979; 6.11 © 1983; 7.1 © 1975; 10.3 © 1985; 11.9 © 1985; 16.1 © 1981; 16.2 © 1981; 20.1 © 1979; 20.2 © 1977; 20.3 © 1979; 20.4 © 1979; 20.9 © 1979; 20.10 © 1979; 23.12 © 1989

University of California Press
12.5 © 1979

University of Chicago Press
11.6 © 1979; 16.3 © 1979; 19.1 © 1979; 19.2 © 1977; 19.9 © 1979; 19.11 © 1979; 19.12 © 1979; 19.13 © 1979; 19.14 © 1979; 19.15 © 1979

University of Georgia
22.4a © 1983; 22.6b © 1983

Wadsworth Publishing Company, Inc.
6.13 © 1978; 8.2 © 1969; 12.6 © 1979; 15.6 © 1979

Westview Press
9.10 © 1981; 9.12 © 1981

John Wiley and Sons, Inc.
9.2 © 1974; 9.3 © 1974; 13.1 © 1977; 13.2 © 1985; 13.3 © 1985; 13.4 © 1985; 13.7 © 1985; 13.8 © 1972